Introduction to
HORTICULTURE
Third Edition

CHARLES B. SCHROEDER

Head, Horticulture Department
Danville Area Community College
Danville, Illinois

EDDIE DEAN SEAGLE

Head, Environmental Horticulture
Abraham Baldwin Agricultural College
Tifton, Georgia

LORIE M. FELTON

Environmental Horticulture Department
Abraham Baldwin Agricultural College
Tifton, Georgia

JOHN M. RUTER

Coastal Plain Experiment Station
University of Georgia
Tifton Georgia

WILLIAM TERRY KELLEY

Extension Horticulturist
University of Georgia
Tifton, Georgia

GERARD KREWER

Extension Horticulturist
University of Georgia
Tifton, Georgia

Introduction to

HORTICULTURE

IST

AgriScience and Technology Series

Jasper S. Lee, Ph.D.
Series Editor

Interstate Publishers, Inc.
Danville, Illinois

Introduction to

HORTICULTURE

Third Edition

Copyright © 2000 by
Interstate Publishers, Inc.

Prior editions 1995 and 1997.

Cover Photo Courtesy, Ron Biondo

Printed in the United States of America.

Library of Congress Catalog Card No. 99-71468

ISBN 0-8134-3170-0

2 3 4 5 6 7 8 9 10   04 03 02 01 00

Order from

Interstate Publishers, Inc.

510 North Vermilion Street
P.O. Box 50
Danville, IL 61834-0050

Phone: (800) 843-4774
Fax: (217) 446-9706
Email: info-ipp@IPPINC.com

PREFACE

Welcome to the Third Edition of *Introduction to Horticulture*! In the few years since the first edition was published, this book has become the leading introductory book in the field of horticulture. Why? The book has a student-friendly format that is easy for teachers to use. Reading ease, outstanding illustrations, and quality content have moved this book to the top. It is also popular because of its subject—horticulture.

Horticulture is an exciting field of study. Horticulture makes the world a better place to live. All people benefit from the diverse products and services of horticulture. This diversity offers many career opportunities. Even with the broad emphasis, horticulture has one central focus—using plants to benefit people.

Horticulture includes art, science, and technology. These have been expanded in the Third Edition. More emphasis is on plants and how they are used in the environment. This reflects a dynamic blending of science and technology. This edition also reflects expansion into horticulture as a business, including a new chapter on horticulture business management.

Introduction to Horticulture is a practical approach to the study of horticulture. The authors have carefully prepared the book to meet the needs of students and teachers. Years of education and experience in the horticulture industry have guided their writing. Curriculum guides, research reports, industry information, and teacher input from throughout the United States have been used.

The author team is likely the best qualified group to ever write a horticulture book. They know the industry and have experienced it all across North America. The team members represent the specializations in many areas of horticulture. The quality of this book is further enhanced by the dedication of the publisher, Interstate Publishers, Inc., to quality educational textbooks and materials for agricultural education.

This book is exactly what is needed in today's horticulture education programs. It is written just for you—the horticulture students and teachers in North America.

Jasper S. Lee
Series Editor

v

ABOUT THE THIRD EDITION

The Third Edition of *Introduction to Horticulture* represents a major step forward for horticulture books in agricultural education. The attractive, student-friendly book has been expanded and made much more appealing. Changes from the prior editions have made this book increasingly useful in the hands of students.

Major changes include:

- Increased technology applications in horticulture are presented.

- More full-color photographs that appeal to the visual sense of students are used.

- Additional line art in full color is used to help convey important concepts.

- Environmentally-friendly approaches have been emphasized.

- Student models are used in many photographs to add a youthful orientation to the book.

- Portions have been re-written to enhance readability and student motivation.

- A new chapter on horticulture business management has been added.

Responses by teachers and students to the prior editions have been tremendous. The changes made in producing the Third Edition will no doubt result in a book with even greater appeal. Classes all across North America will be excited by this edition. Both teachers and students will find this book just what they need.

ACKNOWLEDGMENTS

Many people made important contributions to *Introduction to Horticulture*. Without their help, the work in writing, illustrating, reviewing, and otherwise producing the book would have been impossible. The authors would like to acknowledge horticulture teachers, scientists, government officials, business and industry, and the many others that helped with materials, photographs, and technical and educational reviews for this book. Their help is genuinely appreciated by the authors. Some important individuals are listed below.

- **Ron Biondo,** FCAE Field Advisor in Agricultural Education, Countryside, Illinois

- **Dr. Dianne A. Noland,** Horticulture Instructor, University of Illinois

- **Gail Komoto,** Agriculture Teacher, Vancouver, Washington

- **John Sharber,** Agriculture Teacher, Sapulpa, Oklahoma

- **Dr. Miles (Bud) Smart,** The Turf Science Group, Inc., Cary, North Carolina

- **Dr. Edward W. Osborne,** Agricultural Education, University of Florida

- **Chad Nunley,** Agriculture Teacher, Greencastle, Indiana

- **Dr. Darbie M. Granberry,** Professor of Horticulture, University of Georgia, Cooperative Extension Service

- **Doug Anderson,** Agriculture Teacher, Paxton, Illinois

- **Carl Jensen,** Horticulture Teacher, Lake Stevens, Washington

- **J. Michael Hart,** Grounds Superintendent, Danville Country Club

- **Norman Hammond,** State FFA Advisor, North Scituate, Rhode Island

- **Michael Van Winkle,** Agriculture Teacher, Everett, Washington

- **John Blue,** Horticulture Technician, Danville Area Community College

- **Regina Grubb,** Agriculture Teacher, Granite Falls, Washington

- **James O. Felton, III,** ABAC Grounds Superintendent and former agriculture teacher

- **Dr. Dean Evert,** Research Horticulturist, University of Georgia, Coastal Plains Experiment Station

- **John Tonsor,** Grounds Superintendent, South Side Country Club, Decatur, Illinois

- **Dr. Mike Bader,** Assistant Professor of Agricultural Engineering, University of Georgia, Cooperative Extension Service

- **Julie Taylor,** Horticulture Technician, Abraham Baldwin Agricultural College

- **Dr. Gregory M. Pierceall,** Landscape Architecture, Purdue University

- **Gerry Douglas,** Agriculture Teacher, Lynnwood, Washington

- **M.E. Carter,** Computer Generated Graphics, Interstate Publishers, Inc.

- **Dr. Charles H. Peacock,** North Carolina State University

- **Kim Romine,** Desktop Operations and Design, Interstate Publishers, Inc.

- **Dr. James Leising,** Oklahoma State University

Special appreciation is expressed to **Dr. Jasper S. Lee** for his planning and design in this innovative introductory horticulture textbook. Dr. Lee went above and beyond in his dedication to make this new book the best it can be and to meet the needs of students and teachers in horticulture.

The Authors

CONTENTS

PART ONE—Overview of Horticulture

PART TWO—Science in Horticulture

PART THREE—
Greenhouse Management and Production

PART FOUR—Nursery Management and Production

PART FIVE—Using Floriculture Products

PART SIX—Landscaping

PART SEVEN—The Turfgrasses

PART EIGHT—Food Horticultural Crops

PART NINE—Technology in Horticulture

APPENDIXES

1

EXPLORING THE HORTICULTURE INDUSTRY

People like plants! All of us like the beauty, comfort, and food they provide. Having plants and their products when and where we want them requires special skills. A huge horticulture industry has emerged to help meet demands.

Flowers have special meaning to people. A good example is Valentine's Day. Red roses are used to express human feelings for each other. The demand for roses is greater just prior to February 14 than any other time of the year. Sometimes shortages develop and the price goes up. That's when you need to have roses to sell!

The horticulture industry is more than pretty flowers on a special day. It is a broad industry that meets human needs in many ways.

1-1. Horticulture is concerned with the culture of specific crops for food, comfort, and beauty.

OBJECTIVES

This chapter provides background information on horticulture, including its history and importance. It has the following objectives:

1   Explain horticulture as a science, technology, and industry

2   Describe the three major areas of horticulture

3   Trace the history of horticulture

4   Explain the popularity of horticulture

TERMS

agriculture
botany
floriculture
foliage plant
horticulture
horticulture industry
horticulture science

horticulture technology
interiorscaping
landscape horticulture
nursery
olericulture
ornamental horticulture
pomology

HORTICULTURE COMES TO LIFE

Plants are living organisms. In order to live and grow, the needs of plants must be met. Plants will not grow, and may die, if they don't have the proper care. Humans learned long ago about some of the basic needs of plants.

AN AREA OF AGRICULTURE

Agriculture is the production of plants and animals to meet basic human need. Humans must have food, clothing, and shelter in order to survive. Many of these needs are supplied by plants.

Agriculture includes three important areas of plant science: horticulture, agronomy, and forestry. All three areas deal with growing and using plants to meet human needs. Horticulture has a unique place in plant science.

Horticulture

Horticulture is the culture of plants for food, comfort, and beauty. This includes a wide range of important and useful plants. It does not include the traditional agronomy crops, such as grain and fiber, nor forestry, which focuses on tree production for timber.

The modern notion of horticulture has developed over hundreds of years. Historically, horticulture was "garden culture." The word "horticulture" is derived from the Latin words *hortus,* meaning garden, and *colere,* meaning to cultivate. Today, horticulture is far more than "garden culture."

1-2. The horticulture industry is all of the activities that support meeting the needs of consumers for horticulture products. (Courtesy, Ron Biondo, Illinois)

An Industry

Practicing horticulture is much more than growing certain kinds of plants. People don't work alone. They have a vast network of support for their efforts.

Horticulture industry is all of the activities that support meeting the needs of consumers for horticulture products. The horticulture industry includes the supplies and services that growers use, the actual production of horticultural crops, and the marketing of the crops so that they reach the consumer in the desired forms. Just think about what it takes to get the beautiful flowers into the local florist shop!

Horticulture Science

Horticulture science is the broad field that deals with observed facts and relationships among the facts as related to horticulture. Many of the facts are from areas of plant and soil science. Relationships are carefully studied in research stations, natural settings, and plant production.

The major area of science in horticulture is botany. *Botany* is the study of plants. This includes their life cycle, structure, growth, and classification. Processes occurring in plants and how to promote growth are studied. Today, this is taken further than ever before with biotechnology. Many improvements are being made in horticultural plants using biotechnology methods. Some of this deals with basic plant science, such as genetics and physiology.

The goal of horticulture science is to have more and better plants. This must be accompanied by efficiency and profitability for the producer.

1-3. The growth and appearance of plants can be managed by controlling environmental factors, such as light intensity. This photograph shows black cloth curtains in a partial closed position in a greenhouse. (Courtesy, United Greenhouse Systems, Inc.)

HORTICULTURE TECHNOLOGY

Horticulture technology is applying science in horticulture production. It includes several important areas.

Controlling and managing a plant's environments (macroenvironment and microenvironment) serve as major contributors to success in the horticultural industry. Developing and using the proper cultural practices (mowing, pruning, fertilization, irrigation, pest control, etc.) will impact the visual, functional, and productive qualities of the host plant or community of plants. Combining modern technology with a solid, scientific understanding of plants promotes a more effective management program.

1-4. Water aerification is one of the latest technologies in turf-grass management.

AREAS OF HORTICULTURE

Horticulture is a broad and diverse science divided into three important areas: ornamental horticulture, olericulture, and pomology. In general, horticulture meets the needs of people for food, comfort, and beauty.

ORNAMENTAL HORTICULTURE

Ornamental horticulture describes growing and using plants for their beauty. These plants may be used inside or outside of our homes. Ornamental plants are used for their aesthetic qualities and their appeal to the human

1-5. Floriculture is the production, transportation, and use of flower and foliage plants.

senses. Ornamental horticulture can be further divided into two major areas: floriculture and landscape horticulture.

Floriculture is the production, transportation, and use of flower and foliage plants. Greenhouse growers produce cut flowers, flowering potted plants, foliage plants, and bedding plants. Cut flowers are sold to florists, who arrange them into beautiful bouquets and sell them to the public. Flowering potted plants are sold in the containers in which they grew. Foliage plants are often sold in pots for use as houseplants. A variety of annual flowers are sold as bedding plants that homeowners will transplant into their gardens.

Landscape horticulture deals with producing and using plants to make our outdoor environment more appealing. A *nursery* is a place where plants, shrubs, and ornamental trees are started for transplanting to landscape areas. Nursery managers propagate and grow ornamental plants for installa-

1-6. Landscape horticulture uses plants to make our outdoor environment more appealing.

tion in landscapes. The landscaper designs planting plans, installs plant material, and maintains the plants in the landscape environment. Lawn and turf maintenance is another area of study under landscape horticulture. Golf course management includes maintaining all the grass areas of the golf course which may also be considered a part of landscape horticulture.

Interiorscaping is the use of foliage plants to create pleasing and comfortable areas inside buildings. ***Foliage plants*** are only grown and sold for their beautiful colored leaves and stems. Most offices and businesses use these plants to create an attractive interior environment.

*F*OOD CROP PRODUCTION

Many horticulture crops are grown for use as food. The toppings on your favorite pizza—tomatoes, olives, peppers, and onions—are examples of horticulture plants used as food.

Horticulture crops used for food can be divided into olericulture (vegetables) and pomology (fruits and nuts). With horticulture as broad as it is, some additional areas of production may be important in local communities.

Olericulture is the growing, harvesting, storing, processing, and marketing of vegetables. This is a huge industry in some farming areas. For example, many important vegetable crops are produced in the Salinas Valley of California, including lettuce, celery, tomatoes, broccoli, beans, and sweet corn. Olericulture is found throughout North America, with many families having gardens where their favorite vegetables are grown.

Pomology is the growing, harvesting, storing, processing, and marketing of fruits and nuts. Oranges, apples, peaches, cherries, grapes, strawberries, blueberries, pecans, walnuts, and pistachios are a few examples of pomology crops. Pomology is found throughout North America. Nearly every region produces

1-7. Cauliflower being harvested in Salinas Valley, California.

1-8. Oranges are grown in groves, harvested, graded, and placed in boxes for shipment. (Courtesy, Florida Department of Citrus)

some type of important fruit crops. States, such as Washington, Michigan, and Virginia, grow apples and similar crops. Florida citrus is well known. Georgia is known for peaches and South Texas for citrus. California is known for grapes, strawberries, citrus, and nuts. Most fruit will grow in many locations, but some places have much better growing conditions than others.

HISTORY

Early work in horticulture was often by botanists, who called what they were doing "applied botany." A knowledge of taxonomy (plant classification), anatomy (plant structure), physiology (plant function), entomology (insects), pathology (plant diseases), heredity (plant genetics), and weed control was crucial.

Horticultural science relates to the cultivation of ornamental plants, vegetables, and fruits. It includes the technology modern horticulturists use along with current management strategies.

THE EUROPEAN INFLUENCE

The ancient Greeks were innovative in sciences, including horticulture. The most significant Greek horticulturist was Theophrastus (377–288 BC), who was precise in his work. He speculated that the roots of a plant absorbed nutrients. He observed differences between the leaves of germinating seeds of wheat (monocotyledon) and the leaves of germinating seeds of beans (dicotyledons). He described how root pruning encouraged the flowering and subsequent fruiting of plants.

The Greek influence declined with the rise of the Roman empire, which adopted different techniques in agriculture. The Romans, while holding agriculture in high regard, were not innovative and scientific thinkers. Instead, they adopted and improved some of the Greek techniques of farming. The use of legumes to improve poor soil, manure to improve soil production, and cultivation to impact weed control were started by the Romans.

1-9. Carolus Linnaeus developed a taxonomy for classifying living organisms. (Courtesy, Stock Montage, Inc.)

The techniques of post-harvest storage of fruits was developed by Varro (116–20 BC). He recommended the placement of straw in a cool, dry place, such as a cave, which provided the Romans a means of storing and having fruits during the colder months.

Knowledge of plants reached its peak with the publication of an authoritative book by Dioscorides about 77 AD. This book, *De Materia Medica*, was not the scientific work of Theophrastus, but did serve as the authority for 1,500 years. It had practical information. He described roots, stems, leaves, and sometimes flowers. For centuries, no drug plant was considered genuine unless it could be identified by the descriptions given by Dioscorides.

Approximately 1,700 years passed before the next major authority. C. V. Linnaeus (1701–1778), a Swedish botanist and physician, made a big impact in the sciences. He developed a method of classifying plants by clear and concise descriptions. These descriptions were known as binomial nomenclature. They have endured the trial of time, as modern plant explorers and taxonomists classify plants by the same technique.

Two additional botanists, Charles R. Darwin (1809–1882) and Gregor T. Mendel (1822–1884), were important. In his *On The Origin of Species*, Darwin concluded that "favorable variations would

1-10. Gregor Mendel laid the foundation for the science of genetics. (Courtesy, Stock Montage, Inc.)

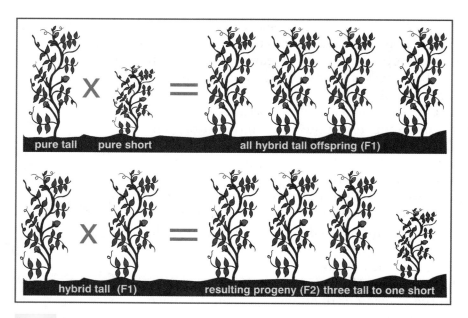

I-11. Gregor Mendel's work with garden peas discovered that crossing pure tall with pure short plants resulted in all tall offspring (F1). He crossed two of these hybrid tall plants and found that segregation took place with the resulting progeny of F2 generation. There was a distinct ratio of three tall plants to one short plant, illustrating the characteristics of dominant and recessive traits.

tend to be preserved and unfavorable ones to be destroyed. The result of this would be the formation of a new species." Darwin was the first to present a scientifically sound explanation of evolution. He wrote another important book, *The Power of Movement in Plants* (1880), in which he referenced the ability of plants to respond to the effects of light and gravity. This was a fundamental recognition of phototropism and geotropism, or the responsive growth of plants to light and gravity.

The work of Gregor Mendel (1822–1884), beginning in 1843, laid the foundation for the science of genetics. With garden peas, Mendel observed seven pairs of contrasting characteristics which became the first step towards analyzing inheritance. The next step was to cross the individuals which had different specific characteristics. He then crossed the resulting hybrids and observed the segregation that had taken place. As an example, a tall pea and a short pea parental line were crossed. From the seed formed (F1 or first filial generation), he observed that all of the progeny (F1 hybrids) were tall. In his next step, he crossed two of these hybrid tall plants and found that segregation took place with the resulting progeny of F2 generation. There was a distinct ratio of three tall plants to one dwarf plant, illustrating the characteristics of dominant and recessive genes.

EARLY U.S. HORTICULTURE

The first commercial nursery in the United States was founded by the Prince family during the early 1730s. Robert Prince established the Prince Nursery at Flushing, Long Island, New York. Under the influence of his son, William Prince, the nursery expanded rapidly until the American Revolution. In 1794, a catalog published by the nursery contained an extensive list of cultivars, including apricots and nectarines. The fruit trees were the foundation of the business and necessary for stocking the country during a period of new and different growth. Ultimately, ornamentals became a significant part of the business. In 1784, the Lombardy poplar was introduced into America, and the Prince Nursery became a major supplier of these trees. The Lombardy poplar became the emblem of a democracy. During the post-Revolutionary era, these trees became the most popular trees in America.

The mere presence or abundance of trees and shrubs does not automatically provide an attractive landscape. As the interest in landscape gardening continued to develop, it was only reasonable that some great, creative people emerge and expand it. Andrew J. Downing (1815–1852) was the first great American landscape gardener. His influence was on developing the simple, natural, and permanent in the landscape, as opposed to the complex, artificial, and temporary that had been promoted by the Italians, Dutch, and French.

Downing's greatest influence was his ability to inspire his pupils. Such inspiration is evident through one of his most renowned learners, Frederick Law Olmstead (1822–1903). Olmstead is now considered to be the father of landscape architecture. He retained Downing's concepts and became the pri-

1-12. Frederick Law Olmstead was the primary landscape architect for Central Park in New York City.

1-13. Roses were initiated in the ancient Egyptian culture.

mary landscape architect for Central Park in New York City, as well as parks in Buffalo, Chicago, and Detroit. Olmstead was a "revolutionary," and his developments were bold, sweeping, and imaginative. The natural, restful settings of his finished projects masked the political turmoil and frustrations that filled his personal life.

Modern horticulturists have also impacted the field. Liberty H. Bailey (1858–1954) is the twentieth century's analogue to the ancient plantsman, Dioscorides. Bailey's many books are often quoted as the standard authority on plant nomenclature, taxonomy, pruning, etc. It is difficult to believe that one man could contribute so much to a particular profession in just one lifetime, yet his credits are valid. Three of his most popular books were *Manual of Cultivated Plants*, *Hortus Second*, and *How Plants Get Their Names*. He established the Bailey Hortorum at Ithaca, New York, during the last years of his life.

Horticulture will continue to impact our lives—as a hobby, a second or seasonal interest, or perhaps a career choice. Whether for food production or ornamentation, the industry will continue to grow and develop.

THE POPULARITY OF HORTICULTURE STUDY

The renewed interest in plant life began in the mid-1960s and continues into the 2000s. From the free spirit attitudes of the 1960s and 1970s to

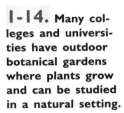

1-14. Many colleges and universities have outdoor botanical gardens where plants grow and can be studied in a natural setting.

the environmental issues of the 1980s and 1990s, the interest continues to increase in our culture—young and old, rich and poor.

Interests and site development have resulted in an increased awareness and job market in horticulture. Plants function as a food source for all animal life, but also provide emotional and psychological gratitude to people. Technology has expanded and career choices abound throughout the field of horticulture.

Land grant universities and two-year college programs in America are experiencing an acceleration in horticultural and agronomic enrollments. This increase is an indicator of the professional interest in horticulture as a career. In addition, the job market for those completing a degree is excellent.

The study of horticultural science and technology has a dominant role in the potential of those electing to find successful careers in this dynamic industry. Two-year degrees emphasizing technology are awarded to those individuals successfully completing prescribed collegiate programs of study in two-year colleges. Bachelor, master, and doctorate degrees emphasizing the horticultural sciences are available at universities. Students must identify the degree requirements and take needed courses for the degree.

REVIEWING

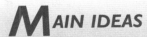

M AIN IDEAS

Horticulture helps meet important needs of humans. Culturing plants for food, comfort, and beauty has resulted in a large horticulture industry. The industry focuses on ornamental horticulture and food crop production, especially vegetables, fruits, and nuts.

Horticulture in North America was influenced by developments in Europe and, a little later, the United States. Early efforts were known as applied botany and dealt with basic plant growth and genetics. Plant classification efforts led to horticulture becoming a science-based activity.

Today, horticulture products are in greater demand than ever before. People are enjoying the beauty of carefully designed and maintained landscapes. The special qualities of a beautiful flower are used to express the gentleness and caring of people.

QUESTIONS

Answer the following questions. Use complete sentences and correct spelling.

1. What is horticulture? How is it distinguished from agronomy and forestry?
2. What is the horticulture industry? Why is it important?
3. What are the three important areas of horticulture? Define each.
4. Briefly trace the history of horticulture in Europe and in the United States.
5. Identify one specific contribution of the following people to horticulture: Carolus Linnaeus, Charles Darwin, and Gregor Mendel.
6. What areas are in landscape horticulture?
7. What are foliage plants?
8. What is the distinction in cut flowers and flowering potted plants?
9. What is horticultural technology?
10. Why is the study of horticulture popular?

EXPLORING

1. Select an important area of horticulture and investigate how it is practiced in your community. Arrange to interview horticulturists about their work and industry. You can select landscaping, floriculture, olericulture, or pomology. Determine the kinds of plants and what they need to grow well.

2. Select an individual who made important contributions to the development of horticulture. Conduct library or computer research to collect details about their work. Give a written or oral report in class.

3. Investigate opportunities for education in horticulture. Contact a local college or university and get information about opportunities in studying horticulture. Prepare a poster or bulletin board that reports your findings.

2

ACHIEVING SUCCESS IN HORTICULTURE

Horticulture is many things. It is an art as old as the Garden of Eden, a science as new as the future, and a technology limited only by resources. It is a hobby to some and a profession to others.

Horticulture deals with the development, improvement, growth, distribution, and use of fruits, vegetables, and ornamental plants including turfgrass. Many people are needed in these areas. Some must have specific education and experience. Choosing a career is one of the most important decisions a person makes. Time and energy are needed to make the best decision. The field of horticulture offers many opportunities for qualified people. If you are interested in a career in horticulture, conduct a thorough study to identify the occupations and training needed.

2-1. Entering a horticulture career will require the appropriate education.

OBJECTIVES

This chapter covers the careers in horticulture and the personal skills needed to enter and advance. Upon completion of this chapter you should be able to:

1   Explain how to prepare for a career in horticulture

2   Describe how to get and succeed in a job

3   List examples of careers in the horticulture industry

TERMS

arboretum

botanical garden

career

career goal

garden center

goal setting

horticultural garden

job

job interview

landscape maintenance

landscape nursery

nursery production

occupation

personal skills

PREPARING FOR A CAREER IN HORTICULTURE

Many different careers are available in the horticulture industry. Every local community in North America has opportunities for career success in horticulture. Education, personal skills, experience, and hard work are needed to advance in these careers.

Horticulture career opportunities are organized around the areas of the horticulture industry. In addition, several specialized careers are found that may not be a part of one of the areas of horticulture. The three major career areas of horticulture industry are: ornamental horticulture, olericulture, and pomology.

MAKING GOOD CHOICES

People want to make good decisions when choosing careers. They need to study information about careers and assess their interests. Part-time work and observing horticulturists will provide information to help make a good decision. School counselors and horticulture teachers also have useful information.

A *career* is the general direction of a person's life as related to work in the field of horticulture. It includes a succession of horticulture occupations and

2-2. Personal communication skills are important in being successful. (Courtesy, Jasper S. Lee)

jobs. Individuals start at a lower level and advance on the basis of interest, productivity, and education.

An *occupation* is specific work that has a title and general duties that a person in the occupation would perform. The occupations are often organized in sequence. The entry-level occupations are available to people without experience. In some cases, more responsible entry-level horticulture occupations may have educational requirements.

A *job* is the specific work that a person in a horticulture occupation performs. It is often at a certain site and performed for remuneration, such as salary and benefits. Jobs are with specific employers. The same occupation can be found at many different employers. People can share the same occupation but not the same job. People sometimes change employers and jobs but keep the same occupational title.

BEING SUCCESSFUL IN A CAREER

People want to be successful. Success often depends on the personal skills of an individual. *Personal skills* are the abilities of an individual to relate to other people in a productive manner. People can develop personal skills. Once a person has the education and a job, they need to go about work in a manner that shows that they can be productive.

Being successful involves setting goals and going about achieving them. A *career goal* is the level of accomplishment you want to make in your work. The first step is to set goals by a process known as goal setting. *Goal setting* is describing what we want to achieve in life. In the process, people must consider their personal interests and what is realistic for them to achieve. As goals are set, steps to achieve each goal must be identified. Ways and means to accomplish the steps and deadlines for reaching each step are set.

Goals can change. People sometimes gain more information and decide that goals they have set need to be modified. It is smart to regularly assess and modify goals. You want to be sure that you achieve what you are capable of doing.

Education and Training

Entering a horticulture career will require appropriate education. The education is available through schools, practical experience, and other special opportunities.

2-3. Education and training are available through schools, practical experience, and other special opportunities. (Courtesy, Jerry D. Gibson, North Carolina)

Horticultural training is offered in high schools, community colleges, and universities. High school horticulture classes may be offered in the agriculture department of the school. School student organizations include the National FFA Organization and NJHA. Community colleges may have one- and two-year programs. Students in two-year programs may join the PAS—a national organization for students in agricultural areas. Universities typically require four years for a baccalaureate degree in horticulture. Masters and doctoral degrees are also offered at some universities. Arboreta and public gardens also offer education in horticulture.

The educational requirements of horticulture occupations will vary. Some will require a university degree; others will require high school or community college education. A few horticulture occupations have no formal education requirements. These are usually the lowest level jobs and offer little opportunity for advancement.

The education to prepare for a horticulture career should provide a good background in plant science and related areas. It should include soils and soil preparation and management. Plant propagation, pest control, fertilization, and other areas should be included in the education. Plant materials identification and use should be a part of the education. In some cases, equipment operation and maintenance will be necessary. High level positions will require education in business management and personnel. Individuals should determine the needs of the occupation that is of interest to them. Some require high levels of specialized education.

GETTING AND SUCCEEDING IN A JOB

FINDING JOB OPENINGS

A person who is not an entrepreneur will likely work for another person or employer. Finding a job is not always easy. Knowing how to go about the job search process is important. Landing a job is part of gaining employment.

Some suggestions for finding job openings are:

- Directly contact a potential employer in person, by telephone, or by e-mail.
- Visit personnel offices of potential employers.
- Visit placement offices to identify job openings.
- Newspapers, magazines, television, and radio may have announcements of job openings.
- Use job postings on the Internet.
- Ask family and friends about possible job openings.
- Use contacts with teachers, counselors, and others at your school.

2-4. A teacher or counselor can help you with the job search process.

APPLYING FOR A JOB

Once a job opening has been found, the next steps are to apply for the job and go through the employment process. This may involve filling out an application form, writing a letter of application, preparing a personal data sheet, and going for an interview. It is important to properly prepare written job application materials. Why? The written material you submit represents you. Always spell words correctly, use correct grammar, give accurate information, and write legibly. Be neat!

A part of applying for a job is going for an interview. A *job interview* is a personal appearance of the job applicant with the employer. It provides information about you to the employer and provides you with information about the employer. Taking a job is an important step. Both you and your employer need to understand each other.

Here are some suggestions in making a job interview:

- Take needed information and materials to the interview.

- Groom and dress appropriately.

- Be on time.

- Speak clearly and confidently and use good manners.

- Conclude the interview on time unless the individuals interviewing you wish to continue.

The employer will likely indicate when a decision will be made about the job during your interview. Give the employer that amount of time. Do not call or visit the employer about the job. When you hear from an interview, it will be to offer you the job or indicate that the job has been offered to another person. In either case, always appropriately thank the individual that contacts you.

If you get the job offer, ask when the work would begin and any other questions you may have. Indicate that you will provide a response within the time frame allowed. You may want to talk with people you trust before making a final decision. If you do not want the position, indicate to the employer that you have carefully considered the offer and feel that it is in your best interest not to accept. If you want the job, indicate that you accept it. In either case, be appreciative of the offer and the courtesy you have received.

GETTING ALONG ON THE JOB

Getting along on a job is essential. People who work for others must show that they will be productive. They need to have important personal skills. Here are a few of the essential personal skills:

2-5. The ability to get along with others is important in a job. (Courtesy, Jerry D. Gibson, North Carolina)

- Work ethic—Work ethic is how a person views work. It reflects a person's attitude toward work. Successful people view work as a natural part of life. They strive to be productive as workers. Showing pride in work, taking care of tools and equipment, and doing extra work are indications of work ethic.

- Getting along with people—The ability to get along with other people is important in a job. You need to get along with those you work with as well as the public. Getting along with customers is essential. Research indicates that more people lose their jobs because they cannot get along with other people than for any other reason!

- Honesty—Honesty is the trait of individuals who have high principles. They do not steal, lie, or cheat. People who are honest are truthful. They do not speak or write inaccurate information. Honest people always do what they say they will do. Honesty includes paying debts and obeying laws and rules.

- Life style—Life style is how a person goes about living. People need a life style that builds their well-being as well as that of the people around them. Keeping the human body in good condition is important. Getting adequate sleep, exercise, and nutrition promotes well-being. Substances that impair the body are to be avoided. A good life style helps a person to be productive when on the job.

- Enthusiasm—Enthusiasm is indicated by the energy that a person demonstrates when talking or moving. They show enthusiasm and go about work in a way to get the job done.

■ Dedication—Dedication is probably best described as loyalty to work. The work has a high priority in a person's life. The individual goes about work to do what needs to be done as efficiently as possible.

■ Education and skill—Education and skill are important in most jobs. People gain education and skill through school and practical experience. Education includes the ability to read, write, and do the necessary mathematics. Computer skills are often needed for a job.

■ Dress, grooming, and personal hygiene—How people dress and present themselves create an image of the person. Being dressed appropriately is important. Never dress in clothes that reflect fads nor groom your hair in a style that is a fad. Leave off inappropriate jewelry. Taking regular baths, using deodorant, and brushing teeth help make a good impression on others.

CAREER OPPORTUNITIES IN HORTICULTURE

The opportunities in horticulture may be as an entrepreneur or as an employee. An individual can be successful in either type of employment. Examples of horticulture career areas with employment opportunities are described here.

LANDSCAPE HORTICULTURE

Production Nursery

Nursery production is the growing of plants in containers or fields. Woody plants are grown from seedlings, liners, or rooted cuttings to marketable sizes for sale on the wholesale market. Propagation is with seed, unrooted cuttings, and by grafting. Many producers specialize in specific types of nursery stock, such as fruit trees, shade trees, container plants, broad-leaved evergreens, or needle evergreens. Here are a few occupations in nursery production:

■ Propagator—produces new plants by selected propagation techniques, such as grafting, budding, layering, and rooting. Supervises the work and training of propagation crews.

2-6. A salesperson may attend trade shows for product promotion. (Courtesy, Agricultural Research Service, USDA)

■ Inventory Manager—maintains records of all plant materials; number of plants by species, variety or cultivar, age, location; and other important data. This information helps the nursery fill orders.

■ Field Supervisor—supervises crews as they work in the production of plants and trains inexperienced workers. Responsible for informing employees of the dangers and risks involved in their job duties. Also, responsible for having each job performed properly, without injury to workers or crops.

■ Manager—coordinates entire nursery operation. Does the hiring, firing, promoting, makes recommendations of pay raises, and makes long-term plans to maintain minimal cost production and insure profits. This may involve purchasing new equipment, developing production budgets, changing to new or different chemicals, determining marketing trends, planting different cultivars of plants, or building or expanding storage facilities.

■ Salesperson—identifies and develops customers in a given geographical region. Is responsible for attending seminars and trade shows for product promotion. This may involve managing a booth at a trade show.

■ Sales Manager—coordinates the sale of nursery products. Trains new salespeople, and assigns them to specific areas. Reports changes in demand trends reflected by orders, and makes recommendations for new products requested by the clientele.

■ Shipping Supervisor (Traffic Manager)—supervises the labeling, packaging, and movement of materials in preparation for shipment.

Landscape Nursery

The *landscape nursery* area is concerned with the preparation of sites for landscaping and the purchase and planting of trees, shrubs, evergreens, vines, and turfgrasses to meet the specifications of the landscape architect or designer. The construction components (structures, walls, terraces, and sidewalks) that are part of the landscape design may be subcontracted. Here are a few landscape nursery occupations:

- Construction Supervisor—is in charge of the actual construction on a site. Is responsible for hiring, training, firing, and retaining employees to do high quality work.

- Construction Superintendent—coordinates the various construction jobs, and maintains the proper staff. Determines job selection for each crew, and makes certain that each crew has the necessary materials and equipment to do their job.

- Designer—prepares landscape designs for homes and commercial buildings matching plant material with the micro-environment of the site and the client's needs. In contrast, the landscape architect designs mainly for shopping malls, industrial complexes, and other large projects.

- Salesperson—keeps informed about building construction and development plans in their assigned area. Makes customers aware of the availability and quality of the company for landscape work. Prepares and submits bids for landscape jobs. Interprets landscape drawings.

2-7. Landscape designers are trained in the art of design and the science of growing horticultural plants. (Courtesy, Ron Biondo, Illinois)

Landscape Maintenance

Landscape maintenance is the care of established landscapes. It involves cultural practices, such as mowing, spraying, pruning, weed control, and fertilization, to maintain a landscape. Because the novice can start a business of this sort on a relatively small scale with a small initial investment, this is often a good way for young people to start a private business. Landscape maintenance is seasonal in most locations. However, expansion and diversity of duties have resulted in year-round work.

2-8. Landscape maintenance technicians are trained in and given charge of maintaining an existing landscape. Note the eye, ear, and hand safety protection. (Courtesy, Husqvarna Forest and Garden Company)

- Crew Supervisor—supervises workers on individual jobs, and trains new workers in handling special equipment, such as sprayers and mowers.

- Superintendent of Operations—coordinates the maintenance jobs and assigns crews to specific jobs in larger firms which have an administrating manager.

- Salesperson—solicits jobs for the company among potential customers, responds to inquiries for services, submits bids, and arranges contracts for services.

- Manager—coordinates all phases of the work, and supervises the maintenance work in small operations. Assigns crews to specific jobs each day, prepares budgets, hires, trains, fires, retains, promotes employees, and recommends pay increases.

Herbaceous Plant and Seed Producers

The seed and young herbaceous plants used by amateur and commercial growers of flowers and vegetables are supplied by large operations. These operations specialize in the development of new cultivars and plant propagation. Seed sold by American firms are usually mass produced for them by independent growers on the west coast, Central America, and South America, where growing and harvesting conditions are very favorable for seed production.

- ■ Plant Breeder—specializes in the development of improved cultivars of either flowers or vegetables.

- ■ Propagator—responsible for seed production. Supervises the fields where the different varieties are being grown and the rooting of cuttings of species, which have to be propagated vegetatively, to supply new plants.

- ■ Independent Grower—produces the seed crops for a seed firm under a contractual agreement.

- ■ Sales Manager—coordinates sales of seeds and plants.

- ■ Salesperson and Dealer—sells to both retail distributors and commercial flower and vegetable growers. A dealer is a locally based salesperson in an area where demand is high. The salesperson performs an educational function for growers by providing current information on cultural practices for various crops.

Garden Center (Retail Nursery)

The ***garden center*** is the retail sector of the nursery industry. It provides consumers with a source for all plant materials and supplies including fertilizers, mulches, chemicals, pesticides, tools, support books, and outdoor furniture. A garden center may be part of an incorporated chain of outlets (re-

2-9. This customer discusses her need for bedding plants with a plant technician. (Courtesy, James Leising, Oklahoma State University)

gional or local), and be managed by a central administrative office. Policies on buying and pricing plants, advertising, and maintenance of central warehouses for hard goods are established for the entire chain from this office. As the number of outlets increase, the need for district managers is recognized. Examples of occupations in garden centers are:

- Buyer—responsible for locating the sources and purchasing the kinds, quantity, and quality of plants and other products that will satisfy customer needs. Makes necessary contacts with suppliers and arranges delivery. Close coordination with district managers and store managers is needed.

- Landscape Designer—develops and prepares landscape designs for small homeowners. Designs are either standard and inexpensive, or elaborate and expensive, according to specifications of the individual site and customer. Analyzes sites and prepares plans without hesitation.

- Plant Technician—specializes in plant problems. Offers recommendations on landscape practices and requirements of specific plants. In smaller garden centers the landscape designer and plant technician may be the same individual.

- Manager—administrates and coordinates the entire marketing operation. Plans, budgets, controls, and evaluates each phase of the business, and makes needed operational changes. Supervises all activities in a small operation.

Arboretum, Botanical, and Horticultural Gardens

Many plant collections are available for study by the general public, students, botanists, and horticulturists throughout the United States and abroad. These collections may be found in arboreta, and botanical and horticultural gardens.

An **arboretum** is a collection of trees arranged in a naturalized fashion. Many arboreta contain shrubs, vines, and flowers throughout the facility. A true **botanical garden** is a plant collection habitat. The **horticultural garden** contains an assortment of plants represented by many horticultural varieties, arranged to achieve a desirable aesthetic effect. It is designed as an environment people may frequent to enjoy the color and beauty of the plants, and to provide a place for specific study of plants.

Some botanical gardens, horticultural gardens, and arboreta offer conservatory greenhouses for the display of the more tender plants and plant propagation. Many of them have an educational program and research studies in the plant sciences. A few occupations are:

■ Writer—prepares educational materials on horticultural subjects.

■ Researcher—conducts selected research projects to determine plant responses for improvement of our environment.

■ Propagator—supervises and propagates new plant material for use in garden areas and greenhouses.

■ Educational Director—develops and supervises educational programs in botany and horticulture.

■ Librarian—establishes and maintains a collection of horticultural books and periodicals for the use by the public and in educational programs.

■ Director—implements the policies of the organization's board of directors, develops programs within budgetary limitations, and coordinates the activities of the various divisions or groups in the organization.

■ Curator—responsible for planning, obtaining, placing, and labeling all plant materials throughout the complex.

■ Greenhouse Manager—responsible for maintaining collections of plants in the greenhouses, and attractively displaying them for public view.

FLORICULTURE

Floriculture is the cultivation, harvesting, storing, designing, and marketing of flowering plants. Most floriculture employment opportunities involve the growing and distribution of cut flowers, bedding, and pot plants and the marketing of florist supplies. This industry directs production towards

2-10. A grower is in charge of floral crop production. (Courtesy, *Greenhouse Product News*)

major peaks of consumption, such as special holidays (Christmas, Easter, Mother's Day, etc.). The floricultural greenhouse industry is concentrated near large population centers or in regions favored by long growing seasons, winter sunshine, and mild temperatures.

Production

Production deals with growing floral plants and materials. Four occupations are:

- Grower—responsible for all the stages of production of a crop. Trains new workers and serves as supervisor for crop production employees. Large firms may have many growers.

- Production Superintendent—supervises growers, and coordinates all production activities. Responsible to the owner for the profitability of the enterprise.

- Marketing Manager—supervises the grading, handling, storage, packaging, and shipment of cut flowers. Programs the sales of cut flowers, responsible for direct sales, and delivery of pot plants to retailers.

- Inventory Controller—schedules the flowering of crops to coincide with the timing of their greatest need and best price. Cooperates closely with marketing manager and growers. Orders supplies and equipment needed for production and marketing.

Wholesale Florist

The wholesale florist merchandises both cut flowers and hard goods. They maintain regular communication with retail flower shops. A few occupations are:

- Manager—responsible for entire operation. After consulting with buyers, makes the decisions involving purchase sources or suppliers to meet the specific demands of retail customers.

- Buyer—investigates potential sources of supplies or producers of cut flowers and hard goods, determines the availability of this supply, and makes recommendations to manager. Keeps in close contact with suppliers and monitors the quality of flowers and supplies being received. Sets up delivery schedule of flowers from suppliers. Individual buyers for large companies may deal only with hard goods or certain flower species.

- Salesperson—markets flowers and supplies to retailers. May specialize in either flowers or supplies.

Retail Florist

The retail florist shop is the common outlet for flowers, houseplants (foliage plants, flowering plants, dish gardens), and terrariums. A large proportion of a florist's income is derived from corsages and floral arrangements for weddings and other special occasions. In some cases, supermarkets and other large stores may have floral departments. Retail florist occupations include:

2-11. A floral designer makes flower arrangements. (Courtesy, Jasper S. Lee)

- Store Manager—responsible for all aspects of the operation and its financial success. In small shops, the manager often handles the special orders, such as weddings and funerals.

- Sales Clerk—must be more knowledgeable than sales clerks in many other retail businesses. Must be able to make immediate suggestions to the customer. In larger shops, the sales clerk may specialize in weddings, funerals, etc.

- Designer—makes various flower arrangements. Must be highly skilled and creative. The most successful designers are artists, and are generally paid extremely well.

FRUITS, VEGETABLES, AND NUTS

Olericulture is the branch of horticulture dealing with the production, storage, processing, and marketing of vegetables. Pomology is the branch of horticulture science involving fruits and nuts.

Tons of produce are shipped fresh or held in refrigerated storage worldwide. Many additional tons are preserved by canning, freezing, pickling, or drying and are shipped throughout the U.S. and global market. Fruit, vegetable, and nut production, preservation, and distribution are highly mechanized, and involve the use of very advanced techniques. Professional horticulturists fulfill responsibilities in both the commercial and residential markets.

Production

Fruit, vegetable, and nut producers vary depending on the nature of the crop, climate, intensity of culture, and the degree of mechanization. Operations vary in size from small to large. A few occupations are:

2-12. This vegetable producer coordinates all stages of production. (Courtesy, Agricultural Research Service, USDA)

- Manager—coordinates all stages of production from the selection of varieties and delivery of the harvested crop for packaging to its final shipment. Responsibilities include cost accounting, budgeting, personnel management, and cultural practices.

- Processing Company Field Technician— many of the fruits and vegetables preserved by processors are grown by independent growers under contract. To insure a sufficient supply of fresh products of high quality, the processor utilizes a field technician (similar to agricultural extension agent) in advising the contracted grower of the newer varieties and improved cultural practices.

Marketing

Considerable effort and money have been invested to improve marketing technology. These improvements include grading and packaging equipment and materials, chemical treatments to reduce spoilage, devices to increase storage life, and methods of shipment. Specialized packing, storage, sales agencies, marketing organizations, and legislation have evolved. Much of this effort requires people with special training in horticulture. Several occupations in fruit and vegetable marketing are:

- Sales Agency—private or cooperative organization that handles the sales, grading, storing, and packing of fruits and vegetables for growers.

2-13. A promoter is selling customary foods at this local Japanese market. (Courtesy, Marshall Stewart, North Carolina)

- Broker—resells wholesale lots of produce bought directly from growers or sales agencies.

- Marketing and Promotional Organizations—active in the promotion of specific fruits or vegetables at the state or national level. Finds new market outlets and promotes the crop. Also, distributes market information to growers.

TURFGRASS (HORTICULTURE AND AGRONOMY)

Sod Production

As the use of turfgrasses increases, sod farms will become increasingly important. More people will be needed in this area. Here are a few occupations in sod production:

- Farm Manager—responsible for producing a marketable, pest-free, non-contaminated commodity. Must be a "people person" and understand budgeting, planning, chemicals, and staffing.

- Assistant Farm Manager—assists the farm manager in fulfilling the responsibilities and meeting the goals of the organization.

- Spray Technician—responsible for planning and implementing all activities involved in spraying chemicals. Must have an understanding of and working knowledge in weeds, insects, diseases, nematodes, equipment, and mathematics.

- Staff Leader—responsible for supervising and assisting labor crews in completing their function in sod production.

Turf Establishment

The establishment of a site with turfgrasses (sod, sprigs, and seed) for their specific purpose (recreation, ornamental, utility, or sports) requires skilled workers. Here are two occupations:

- Construction Superintendent—responsible for preparing a site for its intended use. Must have a working knowledge of people, project, and time management.

- Planting Superintendent—responsible for planting a site with turfgrasses for its intended use. Must have a working knowledge of people, project and time management.

2-14. A turf installation contractor is using a low-compaction crawler unrolling large rolls of sod onto a burlap mat over prepared soil. (Courtesy, Woerner Turf, Inc., Alabama)

Golf Course Design and Maintenance

As the demand for the game of golf continues to increase, the need for trained and educated personnel to manage and maintain such facilities will grow. Here are several occupations:

- Golf Course Architect—designs golf courses to creatively use features of the site.

- Superintendent—responsible for supervising the construction and maintenance of a golf course, servicing and repairing turf equipment, keeping appropriate records, and preparing budgets and reports.

- Assistant Superintendent—directs and participates in the construction and maintenance of a golf course. Supervises the operation, maintenance and repair of turf equipment, and performs support tasks.

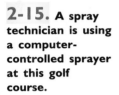

2-15. A spray technician is using a computer-controlled sprayer at this golf course.

- Irrigation Technician—responsible for all irrigation systems on the golf course. Must have an understanding of and working knowledge in irrigation installation and repairs. Must be familiar with pumping systems, irrigation lines and heads, controllers, irrigation hydraulics, and electricity.

- Spray Technician—responsible for planning and implementing all activities involved in spraying chemicals on a golf course. Must have an understanding of and working knowledge in weeds, insects, diseases, nematodes, equipment, and mathematics.

- Landscape Technician—under supervision, responsible for the care and maintenance of the ornamental plants, shrubs, and trees on the clubhouse grounds and selected golf course sites. Also, operates the equipment needed in performing associated tasks.

- Equipment Operator—under direction, operates motorized equipment and trucks used on the golf course and performs related tasks.

OTHER CAREER AREAS

- Horticulture Therapist—helps heal people with health problems or disabilities. Works with physically or mentally challenged individuals, patients in nursing homes and mental institutions, or with individuals having special educational needs. Therapy projects include growing and propagating house plants; forcing bulbs; making corsages and flower arrangements; flower, vegetable, and fruit gardening; and the construction of dish gardens and terrariums.

■ Cooperative Extension Agent or Horticultural Specialist—employed by the Cooperative Extension Service, with offices in local communities and on the college campus. Works with 4-H youth activities; home gardeners in urban areas; or commercial growers, processors, and distributors of horticultural crops. Activities of county extension agents and horticultural specialists are educational. Such activities include disseminating the most recent recommendations of researchers to the people.

■ Consultant—works independently, advising individual growers or other groups in the industry on a regular basis.

■ Communicator in Horticulture—prepares and presents educational presentations on a regular interval through mass media and garden clubs. Writes articles or columns for newspapers or magazines, or serves as an editor of a garden magazine.

■ Teacher—provides instruction in high schools, technical schools, community colleges, two year colleges, universities, arboreta and public gardens.

2-16. **Research scientists checking root development. (Courtesy, Agricultural Research Service, USDA)**

■ Research Scientist and Technician or Assistant—seeks to improve horticultural crops through plant breeding, or by developing more economical and effective techniques for growing, handling, and marketing horticultural crops. Employed by private industry, state and federal government, arboreta and public gardens, and colleges. May initiate, plan, direct, and conduct research projects.

■ Plant Inspector—participates in activities designed to prevent the spread of pests among plants through an inspection service. Polices commercial nurseries, garden centers, and field planting of fruits within their jurisdiction (local, state, or federal) to insure that plants are free of certain pests. Federal plant inspectors concern themselves with plants being shipped from one state or country to another.

Plant inspection positions are civil service. Entrance into this career is as an inspector in training (trainee). A trainee becomes a plant inspector only after a specified time of satisfactory service and performance.

Foreign Assignments

Numerous opportunities exists for people with horticultural training and education, especially in the developing countries. Global marketing through private organizations employs educated and trained people. Global golf course development is one example of numerous employment opportunities.

REVIEWING

MAIN IDEAS

Horticulture is a rapidly growing industry. Many career opportunities are available in horticulture for those who set realistic goals, get the needed preparation, and seek jobs. A number of the careers are based on experience and educational qualifications. Employees often begin in an entry-level occupation and advance to more responsible positions. Students should identify a specific area and make a plan to obtain the needed education and work experience.

Once an individual has the needed preparation, success begins with getting a job. This involves making an application and having a good interview. Personal skills are important. These include good communication, proper dress, and a strong work ethic. On the job, enthusiasm, honesty, productivity, and the ability to get along with people are very important.

Careers in horticulture can be clustered into four main areas: landscape and nursery production; floriculture; vegetable, fruit, and nut production; and turfgrass. In addition, a number of specialized positions are found that do not fit into one of these clusters, such as horticulture therapists and horticulture teachers. Horticultural careers are becoming more technical and science-based but still require a love of plants.

QUESTIONS

Answer the following questions using correct spelling and complete sentences.

1. What are the three major career areas in horticulture? Briefly explain each.

2. What does a person need in order to make a good choice about a horticulture career?

3. What is the distinction between an occupation and a job?

4. What is a career goal?

5. How is the needed education obtained in horticulture?

6. How does a person locate a job?

7. How does a person apply for a job?

8. What are five suggestions in making a job interview?

9. What are the career areas in landscape horticulture?

10. What are the career areas in floriculture?

11. What are the career areas in fruits, vegetables, and nuts?

12. What are the career areas in turfgrass?

EXPLORING

1. Make a list of horticulture businesses in the area. Use the telephone directory to help identify the names of the businesses. Consult an extension agent or horticulture specialist for assistance in this activity. Categorize each business into one of the career areas included in this chapter: landscape horticulture, floriculture, fruits and vegetables, turfgrass, and occupations in education, extension, and research. In some cases, you may have businesses with careers that do not fit into one of these areas. If so, develop your own system for classifying these.

2. Investigate job opportunities in areas of horticulture in your community. Use a variety of sources, including the newspaper, employment offices, Internet postings, interviews with counselors, and visits to personnel offices. Assess which opportunities may match best with your interests and skills. Prepare a report on your findings.

3. Prepare a poster or a bulletin board that depicts a horticulture occupation in the local area. Make photographs and include them in your work.

4. Interview a person who works in horticulture. Determine the nature of their work and the requirements needed for entry. If possible, job shadow them for a day or more. Prepare an oral or written report about your observations.

3

HORTICULTURE AND THE ENVIRONMENT

People want to have a good quality of life; they want to have the things they need to live and enjoy a healthy environment. Horticulture plays a big role in helping people achieve these "wants" by making the earth a better place to live.

Horticulture contributes in many ways to the environment. Plants add beauty, help prevent pollution, and provide oxygen for people and animals. Plants also help conserve and improve the earth's natural resources. Unfortunately, some horticulture practices can create environmental problems. Using pesticides incorrectly can cause pollution, and not using a ground cover can result in soil erosion.

A strong interdependent relationship exists within the environment. Living things depend on other living things and on nonliving features of the environment. The emphasis in horticulture is on living things—primarily living plants!

3-1. Our environment is all of the factors that affect life.

39

OBJECTIVES

This chapter is about the important relationships between horticulture and the environment. It has the following objectives:

1   Explain environment and the issues associated with maintaining a good environment

2   Describe how water resources are related to horticulture

3   Explain how fertilizers are environmental concerns

4   Explain how pesticides pose a threat to the environment

5   Describe the relationships between horticultural practices and wildlife

6   Explain the relationship of wetlands to horticulture practices

7   Describe the role of best management practices

TERMS

coastal wetland
environment
eutrophication
habitat
hydrologic cycle
infiltration
inland wetland
intensive land use
macroenvironment

microenvironment
nitrogen cycle
nonpoint source pollution
plant environment
point source pollution
pollution
wetland
wildlife

KEEPING A GOOD ENVIRONMENT

The **environment** is composed of all of the factors that affect the life of living organisms. It is the surroundings of plants and animals—and people! The environment is made of living and nonliving things. The living things are biotic factors, while the nonliving things are abiotic factors.

Plants exist in close association with each other and their environment. Each part of a natural or artificial environment effects the survival and quality of plants. The **plant environment** is the above and below ground surroundings of a plant. It consists of related and interacting forces that determine the adaptation and growth of plant species. The large atmosphere above a plant is known as **macroenvironment** and the area immediately surrounding a plant is the **microenvironment.**

Horticulture improves the environment, but using the wrong horticultural practices can create problems. The benefits of horticulture are primarily from the plants, and the problems are usually created by the methods people employ when using horticulture. For example, litter results when people improperly dispose of wastes; too much pesticide causes pollution; and land left unprotected erodes. Good horticulturists know the proper practices to use to keep the environment safe.

IMPORTANT ISSUES

Today, people are concerned about the environment. Often, their concerns and the approaches to use are not clear cut. This makes it difficult to deal with the issues.

Studying the issues is important. Practices that maintain high quality plants, preserve the environment, and conserve water resources should be followed. Integrated management strategies help assure the least damage to the environment.

Intensive Land Use

Land is often intensively used. **Intensive land use** involves using large fields. Production practices are used to get top yields. These practices change the natural environment of the land. As a result, wildlife habitats may be destroyed, land may erode, or irrigation may deplete water supplies.

Intensive land use issues are not limited to horticulture. Intensive land use applies to grain, other crops, and forests. The practices of intensive land

3-2. No space is wasted in this intensively-farmed California lettuce field.

use change the natural processes of the environment. Further, construction has a huge impact on the environment. Streets, building sites, and other features require much land to develop. This development makes big changes in how the land is used.

Integrated Pest Management

Integrated pest management (IPM) is using approaches that are environmentally friendly to control pests. It includes biological, mechanical, cultural, genetic, and chemical pest control. IPM reduces the adverse effects of pesticide use on the environment.

The goal of integrated management is to balance effective horticulture systems with good product quality. Costs, benefits, public health, and environmental quality are all considered in IPM. Integrated strategies are also used with water, pests, and nutrient management. These strategies help maintain high quality plants while insuring minimum damage to the environment.

Public Demand

The public demands high quality plant materials. People want attractive homes, businesses, golf courses, sports fields, and parks. People also want fruits and vegetables that are free of insect damage. Meeting these demands requires intense management. Practices must be used that reduce damage to resources and the environment.

Grower Dilemma

Growers are in a dilemma. They must produce quality products at a low cost while also dealing with public concern about the environment.

Large volumes of horticulture products are needed to meet the demands of people. These quantities would be impossible without the use of certain potentially damaging pesticides and other practices. Are people willing to pay higher prices for products grown without pesticides?

3-3. People want plenty of good fruits and vegetables.

Biotechnology

Using biotechnology is an important issue for some people. Genetic engineering has created new kinds of plants. Although these new plants have useful features, some people feel that they are not natural and pose an environmental danger. For example, tomatoes and squash have been genetically altered to achieve certain qualities. Even though these products are of better quality, people are unsure of their impact on the environment.

Emotion

Environmental issues often trigger a lot of emotion. Opinions differ about the use of technology. Soil erosion, contaminated water, and pesticide residues are often emotional issues. Questions about risk and safety are also important.

It is hard for people to make good decisions when emotion is high. Since accurate information helps people make the best decisions, horticulturists need to help provide that information. Helping people understand the issues leads to better choices.

BENEFITS OF HORTICULTURE

Horticulture has many environmental benefits. A few examples are:

3-4. An attractive golf course is a benefit of horticulture.

#3

- plants have recreational and aesthetic value
- roots, leaves, and stems help control soil erosion
- plants absorb atmospheric pollutants
- plants serve as noise and sound buffers
- plants control dust
- plants have a cooling effect on the earth in hot weather
- plants filter fertilizers and pesticides from the environment
- landscaping increases the value of real estate
- plants provide wildlife habitat
- plants increase ground water supply by slowing runoff

Two examples are presented here in more detail.

Maintaining Water Supplies

Turfgrass and groundcover have a big impact on water resources. They hold a large amount of moisture in the upper soil layer. Groundcover reduces runoff and surface soil erosion. Moisture infiltration is increased. Horticultural plants act as filters for pesticides.

Using pesticides properly reduces danger. A good cover of turf "holds" the pesticide and reduces leaching. Only 1 percent of the pesticide used on turfgrass leaches below the zone of the turf. This means that only a very small amount might reach the ground water supply. Other excess pesticide is washed in runoff into streams and lakes, degrading surface water. Always use no more pesticide than recommended.

Maintaining the Atmosphere

Plants help keep the air in good condition. They absorb many contaminants, such as ozone, carbon dioxide, and hydrogen fluoride, without sustaining any damage.

Plants produce oxygen for other forms of life. Plants filter the air and remove smog components, such as smoke and dust. Plants provide recreational and aesthetic areas in urban landscapes. They also reduce glare and noise.

DANGERS FROM HORTICULTURE PRACTICES

Many kinds of production practices are used in horticulture. Some of these practices have the potential to damage the environment. This damage is caused by the actions of people—not from the horticultural plants themselves. Here are a few examples:

- excessive and improper use of pesticide creates pollution
- improper soil management leads to erosion
- horticulture crops can deplete ground water supplies, particularly if irrigation water is pumped from aquifers
- horticultural wastes may be improperly disposed of and create pollution, such as those from a vegetable processing plant

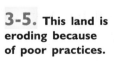

3-5. This land is eroding because of poor practices.

POLLUTION AND ENVIRONMENTAL DAMAGE

Pollution occurs when harmful or degrading materials get into the environment. People are aware of the possible environmental harm that some pesticides can cause, but they are often not as aware of other ways the environment is degraded or made less desirable.

Pollution comes from two sources: point and nonpoint. *Point source pollution* comes from sources that can be readily identified. For example, waste water from a steel mill is point pollution. Both sources are important, but nonpoint source pollution is the most difficult to control.

Nonpoint Source Pollution

Nonpoint source pollution comes from many sources that cannot be specifically identified. Nonpoint sources are diffused in origin, time, and points of discharge. They may relate to climatic conditions, such as rainfall. They can occur naturally or as the result of human activity.

Nonpoint sources in horticulture include:

- leaching and runoff containing nutrients and pesticides may damage the water supply
- soil erosion and runoff losses of sediment and nutrients during construction and adverse weather conditions damage land and water
- beneficial nontarget soil organisms may be killed by pesticides
- wildlife and aquatic organisms may be affected by pesticides
- use of pesticides results in insects and diseases developing resistance to the pesticides
- excessive use of water for irrigation reduces ground water supplies
- some horticulture practices may disturb wetlands and wildlife habitats

Most environmental concerns in horticulture are from nonpoint sources rather than point sources. The concept of nonpoint source pollution can be used to develop an environmentally-friendly horticulture.

Potential Damage

Some horticulture practices can damage the environment. Construction and maintenance work can lead to contamination of surface and runoff water

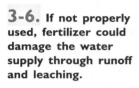

3-6. If not properly used, fertilizer could damage the water supply through runoff and leaching.

by sediment and nutrients. Because of this potential damage, erosion control is often used at construction sites.

Pest populations can develop resistance to chemicals. Using pesticides can have negative impacts on nontarget organisms, including plants and animals. Excessive use of water can create shortages. In some cases, wetlands have been developed into other uses. The benefits of the wetland area on the water cycle are lost.

People can follow steps that reduce the potential for damage. The procedures in this book are sensitive to maintaining a good environment.

WATER RESOURCES AND HORTICULTURE

Water is essential for life. It controls the structure and function of ecosystems. The earth has a limited supply of usable water. The amount of water on the earth does not change, only the location and suitability of that water changes. Nature has important ways of restoring water to good condition so it can be reused.

THE HYDROLOGIC CYCLE

The **hydrologic cycle** is the cycle of water in the environment. Water moves in a series of processes. Precipitation either goes into the soil or runs off. Plants help hold the water from run off so that it can soak into the soil.

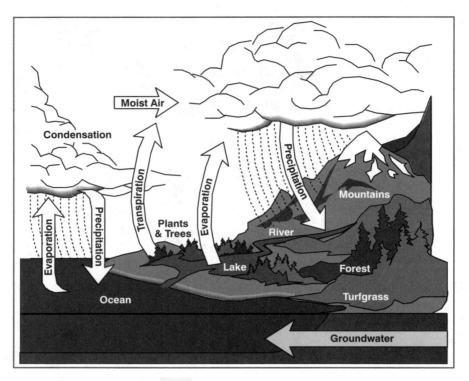

3-7. The hydrologic cycle.

As water moves through the soil, impurities are removed from it, and ground water is restored.

The major sources of water are groundwater, streams, rivers, lakes, and reservoirs. These provide water for drinking, irrigation, industrial use, recreation and natural beauty, and transportation.

How water is used is a concern. Industrial development has been shaped by the location of water. Water for irrigation is often needed for horticulture production. Both industry and agriculture expand where the water supply is adequate.

INFILTRATION

Infiltration is the movement of water into the soil. Horticultural practices influence infiltration. Thatch in turfgrass holds water. Black and hydrophobic layers under the turfgrass cause water to run off. Using synthetic materials, such as black plastic in landscape beds, decreases the rate of infiltration. The quantity and quality of water that becomes ground water depends on how the land is managed.

FERTILIZERS AND THE ENVIRONMENT

Fertilizers are often used with horticulture crops. They provide additional nutrients to help plants grow. Some of the fertilizer may be unused. The unused fertilizer may enter water supplies. Water with high nutrient levels changes the natural ecology of streams and lakes.

Fertilizer should be used only as needed. The rate of application should be appropriate. Excessive fertilizer is wasted and can damage water.

Plants require different amounts of nutrients, with their needs based on the level of management and how they are being grown. Turfgrass on golf greens requires more nutrients than utility turfgrasses on roadsides. An ornamental plant in a high intensity culture requires more nutrients than the same plant in a natural setting.

Eutrophication happens when lakes or streams have too many nutrients. It is caused by excess fertilizer nutrients washing into the lakes and streams. This can create problems, such as excessive algae growth, oxygen depletion, and reduced water clarity, for aquatic life. Oxygen depletion can quickly cause the death of fish and other organisms in the water. Using fertilizer properly will help prevent eutrophication.

NITROGEN CYCLE

An important plant nutrient is nitrogen. Fertilizers often contain high amounts of nitrogen. Excess fertilizer that is not used by plants can potentially damage the environment.

The *nitrogen cycle* is the circulation of nitrogen in nature. Decaying organic matter is converted to ammonium compounds in the soil. These compounds are converted to nitrates through nitrification as a result of bacteria action. The presence of nitrates in drinking water is a major concern.

PHOSPHORUS LOSSES

Phosphorus is lost by surface runoff. Turfgrasses and groundcovers help stop leaching. Soil texture, amount of percolation, fertilizer rate and timing, irrigation, and rainfall affect leaching.

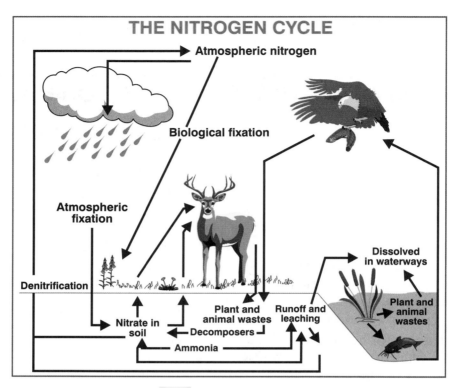

3-8. The nitrogen cycle.

PESTICIDES AND THE ENVIRONMENT

Pesticides are chemical materials used to kill pests. They are widely used in agriculture, forestry, and horticulture. Pesticides help control weeds, insects, diseases, and other organisms that damage plants. Regardless of the benefits, using pesticides is controversial.

Some pesticides leave residues or small traces of the material behind. These residues have been associated with many environmental problems. Here are a few examples:

- a decline in certain bird populations
- the presence of residues in aquatic ecosystems
- the possible effects of carcinogens on human health
- residues in the soil
- contamination of water sources

- destruction of nontarget organisms
- advancement of nonpests to pest status
- the emergence of resistant pest populations

HEALTHY PLANTS TOLERATE LOW PEST POPULATIONS

Keeping plants healthy reduces the need for pesticides. Healthy plants can tolerate low-level pest populations, and can recover more quickly from pest attacks. With severe pest attacks, however, pesticides may be the only alternative.

HOW PESTICIDES GET INTO THE ENVIRONMENT

Pesticides get into the environment in several ways. After being applied, some pesticides change into vapor and enter the atmosphere. Other pesticides are held in union by the soil colloids and thatch. Water can carry pesticides in runoff or when it percolates into the soil. Properly disposing of used pesticide containers prevents materials in the containers from polluting the environment. Decomposition of pesticides often occurs slowly in the environment.

Turfgrasses and groundcovers serve as biotic environmental filters for pesticides. These plant materials hold the pesticide and allow it to degrade. This keeps it from immediately entering the soil or water.

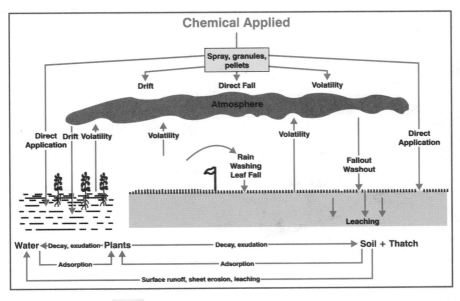

3-9. Fate of chemicals applied to turfgrass.

GUIDELINES FOR USING PESTICIDES

General guidelines for using pesticides should be followed. The guidelines are "environmentally friendly management strategies."

Here are several guidelines for using pesticides:

- ◼ use only chemicals specifically labeled for application
- ◼ properly maintain and operate equipment
- ◼ calibrate equipment to insure even applications of intended volumes
- ◼ properly dispose of unused chemicals and containers
- ◼ use antibacksiphoning devices in chemigation equipment
- ◼ plan the rate and time of the pesticide application relative to precipitation and irrigation
- ◼ maintain a buffer zone between target areas and surface water
- ◼ carefully clean application equipment after use

WILDLIFE AND HORTICULTURE

Wildlife is plant and animal organisms that have not been domesticated. They depend on natural food and habitat. When these are destroyed, the wildlife may not survive. With planning, horticultural installations can be designed to help wildlife.

3-10. Song-birds can be at-tracted with good habitat.

Habitat is the place where wildlife live in nature. The habitat provides food, protection, and other wildlife needs. Golf courses, parks, grounds, production land buffers, and other areas can offer a variety of habitats for wildlife. Ducks can live in ponds. Songbirds can use trees in horticultural areas. Hummingbirds can be attracted to the flowers in a landscape. Areas that have had horticultural improvements support a wide range of wildlife.

CHEMICAL HAZARDS

Wildlife may be exposed to pesticides, fertilizers, and other chemicals. This exposure can come from residues on treated plants and in standing water or wetland areas. In some cases, the wildlife eat chemically-targeted species, such as when a bird eats a poisoned insect. Wildlife also inhale and absorb pesticides. Formulations may have a degree of danger to wildlife. Granular formulations may be picked up and swallowed. Pesticides that are in or on materials wildlife normally eat are especially dangerous. For example, if grain is treated with a pesticide for ants, birds may eat the grain products!

Pesticide applications should be carefully planned. Exposure can be reduced by knowing the activity of various wildlife. Areas that have early morning wildlife activity should be treated later in the day. Areas having nocturnal activity should be treated early in the day. Horticultural areas can be good sources of food, cover, water, space, and nesting habitat for wildlife.

PROJECT INSTALLATION

Projects in or near a wildlife refuge or habitat may run the wildlife away. Allow space for wildlife. Unchanged areas along creeks or in woods provide a safe habitat. Work on one part of a site at a time. This will allow wildlife to move about but remain in the area.

WETLANDS AND HORTICULTURE

Wetlands are swamps, bogs, marshes, mires, ponds, and other places where water often stands. Wetlands are important in the water cycle. Underground water supplies are restored by the actions of wetlands.

Wetlands form between dry uplands and streams or lakes. Wetlands are often difficult to identify because they range from aquatic to terrestrial systems.

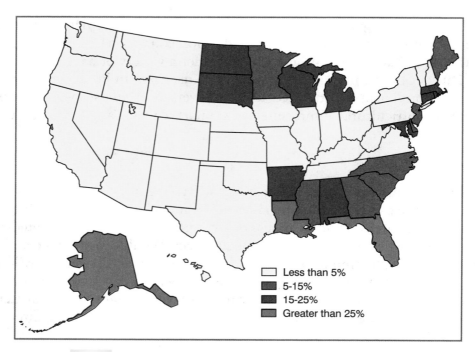

3-11. Percentage of land area in states classified as wetland.

Wetlands were once considered to have little value; thus, many wetlands were destroyed for agricultural production and in urban development. However, with a good understanding of ecological processes and environmental values, wetlands are now viewed as a natural resource. Wetlands provide habitat for wildlife, aesthetic and recreational areas, flood control, improvements in water quality, and shoreline erosion control.

Wetlands are divided into two major groups, coastal and inland wetland ecosystems. *Coastal wetlands* include tidal salt marshes, tidal freshwater marshes, and mangrove wetlands. *Inland wetlands* include freshwater marshes, northern peatlands, southern deepwater swamps, and riparian (stream) wetlands.

WETLAND PROTECTION

Wetlands are now protected by law, and programs to restore previously destroyed wetlands have been established. In some cases, wetland restoration is improving existing wetlands that have been degraded. Wetlands may be reestablished in areas where they were once located.

Wetland creation is forming a wetland in an area where wetlands do not naturally occur or have not recently occurred. Wetland enhancement is changing one or more ways the wetland is used.

WETLAND ECOSYSTEMS

Wetlands are diverse and productive ecosystems. They are the habitat for zooplankton, worms, insects, crustaceans, reptiles, amphibians, fish, birds, mammals, and plants.

Wetlands are homes for water wildlife. Waterfowl, such as ducks and geese, are often found in wetlands. Fish and shellfish live in wetland areas. Muskrats, otters, and beavers make wetlands their home. Wetlands located along coastlines, riverbanks, and lakeshores help control erosion. Wetland vegetation protects the shorelines from erosion by waves, tides, storms, and wind. Inland wetlands serve as buffer zones by slowing water flow in rivers.

Rural and urban wetlands are used for recreation. People enjoy fishing, hunting, canoeing, and bird watching. Wetlands also provide excellent aesthetics for urban retreats. People enjoy the wonders of nature.

3-12. Waterfowl use wetlands for habitat.

MAINTAINING WETLANDS

Here are several suggestions for maintaining wetlands when using horticulture practices:

- mow turfgrass with smaller equipment

- limit fertilizer use

- use slow release fertilizers to avoid nutrient buildup

- direct irrigation and drainage systems away from wetlands

- limit pesticide use to minimize effects on nontarget species

- implement an integrated systems approach to management

- establish buffer zones along the perimeter between wetlands and development areas

- store pesticides where leakage would not get into wetlands

- design paths, streets, and parking lots so that storm runoff does not directly enter the wetlands area.

BEST MANAGEMENT PRACTICES (BMPs)

Properly managing water, nutrients, and pests is a must in having high quality plant materials. Reductions in the use of pesticides, fertilizers, and irrigation water are not always economic issues.

The public sometimes feels that pesticides and fertilizers have a negative effect on the environment. People may oppose their use. Although it is often easy to identify the issues, it is much more difficult to find good solutions.

SYSTEMS APPROACH

A systems approach involves a number of decisions. The proper species and cultivar must be selected. Soil management practices must be used. Nutrient levels, irrigation, and drainage must be managed. Integrated pest management is a part of the systems approach.

Conservation of soil, water, and other natural resources is a primary concern. Best management practices (BMPs) are based on economic thresholds, chemical management, and biological controls. The emphasis is on environmentally friendly approaches.

Integrated systems have many areas. People need much information. Site-specific strategies are used with horticulture programs. Realistic uses are made of cultural practices, water, and nutrients. Thresholds are set for chemical use. Careful monitoring is used.

UNDERSTANDING NATURAL PROCESSES

Plants and animals have certain needs. Each species has requirements for food, water, shelter, and space. Horticultural activities should be planned so that food, water, and shelter are provided close by so that animals can meet their needs without going very far.

The "edge effect" is when one vegetation type meets another type. The border between a marsh and meadow is an edge. The border between a woods and a nursery is also an edge. The area between a golf hole and a body of water provides an edge effect. The "edge" supports more diversity in wildlife than the pure area.

Best management practices involves three things: a habitat inventory, an animal or resident inventory, and an interspersion analysis. Such an analysis locates the present habitat and identifies needs for a desired species. This also helps in identifying what is needed to discourage unwanted wildlife. An example of unwanted wildlife is geese leaving their habitat near a golf course and walking on the golf courses. The manure they leave reduces the quality and appearance of the golf green, as well as being undesirable to people. A more detailed presentation of best management practices (BMPs) is included in Chapter 8.

REVIEWING

*M*AIN IDEAS

Horticulture can help maintain a good environment. Issues are sometimes associated with land use for horticulture. Many of these issues are not horticulture specific. Intensive management of land for crops, forests, and urbanization impact the environment. The use of integrated management systems reduces environmental impact. Pest control measures include biological, mechanical, cultural, and chemical.

The public wants high quality plant materials. This includes golf courses, sports fields, parks, and residential sites. As a result of this demand, a production intensity must be used to assure that people get what they want.

Horticulture systems offer considerable environmental benefits. Benefits include erosion control, chemical filters, dust control, wind control, cooling effects, noise reduction, desirable wildlife habitat, and appreciation of real estate value.

The management of water, nutrients, and pests is important in having high quality plant materials. Reducing the use of pesticides, fertilizers, and irrigation water are not always economic issues. Using natural conditions can help control pests. Natural repellents in plants, insect-eating birds, toads, frogs, and other wildlife can help control pests.

QUESTIONS

Answer the following questions. Use complete sentences and correct spelling.

1. What is the environment and why is it important to people?

2. Why are growers in a dilemma? Explain.

3. What are the benefits of horticulture to the environment?

4. What are the major dangers to the environment from horticultural practices?

5. What is pollution? What is nonpoint source pollution?

6. How does horticulture impact the water cycle and infiltration?

7. What is eutrophication? Why is it important?

8. What are pesticides? Why are they used? What can be done to reduce reliance on pesticides?

9. What are wetlands? How does horticulture relate to wetlands?

10. What is integrated pest management? What is best management practices?

EXPLORING

1. With the assistance of your teacher or county extension agent as resource people, select a turfgrass or horticultural area in your community for study. Identify the management issues that currently impact the environmental quality on that site and the surrounding areas. Place these issues in order of priority based on local community laws and standards. Decide what issues might surface in the next five to ten years. How similar are these lists and how do they differ? Why?

2. Select a wetlands area in your community and conduct an in-depth study of a representative sample of the site. Conduct an integrated resource management study. What impact does the surrounding land use have on the environmental quality of the wetlands? Is there a buffer zone surrounding the wetlands, and did you observe the "edge effect?" Explain your findings. Use your teacher and the local county extension agent as resource people.

4

PLANT ANATOMY

Most often horticultural plants are thought to be pretty flowers. However, they play a major role in the everyday lives of humans. Not only do they provide a food source as in fruits, nuts, and vegetables but they are also a major producer of oxygen in and around our homes and workplace. In addition, horticultural plants beautify our environment by providing a variety of colors, textures, and patterns in the landscape. They enhance architectural structures, frame views, and add fragrances and sounds. Plants also engineer our environment to reduce solar radiation, control erosion, reduce noise pollution, absorb heat, and reduce injuries from sports fields. An added benefit of plants in the environment is that they provide food and shelter for wildlife.

Plants are complex organisms. Plants are made of organs consisting of tissues and cells. Plant organs include leaves, stems, roots, and flowers. Understanding plant growth and the function of plants is very important in horticulture. Horticulturists apply their knowledge of plant anatomy of the different plant organs to promote growth and high quality crops.

4-1. An understanding of the various plant parts and their functions is important in identification, plant care, and producing more when working with horticultural plants.

OBJECTIVES

This chapter provides information on the plant structures and their functions and has the following objectives:

1   Explain different ways plants are classified

2   Describe the differences between annuals, biennials, and perennials

3   Explain the processes of photosynthesis and respiration

4   Identify and describe the functions of the vegetative plant parts

5   Discuss the differences between simple and compound leaves

6   Describe the parts found on a plant stem

7   Explain the structural differences between dicot and monocot stems

8   Describe the differences between taproot and fibrous root systems

9   Identify and describe the reproductive structures of plants

10   Describe the two types of fruits

TERMS

adventitious roots	cultivar	imperfect flower	root hairs
annual	deciduous plant	incomplete flower	scientific name
apical meristem	dicot	inflorescence	secondary root
axillary bud	dormancy	monocot	seed
biennial	endosperm	morphology	seed coat
botanical nomenclature	evergreen	narrowleaf	seed embryo
botanist	fertilization	perennial	simple leaf
broadleaf	fibrous root system	perfect flower	stem tubers
calyx	flower	phloem	stomata
cambium	fruit	photosynthesis	taproot system
chlorophyll	germination	pith	tender plant
complete flower	hardiness	pollination	transpiration
compound leaf	hardy plant	primary root	vegetative phase
cotyledons	herbaceous plant	reproductive phase	woody plant
cross-pollination	hybrid	root cap	xylem

PLANT CLASSIFICATION AND NAMING

Early classifications of plants were based upon segregating those plants that were harmful from those that were useful. They were further divided by their specific uses. People named plants they could eat or that were used for medicine. They named poisonous plants so others could identify which plants not to eat.

CLASSIFYING PLANTS

Scientists use the similarities of plants to classify them into groups. This makes plant identification easier for scientists and horticulturists. Kinds of stems, size, stem growth form, kind of fruit, life cycle, and foliage retention are examples of natural characteristics.

Kinds of stems describes the type of plant, such as a herb (a plant with a soft, nonwoody stem with primarily vegetative parts), a shrub (no main trunk), or a tree (one main trunk). In this usage, the term herb is not describing a spice. Size classification refers to the average mature size of the plant, such as a dwarf shrub or a large tree. Stem growth form describes how the stem stands in relation to the ground, such as erect, creeping, climbing, or decumbent. Fruits can be classified as fleshy fruits or dry fruits.

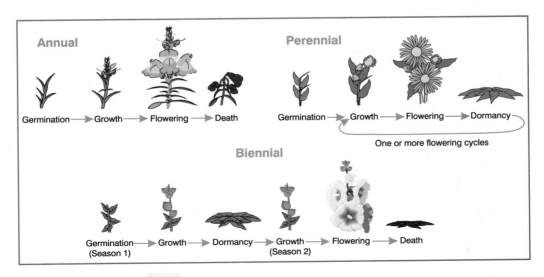

Annual

Germination ⟶ Growth ⟶ Flowering ⟶ Death

Perennial

Germination ⟶ Growth ⟶ Flowering ⟶ Dormancy

One or more flowering cycles

Biennial

Germination (Season 1) ⟶ Growth ⟶ Dormancy ⟶ Growth (Season 2) ⟶ Flowering ⟶ Death

4-2. Plants can be classified by their life cycle.

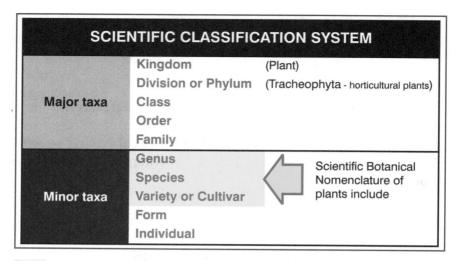

4-3. Horticultural plants in the plant kingdom are contained in a single division, Tracheophyta.

Ornamental plants are often classified by stem type, foliage retention, or life cycle. *Herbaceous plants* have stems that are soft and not woody, such as herbs, certain vines, and turfgrasses, that die back to the ground each year. *Woody plants* include any shrubs, trees, or certain vines which produce wood and have buds surviving above ground over the winter. They are self-supporting. Foliage retention refers to *deciduous plants* which are leafless during a portion of the year (usually winter) and *evergreens* which hold their leaves all during the year. Scientists further classify evergreen plants as narrowleaf (needle-like leaves) or broadleaf (flattened leaf blade) evergreens.

Another means of classifying plants is in their life cycle (vegetative, reproductive, and senescence or dormancy). Annuals (winter and summer) complete their life cycle in less than one year. Biennials complete their life cycle in two years. Perennials may be herbaceous or woody and will grow indefinitely from year to year.

The ability of a plant to withstand colder temperatures is known as *hardiness* (adaptation). A *tender plant* is more sensitive to temperature extremes. A *hardy plant* is less sensitive to temperature extremes. Another classification is whether the plant is edible (fruits, nuts, and vegetables) or ornamental (trees, shrubs, groundcovers, flowers, and turfgrasses).

Dicots are plants characterized by two cotyledons (seed leaves) in their seedling stage. They have flower parts in fours or fives or multiples of these numbers and pinnate or palmate leaf venation. *Monocots* are plants characterized by one cotyledon in their seedling stage, flower parts in threes or multiples thereof, and parallel leaf venation.

The dicots, numbering about 200,000, include most broadleaf herbs, shrubs, and trees. The 50,000 monocots are grouped into such orders as the Liliales (lilies), the Palmales (palms), and the Graminales (grasses and sedges).

SCIENTIFIC CLASSIFICATION AND NAMES

No two plants have the same scientific name. The scientific name helps the scientists and horticulturists identify plants and recognize their characteristics. Common names are often confusing. One plant may have many common names or two different plants may have the same common name. Common names are often regional.

Scientific classification is based on the morphology of plants. **Morphology** deals with plant form and structure. The plant kingdom is separated into about a dozen major phyla or divisions, and horticultural plants are contained in a single division (Tracheophyta). The categories from kingdom to family are called the major taxa, such as Kingdom, Division, Class, Order, and Family. The Genus, species, variety, form, and individual are labeled the minor taxa. **Botanical nomenclature** is the scientific classification of plants. It includes the Genus, species, and variety or **cultivar** (cultivated variety that retains its features when reproduced).

Scientists and horticulturists use scientific names (botanical names) to name plants. The **scientific name** is the Latin name of the plant written using the Roman alphabet which includes the genus and species names. The first name is the genus (plant group name). The first letter of the genus is always capitalized and the species is lower case, both italicized. An examples would be *Cornus florida* (flowering dogwood) or *Cynodon dactylon* (common bermudagrass). Plants in the same genus have similar characteristics and relate to each other. The second name is the plant's species. A species is made up of plants that often exhibit many more morphological similarities than do members of a genus. Each plant in the same genus has a different species name. A plant's scientific name is the same throughout the world.

The species has always been the basic unit of all taxonomic work, from Linnaeus to modern times. It often gives important information about the plant. Some species names describe a specific characteristic of a plant and others give geographical information that describe the origin of the plant or where it grows in nature. Some species examples that give information would be—*alba* (white), *parvifolius* (small leaves), *altus* (tall), *reptans* (creeping), *americana* (of America), *chinensis* (of China), and *grandiflora* (large, showy flowers). The naming of plants is governed by the International Code of Botanical Nomenclature, and strictly adhered to by botanists globally. A **botanist** is a scientist who studies plants.

When one or more of the populations of plants in a single species is sufficiently different in appearance from the remaining members of the species, it is often given varietal status. A population deviating from other members of the species, but not enough to be called variety, is called a form. In horticultural terminology, cultivar refers to a named group of plants within a particular cultivated group of plants that is distinguished by a character or group of characters.

Scientific classification integrates and summarizes our knowledge of plants, including morphological, genetic, ecological, or physiological characteristics. Classification knowledge assists the horticulturist in predicting the cultural requirements of a plant, plant propagation techniques, and new plant development.

PLANT LIFE CYCLES

Life cycles of plants refer to the length of a plant's life. Plants generally fall into three groups—annuals, biennials, and perennials. Typically, plants pass through three phases of development. The **vegetative phase** begins when a plant seed germinates and grows producing leaves, stems, and roots. Some plants remain in a vegetative phase briefly while others continue growing vegetatively for years. The common century plant (*Agave americana* L.) may remain in the vegetative phase for up to 100 years before maturing to the reproductive phase.

The **reproductive phase** is when a plant flowers and produces fruit. The type of plant and environmental conditions can determine when a plant enters into the reproductive phase. Often, plants growing in favorable environmental conditions, such as adequate temperature, moisture, and nutrient levels, will delay the plant from flowering. On the other hand, placing a plant under unfavorable environmental conditions can cause plants to enter into the reproductive phase.

During the life cycle of plants, most will have a phase of slowed or inactive growth. This is known as **dormancy**. Dormancy can occur in seeds as well as mature plants. Plants enter dormancy usually when adverse growing conditions are present, such as cold, drought, or short daylight periods. They remain dormant until conditions are favorable for growth to resume. For example, many trees in North America will drop their foliage and slow or even stop growing during the winter season as they remain dormant.

ANNUALS

Annuals are plants that germinate from seed, grow to maturity, flower, and produce seed in one growing season. In annuals, flowering, then production of seed, is a signal that the plant has entered into the final stage of growth and death will soon occur. Often times gardeners will remove dead or dying flowers from annuals so that the plant will continue to live and bloom for a longer period of time. This process is known as dead-heading.

Annuals are often divided into groups according to what season or climate is best suited for their growth. Summer annuals are sensitive to cold temperatures and are killed by frost. These plants are normally planted in the spring after the last frost and grown until fall. Petunia, marigold, and tomatoes are examples of summer annuals.

Winter annuals, on the other hand, are planted in the fall, grow through the winter, and mature in the spring. Broccoli, spinach, and pansy are examples of winter annuals.

4-4. Petunias are an annual. They complete their life cycle in one growing season.

BIENNIALS

Biennials are plants that complete their life cycle in two growing seasons. During the first season, usually summer, the plants grow vegetatively then become dormant in the winter. The following spring the plants produce flow-

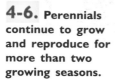
4-5. Biennials complete their life cycle in two growing seasons.

ers and fruit then die. Hollyhock, sweet william, cabbage, and beets are common biennials.

PERENNIALS

Perennials are plants that may be herbaceous or woody and live for more than two seasons. Perennials may be placed in subgroups of herbaceous or woody types. Above ground portions of herbaceous perennials generally die in the winter but grow new shoots and leaves the following spring from the below-ground portions of the plant. Strawberries, asparagus, and daffodils are examples of herbaceous perennials. Woody-type perennials remain alive during the winter season, but growth is slow or the plants become dormant. Trees, shrubs, and some vines are woody perennials.

Woody perennials may be either deciduous or evergreen. Evergreen plants

4-6. Perennials continue to grow and reproduce for more than two growing seasons.

retain leaves at all times. They do drop some of their leaves, but not all at once. Pine trees, junipers, and hollies are classified as evergreens. As deciduous plants become dormant, they will drop all their leaves after the fall season. Apple trees, maples, and flowering quince are classified as deciduous.

PHOTOSYNTHESIS AND RESPIRATION

Plants are living organisms. They have complex chemical processes that direct growth and development. Photosynthesis and respiration are two major processes. These and other life processes are regulated by hormones and conducted by enzymes.

PHOTOSYNTHESIS

Green plants have the unique ability to produce their own food. The main function of leaves is to manufacture food for the plant through a process known as *photosynthesis*. This process is a series of chemical reactions in which carbon dioxide and water are converted in the presence of light to sugar and oxygen. *Chloroplasts* are specialized structures within individual leaf cells. As light enters the chloroplast, chlorophyll will absorb this energy.

Chlorophyll is the green pigment contained in the chloroplast of plant cells involved in the manufacture of food made by the photosynthetic process. This is a complex process that takes place in two successive steps—the light phase (Hill reaction) and the dark phase (Calvin cycle). The light phase of photosynthesis is not influenced by temperature. However, the dark phase reaction rates will increase as the temperature increases and slow as the temperature decreases.

Light strikes chlorophyll molecules and initiates the conversion of light energy into chemical energy. Energy storing molecules ATP and energy-carrying molecules $NADPH_2$ are created in the light phase of photosynthesis. During the Hill reaction, water molecules

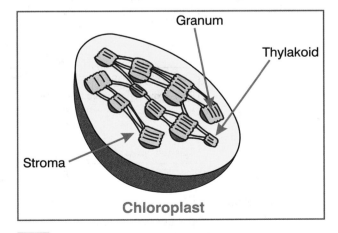

4-7. Chloroplasts are structures within individual plant leaf cells that contain chlorophyll.

are split into hydrogen and oxygen atoms. Two oxygen atoms bond together and form O_2. This gas escapes through the leaf's stomata into the atmosphere. **Stomata** (singular: stoma) are tiny pores in the epidermis of a leaf.

In the dark phase of photosynthesis, carbon dioxide and hydrogen atoms combine with a carbon-acid to produce glucose and fructose, simple sugars. Some of these sugars are recycled, while others are stored in plant tissue.

Plants are divided into C_3 or C_4 groups. Horticultural plant examples are listed in Table 4-1. These groups are based upon whether a plant produces a three-carbon or four-carbon molecule during the Calvin cycle. C_4 plants have a higher relative photosynthetic rate than C_3 plants. They are able to continue photosynthesis under higher temperatures and during higher light conditions. C_4 plants use carbon dioxide more efficiently and are able to manufacture sugar when the stomata are only partially open during hot sunny days. Generally, C_4 plants are adapted to more tropical conditions while C_3 plants usually grow in a temperate climate.

Table 4-1. Relative Photosynthetic Rate for Horticultural Plants	
C_3 Plants	
Orchids and desert plants	1–10
Woody evergreen plants	5–15
Perennial flowers	5–20
Woody deciduous plants	15–30
Kentucky bluegrass	15–30
Annual flowers	15–35
C_4 Plants	
Crabgrass and other grass-type weeds	60–70
Fast-growing broadleaf weeds	60–75
Sweet corn	50–80
Zoysiagrass	60–80
Bermudagrass	80–90

Glucose and fructose fuel plant growth. They are the source of energy for the plant's life processes. All live cells in the plant benefit as the sugars are transported through conductive tissue to the rest of the plant. The simple sugars can also be combined to form a more complex sugar, sucrose.

The sugar molecules may be processed further to form starch and cellulose. These are huge molecules resulting from the bonding of thousands of glucose molecules. Starch serves as the principle way in which food is stored for plants. When needed, it is easily broken down into glucose or converted into other plant products by plant enzymes. The cellulose is applied to cell walls for strength and rigidity. Once sugars are converted to cellulose, they are not recycled for other purposes.

A simple chemical equation for photosynthesis follows:

$$6CO_2 + 6H_2O \rightarrow C_6H_{12}O_6 + 6O_2$$

This equation is interpreted as six molecules of carbon dioxide (CO_2) and six molecules of water (H_2O) combined together will yield one molecule of a carbohydrate ($C_6H_{12}O_6$) and six molecules of oxygen (O_2) which is released into the atmosphere.

High rates of photosynthesis contribute to healthier plants and crops develop more rapidly. Growers strive to maintain optimum levels of limiting factors, such as water and light. They carefully monitor watering and supplemental lighting may be used. Also, some greenhouse operations are equipped with carbon dioxide burners to raise CO_2 levels and speed plant growth.

CELLULAR RESPIRATION

The chemical process known as cellular respiration is the reverse of photosynthesis. In **cellular respiration**, sugars made in photosynthesis are broken down into simpler molecules. In the process of breaking the chemical bonds, energy is released. The energy is applied towards growth and development of the plant.

Cellular respiration is extremely important in the growth and development of plants. All living plant cells use energy to live and function. To germinate, a seed needs to break down the food that has been stored in the cotyledons or endosperm. For root cells to function, they need to convert sugars to energy.

Cellular respiration involves sugars produced in the photosynthesis process along with oxygen and water. In fact, oxygen is a critical ingredient to the reaction. This explains why plant roots need a medium that is well aerated. In the reaction, chemical energy is released when the molecular bonds of the sugar molecules are broken. The extracted energy, ATP, drives a variety of chemical reactions in the cell. By-products of the reaction include carbon dioxide and water.

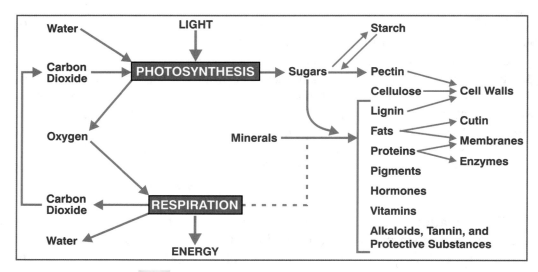

4-8. Diagram of photosynthesis and respiration.

A simple equation for cellular respiration follows:

$$C_6H_{12}O_6 + 6O_2 + 6H_2O \rightarrow 12H_2O + 6CO_2 + Energy$$

Actual plant growth, fueled by cellular respiration, takes place primarily at night when photosynthesis is shut down. With signals from hormones, enzymes or chemical activators are produced. Each enzyme has a specific job. They break down sugars and recombine them with nitrogen and other minerals. Many complex molecules are produced. The speed at which chemical reactions occur is influenced by temperature. Plant metabolism is slowed in cool or cold temperatures and is more rapid in warmer temperatures.

VEGETATIVE STRUCTURES OF PLANTS

The principle structures of a vegetative plant are: leaves, stems, and roots. These structures or organs work together to carry out the plant's growth processes.

LEAVES

The obvious and more diverse structures of plants are the leaves. The most common color of leaves is green. However, they may be other colors such as red, yellow, or purple. Leaves may be large, smooth, and flat, or small, hairy, and fleshy. Whatever the color, size, or shape, leaves are very helpful in the identification of the plant. But that is not their only purpose. Leaves carry out several important plant growth processes which are essential to the plant's survival. Green plants have the unique ability to produce their own food through photosynthesis.

Leaf Functions

Leaves are thin organs containing specialized cells. There is an upper and lower epidermis which is a protective layer of cells. The epidermis cells have an exterior coating that is waxy. The waxy coating, called the **cuticle**, serves to prevent excessive water loss from the leaf tissues.

Leaves have the capability to "breath" through tiny pores in the

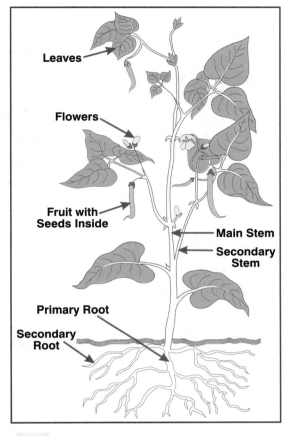

4-9. Principle parts of a plant are shown here.

epidermis called stomata. These openings allow the exchange of oxygen, carbon dioxide, and water vapor. Most stomata can be found on the underside of leaves. The loss of water from the plant through the leaves in the form of water vapor is referred to as **transpiration**. The stomata are opened and closed by a pair of guard cells. When water is plentiful and light is shining, the guard cells are pumped full of water. This causes the cells to push apart, creating an opening to allow an exchange of gases.

The stomata close when water leaves the guard cells, causing the cells to collapse. Stomata close at night. They also close when the plant is experiencing water stress. If the medium dries out or if the roots cannot replace the water lost by transpiration, the plant tries to conserve water. The plant might even wilt or become limp. When the stomata close, carbon dioxide is no longer available, so photosynthesis stops and the plant may wilt.

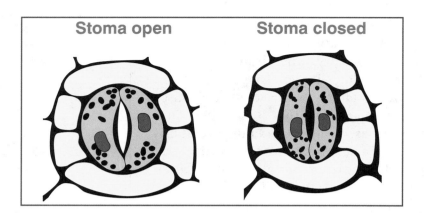

4-10. Tiny pores in a leaf which allow a plant to breath are called stomata (stoma singular).

Parts of Leaves

Leaves come in a great variety of shapes, sizes, and colors. Leaves are also useful when trying to identify plants. There are different parts to a leaf. The large broad part of a leaf is called the leaf blade. The leaf blade provides a large surface area that increases the amount of solar energy absorbed for photosynthesis. The edge of the leaf blade is referred to as the margin. The margins can be one of many forms including wavy, toothed, lobed, entire, or smooth. The leaf blade is connected to the stem by the petiole or leaf stalk. Water and minerals flow through the xylem in the petiole to the cells in the leaf blade.

Leaves consist of several parts which help us to identify them. The major parts of a simple leaf are:

■ **Petiole**—The leaf stem or stalk.

■ **Blade**—The blade is usually a large flat structure that is designed to capture the greatest amount of light.

■ **Midrib**—The largest vein in the center of the leaf which may have smaller lateral veins branching from it.

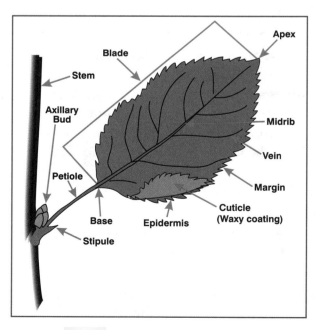

4-11. Parts of a simple dicot leaf.

■ **Veins**—Tiny tubes that form patterns in the leaf blade. The veins move water, minerals, and nutrients in and out of the leaf blade. The arrangement of these veins, called venation, is used as an identifying characteristic of plants.

■ **Leaf margin**—The leaf margin is the outer edge of the leaf blade. The margin may be smooth, toothed, lobed, or in various combinations.

■ **Leaf apex**—The leaf apex is the tip of the leaf. The apex can be rounded, pointed, indented, or various other shapes.

■ **Leaf base**—The leaf base is the part of the blade that is attached to the petiole. The leaf base can be rounded, triangular, heart-shaped, or various other shapes.

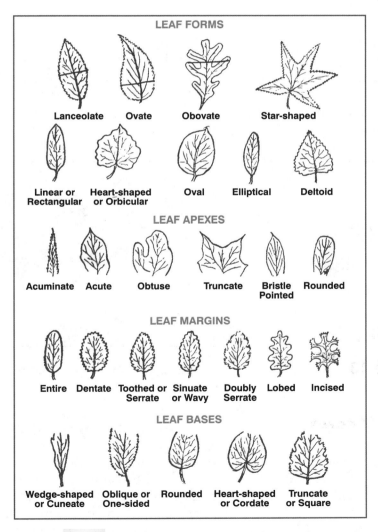

4-12. Leaf forms, apexes, margins, and bases.

- **Leaf covering**—The leaf covering can be waxy, which prevents water loss, or hairy. The hairs on the leaf surface may be few or many.
- **Stomata**—The tiny pores through which gases and water vapor pass in or out of the leaf.

Leaf Venation Pattern

A leaf blade contains many veins. The largest vein in the center of a leaf is the midrib. The smaller veins that branch from the midrib are lateral veins. *Venation pattern* is the arrangement of veins in a leaf. Leaf venation patterns differ among plants. Most plants have parallel, pinnate, or palmate venation patterns.

The major veins in a leaf with parallel venation are parallel to the midrib and are nearly equal in size. They extend the length of the leaf. For example, all true turfgrasses have parallel venation. A leaf with pinnate venation has the midrib with smaller, lateral veins branching from it. Palmate venation has three or more major veins that extend from the base of the leaf blade. Smaller veins branch from these main veins.

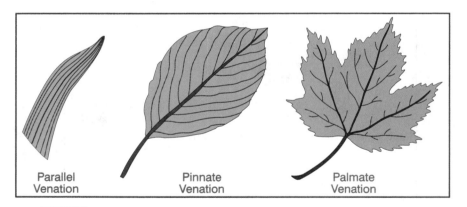

Parallel Venation   Pinnate Venation   Palmate Venation

4-13. Leaf venation patterns include parallel, pinnate, and palmate.

Types of Leaves

Leaf complexity describes the form of a leaf. Leaves may be simple or compound. A *simple leaf* consist of a single leaf blade and a petiole. A *compound leaf* is made of a petiole and two or more leaf blades called leaflets. To determine the type of leaf, one must first look for the position of the axillary bud. The *axillary bud* is a bud located at the base of the entire leaf. Leaflets

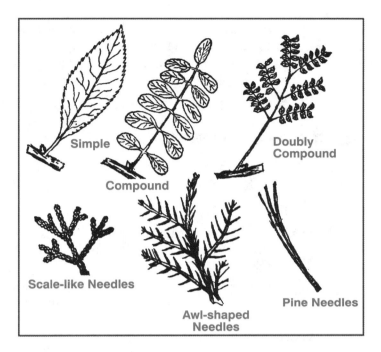

Types of leaves.

do not have axillary buds. English ivy, oak trees, and corn have simple leaves, while poison ivy, hickory trees, and roses have compound leaves.

Compound leaves may be pinnately compound or palmately compound. On a palmately compound leaf, all the leaflets come from a point at the tip of the petiole. A pinnately divided leaf has leaflets arranged along both sides of the petiole. If no leaflet occurs at the end of the petiole, it is even pinnate. If it has a leaflet at the end of the petiole, it is odd pinnate.

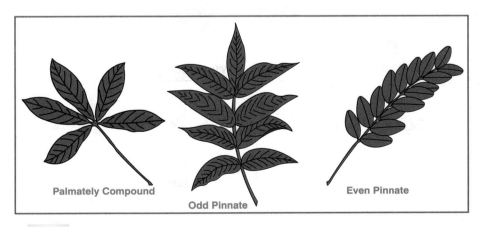

4-15. Compound leaves may be pinnately compound or palmately compound.

Foliage is usually considered as broadleaf or narrowleaf. **Broadleaf** plants have flattened leaf blades, while **narrowleaf** plants have awl-like, scale-like, or needle-like leaves. Plants such as oak, hydrangea, and holly are broadleaf plants. Narrowleaf plants are usually evergreen cone-bearing plants such as pine, juniper, fir, spruce, and cypress.

Leaf Arrangement on Stems

Leaves are attached to stems in patterns. These different patterns are used to aid in the plant's identification. Three types of leaf arrangements most commonly found are:

- **Opposite**—Opposite leaf arrangement occurs when the two leaves and buds are directly across from each other on the stem. Maple trees have an opposite leaf arrangement.
- **Alternate**—Alternate leaf arrangement occurs when leaves and buds are alternated or staggered along the stem. Oak trees have an alternate leaf arrangement.
- **Whorled**—Whorled leaf arrangement occurs when three or more leaves and buds arise from the same point on the stem. Many of the clump-type grasses, such as pampas grass, have a whorled leaf arrangement.

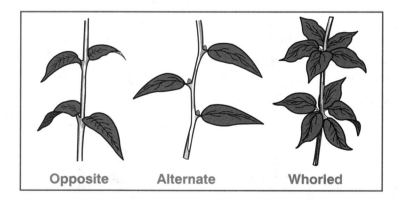

Opposite   Alternate   Whorled

4-16. Three types of leaf arrangements on stems—opposite, alternate, and whorled.

Modified Leaves

There are several modifications of the common green leaf. Some leaves may be thought of as flower petals but are actually leaves located just below the flower. These are called bracts. What is often thought of as the petals of a poinsettia or dogwood are actually the bracts. Also, tendrils of cucumbers or peas are modified leaves used to climb.

STEMS

Stems arise from germinating seeds or from other stems. They grow in length and increase in width. They vary in appearance from one type of plant to another. With practice, one can identify the plants by just observing their stems without leaves.

Stem Functions

Plant stems serve several functions. One is support of leaves, flowers, and fruit. Another function is the transportation of water, minerals, and manufactured food throughout the plant. Green stems also manufacture food just like leaves. In some cases stems function as storage organs of food, as in the Irish potato.

Parts of Stems

Stems have structures called buds. Buds contain undeveloped leaves, stem, or flowers. Bud scales cover and protect these undeveloped parts during dormancy. On some plants, the buds can be difficult to see. Buds tend to produce either vegetative or flowering parts of the plant. Vegetative buds contain immature leaves. Flower buds hold immature flowers. Usually, flower buds are visibly larger than vegetative buds.

The large bud at the tip of a stem is referred to as the terminal bud. The terminal bud is important because it contains the *apical meristem* or the primary growing point of the stem. Hormones, called auxins, are produced in the apical meristem. The auxins migrate down the stem and influence the growth of the stem.

Buds located along the sides of a stem are axillary or lateral buds. Lateral buds can be found where the leaf petiole is attached to the stem. Lateral buds

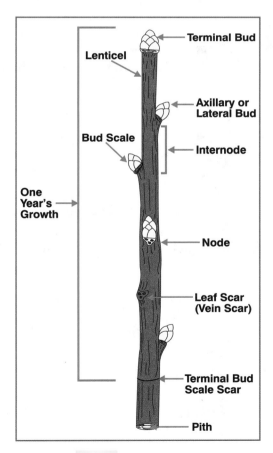

4-17. A typical stem.

closest to the apical meristem are inhibited from developing by the concentration of auxins. The apical bud exerts its dominance over the lateral buds. The term to describe this is apical dominance. Further down the stem, lateral buds develop as the influence of the auxins becomes weaker. A bud also will develop if the terminal bud is removed, as is the case when plants are pruned, pinched, or damaged. This is known as an adventitious bud. Fruit trees have another type of bud known as a fruit bud. Fruit buds are small buds that contain undeveloped flowers and occur next to the axillary buds.

In the winter season parts of stems of deciduous plants are noticeable. Some of the parts of a stem are as follows:

- **Terminal bud**—A bud that contains cells of undeveloped leaf, stem, flower, or mixture of all. A terminal bud is positioned at the tip of the stem. This sometimes forms the central leader or main trunk of plants

- **Bud scale**—Bud scales are tiny leaf-like structures that cover the bud before it opens and begins to grow. These bud scales can have different shapes, sizes, and colors on each type of plant species.

- **Terminal bud scale scar**—Bud scale scars are ring-like scars that can be found inches, or up to a few feet back from, the terminal bud. The scar is left when the terminal bud begins to grow in the spring. One year of growth is represented from one terminal bud scale scar to another. A single stem can have several years of growth on it.

- **Axillary** or **lateral bud**—An axillary or lateral bud which will produce a new leaf or stem is located along the side of the stem.

- **Node**—A node is a point along a stem where leaves or other stems are attached.

- **Internode**—An internode is the area between two nodes.

- **Leaf scar**—The leaf scar is a scar that is left when a leaf drops from a stem. They are different shapes and sizes and useful in identification.

- **Lenticel**—Lenticels are tiny pores located on the stem that allow for gas exchange between the plant and the environment. Some plants, like Cherry trees, have conspicuous lenticels.

Some plants have adaptations in the structures of their stems. These adaptations provide protection for the plant. Thorns are an example of stem adaptations which are sharp-pointed, woody stems.

Internal Structure of Stems

Water, minerals, manufactured food, and other substances are transported in conductive tissues within the stem. The conductive tissue in the

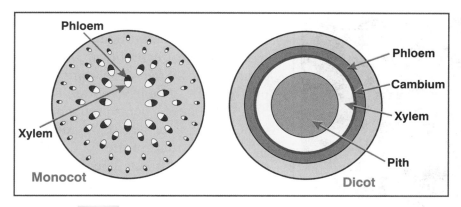

4-18. Internal stem structures of a monocot and dicot.

stem that transports water and minerals from the roots to the leaves is called the *xylem*. Xylem cells also have stiff walls that provide structural support for the plant. Xylem is located in the roots, stem, leaves, and flowers.

Food made in the leaves is transported to the rest of the plant through the *phloem* tissue. Phloem, like the xylem, is found throughout the plant. The roots would die if the plant lacked the ability to move food to the root system.

The location of xylem and phloem within a stem differs with the type of plant. Monocot plants have xylem and phloem scattered in bundles throughout the stem. Monocot plants include narrow leaf plants. Turfgrasses, corn, Easter lilies, iris, and spider plants are examples of monocots. Dicot plants have xylem and phloem in a ring within the stem. Annual rings found in the wood of stems occurs from the growth of the xylem. The *pith*, which stores food and moisture, is the center portion of the stem. Dicots also have a layer of cells called the *cambium* where cell division takes place. These dividing cells become either xylem or phloem cells depending on which side of the cambium they are located. Examples of dicots are trees, most shrubs, African violet, and chrysanthemum.

Specialized or Modified Stems

Stems can become specialized as in the case of bulbs, corms, rhizomes, and stem tubers. These structures serve as underground food and water storage organs. Bulbs are short, flattened stems that bear fleshy food storage leaves. At the base of each fleshy leaf there is a bud. Examples of bulbs include the onion, lily, tulip, and Narcissus. Corms, although they look like bulbs, differ in structure. A corm is a globe-shaped, fleshy underground stem.

4-19. Gladiolus have a modified stem. They produce corms—globe-shaped, fleshy underground stems.

Crocuses and gladiolas are corms. A rhizome is an underground horizontal stem. Iris, asparagus, calla lilies, and most ferns are rhizomes. *Stem tubers* are swollen tips of a rhizome. The Irish potato is a stem tuber. The eyes of a potato tuber are its nodes which consist of leaf scars and buds.

A stem structure similar to a rhizome is the stolon or runner. The difference is the stolon is a stem that grows horizontal above the ground and may produce roots at its tip or at nodes. Strawberries and spider plants produce stolons.

Roots

Roots are the vegetative part of the plant that is not seen and sometimes forgotten. Most living roots are whitish or tan in color and make up about one-half or more of the entire plant body.

Root Functions

The chief function of plant roots is to absorb water and nutrients from the soil then transport them to the above-ground portion of the plant. Water and mineral absorption occurs mostly in the tips of young actively growing roots and transported through the xylem. Next, roots serve to anchor and support the top portion of the plant. Lastly, roots in some plants can store carbohydrates to be used later for a source of energy by the plant and in some cases by animals.

Parts of a Root System

The first structure to emerge from a germinating seed is a root which immediately begins absorbing water and minerals for growth. This root develops into the *primary root*. The primary root continues to grow and branch. Unlike stems, roots do not have nodes or buds, therefore they have an irregular branching pattern.

A root that arises from the primary root is called a **secondary root.** Seedlings or cuttings transplant best when secondary roots have formed. Both primary and secondary roots have **root hairs** found near the growing tip of the root. These are single cell roots that are located a few millimeters back from the root tip. The greatest amount of water and mineral absorption occurs through the root hairs which is transported in the xylem tissue throughout the plant. Since root hairs are so small, they can be easily damaged through improper handling of the plant. Damage to root hairs can occur when one lifts container plants by the stems and leaves.

As roots grow, their tips are protected from coarse soil. This is accomplished by a mass of cells, called the **root cap**. The area directly behind the root cap is where new cells are formed. When a root cap comes in contact with an object—for example, a stone—it will either grow around it or branch.

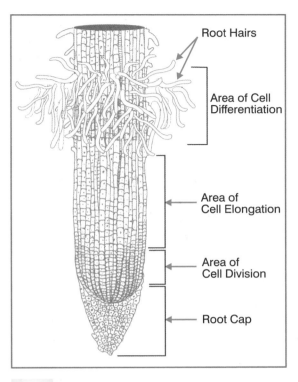

4-20. A microscopic view of a root tip showing root hairs and the root cap.

Types of Root Systems

Plant root systems are grouped according to their growth habits. A **taproot system** is one in which the primary root grows down from the stem with some small secondary roots forming. Walnut trees, dandelions, and carrots have a taproot system.

Plants that branch into a number of small primary and secondary

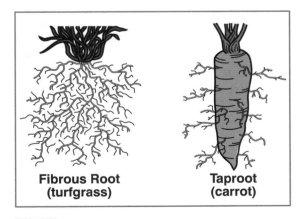

4-21. The plant on the left has a fibrous root system, while the one on the right has a taproot.

roots form a *fibrous root system*. Generally, plants with fibrous root systems grow shallow near the soil line and are subjected to drought and mineral deficiencies. However, plants with fibrous root systems are easier to transplant than ones with taproots. Most landscape plants have a fibrous root system.

Modified Roots

Not all roots begin growth from root tissue. Some roots begin growth from the stem or a leaf. These roots are called *adventitious roots*. Adventitious roots can be found on climbing plants, such as ivy and heart-leaf philodendron. Many greenhouse crops are started or propagated by cuttings. In these cases, a stem or leaf is removed from a parent plant and placed in a growing medium. Given the right conditions, adventitious roots grow from the stem or leaf.

Roots of certain plants can store carbohydrates for a food source. Many of the vegetable root crops, such as sweet potatoes, are not tubers but tuberous roots.

REPRODUCTIVE STRUCTURES OF PLANTS

The reproductive structures of plants consist of the *flowers,* which then produce the fruit and seeds of the plants.

FLOWERS

Flowers are produced commercially for their beauty and fragrance. In some countries, flower petals are collected and used in the manufacturing of perfumes. However, flowers are plant reproductive organs that produce sex cells. Flowers are quite diverse in size, shape, and color. They may be designed to attract certain animals, such as bees, butterflies, bats, or birds. Some may even be designed to take advantage of the wind or rain in order to reproduce. Botanists and horticulturists use flowers to positively identify the plant species.

Parts of a Typical Flower

The exterior of the flower consists of green, leaf-like structures called sepals. The sepals fold back as the flower opens. Collectively, the sepals of a

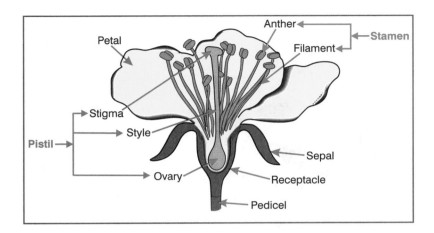

4-22. Parts of a perfect flower.

flower are called the calyx. Petals are located just inside the sepals. Petals appear leaf-like and are often very colorful. The function of brightly colored petals is to attract pollinators, such as insects and birds.

The stamens or male reproductive parts of a flower are arranged around the female parts of the flower. The stamen consists of a stalk called the filament and an anther. The anther produces and holds the pollen. The pollen grains contain the male sex cells.

The female part of the flower is the pistil. The pistil has three main parts, the stigma, the style, and the ovary. The stigma is found at the end of the pistil and it has a sticky surface on which pollen can be caught. The neck of the pistil is referred to as the style. The third part is the ovary, which contains one or more ovules. The eggs are produced and seeds develop within the ovule. As the seeds form, the ovary becomes a fruit.

A typical flower consists of four major parts:

- ■ **Sepals**—The sepals are the green leaf-like structures beneath the petals. The *calyx* is made up of all of the sepals on one flower. They form the protective covering of the flower before it opens. Some sepals have spines or chemicals that protect the flower bud from insects and other animals.

- ■ **Petals**—The petals are usually brightly colored and serve to attract pollinators, such as insects.

- ■ **Stamens**—The male reproductive part of the flower is the stamens. A stamen consists of a filament or stalk which supports the anther. The anther produces the pollen or male sex cells. Depending on the type plant, pollen can be large and easily seen or microscopic.

- ■ **Pistils**—Pistils are the female reproductive parts of the flower. A pistil consists of a stigma, which usually has a sticky surface for capturing pollen; style, which is tube-like and connects the stigma with the ovary; and an ovary, which contains ovules or eggs.

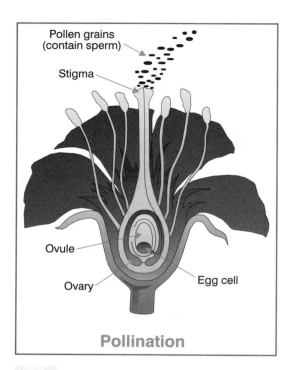

Pollen grains
(contain sperm)

Stigma

Ovule

Ovary

Egg cell

Pollination

4-23. **The transfer of pollen to the stigma is called pollination.**

Pollination occurs when pollen grains are transferred from the anther to the stigma. In nature, flowering plants accomplish pollination a number of ways. Colorful, scented flowers attract birds, insects, bats, and other animals. These creatures unknowingly pick up pollen from the anthers and, when they visit another flower, deposit the pollen on the stigma. The plant rewards pollinators with sugary nectar. Other plants rely on the wind to transfer pollen to the stigma. Since there is no need to attract pollinators, these plants do not produce colorful flowers with large petals, scents, or nectar.

The pollen grains then germinate and a pollen tube grows down the style to the ovary. The cell within the grain of pollen divides to form two sperm nuclei. These travel down the pollen tube to an ovule that holds the egg. *Fertilization* occurs when one sperm nucleus fuses with the egg cell nucleus, forming a zygote that will become a seed. The other sperm nucleus fuses with two nuclei in the ovule to create the endosperm.

When the pollen of a plant pollinates a flower on the same plant, it is called self-pollination. Some plants have this ability; others do not. When the pollen of a plant pollinates the flower of another plant, it is said to be cross-pollination.

Controlled cross-pollination is an important technique used by plant breeders in developing new cultivars. *Cross-pollination* occurs when pollen grains from the flowers on one plant transfer to the stigmas of flowers on another plant. Cross-pollination occurs between closely related plants. Most cultivars are hybrid plants that are the result of cross-pollination. A *hybrid* is the offspring resulting from cross-pollinating two different varieties of a species.

Many cultivars exist for vegetables and flowering plants. The goal of cross-pollination is to create hybrid cultivars that have improved traits or characteristics. These traits or characteristics could include greater tolerance to environmental conditions, increase in size, faster speed of growth or ripening, and resistance to pests.

Types of Flowers

A flower that has all four major parts (sepals, petals, stamens, and pistils) is called a *complete flower*. Apple, lily, and pea are examples of plants with complete flowers. Some flowers lack one or more of the major parts, such as their petals. These are called *incomplete flowers.*

Types of flowers can be grouped as to the presence of stamens and pistils. A flower containing both stamens and pistils is known as a *perfect flower.* Most flowers used in the horticultural industry have perfect flowers. A flower that lacks either stamens or pistils is called an *imperfect flower.* Corn is a common example of an imperfect flower. The male flowers, or stamens, are the tassels, which are located near the top of the plant; while the female flowers, or pistils, produce the ears. The silks in an ear of corn are actually the styles of the pistil. Other examples of imperfect flowers include holly, squash, walnut, and willow.

Flower Inflorescences

To help identify the plants, flower forms are grouped as to their position or arrangement on a stem. Some plants have solitary flowers, such as the tulip, narcissus, and rose. Other plants have flower clusters, known as an inflorescence. A flower *inflorescence* is the arrangement of flowers on a stem. Some flowers, like daisies, are clustered at the tip of the stem, while others may be in spikes, such as gladiolas.

Flower inflorescence types include:

- **Cyme**—A cyme takes on several forms, although they are usually a flat-topped inflorescence. Baby's breath is a cyme.
- **Spike**—A spike is an elongated inflorescence with a central axis along which are sessile flowers. Sessile flowers are attached directly without a stem or stalk. Gladiolas and liatris are spike inflorescences.
- **Raceme**—A raceme is an elongated inflorescence with a central axis along which are simple pedicels of more or less equal length. Examples of racemes include snapdragons, delphiniums, Scotch broom, and stock.
- **Panicle**—A panicle is an elongated inflorescence with a central axis along which there are branches that are themselves branched. Astilbe and begonias have a panicle inflorescence.
- **Corymb**—A corymb is a flat-topped inflorescence having a main vertical axis and branches of unequal length. Yarrow is an example of a corymb.
- **Umbel**—An umbel is an inflorescence having several branches arising from a common point. Queen Anne's lace and amaryllis have umbels.

4-24. A flower inflorescence is the arrangement of flowers on a stem. Shown is an anthurium which has a spadix inflorescence.

- **Spadix**—A spadix is a spike with a thickened, fleshy axis, usually enveloped by a showy bract called a spathe. Calla lily and anthurium are spadix.

- **Catkin**—A catkin is a spike, raceme, or cyme composed of unisexual flowers without petals and falling as a unit. Catkins are found on willows, alders, and birch.

- **Head**—A head is a rounded or flat-topped cluster of sessile flowers. Some plants that have a head inflorescence are chrysanthemum, sunflower, marigold, and dahlia.

*F*RUITS

After pollination and fertilization, the flower petals begin to drop and the ovary and other surrounding parts enlarge and develop into a *fruit*. The fruit is the seed-bearing organ. There are two types of fruits:

- **Fleshy fruit**—The mature fleshy fruit is composed of a soft and fleshy material with seed or seeds enclosed. Blueberry, peach, tomato, and watermelon are all fleshy type fruits.

- **Dry fruit**—A dry fruit consists of seeds enclosed in a fruit wall that is hard and brittle when mature. Pea, sunflower, oak, and elm produce dry fruits.

The fruit in some plants help to distribute the seeds. Such as in the case of mistletoe, a parasitic plant that grows in the fork of tree branches. Birds eat the mistletoe fruit, then fly to another tree and deposit the seed. Also, the fruit in most plants serves to protect the seeds until they germinate.

4-25. The fleshy fruit of a watermelon (left) and the dry fruit of a pea (right).

Seeds

Seeds are the mature, fertilized ovules, or eggs, that are contained in the fruit. Seeds can range in size from a few millimeters to several centimeters. They can be flat, cylinder-shaped, or rounded. Some seeds may be distributed to different locations by animals, while other seeds may be caught in wind currents and travel several miles. Whatever the size, shape, and method of dispersal, all viable (living) seeds are capable of producing new plants.

Internal Parts of Seeds

Seeds contain three primary parts. First, is the *seed embryo* which is a complete miniature plant in a resting stage. The embryo has a root, stem and

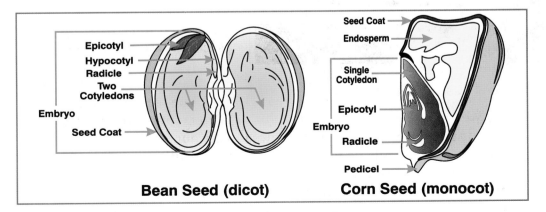

4-26. The major parts of dicot and monocot seeds are shown here using a bean and corn kernel.

one or two seed leaves called *cotyledons.* Monocots have one cotyledon while dicots have two cotyledons.

The second part of the seed is the stored food which contains sugars, proteins and fats or oils. This stored food is used by the plant in its first stage of growth and development. The food may be stored in specialized tissue called the *endosperm.* Plant seeds such as corn have food stored in the endosperm. Others, like beans, have the majority of the food stored in the cotyledons.

The third part of the seed is the *seed coat.* This tissue that surrounds the embryo and endosperm functions to protect the seed from moisture loss, injury, or other unfavorable conditions.

Seed Germination

Germination is a process of events whereby the seed embryo goes from a dormant state to an active growing state. It starts when the seed begins to absorb water and swells. The first to emerge in a germinating seed is the embryonic root-like structure that will soon develop into the primary root. Next, the stem or shoot emerges from the seed. The process ends when the cotyledons begin manufacturing food.

4-27. Seeds germinating in a flat.

Moisture, oxygen, optimal temperature, and sometimes light are environmental conditions that must exist for viable seed germination. The seed must have a sufficient and continual supply of water throughout the germination process. Often times, the seed is given an initial supply of water, but then allowed to dry out, thus the seed dies. Oxygen must also be available to the seed. Seeds have different soil temperature requirements. For example, seed of winter annuals prefer a cooler soil temperature than the seed of summer annuals. In some plants, like celery, light is needed for seed germination.

REVIEWING

*M*AIN IDEAS

The principle structures of plants are the leaves, stems, roots, flowers, fruits, and seeds. Identifying these structures and determining their function is important in naming, maintaining, and reproducing the plants.

Plants go through three phases of growth and development. After they germinate from seed, they produce vegetative growth, which consists of the leaves, stems, and roots. They then enter into the reproductive phase in which they produce flowers that form the fruits. Within the fruits are the plant's seeds. At some time during the plant's life cycle it will become inactive or dormant. Plant dormancy can be brief, as deciduous trees are in the winter season, or extend for several years as dormant seeds lie on the forest floor.

Plants are classified according to their life span. Annuals are plants that grow, flower, and die in one growing season. Biennials live two growing seasons. The first season, biennials grow vegetatively and then the next season they reproduce. Perennials are plants that live for more than two growing seasons.

The vegetative structures produced by plants are the leaves, stems, and roots. The leaves are the most conspicuous part of the plant. Their main function is to produce food or carbohydrates through a process called photosynthesis. Leaves are various shapes, sizes, and colors. Simple leaves have one blade per petiole, while compound leaves have many blades or leaflets.

Stems support leaves, flowers, and fruits. They also transport water, minerals, and manufactured food throughout the plant. Buds, or growing points, are located at the nodes along the stems or at the tip. They can produce new stems, leaves, or flowers.

Roots are the underground portion of the plant. Depending on the types of plants, roots can have either a taproot system or fibrous root system. The main function of roots is to absorb water and minerals. This occurs mostly through the root hairs.

The reproductive structures of the plants are the flowers, fruits, and seeds. The flowers, like leaves, are quiet variable. A complete flower consists of sepals, petals, stamens (male parts), and pistils (female parts). An incomplete flower lacks one or more of these parts. Pollination occurs when pollen, the male sex cells, are transferred from the stamen to the pistil. Fertilization occurs after the pollen grains grow down the styles, a part of the pistil, and fuse with the egg located in the ovary.

Once fertilization occurs, the fruit develops. The fruit is a structure that contains the seeds. Fruits can be fleshy or dry. They protect the seeds inside and aid in the dispersal of them.

Seeds are the mature, fertilized ovules or eggs. They consist of a seed coat, endosperm, and embryo. Seed germination is a process that begins when seeds absorb water. Besides water, seeds also need oxygen, proper temperature, and sometimes light for germination. The germination process is complete when the seedling can manufacture food through photosynthesis.

QUESTIONS

Answer the following questions using correct spelling and complete sentences.

1. What are the three phases of development a plant passes through in its life cycle?
2. What is the difference between annuals, biennials, and perennials?
3. What is the major function of a leaf?
4. What are the major components in the photosynthesis process?
5. What are the parts of a leaf? (Sketch a leaf and label its parts.)
6. What is the difference between a deciduous plant and an evergreen plant?
7. What are three types of leaf arrangements most commonly found? (Sketch and label the three arrangements.)
8. What are the major parts of a stem? (Sketch a stem and label its parts.)
9. What are modified stems? Give three examples.
10. What are the major parts of a root system?
11. How are taproot and fibrous root systems different?
12. What are the four major parts of a flower?
13. How are perfect and imperfect flowers different?
14. How are fleshy and dry fruits different?
15. What are the three primary parts of a seed?
16. How is germination accomplished?

EXPLORING

1. Make a list of the fruit, vegetable, and ornamental plants grown in your area. Classify them as annuals, biennials, or perennials.

2. Make a collection of leaves of the common plants in your area. Observe their parts. Classify the leaves as simple or compound and how they are arranged on the stem.

3. Obtain a cross-section of stems from a dicot such as an oak tree and a monocot such as corn. Observe the differences. Note the locations of the phloem, cambium, and xylem.

4. Visit a local nursery, garden center, or botanical garden in your area. Study the vegetative and reproductive structures and become familiar with their parts.

5. Dissect a complete flower. Sketch and label the reproductive structures.

6. Germinate some large seeds, like corn or beans, by rolling them up in moist paper towel for 7 to 10 days in a warm place. After they have germinated, observe their parts and examine their root systems under a magnifying glass or dissecting scope.

5

PLANT PROPAGATION

At first, horticulturists could only copy nature to propagate plants from the wild or plants with desirable characteristics. They gathered seeds or divided plants and hoped they would grow. However, this method was sometimes slow. Often they did not get the number of plants when and where they needed them. Eventually, they learned to artificially propagate from cuttings, air layering, grafting, and budding. Today, hundreds of plants can be propagated in laboratories by using small sections of plant tissue.

The various propagation techniques used now allow horticulturists to produce new and better breeds of plants at a faster rate. For example, the wild species of tomato plants, a native species of North America, is spindly with very small, acid-tasting fruit. Over the years plant breeders have bred and propagated desirable characteristics of the tomato plant. So now people can have a large juicy slice of tomato on their hamburger and ketchup on their French fries.

5-1. Dozens of improved varieties of fruits, nuts, and vegetables continue to emerge from greenhouses and laboratories. (Courtesy, Agricultural Research Service, USDA)

OBJECTIVES

This chapter focuses on plant propagation techniques used by horticulturists to reproduce plants and has the following objectives:

1   Discuss the importance of plant propagation

2   Explain the difference between sexual and asexual propagation

3   Describe how to successfully direct seed outdoors

4   Describe the factors involved in planting seeds for transplanting

5   Discuss the various methods of stem cutting propagation and identify each method

6   Describe the various types of growing media used for cuttings

7   Describe grafting and identify three common methods

8   Explain the difference between separation and division in plant propagation

9   Discuss the importance of tissue culture

TERMS

agar
asexual propagation
budding
callus
callus tissue
clone
damping-off
direct seeding
division
explants
genetic engineering

grafting
harden-off
indirect seeding
layering
leaf-bud cutting
leaf cutting
micropropagation
percent germination
plant crown
plant propagation
plantlet

root cuttings
scarification
scion
seedling
separation
sexual propagation
stem cuttings
stratification
tissue culture
understock

PLANT PROPAGATION METHODS

One of the most interesting and important areas of horticulture is plant propagation. All organisms, including plants, reproduce. ***Plant propagation*** is the term used to refer to the reproduction of new plants from seeds or vegetative parts of a plant. Plants reproduced from seeds are reproduced through sexual propagation. Asexual propagation is plant reproduction using vegetative parts of a parent plant like leaves, stems, or roots. Tissue culture is the most recent method of asexual propagation. The methods of propagation and techniques vary with the plant.

SEXUAL PROPAGATION

Sexual propagation is the reproduction of plants with the use of seeds. Since most plants reproduce naturally from seeds, this method is often the easiest and least expensive. Plants grown from seeds are called ***seedlings***.

Seedlings are used for the underground portion of fruit trees and roses. They are also used when a large number of plants are needed, for instance, a lawn or a large bed of summer flowering annuals. In addition, plants grown from seeds are genetically different from their parents. Sexual propagation allows a plant breeder to create new plants.

Creating new and different plants normally takes patience and years of research. Plants that are bred must be genetically similar. In most cases they belong to the same species. The parent plants that are chosen to be bred are cross-pollinated. The pollen is taken from the stamens of one parent and placed on the style of another. The seedlings that are produced have char-

5-2. This horticulturist has developed a propagation technique of taking tiny lily bulbs (bulbils) and, with sequential temperature changes, bring them to flower in 280 days in the greenhouse. No longer must they grow in fields for two to three years. (Courtesy, Agricultural Research Service, USDA)

acteristics which are more desirable, like producing larger flowers or fruit or being more disease resistant.

Seed PRODUCTION

Seed production is widespread. A seed producer can be one who saves seeds from the garden to start plants for next year or one who grows hundreds of acres of plants for seeds. Generally, commercial seed producers locate in areas with an arid climate because of less plant disease problems. A large number of commercial seed producers are located in the Pacific Northwest.

Harvesting and Collecting Seeds

One of the most important parts of harvesting seeds is knowing when the seeds are ripe. If they are harvested when immature, the seeds will never germinate. If they are harvested too late, seed germination will be poor. Only through careful observations can one determine when seeds are ripe. Often, fruit will change color or the seeds will no longer increase in size. Commercial seed producers keep accurate records of the weather and the seeds' development to help them determine the best time to harvest them. Mechanical and hand harvesting are the two methods used to collect seeds. The growth habit of the plant and the type of seed determine the method used. Mechanical harvesting is the fastest, but damage to seeds can sometimes occur. Plant seeds that mature at intervals must be picked by hand. Hand picking reduces the amount of cleaning.

5-3. Hand harvesting seeds from Heavenly Bamboo is a time consuming procedure.

Cleaning Seeds

The type of cleaning method will depend on the type of fruits. In large commercial seed operations, professional cleaning machines clean the seeds. For smaller operations, the dry fruits are gathered and spread to dry. The seeds are then threshed from the pods or capsules.

Cleaning fleshy fruits is a messy task. Large fruits, like tomatoes, are allowed to soften and ferment. The pulp is then washed from the seeds, followed by drying and storing. Smaller fleshy fruits with tough fruit walls, like Magnolia, are cleaned by soaking in water for 8 to 24 hours. In some cases, one teaspoon of baking soda per quart of water is used for soaking to further soften the hard pulpy fruit. After the fruits have softened they are washed to remove the pulp. The remaining seeds are spread to dry and then stored.

Storing Seeds

Seed storage is dependant upon the type of seeds. Most of the dry seeds of annual flowers and vegetables can be stored in a cool, dry place until the next growing season. The best place to store these types of seeds is in a refrigerator with a temperature of 40°F. The seeds need to be placed in air-tight polyethylene bags or glass jars with lids.

Some seeds of trees and shrubs have to be sown immediately. Maple, elm, and many of the tropical plants are short-lived and are sown as soon as they have ripened. Other seeds, like apple, oak, and pecan, are stored in a moist, cool place. These types of seeds are kept in a refrigerator with 40 to 90 percent relative humidity. Most seeds are stored for one year. Those that are stored longer lose their ability to germinate. However, some weed seeds have been able to germinate after remaining in the soil for 50 or more years.

5-4. Proper storage maintains seed viability and vigor. (Courtesy, Ball Horticultural Co.)

Testing and Labeling of Seeds

Using the best seeds saves time and money. Many federal and state laws regulate the shipment and sale of seeds. These laws require that seeds be tested and labeled for trueness of name; origin; germination percentage; pure seed percentage; and percentage of other ingredients, such as weed seeds and inert materials.

After seeds are stored for any length of time they should be retested for percent germination. The **percent germination** is the percentage of seeds that will sprout and grow. Seeds can be tested for percent germination by placing a sample or small portion of the seeds in an environment favorable for growth. For example, 100 tomato seeds are placed on a moist paper towel. They are kept moist and at a temperature of 85°F for 7 to 10 days. After that time, the number of sprouted seeds are counted and divided by 100. This will give the percent germination.

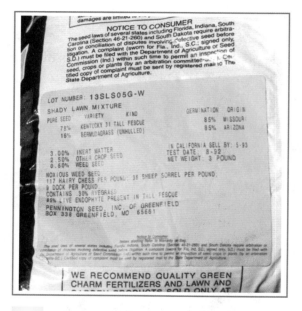

The percent germination will effect the number of seeds that will need to be sown. If one needs 1,000 tomato plants, and the percent germination is 90 percent, then an additional 100 seeds will have to be sown.

5-5. An example of a seed label which is required by law.

PLANTING SEEDS OUTDOORS

Many vegetable and grass seeds are planted directly into the soil outdoors. This method is called **direct seeding.** Plants that are difficult to plant or easily germinate and grow from seeds are planted by direct seeding. In large areas or fields machines are used to plant seeds, while in small areas seeds are planted by hand.

Major factors that will determine the success of the plants grown from the direct seeding method are:

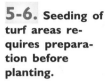 **5-6.** Seeding of turf areas requires preparation before planting.

- **Site Selection**—The site needs to have sufficient sunlight for the plants to grow. Soil drainage is also important. Water should drain from the soil surface after a rainfall.

- **Seedbed Preparation**—The soil needs to be loose, fine textured, and not compacted to allow for adequate moisture and aeration in seed germination and growth. Seeds need to have direct contact with the soil particles to absorb moisture. At the same time, seeds need oxygen to carry out the growth process. In addition, seedbeds need to be free of weeds which compete with the seeds for moisture and sunlight.

- **Planting Date**—The date when seeds are planted is determined by the germination temperature required by the seeds. Soil temperature requirements of seeds vary. If the seeds are planted too early, they will fail to germinate. Also, the planting dates will determine harvesting dates of vegetables and peak bloom dates of flowers.

- **Planting Depth and Spacing**—Seeds need to be sown at recommended planting depths and spacing. The depth and spacing will effect the germination rate as well as the survival of the seedlings. A general rule, if the planting depth is unknown, is to plant seeds at a depth of three to four times their width. The percent germination, environmental conditions, and mature size of the plant will determine the spacing.

- **Moisture**—Seeds need to be irrigated when first planted to supply sufficient moisture for seed germination to begin. The soil should not be allowed to dry out during the germination period or the seeds will stop germinating and die. On the other hand, soil should not remain wet during germination. This will cause the seeds to rot.

PLANTING SEEDS FOR TRANSPLANTING

Seeds can be planted indoors or in a greenhouse in flats or containers with a germinating medium. Once they grow into seedlings, they are transplanted to larger containers or to a permanent location outdoors. This is called **indirect seeding.** Many trees, shrubs, and especially annual flowers and vegetables are grown in this manner. By starting seeds in a greenhouse early, growers can extend the length of the growing season.

The seedling stage begins when plants have produced a set of true leaves. At this stage, seedlings need to have an adequate supply of moisture, nutrients, and light. To provide these requirements to the seedlings, certain cultural practices must be followed. These practices are discussed in later chapters.

5-7. Automated seeders sow seeds directly into plug trays. (Courtesy, *Greenhouse Product News*)

Overcoming Seed Dormancy

Many seeds have no dormancy period and can be planted whenever needed. However, some seeds of woody plants have one or both types of seed dormancy. They are as follows:

Seed Coat Dormancy—Seeds of some plants, like redbud seeds, have a hard seed coat which will not allow the absorption of moisture. The seed coat must be broken or softened. This process is called **scarification.** When scarification is used, care should be taken not to damage the seed embryos.

Embryo Dormancy—Some seed embryos, like in elm seeds, must go through a chilling period before they will grow. Growers overcome embryo dormancy by placing seeds in a moist soil medium at temperatures between 32° and 50°F for a certain period of time. This is called *stratification.* The time period varies depending on the type of seeds.

Germination Media

Flats, plastic containers with drainage holes, and compressed peat pots are used for holding the germination or growing media. The best medium for germination is one that provides proper drainage and aeration, while still retaining adequate moisture. It must be in a favorable pH range and provide essential nutrients for growing plants. The medium also must be free of insects, weeds, and disease organisms.

Germination medium is prepared by mixing a combination of peat moss, perlite, and/or vermiculite. Many growers find that a commercially prepared medium which is made especially for germinating seeds is more economical. Germination and growing media are discussed in more detail in Chapters 6 and 11.

Seed flats or containers are filled with the germination medium and leveled to provide a uniform seedbed. The medium is watered and allowed to drain before planting the seeds.

Planting in Rows or Broadcasting

Seeding machines or hand sprinkling of the seeds are used to plant seeds in the medium. Seeds are planted at a recommended depth found on the seed package or catalog. They may be evenly distributed over the surface of the medium, which is known as broadcasting; or, seeds may be planted in rows. Planting in rows requires more space, but allows for easier transplanting and reduced chances of disease. Seeds are spaced to allow for development. Around 750 to 1,000 seeds are normally planted in a standard flat (11" × 22").

5-8. When using indirect seeding methods, seeds can be planted in rows or broadcasted evenly over the surface.

Moisture, Temperature, and Light

Water is applied after the initial planting of seeds by misting. Thereafter, the flats are misted when the medium begins to dry. Some seed propagators cover the flats with clear plastic or glass to keep the soil moisture from drying out. In these situations, care should be taken to avoid direct sunlight so excess heat will not damage the seeds as they germinate.

The medium temperature is dependant upon the type of plants that are grown. Seed catalogs and other resources contain information for the specific temperature. In general, seed flats are placed on greenhouse benches with heating cables or mats that maintain a temperature of 70° to 80°F.

Many seeds do not require sunlight for germination. Once the seedlings emerge, direct sunlight is essential. If seedlings do not receive enough sunlight, they are weak, spindly, and of poor planting quality.

Damping-off

Germinating seeds and seedlings are susceptible to many fungal diseases. One common disease which causes the stems to rot at the soil line is *damping-off.* The organisms which cause this disease are present in unsterilized media, containers, tools, and benches. The most effective control measures are proper sanitation practices and avoiding conditions which favor the disease, such as a warm, wet medium. Fungicidal drenches can be used, but mostly as a preventive measure.

Transplanting

Transplanting to larger containers or a permanent location is done when the seedlings have developed their first true leaves. Some growers transplant marigold and tomato seedlings when the cotyledons have developed fully and have not produced their first set of true leaves.

Transplanting is a shock to seedlings. Prior to transplanting, seedlings need to be harden-off. Seedlings that are *harden-off* are placed in areas of cooler temperatures with less frequent watering. The harden-off period varies according to the type of plants and their new location. For example, two or three days of a harden-off period is sufficient for most summer flowering annual seedlings. Other seedlings may require one to two weeks.

A few days before transplanting, knives are used to cut between each row of seedlings in both directions. This confines the roots of plant to their own area and encourages them to branch.

When transplanting, care should be used in handling seedlings. Seedlings should be held by their leaves or cotyledons—not stems.

Holes are made with a dibble, stick, or forefinger in their new location and the seedlings are placed in the holes slightly below or at the same depth they were grown in the seedling flats. After transplanting, the seedlings need to be watered to avoid wilting. This also encourages the root system to be established sooner in the growing medium. Shading the seedlings after transplanting also helps reduce the likelihood of wilting.

5-9. Care should be taken when transplanting seedlings.

ASEXUAL PROPAGATION

Asexual propagation is the reproduction of new plants from the stems, leaves, or roots of a parent plant. In this method of plant production no seeds are used—just portions of the parent plant which are placed in soil, soilless media (potting soil), or even in test tubes containing nutrient-rich media. The discovery of asexual propagation allows propagators to produce more plants faster, especially in cases when seeds are difficult to germinate or plants produce seeds that are not viable. One of the most important benefits of asexual propagation is that the plants produced are genetically identical to the parent plant; therefore, they have the same traits as the parent plant. This type of plant is known as a *clone*.

5-10. These rooted chrysanthemum cuttings are newly potted. (Courtesy, Ron Biondo, Illinois)

5-11. The most widely used asexual propagation method for floriculture crops is stem cuttings.

CUTTINGS

The most common and often used method of asexual propagation is the use of cuttings. Cuttings are detached portions of the plant, such as stems, leaves, leaf-buds, or roots that form missing parts to grow into complete new plants. Cuttings are classified as stem cuttings, leaf cuttings, leaf-bud cuttings, and root cuttings.

Stem Cuttings

Portions of the stem that contain terminal or lateral buds are used for *stem cuttings.* Stem cuttings are placed in growing medium in hopes that they will produce roots. The time of year, stage of growth, and type of stem are factors that will determine if the cuttings will produce roots. Stem cuttings are grouped as follows:

■ **Softwood Cuttings**—These are cuttings taken from soft, new spring growth of woody plants. Softwood cuttings root easy but require more attention. Cuttings are usually 3 to 5 inches in length with 2 or 3

5-12. Stem cuttings can be placed on misting benches in the greenhouse.

nodes. The stems are cut at a 45° angle approximately ¼ inch below a node. Then, to reduce water loss through transpiration, one-third of the lower leaves are removed. Large leaves can be cut in half to reduce wilting and thus increasing the chances of rooting. All flower buds should be removed from the cuttings.

■ **Herbaceous Cuttings**—Herbaceous cuttings are also considered softwood cuttings. Numerous succulent greenhouse plants are propagated by herbaceous cuttings. These cuttings are 3 to 4 inches long and contain several leaves. Many species root easily. Examples include geranium, carnation, chrysanthemum, and coleus.

■ **Semi-hardwood Cuttings**—These are cuttings prepared from woody, broadleaved plants. The cuttings are taken in the summer from new shoots that have partially matured. The cuttings are 3 to 6 inches in length and are handled in the same manner as softwood cuttings.

■ **Hardwood Cuttings**—These are cuttings prepared during the dormant season from either deciduous or evergreen plants. Cuttings are usually 6 to 8 inches in length and taken from one-year-old wood. The cuttings are allowed to form callus tissue during the winter. The *callus tissue* is white tissue that forms over the wounded area or base of the cuttings. This is where new roots sprout. In the spring, the callused cuttings develop roots.

■ **Conifer Cuttings**—Conifer (narrowleaf) evergreens are propagated by hardwood cuttings obtained from plants in early winter. They may require several months to produce an adequate root system. Juniper, yew, spruce, and pine are examples of conifers that can be propagated with this method.

5-13. Mother-in-law's tongue is one of a relatively small number of plants that have the ability to reproduce with leaf cuttings.

Leaf Cuttings

A *leaf cutting* consists of a leaf blade or a leaf blade with a petiole attached. Leaf cuttings are used when the plant material is scarce and a large number of new plants are needed. Houseplants or foliage plants are often propagated by this method. Sansevieria, Rex begonia, and African violet are commonly propagated by using leaf cuttings.

Sansevieria leaf blades are cut into pieces approximately 3 to 5 inches long and inserted into a growing medium. The bases of the leaf cuttings will produce new roots and shoots. Leaf cuttings taken from variegated forms of Sansevieria will not produce variegated plants due to their type of cell arrangement. This type of variegated plant can be propagated by division, which will be discussed later in this chapter.

Rex begonia plants can be started by using the leaf blade. The primary veins are cut and the leaf cutting is laid flat on top of the growing medium. The leaf, with the upper side exposed, is pinned down, usually with some type of pins,

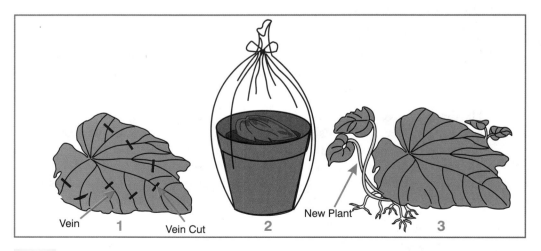

5-14. Leaf cutting of Rex begonia with veins cut is anchored to the medium. The cutting is enclosed in a plastic bag. After several weeks, new plants develop.

such as florist greening pins. New plants are formed at each point where the veins were cut.

African violet is commonly propagated by using the leaf blade with the petiole attached. The petiole is inserted in the growing medium. After a time, a new roots and shoots will emerge from the base of the petiole. The old leaf is usually pinched off and discarded. The new plants are separated and re-planted.

*L*eaf-bud Cuttings

A *leaf-bud cutting* consists of a leaf, petiole, and a short piece of stem with the lateral bud. The leaf-bud cutting method is extremely valuable to the propagator, especially when the woody plant material is scarce and many new plants are needed. Leaf-bud cuttings should only be made from plant material having well developed buds, and healthy, actively growing leaves. The stem piece of the cutting is treated with a root-promoting substance and then inserted in the growing medium with the lateral bud just below the medium surface. A new plant will develop from the lateral bud. Plants such as rhododendron, camellia, maple, and magnolia can be propagated using this method.

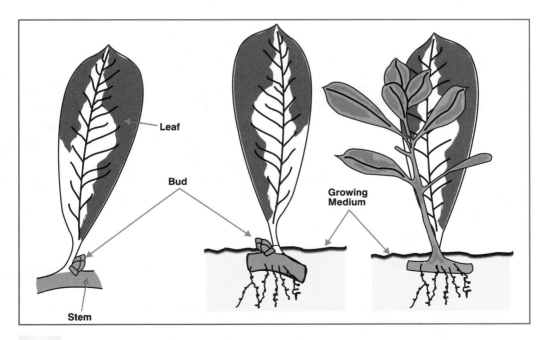

5-15. A leaf-bud cutting consists of a leaf, petiole, and a short piece of stem with the lateral bud. Leaf-bud cuttings are used when plant material is in short supply.

Root Cuttings

Root cuttings are made from root pieces of young plants during late winter or early spring. Roots are dug, cleaned, and treated with a fungicide. Root pieces are cut 2 to 6 inches in length. They are either placed horizontally, approximately 2 inches deep in the growing medium, or vertically, with the root end that was closest to the crown of the plant at the top. To avoid planting the roots upside down, propagators will cut the root closest to the main stem with a straight cut and the root portion further from the main stem with a slant cut.

Plants that naturally produce suckers are easily propagated by root cuttings. However, propagators find that preparation of root cuttings are quite laborious and therefore will use other propagation methods. Raspberries are commonly propagated with root cuttings.

CARE OF CUTTINGS

Timing, collecting, preparing, and providing the proper environment are important factors that determine the success of rooting of cuttings. Cuttings should always be taken from healthy plants which have been watered properly.

Timing and Collecting

The time of year in which the cuttings are taken will effect the percentage of cuttings that root. One should research the particular plant species to determine which month and type of cutting are best for rooting. Cuttings should be taken early in the morning when the water content of the plant is at its highest. Hands and tools should be clean so as to reduce chances of diseases. After cuttings are taken great care should be used to insure that the cuttings do not wilt before they are placed in the growing medium. If cuttings cannot be placed in the medium immediately, they need to be stored in a plastic bag under refrigeration.

Preparing Cuttings

Cuttings should be prepared according to the type method used. Stem cuttings can be lightly wounded at the base by gently scraping the bark on either side of the stem with a knife. This will encourage the formation of roots. The base of each stem cutting should be treated with a root-inducing substance. These root-inducing powders or liquids are available from nursery supply dealers or local garden centers. They contain a synthetic root-promoting substance which hastens the root initiation as well as increase the number of roots formed. Many commercially prepared root-inducing substances contain fungicides which are important in preventing root rot. Some herbaceous plants, like coleus and Swedish ivy, will root without a root-inducing substance. Hardwood cuttings may require a higher concentration of root-inducing substance than softwood cuttings.

Root-inducing Environment

Since many of the types of cuttings have leaves but no roots, it is very important to keep them in a high humidity environment to reduce water loss through transpiration. High humidity around the cuttings can be achieved by simply enclosing the cuttings which have been placed in the growing medium in a plastic bag. Many of the large commercial growers maintain the high humidity around the cuttings by placing them on misting propagation benches in greenhouses. Sunlight is important in the growth and development of cuttings, but direct sunlight increases transpiration. Therefore, cuttings need indirect sunlight, which is achieved by placing shade fabric over the greenhouse.

The type of growing medium used for rooting cuttings varies depending on the type of cutting and the cultural practices. Many different types of media can be used to root cuttings. Whatever type is used needs to hold moisture, provide good aeration and drainage, as well as be disease and weed free. Some of the common types of growing media used for rooting cuttings are:

- **Peat moss and perlite mixture**—This type of medium in a 1 to 1 ratio will hold moisture while providing adequate aeration.

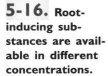

5-16. Root-inducing substances are available in different concentrations.

5-17. High humidity around cuttings reduces the transpiration rate.

■ **Vermiculite**—Vermiculite is sterile and has a high moisture retention. However, large cuttings tend to fall over when straight vermiculite is used.

■ **Sand**—Sand provides aeration but does not retain moisture well. Therefore, this type of medium will require more frequent watering.

■ **Sand and peat moss mixture**—This medium in ratio of 1 to 1 or 2 to 1 increases moisture retention but care must be taken not to allow the medium to become to wet.

Prior to placing cuttings in the growing medium, the medium needs to be moistened. Fertilization of cuttings is not necessary until roots have emerged.

Proper growing medium and air temperature will influence the rooting of cuttings. Temperature will vary with the method of propagation and the type of plants being reproduced. Medium temperature of most stem cuttings needs to be maintained between 70° and 80°F. Propagation heating mats or cables with thermostatic controls can provide the proper temperature. Methods of propagating common nursery plants by species are listed in Table 5-1.

GRAFTING

Grafting is the process of connecting two plants or plant parts together in such a way that they will unite and continue to grow as one plant. Grafting is a complicated process which takes much practice. A plant that has been

Table 5-1. Methods of Propagating Common Nursery Plants			
Plant	**Method**	**Time**	**Root-inducing Substance/ Comment**
Abelia	Softwood cuttings Semi-hardwood cuttings	April–May June–Aug.	
Aucuba	Semi-hardwood cuttings	Summer	
Azalea	Semi-hardwood cuttings	Early July	Rootone/Hormodin #1

Barberry
  Deciduous
  Evergreen
	Semi-hardwood cuttings
	Hardwood cuttings
		July–Sept.
		Nov.–Dec.
			Hormodin #2
			Hormodin #2

Boxwood	Semi-hardwood cuttings Hardwood cuttings	July Nov.–Dec.	Hormodin #2
Camellia	Semi-hardwood cuttings Seeds	May–July	Give bottom heat in late summer Soak in warm water
Cotoneaster	Semi-hardwood cuttings	July	Hormodin #2
Crape myrtle	Seeds	Sept.–Nov.	Germinates in spring
Dogwood	Softwood cuttings Hardwood cuttings	April–June Nov.–Jan.	Use new growth after flowering
English ivy	Hardwood cuttings	Nov.–Dec.	Rootone
Euonymus	Hardwood cuttings	Nov.–Dec.	
Fig	Hardwood cuttings	Nov.–Dec.	
Forsythia	Hardwood cuttings Semi-hardwood cuttings	Nov.–Jan. May–Aug.	Very easy to root
Gardenia	Semi-hardwood cuttings Hardwood cuttings	May–June Sept.–Jan.	
Hawthorn	Semi-hardwood cuttings	July–Aug.	
Hibiscus	Hardwood cuttings	Nov.–Dec.	Varieties not true from seed
Holly	Semi-hardwood cuttings	May–Aug.	Use under glass
Hydrangea	Hardwood cuttings	Nov.–Jan.	
Junipers	Hardwood cuttings	Nov.–Dec.	Rootone
Magnolia	Seeds Softwood cuttings	Fall Spring	Soak few days, remove pulp
Oleander	Semi-hardwood cuttings	Aug.–Sept.	
Pittosporum	Semi-hardwood cuttings	May–Aug.	

(Continued)

Table 5-1 (Continued)			
Plant	**Method**	**Time**	**Root-inducing Substance/ Comment**
Privet	Hardwood cuttings Softwood cuttings	Nov.–Jan. May–July	Very easy to root
Pyracantha	Softwood cuttings	Late spring	
Quince	Semi-hardwood cuttings Hardwood cuttings	May–Aug. Oct.–Dec.	
Redbud	Seeds	Late spring	Scratch, acid bath
Spirea	Hardwood cuttings Softwood cuttings	Sept.–Dec. May–Aug.	Easy to root
Sweet olive	Semi-hardwood cuttings	Late summer	
Wisteria	Root cuttings	Spring	

grafted consists of the *scion,* which is a short piece of stem with two or more buds, and the *understock,* sometimes called the rootstock, which is the lower portion of the graft. The scion develops into the top of the plant and the understock develops into the root system.

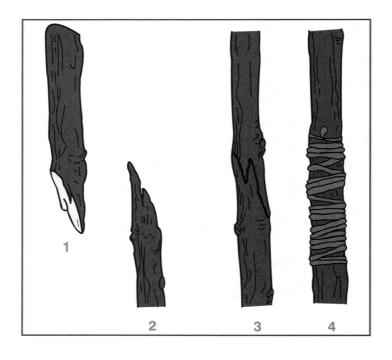

5-18. The scion and understock of the whip-and-tongue graft are prepared (steps 1 and 2). The cambiums are matched (step 3) and the graft is tied to prevent drying (step 4).

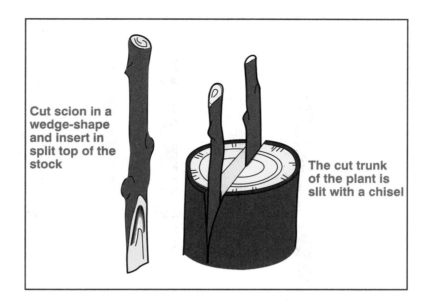

5-19. The cleft graft is a popular way of grafting young trees.

Cut scion in a wedge-shape and insert in split top of the stock

The cut trunk of the plant is slit with a chisel

The main reason for grafting is to asexually propagate plants that are difficult by other methods. Grafting is used quite often in the production of orchard tree, shade tree, and roses. The grafting methods commonly used are:

■ **Whip-and-Tongue Grafts**—This method joins small scion and understock parts together, usually under an inch in diameter. During the winter, many fruit trees are grafted using the whip-and-tongue method.

■ **Cleft Grafts**—Cleft grafts are used to join small scion parts to larger diameter understocks. Cleft grafting is usually done in late winter to top graft new cultivars on to existing tree limbs or rootstocks. It is important in cleft grafting to match the cambium layer of the scion with the cambium layer of the understock.

■ **Bark Grafts**—Bark grafts are similar to cleft grafts in that they join smaller scion wood to larger diameter understock. This method is done in the early spring when the bark easily separates from the wood along the cambium.

In order for graft unions to be successful, several important factors should be considered. First, the scion and understock must be compatible such as grafting an apple onto an apple. Second, the cambium of the scion must be in close contact with the cambium of the understock. Third, grafting must be done at the proper time of year. Last, immediately after grafting, all cut surfaces must be thoroughly covered with grafting wax to prevent the surfaces from drying out.

Shaping the scion to fit the stock   Split bark

Place scion so cambium layers match stock   Scion is secured

5-20. The process of making a bark graft involves several steps. The scion is prepared by shaping to fit the stock, the stock bark is split, the scion is placed so that cambium layers match with the stock, and the scion is secured in position to begin growing.

BUDDING

Budding is similar to grafting except that the scion is reduced to a single bud with a small portion of bark or wood attached. The single bud scion is joined with the understock to form one new plant. Budding is done in the spring or fall when the bark separates easily from the wood of the understock. Plants that have been budded are treated in the same manner as grafted plants. The common methods of budding are:

■ **T-Budding**—This method involves taking buds from one plant and inserting them under the bark of understock. The buds are then tied into place with

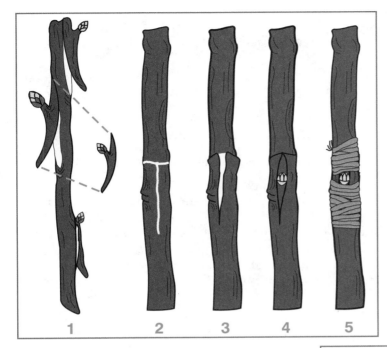

5-21. The scion and understock of a T-bud is prepared (steps 1 and 2). The bark is raised and the bud is inserted (steps 3 and 4). The T-bud is then tied with a budding rubber.

elastic bands called budding rubbers. T-budding is named for the shape in which the bark of the understock is cut. T-budding is used mostly on fruit trees and roses.

■ **Patch Budding**—Patch budding is used when the plant's bark is thick, such as on pecans and walnuts. In patch budding, special knives are used so that the bud patch is the same size as the patch opening on the understock. Patch budding is done when the bark separates from the wood on the scion as well as the understock.

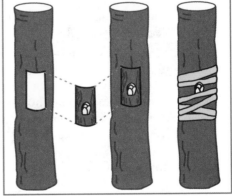

5-22. In patch budding, the opening on the understock should be the same size as the scion.

LAYERING

Layering is a simple method of asexual propagation in which roots are formed on a stem while it is still attached to the parent plant. Larger new plants can be produced using this method. However, only a few plants can be produced from one parent. Some of the types of layering are:

5-23. Air layering is a method that produces large plants.

- **Simple Layering**—Simple layering is accomplished in the spring by bending branches to the ground and covering portions of the branches with 3 to 6 inches of soil. The terminal ends of the branches are left exposed. The portions which are covered by the soil should be slightly cut or wounded. By the end of the growing season new roots should form and the plants are removed from the parent. Many deciduous shrubs and vines can be propagated using this method.

- **Trench Layering**—To propagate by trench layering, a shallow trench is dug near the plant. A stem attached to the plant is placed in the trench and covered with 2 to 5 inches of soil. After a few weeks, roots will develop along the stem and new shoots will form at each node. When the new plant reaches the desirable size, the grower separates it from the parent plant.

- **Mound Layering**—To perform mound layering, the grower prunes the parent plant during its dormant season so it is only 2 to 4 inches above the ground. In the spring, the vigorous new shoots produced by the plant are cut back 2 to 4 inches. After new shoots develop again during the summer, soil is mounded over half of the new shoots. The new shoots will begin to develop their own roots in the mound of soil. The newly rooted plants may then be separated from the parent plant. This method of layering is used for propagating apple and pear trees.

- **Air Layering**—Air layering involves removing a portion of the bark on the stem and placing moist, unmilled sphagnum moss over the exposed area. Then plastic is wrapped and tied around the moss. Root-inducing substance is often applied after the cut is made. After roots develop, the top part of the plant is cut just below the rooted area. The new plant is then potted to grow on its own. Many houseplants can be air layered.

SEPARATION AND DIVISION

Some plants produce vegetative structures which can be separated or divided from the parent plant as a natural means of reproducing. *Separation* is a propagation method in which these natural structures are removed from the parent plant and planted to grow on their own. Whereas, *division* is a method in which parts of the plants are cut into sections that will grow into new plants naturally.

Tulips, daffodils, lilies, and amaryllis reproduce by natural division. Bulbs are produced off the main bulb. These are separated and planted. Corms, including crocus and gladiolas, can be cut into smaller pieces. Each piece of the corm must have a bud that is capable of developing into the stem. Corms also develop small corms called cormels. These miniature corms can be separated and planted.

Some plant structures that can be separated or divided are:

5-24. Lifting and dividing the crowns of perennials can be done in early fall.

■ **Bulbs**—After several growing seasons, bulbs of daffodils, tulips, lilies and amaryllis produce offset bulbs which can be separated from the bulb of the parent plant.

■ **Corms**—Gladiolus and crocus are examples of plants that produce cormels, which are small corms, that can be separated.

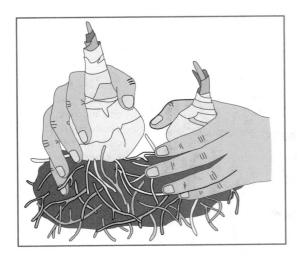

5-25. Daffodil bulbs produce side bulbs that are separated and planted.

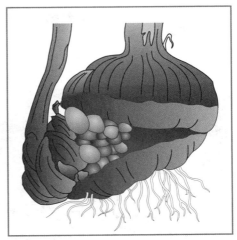

5-26. Gladiolas produce cormels that are separated and planted.

5-27. An Irish potato with eyes (left) can be divided (right) and planted to produce a new plant.

- **Rhizomes and Tubers**—These underground structures can be lifted from the soil, then cut or divided into pieces that will produce new plants. Iris, which produce rhizomes, and Irish potato, which produce tubers, can be lifted and divided into sections or pieces.

- **Plant Crowns**—The *plant crown* is the part of the plant at the soil surface from which new shoots or leaves are produced. Many herbaceous perennials and houseplants are lifted from the soil and the crown is divided into sections which become new plants.

TISSUE CULTURE

The most recent method of asexual propagation is tissue culture. It is also known as *micropropagation.* This highly technical method involves

5-28. Micropropagation is the asexual reproduction of cells or tissues in a closed container. (Courtesy, Agricultural Research Service, USDA)

5-29. This shows plantlets being divided and transferred to jars for continued growth.

taking small tissue or a group of cells from the parent plant and growing it on an nutrient-rich agar gel or liquid in a sterile environment to produce hundreds of new plants. *Agar* is a sugar-based gel derived from certain algae. This agar substitutes for sugars that would normally be produced in a plant's leaves through photosynthesis.

Tissue culture is defined as a method of growing pieces of plants, called *explants*, on an artificial medium under sterile conditions. The explant ini-

5-30. Peach tree tissue culture. (Courtesy, Agricultural Research Service, USDA)

tially forms a *callus*, an undifferentiated mass of cells. The use of certain media can cause this callus to then produce roots, shoots, and other differentiated cells. The plant that was created from the explant and grown by tissue culture, has tiny leaves, stems, and roots that have not yet developed into normal-sized parts of the parent plant. It is called a *plantlet*.

A number of tissue culture methods are used in research and commercial production. Some methods include callus culture, cell suspension culture, embryo culture, meristem culture, and anther culture.

Commercial tissue culturing requires special laboratory facilities and highly trained technicians which can be quite costly. However, this method enables mass production of a large number of plants from a very small amount of parent plants in a short period of time. Also, tissue culturing allows propagators to produce virus-free plants from parent plants. Some of the plants that are now being commercially propagated through tissue culture are orchids, ferns, chrysanthemums, maples, and Kiwi vine.

5-31. A research technician checks on the progress of tiny peach and apple trees being grown from cells which have received new genes through genetic engineering. (Courtesy, Agricultural Research Service, USDA)

Genetic Engineering

Plant breeding practices are being dramatically changed with new discoveries in biotechnology and genetic engineering. Recent discoveries in science have enabled scientists to select and move genetic material from one plant to another. Genes can even be transferred from one species to another. This process, called *genetic engineering*, holds great potential for improving horticulture crops. Desired results can be obtained much more quickly than with traditional breeding methods. There is also greater control over what characteristics will be expressed in the offspring.

Genetic engineering has already been applied to vegetable crops, turfgrass, and grains. A tomato that has a shelf life several weeks after harvest as opposed to several days was first approved in 1994.

Genetically engineered squash that is resistant to viral infection has been developed. Strawberries have been altered to be more resistant to frost damage. Turfgrass has been improved to be toxic to turf insects and diseases. The aesthetic value of floriculture crops could be improved in the future. A true blue rose is possible. Cut flowers that have a vase life of a month could be a reality. Crops could be engineered to be resistant to soil-borne diseases, cutting costs and labor involved in disease control. Poinsettias could be engineered to resist white fly infestations. The possibilities are endless.

REVIEWING

MAIN IDEAS

The various methods used in propagation are important to all areas of horticulture. Plants can be propagated either by sexual or asexual means.

Sexual propagation is the reproduction of plants by seeds. For good germination, seeds must be properly collected, cleaned, stored, and tested. Seeds then can be direct seeded outdoors or indirect seeded indoors. Many environmental factors are involved in the success of the germination, growth, and development of seedlings.

Asexual propagation is the propagation of plants from stems, leaves, or roots of parent plants. Asexual propagation yields plants which are genetically identical to the parent. These types of plants are called clones.

Cuttings are the most popular form of asexual propagation. Stem, leaf, leaf-bud, and root cuttings are techniques that may be used. Cuttings should be taken early in the day and good sanitation practices should be followed. Cuttings are often treated with a root-inducing substance and placed in an environment that favors root initiation and development.

Grafting and budding are methods of joining two plants together so that they will unite and grow as one plant. Different techniques of grafting and budding are used in the production of roses, orchard trees, and shade trees.

Layering is a simple form of propagation in which the plants are produced while still receiving water and nutrients from the parent. Layering produces larger plants than other methods, but only a small number can be produced from one parent.

Separation and division of plants is a form of propagation that is used in the production of bulbs, corms, rhizomes, and tubers. Also, crown division of herbaceous perennial plants can used to produce new plants.

Tissue culture is the most recent form of propagation. Plant tissue placed in a sterile environment can yield millions of new plants in a short period of time.

QUESTIONS

Answer the following questions using correct spelling and complete sentences.

1. What is the purpose of plant propagation?

2. Describe the difference between sexual and asexual propagation.

3. What is percent germination and how can one determine it?

4. Describe the major factors that will determine the success of plants grown from the direct seeding method.

5. What is the difference between scarification and stratification?

6. What is damping-off and how can it be avoided?

7. What are the three types of stem cuttings and how do they differ?

8. When are leaf and leaf-bud cuttings used?

9. How are root cuttings prepared?

10. What are some of the common types of growing media used for rooting cuttings?

11. Name the three types of grafts and draw a simple illustration of each.

12. What is budding and why is it done?

13. Name and describe the two types of layering methods.

14. What is the difference between separation and division of plants.

15. What is tissue culture and why is it so important?

EXPLORING

1. Prepare a bed of flowering annuals by using the direct seeding method.

2. Sow several types of seeds in flats. Compare their percent germination, germination time, and rate of growth. Once developed, transplant the seedlings to larger containers.

3. Collect and propagate various plants from around the community. Observe the propagation method used and rooting time. Prepare a report of your findings.

4. Have a professional nursery manager visit the class and demonstrate the techniques used in grafting and budding. Practice the various methods discussed.

5. Visit a greenhouse operation and observe the different methods of propagation. Prepare a report of your observations.

6

MEDIA, NUTRIENTS, AND FERTILIZERS

Plants need a good environment to grow. Some people only think of the environment for the parts of a plant that can been seen, such as the stems, leaves, and flowers. Although these parts are important, a large portion of a plant is also out of sight.

Plants have an underground environment. This environment is where their roots grow. This environment is very important to the overall health of a plant. A good aboveground environment pays off only if the below-ground environment is in good condition.

Much of the work in horticulture involves preparing a good underground environment. People must know what plants need and how to provide it.

6-1. Nearly half of a plant grows under the ground. (Courtesy, Agricultural Research Service, USDA)

OBJECTIVES

This chapter covers the media in which plants grow, including soil, testing, and fertilization. It has the following objectives:

1 Explain and identify growing media

2 Describe soil materials and structure

3 Name the nutrients needed for plant growth

4 Describe pH and how it is modified

5 Explain fertilization and fertilizer

6 Explain the association between soils and nutrients

7 Describe the use of fertilizer

TERMS

active ingredient
anchor
chlorosis
complete fertilizer
edaphology
elemental fertilizer
fertigation
fertilizer
fertilizer analysis
fertilizer ratio
filler ingredient
growing medium

hydroponics
inert ingredient
leaching
loam
macronutrient
micronutrient
mineral material
nutrient
nutrient solution
organic matter
parent material
pH

pore spaces
soil
soil aeration
soil amendment
soil compaction
soilless medium
soil profile
soil structure
soil test
soil texture
thatch
wear

PLANT GROWING MEDIA

Plants need a source of nutrients and something to hold them in place. The growing medium does this. *Growing medium* is the material in which the roots of plants grow. The medium must be appropriate for the species of plant being grown. (Note: Media is the plural of medium.) *Edaphology* deals with the influence of soil and media on the growth of plants. The environment created consists of synthetic materials, native soil, organic residues, or any combination of these.

6-2. The growing medium provides plant support, water, and nutrients for the plant to grow. (Courtesy, *Greenhouse Product News*)

ROLES OF MEDIA

The role of a growing medium is to provide a plant with nutrients and a place to anchor itself. A growing medium consists of many particles. Moisture and nutrients collect on these particles and in the spaces between them. Plant roots surrounding the particles absorb the moisture and nutrients. Providing the medium that best suits the plant's needs is essential. Without a good medium, plants will not grow.

A good medium provides the nutrients that plants need to grow. *Nutrients* are the substances that roots absorb from the medium and water. If the medium does not contain a needed nutrient, a plant will not grow as it should. It also provides an appropriate place for the roots of a plant to grow. The roots hold the plant in place. This provides an *anchor* for the plant which keeps the plant from falling over and moving about. Plants need oxygen for life processes. The *pore spaces* are the air holes between the growing medium particles which allow oxygen to reach the roots of the plant. The growing medium also provides a favorable environment for microorganisms to grow.

KINDS OF MEDIA

Several kinds of media are used in horticulture. Many involve soil; others are artificial materials, such as vermiculite and perlite. Growers may use several different types and combinations of materials to create growing media mixes for horticultural crops. Any material worked into the soil or applied to the surface to improve plant growth is known as a **soil amendment**. In some cases, plant roots are in a liquid medium or air without soil.

Soil

Soil is the top few inches of the earth's surface that provides for plant growth. Soil is the natural medium in which most landscape plants grow. Many kinds of soils are found in North America. Success with horticulture crops requires a good understanding of soil.

Soilless Medium

A **soilless medium** is one that contains no topsoil. The materials that make up the medium perform a role similar to soil. Soilless media are often made from vermiculite and perlite. Organic components that have been sterilized can be mixed into growing media. Examples of organic components include peat moss, sphagnum moss, leaf mold, and bark. Organic components in a growing medium function to loosen the medium and create larger air spaces between the particles.

Inorganic components result from rocks and minerals, broken down as a result of the physical and chemical process of weathering. Vermiculite is heat-treated mica that has been specially prepared for use as a medium. It is often mixed with other materials to help hold moisture. Perlite is made from volcanic materials. It is used in a medium to provide drainage and aeration. The particles of perlite are somewhat larger than other medium materials.

Growers use various types of growing media. The growing media used depends on the type of crop to be produced. Growers may purchase premixed media or mix their own.

Hydroponics

Hydroponics is a method of growing plants in which the nutrients needed by the plant are supplied by a nutrient solution. A **nutrient solution** contains water with dissolved nutrient salts. Since the roots cannot anchor the plants in the nutrient solution, another method of anchoring must be used. Placing a plant in styrofoam materials which float on the solution is one method of support.

Hydroponic growing systems offer several advantages over soil culture. (1) Plant nutrition is completely controlled through prepared nutrient solutions. (2) Yield per unit area is greater, since plants may be placed closer together. (3) Roots do not spread as much as they do in the soil because water and nutrients are pumped directly to the plant. (4) The need for weed, disease, and insect control is greatly reduced due to the absence of soil. Plant disease in soilless culture is discussed further in Chapter 8.

Hydroponic systems are classified according to the use of substrate materials.

6-3. Terrestrial plants can be grown without soil! Nutrients are provided in a water solution.

Substrate systems:

- **Sand culture**—involves growing plants in sterilized sand and with individual drip irrigation.

- **Gravel culture**—involves irrigating plants grown in gravel for mechanical support.

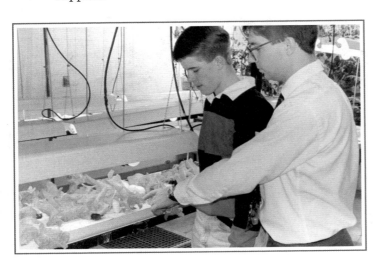

6-4. Lettuce being grown using hydroponics. (Courtesy, Chad Nunley, Indiana)

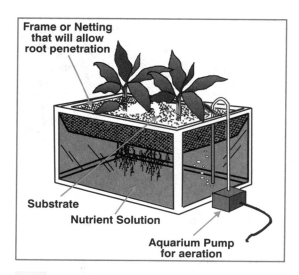

6-5. A simple hydroponic unit can be made from an aquarium.

■ **Bag culture**—utilizes plastic bags that are filled with substrate, such as rockwool, peatlite, and sawdust. Drip irrigation is commonly used to supply the nutrient solution.

Bare root systems:

■ **Aeroponic system**—involves plant roots suspended in air with a fine mist of oxygen-rich nutrient solution sprayed on them at regular intervals.

■ **Continuous flow system**—involves using shallow pools with panels containing plants floating on the surface.

■ **Nutrient film technique (NFT)**—involves using a recirculating, shallow stream of nutrient solution that moves through channels in which the plants grow. The roots are usually covered with a plastic sheet.

Hydroponics is often used commercially to grow high-value crops in greenhouses during the winter. This allows production at a time when the crops could not normally survive outside. For example, tomatoes may be grown in the winter using hydroponics. Individuals can also use hydroponics for growing vegetables, flowers, herbs, and spices.

6-6. These plants are growing in an aeroponic chamber, where nutrient needs are provided by a mist. (Courtesy, Agricultural Research Service, USDA)

Mixtures

Various materials may be mixed to prepare a good medium. Some soils may be improved by the addition of perlite or vermiculite. Peat moss and sphagnum moss may also be used in mixtures. Peat moss and sphagnum moss are the remains of plants and are often found in bogs and swamps. Mixtures may also include decomposed manure, such as from cattle feedlots.

The materials used in preparing a medium must be free of disease, weed seed, dangerous residues, and other materials. Mixtures are often sterilized to kill weed seed and organisms before being used as a medium. Certain kinds of horticulture businesses are inspected to assure that the medium will not transport a plant pest.

SOIL: CONTENTS AND STRUCTURE

Soil is a naturally occurring material. It is formed from rocks, minerals, decaying plants and animals, and other materials. The materials that form soil give it nutrients and structure.

MINERAL MATERIALS

Soil contains several kinds of mineral materials. *Mineral materials* come from inorganic sources. This means that these materials do not come from formerly living plants and animals.

Three kinds of mineral materials are found in soil: sand, silt, and clay. Some soils have a higher mineral content than others. Soils with high mineral content are known as mineral soil.

Sand

The largest mineral particles in soil are sand. The amount of sand

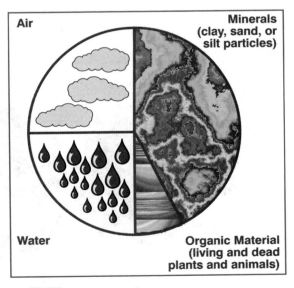

6-7. The four principal components of soil.

in soil varies. Soils with more than 85 percent sand are classified as "sand," not soil. Sand is beneficial in that it assists soil drainage; however, soils high in sand content may not hold enough moisture for plants to grow properly.

S_{ilt}

Silt particles are smaller than sand. These particles are formed by the action of water and other forces that break minerals into silt. Areas near rivers are often high in silt deposits as a result of overflowing water.

C_{lay}

Clay is the smallest size particle in soil. Clay particles fill the spaces between sand and silt particles. Clay holds water and keeps soil moist; however, soils that are high in clay may be slow to dry after a rain. As a result, the soil may form a hard, compact surface, or, when tilled, may form hard clods as the soil dries.

$O_{RGANIC\ MATTER}$

Organic matter is the decayed remains of plants and animals. Decaying leaves, bark, manure, and other plant materials increase organic matter. Organic matter is high in nutrients which are released for plant growth. Since many horticultural crops like soils that have a high organic matter content, crop residue is often left on fields and beds to increase organic matter.

6-8. Three soil components (from left to right) organic matter, clay, and sand.

SOIL TEXTURE

Soil has solid particles and pore spaces. The solid portion includes the inorganic material (sand, silt, clay) and the organic material (peat, humus, etc.). The pore spaces include the large spaces containing air (with various atmospheric gases) and the small spaces containing water and dissolved substances. The relative proportions of these determine the suitability of soil as a growth medium.

Soil texture is the proportion of sand, silt, and clay present in soil. The amount of the mineral materials in soil varies and makes some soils better suited for certain uses than other soils. The most favorable soil has 45 percent inorganic materials, 5 percent organic materials, 25 percent water pore space, and 25 percent air pore space.

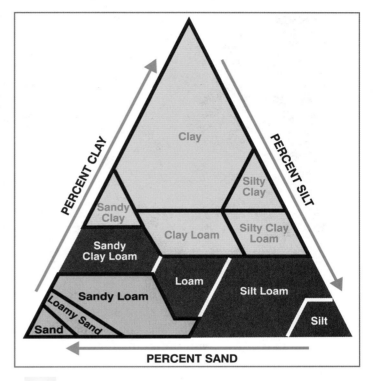

6-9. The soil triangle is used to describe the texture of soil.

A soil triangle is used as a method of classifying soil on the basis of mineral content (texture). Most horticultural crops prefer "loamy" soils. A *loam* has particles of intermediate size and is higher in silt and lower in sand and clay. Loamy soils often have high organic matter.

SOIL STRUCTURE

Soil structure is the physical arrangement of soil particles. Sands constitute lighter soils and clays form heavy soils. Sandy soils are more aerated than clay soils. *Soil aeration* is the movement of atmospheric air into the soil. Plant roots need air to live and grow. The aeration process involves diffusion. Clay soils are more likely to have severe compaction problems.

6-10. Golf course greens are subject to soil compaction due to high traffic in a limited space. Shown here is the famous 18th hole at the Pebble Beach Golf Course in California.

Wear and compaction are two important areas in turf management and landscaping. *Soil compaction* occurs when soil is compressed into a relatively dense mass. The weight of vehicles and people presses on the soil. Plants can't grow well on compacted soil. The soil is hard and unproductive. Vegetable fields may develop hard pans. These are layers of compacted soil just below the depth of plowing. Heavy tractors and harvesting equipment compact the soil. A hard pan restricts plant root growth and water movement in the soil. Subsoiling plows can be used to break up hard pans.

Plants respond to compaction in different ways. Limbs and other plant parts can be broken. Root growth can be reduced. Water and nutrient uptake can be reduced. Plants growing in compacted soil have less tolerance to other environmental forces and greater susceptibility to pests. Plant communities may respond to compaction by the growth of weeds. Some weeds are tolerant of compaction, such as knotweed, annual bluegrass, goosegrass, and white clover.

Wear is the physical deterioration of a plant community resulting from excessive traffic. For example, a path across a lawn is the result of wear. Many people walk in the same place, crushing and destroying the grass and compacting the soil.

SOIL PROFILE

A *soil profile* is a vertical section of soil at a particular location. The profile typically shows a layered pattern of the materials from the surface of the soil down to the bedrock. These layered patterns are divided into horizons.

The horizons include the organic residue layer above the soil surface, the surface soils or topsoil (A), the subsoil layer which contains an accumulation of materials leached from the surface (B), the substratum consisting of partially weathered parent material (C), and the bedrock or parent material (R). *Parent material* is the unconsolidated mass of rock material or peat from which the soil profile develops.

Soil is important as a plant growth medium. It provides nutrients, water, gas exchange, and physical support of plant roots. The layer of organic residue above the surface is *thatch*, a layer of dead, but undecomposed plant material between the soil surface and the living tissue of the host plant.

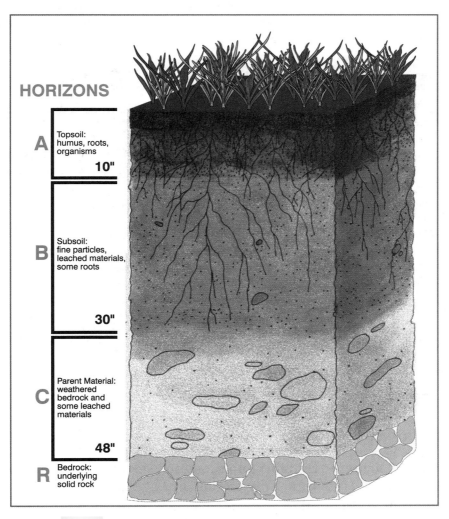

HORIZONS

A Topsoil: humus, roots, organisms
10"

B Subsoil: fine particles, leached materials, some roots
30"

C Parent Material: weathered bedrock and some leached materials
48"

R Bedrock: underlying solid rock

6-11. A soil profile illustrating the location of soil horizons.

PLANT NUTRIENTS

Plants need a constant supply of nutrients to grow and develop properly. Seventeen elements are needed to provide the nutrients necessary for plant growth. Each element serves an important function.

NUTRIENT NEEDS OF PLANTS

Although plants are able to use a few nutrients from the air (carbon, hydrogen, and oxygen), most of the nutrients that a plant needs must be present in the growing medium. Some nutrients are needed in larger amounts than others. Successfully growing plants requires a knowledge of which nutrients are needed by plants and in what amounts.

Table 6-1. Essential Elements for Plant Growth

Name	Symbol	Ionic Form	Ionic Name
Macronutrients			
Primary			
Nitrogen	N	NO_3^-, NH_4^+	Nitrate, ammonium
Phosphorus	P	NPO_4^{-2}, $H2PO_4^-$	Orthophosphates
Potassium	K	K^+	—
Secondary			
Calcium	Ca	Ca_{+2}	—
Magnesium	Mg	Mg_{+2}	—
Sulfur	S	SO_4^{-2}	Sulfate
Micronutrients			
Boron	B	BO_4^{-2}	Borate
Copper	Cu	Cu_{+2}	—
Chlorine	Cl	Cl^-	Chloride
Iron	Fe	Fe^{+2}, Fe^{+3}	Ferrous, ferric
Manganese	Mn	Mn^{+2}	Manganous
Molybdenum	Mo	MoO_4^{-2}	Molybdate
Nickel	Ni	Ni^{++}	—
Zinc	Zn	Zn^{+2}	—

Macronutrients

Macronutrients are the most important nutrients in plant growth. These are major nutrients and must be present in large amounts.

The macronutrients that need to be applied in specialized fertility programs include nitrogen (N), phosphorus (P), potassium (K), calcium (Ca), magnesium (Mg), and sulfur (S). The three nutrients that are needed in the largest quantities are nitrogen, phosphorus, and potassium—the primary nutrients. The three nutrients needed in moderate amounts are calcium, magnesium, and sulfur—the secondary nutrients.

Micronutrients

Micronutrients are essential for plant growth, but they are needed in smaller amounts than macronutrients. Micronutrients are also called trace elements. They include iron (Fe), manganese (Mn), zinc (Zn), copper (Cu), boron (B), molybdenum (Mo), chlorine (Cl), and nickel (Ni).

SOIL TESTING

Soil tests determine what nutrients are present in the soil. This analysis is commonly done in one of two ways—in a laboratory or by the grower.

When using a laboratory, the soil samples must be carefully taken and sent for testing. Soil laboratories are highly accurate, and a specialist may prepare a recommendation based on the test results.

Growers need to be careful in doing their own testing. Most growers buy a commercially available soil testing kit. This kit has all of the materials and instructions needed to perform a soil test, but people need to be sure that the results of the analysis are accurate.

Some laboratories may not test for nitrogen as part of a routine soil test. The recommendations for nitrogen are based on professional judgement by a

6-12. Testing soil samples in a laboratory. (Courtesy, Terra International, Inc., Professional Products)

6-13. Taking a soil sample in turf using a soil probe.

specialist. Nitrogen moves about and is readily leached (washed out) from the soil. Soil and plant tissue analyses should provide similar results.

Most soil tests are for pH, phosphorus, potassium, calcium, magnesium, and sulfur. Calcium, magnesium, and/or sulfur levels are usually taken care of along with the pH. Lime contains high amounts of calcium.

Proper soil testing requires a good sample. The sample tested must be representative of the area. Samples should be collected from several places in the field, bed, or lawn.

Use a soil probe or small shovel to collect seven to nine samples from throughout the site. Place the samples in a clean container. (A dirty container will contaminate the sample.) The samples should be mixed thoroughly together before being tested or shipped to a testing laboratory. When sending the sample to a laboratory for testing, place the sample in a plastic bag and put it in a soil sampling box for shipping. Be sure to label the sample and indicate the plant(s) that are to be grown in the soil. You may also request that a fertilizer report be included with the soil test results.

NITROGEN

Nitrogen is the key element in growing vegetative plant tissue. Levels of nitrogen determine greenness color and density in plants. Chlorophyll, the green pigment contained in the chloroplast of plant cells, must be present for photosynthesis to take place. The dark green color of corn is a result of nitrogen. The amount of vegetative growth is also tied to the amount of nitrogen available to plants.

All plants must have nitrogen to help them recover from damage from elements such as wind, animals, and cultivation. Nitrogen improves a plant's ability to resist disease and to tolerate the effects of heat, cold, and drought.

Nitrogen is taken into a plant as a nitrate (NO_3^-) or ammonia (NH_4^+) ion. The plant then converts the nitrogen into the amide (NH_2) ion. Nitrogen is translocated (moved) to leaf tissue in some plants, such as turfgrasses and herbaceous plants, within 15 to 24 hours.

N	P	K
Functions		
▪ Constituent of amino acids (and thus, proteins and enzymes) ▪ Constituent of chlorophyll (four N atoms in each molecule) ▪ Stimulates carbohydrate utilization ▪ Stimulates root growth and development ▪ Regulates uptake and utilization of other nutrients	▪ Component of DNA and RNA ▪ Affects cell division, root development, maturation, flowering and fruiting, and overall crop quality ▪ Component of high-energy bonds in plant cells, necessary for the release of energy for plant processes	▪ Activates enzymes ▪ Regulates opening and closing of stomata ▪ Regulates water uptake by root cells ▪ Essential for photosynthesis, starch formulation, and translocation of sugars
Deficiency Symptoms		
▪ Activates enzymes ▪ Regulates opening and closing of stomata	▪ Purpling of the stem, leaf, or veins in the underside of leaves	▪ Burn or scorch of margins of leaves, particularly older leaves

6-14. Functions and deficiency symptoms of primary macronutrients.

Nitrogen is used by the plant to make chlorophyll molecules. It is important in amino acid formation, protein synthesis, and nucleic acids. Nitrogen also aids in the transfer of plant characteristics and hastens metabolic reactions with enzymes and vitamins.

When nitrogen is deficient, the proteins in older leaf tissue are transported to younger leaf tissue. This causes older leaves to change color—dark green becomes light green, followed by yellow and copper yellow—and, eventually, necrosis (death) of the leaf blade. The yellowing of the green parts of a plant is called ***chlorosis.***

In tissue analysis, new leaves are used to determine the nutrient content of the plant. The most accurate measure of the nitrogen supply in a plant is tissue analysis. In turfgrasses, the quantity of clippings is the second most accurate measure. Changes in color and density of plants are also good indicators of nitrogen supply.

Denitrification is the process whereby nitrogen is lost from the soil to the atmosphere. In this process, nitrate is converted by certain soil bacteria to various gaseous compounds of nitrogen. Then as these gaseous compounds are lost with soil, the crop-producing nitrogen is also lost.

PHOSPHORUS

Phosphorus helps plants hold and transfer energy for metabolism. Phosphorus is very mobile in the plant and is used repeatedly wherever needed.

Large amounts of phosphorus are found where cell division is taking place in plant tissue. A phosphorus deficiency results in reduced growth and dark to reddish leaf colorations.

POTASSIUM

Potassium ranks just behind nitrogen in the amount needed by plants. Potassium is important in plant life processes. Increased respiration and transpiration, reduced environmental stress tolerance, increased disease, and reduced growth are effects of a potassium deficiency.

Most of the potassium present in fertilizers comes from potassium salt deposits that are mined to make muriate of potash. The word "potash" comes from the term "pot ashes"—ash is high in potassium salts, primarily potassium carbonate.

SECONDARY NUTRIENTS

The secondary nutrients (calcium, magnesium, and sulfur) are found in the organic compounds in plants. Calcium is found in large quantities in the growing regions of plants. It influences the absorption of potassium and magnesium by the roots. It also increases pH and influences the availability of other nutrients. In the soil, calcium is chemically active due to its electrical attraction. Magnesium is a component of the chlorophyll molecule. Sulfur is a component of amino acids and plant vitamins.

The term micronutrients does not refer to nutrients of less importance; instead, it refers to nutrients that are needed in lesser quantities. These nutrients are also called trace elements. Except for iron and manganese which are important as catalysts in enzyme reactions, application of micronutrient fertilizer is seldom needed. Except for molybdenum and chlorine, micronutrients in alkaline soils become highly insoluble. In acid soils, their solubility is so high that toxicity may occur.

SOIL pH

Many important chemical reactions take place in the soil. These reactions may make the soil acidic or alkaline. The degree of acidity or alkalinity of a soil controls the availability of nutrients to plants. Plants grow better in soil that meets their needs.

pH is used to measure the amount of alkalinity or acidity in soil. It is based on the hydrogen ion concentration in the soil. Chemists use a 14-point scale to explain pH. On the scale, 7.0 is neutral. Soil with a pH below 7.0 is acidic. A pH above 7.0 indicates that the soil is alkaline or basic. The greater the distance from 7.0, the greater the extent of acidity or alkalinity. For example, a pH of 4.0 is more acidic than a pH of 5.0.

Samples of soil are tested to determine pH. This test is usually done when the soil is tested for nutrients.

MODIFYING pH

Materials can be added to the soil to modify the pH. Lime (ground limestone known as calcium carbonate, $CaCO_3$) is added to raise the pH. Materials containing sulfur are used to lower the pH.

Most plants grow best in soil with a pH range of 5.5–8.0. Some horticulture plants require a specific pH. If necessary, materials should be added to

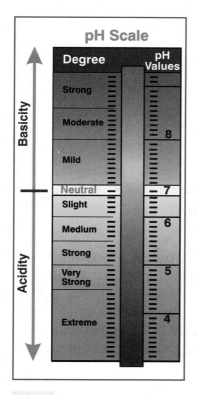

6-15. Soil pH is related to degree of acidity and alkalinity.

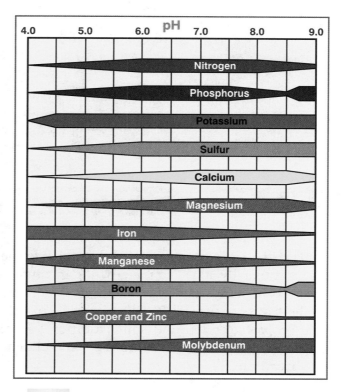

6-16. Nutrient availability as influenced by soil pH.

the soil to change the pH to a level that better meets the needs of the plants currently being grown there. This should only be done on the basis of soil test results.

The color of flowers on some plants is related to pH. The hydrangea, for example, may have blue or pink flowers depending on pH—pink flowers in alkaline soils and blue flowers in acidic soils.

Table 6-2. The pH Range Most Suitable for Some Plants			
pH 5.0–5.5 Strongly Acid	pH 5.5–6.0 Medium Acid	pH 6.0–6.5 Moderately Acid to Slightly Acid	pH 6.5–7.5 Slightly Acid to Slightly Alkaline
Azaleas	Bermudagrass	Apples	Cabbage
Camellias	Fescue, tall	Cantaloupes	Spinach
Irish potato	Hollies	Kentucky bluegrass	Pyracanthea
Watermelon	Kudzu	Lespedeza (annual and Sericea)	
	Peaches	Lima beans	
	Pecans	Peanuts	
	Sweet potatoes	Peas	
	Strawberries	Pepper	
		Sweet corn	
		Tomatoes	
		Turnips	

FERTILIZERS

Fertilizer is any material used to provide the nutrients plants need. Fertilizer is usually added to the growing medium so that it can be absorbed by the roots of plants. However, a few fertilizer materials may be applied directly to the leaves of plants. Fertilizer may be in solid, liquid, or gaseous forms. The proper selection and application of a fertilizer is essential. Money is wasted if the wrong fertilizer is used, and plants can be damaged if too much fertilizer is used.

There are several different types of materials which can be used to improve soil fertility. Many kinds of commercial fertilizers are available to growers. Manure and cover crops, as well as trash from processing crops and other residue, may also be used as fertilizer.

Fertilization is the cultural practice of adding plant nutrients to soil or other growing media. Recommended fertilizer treatments may be analyzed per plant, per one hundred square feet (flower beds), per one thousand square feet (lawns), or per acre (fairways, sod farms, etc.).

FERTILIZER FORMULATIONS

Fertilizers may be complete or elemental. A complete fertilizer contains several plant nutrients. An *elemental fertilizer* provides only one plant nutrient.

Complete Fertilizer

A *complete fertilizer* will contain all three of the primary fertilizer nutrients (nitrogen, phosphate, and potash) and may have select micronutrients included. An incomplete fertilizer will be lacking in one or more of these elements.

6-17. Bag of complete fertilizer showing analysis.

Active Ingredient—The *active ingredient* is the total percentage of nutrients that is being applied. The *inert ingredient* is the carrier or *filler ingredient* that allows deposition of the fertilizer. For example, a complete fertilizer containing 16 percent nitrogen, 4 percent phosphate, and 8 percent potash is marketed as a 16-4-8 analysis fertilizer. The total active ingredient is 28 percent and the total carrier/filler is 72 percent. The 16-4-8 is a commercial formulation. Fertilizers that are medium to high in nitrogen are applied to the surface of the soil. Those fertilizers that are low in nitrogen are incorporated into the soil.

Fertilizer Analysis—*Fertilizer analysis* is the composition of the active ingredients in the fertilizer formulation. A fertilizer grade is the minimum guaranteed analysis of the fertilizer. For example, a 16-4-8 analysis will contain at least 16 percent nitrogen (not 15.99), 4 percent phosphate, and 8 percent potash.

In a complete fertilizer, such as 16-4-8, the 16 percent represents elemental nitrogen, the 4 percent represents phosphate (P_2O_5) (not elemental phosphorus), and the 8 percent represents potash (K_2O) (not elemental potassium). To convert phosphate to elemental phosphorus, multiply by 0.44; to convert phosphorus to phosphate, multiply by 2.29; to convert potash to elemental potassium, multiply by 0.83; and to convert elemental potassium to potash, multiply by 1.20.

Fertilizer Ratio—A *fertilizer ratio* is the least numerical representation of the fertilizer. For example, a 16-4-8 fertilizer has a 4-1-2 ratio. Several analyses may be marketed with the same ratio. For example, a fertilizer with a 1-1-1 ratio may be an 8-8-8, 10-10-10, 18-18-18, or 20-20-20 fertilizer. Also, a 1-2-3 ratio fertilizer may be a 4-8-12, 5-10-15, or 10-20-30 fertilizer.

Fertilizer rates are based on active ingredient. Application is in terms of a commercial formulation.

Elemental Fertilizer

Individual elemental fertilizers are available for nitrogen, phosphorus, potassium, and select micronutrients. Nitrogen fertilizers may be fast release (water soluble), such as ammonium nitrate, or slow release (water insoluble), such as Milorganite.

Other Forms

There may be other materials in fertilizer. Some fertilizers contain herbicides (plant killers). Fertilizer-herbicide combinations should be used carefully; plants can be killed if proper calibration of application equipment is not exercised.

LIME

Lime is used to raise the pH of soil. It is often applied to acidic soils to help increase the pH to a more alkaline range to meet the needs of certain crops. The composition of lime may vary with the source.

Lime can also be used as source of nutrients. Agricultural limestone contains calcium, while dolomitic limestone contains calcium and magnesium. These limes are available in powder or granular form.

SULFUR

Materials containing sulfur are used to lower the pH of soils that are too alkaline. It helps bring the soil to a near neutral or acidic level. Either elemental sulfur or sulfur compounds can be used to lower the pH level.

MICRONUTRIENTS

The micronutrients that are most often deficient in crops are iron and manganese. Iron fertilizers are available in the chelated (claw-like) form which suspends the nutrient in the soil moisture for uptake by the plant.

EFFECTS OF FERTILIZER ON SOIL

Some fertilizers change the pH of soil. These types of fertilizers are said to be acidifying or alkaline forming. Nitrogen fertilizers are the most important acidifying and alkaline forming fertilizers. Fertilizers containing the ammonium form of nitrogen are acidifying, while those containing the nitrate form are alkaline forming. For example, long-term use of ammonium nitrate fertilizer on soil will lower the pH, making it more acidic. Lime may then be needed to correct the acidity.

The acidifying effect of fertilizers is expressed in pounds of calcium carbonate necessary to neutralize one ton of the fertilizer. Phosphorus and potassium sources do not impact soil pH. The pH of the soil influences the availability of certain nutrients. For example, aluminum, iron, manganese, and zinc are very available at a pH of 4.5. Also, calcium and molybdenum are abundant at a pH of 8.0. The availability of most nutrients is best at a pH of 6.0 to 7.0. Most plants grow within this range.

HOW SOIL AND FERTILIZER REACT

Nutrients applied to the soil may be used by plants, lost to leaching, or tied up in chemical reactions in the soil. *Leaching* is the downward pulling of materials through the soil by percolating water. Growers want to use a fertilizer that will be readily available for use by the plants. Other additives may be needed to allow the soil to provide the nutrients to plants in an available form. The fate of nutrients deals with what happens when they are applied in the form of a fertilizer. Some nutrients get "tied up" in the soil and aren't available to plants.

6-18. Excessive rain and flooding wash nutrients away. (Courtesy, U.S. Soil Conservation Service)

AVAILABLE NITROGEN

Nitrogen is added directly to the plant-available nitrogen pool. This pool decreases as plants use nitrogen, and microbes tie it up for use in decaying organic matter. Nitrogen is also lost by leaching as nitrate, clipping and leaf removal, gaseous loss by volatilization of ammonia, and denitrification to N_2 and N_2O. Increases in the pool reserves result from fertilization, deposits of nitrogen from the atmosphere, and organic matter decomposition.

AVAILABLE PHOSPHORUS

Phosphates are immobile in the soil and do not leach readily. Much of the phosphorus in the soil may be tied up in insoluble forms. Phosphate ions ($H_2PO_4^-$) combine with iron and aluminum cations, especially in acidic soil conditions. The available pool is depleted by plant use or by it becoming insoluble.

Plants will generally take up more phosphorus than is needed. Growers can help regulate plant use by applying smaller amounts at a time. Several smaller applications are better than one large application.

AVAILABLE POTASSIUM

Potassium salts are readily soluble in water. Leaching losses can be high in sandy soils. In chemically active soils, potassium is in an unavailable form. Depletion results when plants use up the potassium or when it changes to an unavailable form.

AVAILABILITY OF OTHER NUTRIENTS

Calcium minerals are soluble. This means that they are dissolved in water and may be leached. Calcium is found in the soil as exchangeable ions (Ca^{++}), carbonates, silicates, and a component of organic matter. Calcium minerals are so widespread that deficiencies are rare, except in sandy soils or where acidity is a result of intensive leaching.

Magnesium is subject to leaching but at a slower rate due to its reduced solubility. It is absorbed on exchange sites of the soil colloids. High rates of calcium can cause magnesium deficiencies in plants. Regardless, deficiencies of magnesium are rare.

In contrast to calcium and magnesium, sulfur is found in much smaller amounts. Most sulfur is found in the organic matter. Its availability to the plant is dependent upon the rate of organic matter decomposition. Sulfur may also be made available through sulfur emissions in the atmosphere that are washed to the surface by rainfall or directly absorbed by the leaves as (SO_4^-). The sulfate ion is readily leached, thus deficiencies can occur on sites where clippings and plant sections are removed, in soils that have extensive leaching, and in areas where the atmospheric source is not common.

FERTILIZER APPLICATION

The way fertilizer is applied and the time at which it is applied are important aspects of beneficial fertilizer application. Fertilizer should be applied using a method that does not lead to waste and at a time that plants can most readily use the nutrients the fertilizer provides. Growers select a method that is practical, effective, and cost efficient. Common methods include sidedressing, topdressing, broadcasting, and fertigation.

Sidedressing describes placing fertilizer near the soil surface in bands along both sides of a crop's rows. Topdressing involves placing fertilizer uniformly into the top 1 to 2 inches of the soil or media around the base of the plant. Broadcasting covers the entire area with fertilizer. Fertigation is the application of a soluble fertilizer through an irrigation system.

The frequency and rate of fertilizer application and the way it is applied depend on plant and environmental factors. The kind of plant or plant community and the atmospheric, edaphic (soil condition), and biotic environments determine the specific fertilization program.

6-19. A drop spreader (left) and rotary-type spreader (right). Both are push-type spreaders used for broadcast application.

FERTILIZER RATE

The amount of fertilizer should be adjusted to the natural fertility of the soil, the length of the growing season, and the amount of foot traffic. Turfgrass growing on infertile or sandy soil needs extra fertilizer. The longer the growing season, the more fertilizer needed.

The rate or amount of nitrogen fertilizer that should be applied at one application is about 1 pound per 1,000 square feet. A common problem for a landscaper is determining how much fertilizer is needed to give the right amount of nitrogen to a specified area. A fairly simple equation is helpful and is explained through the following example:

Equation:

$$\frac{\text{pounds nitrogen to apply per 1,000 square feet}}{\text{percent nitrogen in the fertilizer}} \times 100 = \frac{\text{pounds of fertilizer to apply}}{\text{per 1,000 square feet}}$$

Sample problem:

The landscaper has been asked to provide a proposal for maintaining 16,000 square feet of turfgrass for a homeowner. Their selection of fertilizers has an analysis of 32-4-8. How much fertilizer needs to be spread on the lawn to provide the desired 1 pound nitrogen per 1,000 square feet?

$$\frac{\text{I pound nitrogen to apply per 1,000 square feet}}{\text{32 percent nitrogen in the fertilizer}} \times 100 = \frac{\text{3.125 pounds of 32-4-8 fertilizer}}{\text{to apply per 1,000 square feet}}$$

16 multiplied by 3.125 equals 50 pounds of 32-4-8 fertilizer to apply to the 16,000 square foot of lawn.

USE PROFESSIONAL RECOMMENDATIONS

Fertilization formulations and rates should be based on a professional recommendation for the plant and soil test results. Soil testing should be done at frequent intervals to insure that your formulation remains effective.

FORM OF FERTILIZER

Liquid, granular, and gaseous forms of fertilizer are available, but liquid and granular are most commonly used. Granular fertilizer is available in the form of small pellets or particles. Most granular fertilizers do not easily dissolve when mixed with water. They dissolve over a period of time releasing nutrients into the growing media. Liquid fertilizer may be purchased in powder form that, when mixed with water, forms a liquid solution. This form of fertilizer is called water soluble. Growers use the liquid solution to fertilize and water plants at the same time. The gaseous form includes anhydrous ammonia, which has declined in use in recent years.

Liquid and Granular

The two primary types of application equipment are sprayers—for liquid application of soluble or suspended fertilizers—and spreaders—for dry application of granular fertilizers.

Low volume spraying techniques in fertilization are called foliar feeding, and higher volume sprayer applications are called liquid fertilization.

Granular fertilizers are applied by either a drop (gravity flow), rotary (centrifugal), or oscillating shoot unit. The gravity flow distributes the fertilizer vertically to the ground surface. The rotary and oscillating units distribute the fertilizer in a horizontal pattern before dropping to the ground. Each of these units should be calibrated for the subject fertilizer. Keep records and recalibrate as often as needed. Many data sheets are available from specific suppliers on calibration of spreaders, but the internal practice of calibrations

6-20. A PTO operated rotary broadcasting spreader is often used for rapid coverage on golf courses. (Courtesy, J. Michael Hart)

minimize critical errors from misinterpretation. Granular application systems are most widely used. They are practical for most horticultural situations.

*F*ertigation

Fertigation is the application of fertilizers through an irrigation system. This allows growers to water and fertilize plants at the same time. Growers use fertigation for fertilizing both greenhouse and nursery crops.

6-21. Fertilizer injectors deliver exact amounts of nutrients to plants. (Courtesy, Ron Biondo, Illinois).

In large horticultural operations, growers prepare fertilizers in concentrated solutions. The concentrated solution passes through a fertilizer proportioner or injector attached to the irrigation system. The proportioner continuously mixes a small amount of concentrated fertilizer into the irrigation system. The water reaching the plants contains the proper, diluted amount of fertilizer. These fertilizers often contain a blue or green dye to change the color of the water which lets the grower know the water contains fertilizer.

REVIEWING

*M*AIN IDEAS

A plant's growing medium has many functions in a plant's growth and development. Growing medium is the material in which the roots of plants grow. It provides nutrients, holds moisture, and anchors the plant. It also allows for gas exchange and provides a favorable environment for microorganisms. Growers use different types and combinations of organic and inorganic components to create growing media mixes. The growing medium used depends on the type of crop produced. Growers may purchase premixed media or mix their own.

Soil is made of three important mineral materials: sand, silt, and clay. Soil also contains organic matter and other materials. The texture of soil depends on the proportion of different materials it contains. A soil triangle is used to show the texture of soil as related to composition.

Nutrients are the substances plants need to grow. Seventeen nutrients are needed by plants. Some are needed in much larger amounts than others. Nitrogen, phosphorus, and potassium are three nutrients needed in larger amounts. Soil tests are used to determine the nutrients in the soil medium and the pH. pH is the acidity or alkalinity of the soil medium.

Fertilization is used to add the proper quantities of nutrients to the growing media of plants. Complete or elemental fertilizers may be used. Granular, liquid, and gaseous forms are available. Growers use several different methods to apply fertilizer. They select methods that are practical, effective, and cost efficient. Frequency, timing, and rate are important in fertilizer application.

*Q*UESTIONS

Answer the following questions using correct spelling and complete sentences.

1. What is media? What is the function of media? What kinds are used?

2. What is soil? What are the major ingredients in soil?

3. What determines soil texture?

4. What are the nutrient needs of plants? Distinguish between macronutrients and micronutrients.

5. What is soil testing? Why is it important?

6. What is soil pH? How is pH modified?

7. What is a fertilizer? Distinguish between a complete and an elemental fertilizer.

8. How does nitrogen fertilizer affect soil pH?

9. What effect does soil have on fertilizer?

10. How is fertilizer applied?

EXPLORING

1. Visit a local high school football field. Study the site. With permission, carefully take a soil sample and have it analyzed. Develop a fertilizer schedule for the field for a year. Determine the amounts and types of fertilizers needed for the year.

2. Select a turfgrass site, such as a golf course or athletic field. Talk with the superintendent and get permission to conduct a study under his/her supervision. Take soil samples and have them analyzed. Apply specific fertilizers at specified rates. Set aside an area that does not get fertilizer. Make visual checks of the differences between the untreated and fertilized plots. Also, conduct a soil test to determine further differences. Report findings to your class.

3. Make a field trip to a local garden supply store. Determine the kinds of media and fertilizer that are for sale. Prepare a written report on your observations.

7

PLANT GROWTH REGULATORS

Plants grow the way they do for one or more reasons. When people know what these "reasons" are we can get more of the kind of growth that is desirable. We can cause plants to grow rapidly or stop growing without dying. We can help plants produce more flowers and fruit. Regulating plant growth has many advantages.

Plant growth is a complex process. Careful research to learn what causes plants to grow is essential. Developing materials that can be used with plants requires even more research. Once developed, people must know how and when to use the growth regulators properly. Applying the wrong thing to a plant could result in a crop failure!

7-1. Synthetic plant growth inhibitors can be used to produce more compact flowering plants.

149

OBJECTIVES

This chapter covers the meaning, importance, and use of plant growth regulators. It has the following objectives:

1 Name the two groups of plant growth regulating compounds

2 Differentiate between synthetic growth regulators and plant hormones

3 Describe the control of apical dominance by plant hormones

4 Describe the benefits of using plant growth regulators

5 List the different types of rooting compounds

6 Discuss using plant growth regulators to control the size of flowering plants

7 Explain the impact of growth regulators on turf maintenance companies

8 Discuss timing of growth regulator application

9 Explain gibberellins and anti-gibberellins

TERMS

abscisic acid (ABA)
anti-gibberellin
apical dominance
auxin
cytokinin
ethylene

geotropism
Gibberella fujikuroi
gibberellin (GA)
hormone
leggy
Maleic hydrazide
   (MH)

phototropism
plant growth
   regulator (PGR)
senescence
shatter
sleepy

GROWTH PROMOTERS
AND INHIBITORS

Plant growth regulator (PGR) compounds are divided into two groups—natural (hormones) and synthetic (human-made). Both natural and synthetic growth substances regulate or influence cell division, cell differentiation, root and shoot growth, flowering, and senescence (plant aging). These substances are usually separated into groups as either promoters (cause faster growth) or inhibitors (reduce growth) of plant growth. Auxins, gibberellins, cytokinins, abscisic acid, and ethylene represent the major different classes of plant growth regulators.

Plants manufacture organic compounds that regulate growth. These plant produced natural compounds, called hormones, can inhibit or promote plant growth. *Hormones* are chemical messengers that are produced by plant tissue and then transported to other plant parts where they control plant growth.

Synthetic plant growth regulators are applied to the plant to regulate or control plant growth. Some of these chemicals stimulate, reduce, or stop the

Table 7-1. Characteristics of Common Growth Substances

Group	Chemical Example	Description
Auxin	Indole-3-acetic acid (IAA)	A natural substance that promotes growth through cell division and elongation. Important in phototropism, geotropism, apical dominance, and root formation.
Gibberellin	Gibberellic Acid, GA_3, GA_1	Promotes stem elongation through cell division and cell elongation. Synthesized by young developing tissue and developing seeds.
Cytokinin	Zeaton®	Promotes cell division and delays leaf aging process. Synthesized in root tips and developing seeds. Used as a growth promoter in tissue culture.
Abscisic Acid	Single compound	Inhibitor of growth. It closes stomates of plants under water stress and counteracts the effects of auxins and gibberellins. Synthesized in mature leaves under stress.
Ethylene	Single compound	A gas that forms in tissue undergoing stress. It is important in fruit-ripening process and early petal drop of flowers.

7-2. Early fruit drop is prevented by applying auxin spray.

growth of shoots and branches. Other growth regulators delay or prevent flower bud growth and development. A few growth regulators either increase or decrease fruit production. Chemicals that affect and control plant growth have been studied for over 50 years. However, in the past 20 years these chemicals have been taken from the laboratory to practical applications in horticulture.

PLANT HORMONES

Auxins are growth promoting plant hormones. They are produced by the meristematic cells in terminal buds. These compounds move down the stem and control certain plant responses. Auxins also regulate plant cell division and growth. The primary plant auxin is indoleacetic acid (IAA).

High concentrations of IAA, produced by terminal buds, can prevent the development of side buds below the terminal bud on the stem. Greenhouse growers pinch or remove the terminal buds on carnations, petunias, and poinsettias to eliminate this apical dominance. **Apical dominance** describes the control terminal buds have in preventing the development of axillary (lateral) buds on a plant stem. The reduction of IAA in the stem allows the axillary buds to develop and produce a bushy plant with several new stems.

Auxins also control geotropism and phototropism in plants. Plant roots grow down and shoots grow up in response to different concentrations of auxins in the stem and root cells. This is known as **geotropism**, which is plant growth in response to gravitational forces. Plant growth toward a light source is called **phototropism**. A high concentration of auxins on the shady side of the stem causes cell elongation and a gentle curving of the stem toward the light source.

IAA encourages the development of adventitious roots on cuttings. Adventitious roots develop along the stem where no root cells exist. Many herbaceous greenhouse plants produce enough IAA to promote the development of a good root system on cuttings. However, naturally occurring IAA breaks down easily and cannot be applied to other plants to improve rooting. Synthetic auxins can increase root production on woody plant cuttings. Indolebutyric acid (IBA) and naphthaleneacetic acid (NAA) are two common synthetic forms of auxins.

Gibberellins (gibberellic acid or GA) stimulate stem growth even more dramatically than auxins. They also promote cell elongation, stimulate cell division, and control enzyme release. Applying GA to a dwarf plant

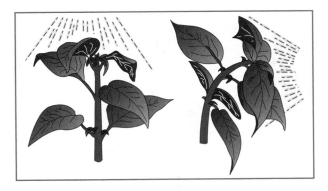

7-3. Auxins are responsible for phototropism, the ability of a plant to bend toward light.

can cause it to grow into a normal size plant. Over 70 different Gibberellins (labeled GA_1, GA_3, GA_7, etc.) have been discovered in various plant tissues. Commercial supplies of GA are obtained from the fungus *Gibberella fujikuroi*. Gibberellic acid is commonly used in the grape industry to thin blossom clusters and dramatically increase berry size. Applying GA to celery increases the size and crispness of the stalks.

Cytokinins are organic compounds that affect the enlargement and division of cells, tissue differentiation, dormancy, leaf aging, and leaf drop. Zeaton® is the name of a synthetic cytokinin growth regulator. Cytokinins produced in the plant root tips and developing seeds will only move upward toward the shoots. They are most abundant in seeds, fruits, and roots. Cytokinins slow the process of *senescence* (biological aging) by preventing the breakdown of chlorophyll in the leaves.

Abscisic acid (ABA) is the only natural plant growth inhibitor. All other growth inhibitors are synthetic. Scientists believe that ABA prevents seeds

7-4. A combination of cytokinins and gibberellins (GA) is used to enhance the elongated shape of some apple varieties.

7-5. Abscisic acid concentrations in seeds prevent germination. (Courtesy, Edward W. Osborne)

from germinating while they are still in the fruit. It also plays an important role in maintaining water supplies within the plant tissue. When a plant becomes water stressed, greater amounts of ABA are produced by the plant tissues. ABA has the net effect of shutting down the metabolic processes in plant cells. Scientists discovered that ABA levels must be reduced before seeds can germinate and buds break dormancy. B-Nine® and Cycocel® are two synthetic ABA type growth inhibitors commonly used in greenhouse plant production.

Ethylene is a water-soluble gas that moves readily throughout the plant. Ripening fruits, old flowers, and plant meristems produce ethylene gas. This ethylene can cause the ripening of other fruits that are in the same container. Thus, one spoiled apple can spoil the entire basket of apples. Orchard operators control the concentrations of ethylene gas in cold storage facilities and regulate the fruit ripening process. Ethylene gas can also cause flowers to drop their petals.

GREENHOUSE GROWTH REGULATORS

Florist and nursery managers apply growth regulators to promote and accelerate root formation on cuttings. The most popular rooting auxins are indolebutyric acid (IBA) and napthaleneacetic acid (NAA). Applied as dusts or solutions to the ends of cuttings the auxins reduce the time needed for good root initiation and development. Both IBA and NAA are considered more effective in inducing root initiation than the naturally occurring indoleacetic acid (IAA).

IBA applications tend to produce strong, fibrous root systems while NAA applications usually produce bushy, stunted root systems. Most greenhouse managers buy IBA and NAA mixed with talcum powder. In this method, the base of the cuttings are treated with the growth regulator mixed with talcum powder. Make a fresh cut at the base before treating the cutting to insure greater absorption of the rooting compound. Another application method is called the quick dip method. Dip the basal end of the cuttings in a concentrated solution of the rooting compound for 5 to 15 seconds. After treating

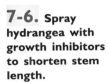

7-6. Spray hydrangea with growth inhibitors to shorten stem length.

with the rooting compounds, the cuttings are placed in the rooting medium for four to eight weeks, where root initiation and development will take place.

Under normal growing conditions, some greenhouse plants grow too tall *(leggy)* to make desirable flowering potted plants. Poinsettias, Easter lilies, chrysanthemums, and hydrangeas can become "leggy" if started too early or grown in a warm greenhouse. The application of growth regulators to these plants can help control their size. Cycocel applied to potted azaleas reduces stem growth, making a more compact plant, while increasing the number of flower buds formed.

Table 7-2. Commercial Growth Regulators Used in Producing Flowering Pot Plants

Product Name	Greenhouse Crop	Effect on Plant
Arest®	Chrysanthemums, poinsettias, Easter lilies	Shortens stem length
B-Nine®	Chrysanthemums, poinsettias, hydrangeas, azaleas	Shortens stem length, increases flower set on azalea
Cycocel®	Poinsettias, azaleas	Shortens stem length, increases flower set on azalea
Florel®	Poinsettias, carnations, roses	Shortens stem length of poinsettias, promotes branching of carnations, promotes new growth from base of roses
Phosfon®	Chrysanthemums, Easter lilies	Shortens stem length
IBA and NAA	Many plants	Improved rooting

FLOWER SHOP GROWTH REGULATORS

7-7. Florists use several chemical preservatives to improve the life of cut flowers.

7-8. Keep flower refrigerators clean and free of ethylene to extend flower shelf life.

Florists use several chemical preservatives to improve the longevity and quality of cut flowers. Some cut flower preservatives contain a chemical from the cytokinin group of growth regulators. When absorbed by stems, cytokinins extend the vase life of flowers and help maintain pigment colors, especially chlorophyll in the leaves. They also can prevent leaf and stem yellowing in many other cut flowers.

Flowers, foliage, ripening fruits, and vegetables naturally produce ethylene. Car exhaust and smoke also contain large amounts of ethylene gas. Flowers exposed to ethylene gas appear wilted or *"sleepy."* Carnations and roses are especially sensitive to small concentrations of ethylene. Snapdragons, delphinium, and larkspur will *"shatter"* or drop their individual flower florets. The effect of ethylene cannot be reversed, but the amount of ethylene present in the atmosphere surrounding fresh flowers can be reduced.

Remove old flowers from the cooler, as decaying or wilting plant tissue produces large quantities of ethylene. Keep apples, other fruits, vegetables, and food out of the flower cooler and away from fresh flowers. It is essential that buckets, containers, tools, and storage areas be kept clean. Use a disinfectant to sanitize the entire storage area weekly. Remove ethylene that the flowers naturally produce by ventilating the storage area.

TURFGRASS GROWTH REGULATORS

Plant growth regulators are widely used in turfgrass management. ***Maleic hydrazide (MH)*** has several trade names, including Slo-Gro, De-Cut, and Retard. In low maintenance areas, such as highway and airport runway turf, MH has been used extensively in controlling vegetative growth and reducing mowing frequency. The negative effect from root pruning has limited its use on home lawns, golf courses, and other high quality turf areas. It is translocated through the plant and causes growth retardation by inhibiting cell division but not cell elongation.

The most common trade name of Chlorflurenol is Maintain CF 125. Maintain CF 125® is a foliar-absorbed growth regulator that is often used in combination with MH to reduce turfgrass growth. It accumulates in the growing tips and inhibits cell division. Maintain CF 125 effectively controls broadleaf weeds besides reducing turfgrass growth.

Paclobutrazol, sold by the Scotts Company as TGR®, is another popular plant growth regulator commonly used on golf courses. Root absorption of TGR is effective because it is transported in the xylem tissue to actively developing and dividing cells. It inhibits plant growth by interfering with the cell elongation process. It is effective in controlling vegetative growth, but does little to suppress seedhead formation. TGR applied to golf putting greens will suppress the growth of annual bluegrass, a troublesome weed on putting greens. TGR has little effect on creeping bentgrass.

7-9. Plant growth regulators are used in turf management in low maintenance areas, such as near highways. (Courtesy, Mollee Thomas, Illinois)

Cutless® is the common trade name for Flurprimidol, a growth regulator, absorbed by foliage, stems, and roots. For best results, always "water-in" Cutless after applying. It reduces vegetative growth of both cool and warm season turfgrass, but it is not effective on seedhead reduction. Today both Cutless and TGR are applied on golf putting greens to suppress annual bluegrass growth.

Mefluidide, sold as Embark®, is absorbed primarily by the grass leaf. It inhibits cell elongation that results in growth suppression. Embark also suppresses seedhead formation on annual bluegrass; however, it does not control the vegetative growth of annual bluegrass. Apply Embark after spring "green-up," but before seedhead emergence. Non-target effects should be considered when using Embark as it can suppress the growth and discolor Kentucky bluegrass.

Trinexapac-ethyl, a turfgrass growth regulator released in the early 1990s, has become popular with home lawn care companies. Trinexapac-ethyl, sold as Primo®, increases the time between mowing turf, thus reducing the total number of mowings per year. This is important to lawn care operators, cemetery managers, and golf course superintendents; less mowing lowers labor costs. Primo can regulate turf growth to varying degrees; however, the standard rate reduces mowing by 50 percent over a period of four to six weeks.

Table 7-3. Growth Regulators for Turfgrass

Generic Name	Product Name	Effects
Flurprimidol	Cutless®	Foliar absorption; suppresses vegetative growth but poor at suppressing seedhead formation
Paclobutrazol	TGR®	Foliar absorption; decrease in cellular elongation and internode length; slows turfgrass growth
Trinexapac-ethyl	Primo®	Foliar absorption; decrease in cellular elongation and internode length
Chlorflurenol	Maintain CF 125®	Inhibits cell division; foliar absorption; translocated up and down plant tissue
Maleic hydrazide	De-Cut® Retard® Slo-Gro®	Inhibits cell division in shoots, buds, and roots; absorbed by foliage; translocated to actively dividing cells
Maleic hydrazide plus Chlorflurenol	Posan®	Suppresses growth and formation of seedheads; foliar absorption; translocated up plant tissue

Primo is absorbed directly into the leaf upon application. It requires no watering-in and is rain-fast within one hour. It results in a decrease in cellular elongation and internode length, but does not in any way stunt the growth of the plant in the long-term. Since it does not inhibit production of plant cells, the process can be reversed with the application of gibberellic acid.

NURSERY AND LANDSCAPE GROWTH REGULATORS

Sprouts on the trunks of trees and suckers from roots are a nuisance to nursery workers, fruit growers, and arborists. Sprouts usually develop after pruning major branches from trees. In many species, pruning the sprouts from the trunk does not eliminate the problem and may even encourage the development of more sprouts. Applying plant growth regulators to the cut surface can reduce or eliminate unwanted sprouts and suckers. A 50 percent reduction in the number of sprouts and suckers can be obtained by applying an asphalt tree wound paint containing Maintain A®.

A new group of synthetic growth regulators, called anti-gibberellin, can reverse the effect of naturally occurring gibberellins in plants. *Anti-gibberellins* counteract the effect of naturally occurring gibberellins in plant tissue; therefore, they reduce the rate of growth instead of increasing it. Clipper®, a commercial anti-gibberellin, reduces internode elongation and thus reduces the total growth rate when applied to shade trees. These growth regulators can remain active within a tree for up to four years. Weak or stressed trees should not be treated with growth regulators. The most widely used method for applying growth regulators to trees is by trunk injection.

Drill $3/_{16}$-inch diameter holes 2 inches deep at the base of the tree trunk. Then inject Clipper at 70 psi pressure into the holes for three

7-10. Applying plant growth regulators to tree wounds after pruning can eliminate or reduce unwanted sprouts and suckers.

minutes. Seal each hole with a vinyl plug to prevent back flushing after injection. For best results, inject Clipper during the growing season before the trees are pruned. Utility companies spend millions of dollars each year trimming trees away from power lines. Although most of the earlier work with tree growth regulators was done on trees under utility lines, recent developments are opening a potential for using growth regulators on trees in the general landscape. Research is needed on the long term effect of these plant growth regulators.

Plants that do not produce flowers cannot make pollen. Applying growth regulators to trees to prevent flowering can provide relief to people who are allergic to tree pollen. Ethephon, sold as Florel®, inhibits the formation of flowers on many trees. Spraying sweet gum trees as the flowers develop will eliminate the messy gum ball fruit production. Female Ginkgo trees have a very bad smelling fruit. Florel sprayed during the flowering stage will prevent the development of the fruit.

Plant growth regulators can be used to control the growth of shrubs and hedges. Off-Shoot-O® was the first material used to chemically prune shrubs. This fatty acid material destroys the meristematic tissue of the shoot apex. This inhibits shoot elongation and promotes lateral branching the same way that manual pinching or pruning does. Some foliar burn can occur when using Off-Shoot-O. A new, safer material, called Atrinal® or Atrimmec, is now available. Both nursery managers and landscape plant managers use Atrinal. In the landscape it is generally used after pruning or shearing to maintain the plant in the desired shape through the growing season. Apply as a foliar spray after trimming. Use Atrinal to suppress flowering and fruit development of certain plants such as ornamental olive (*Olea eurpaea*) and glossy privet (*Ligustrum lucidum*).

Table 7-4. Commercial Growth Regulators Used in the Landscape Industry		
Product	**Landscape Plants Used on**	**Effect on Plants**
Clipper	Shade trees	Reduces internode longation, chemical pruning
Florel	Flowering trees	Reduces flower and fruit development
Off-shoot	Many shrubs	Destroys meristematic tissue and chemically pinches plants
Atrinal	Ornamental olive and glossy privet	Chemical pruning
Maintain A	Trees and shrubs	Reduces sucker production after pruning

REVIEWING

*M*AIN IDEAS

Plants manufacture organic compounds, called hormones, that regulate growth. Hormones are transported to different plant parts where they control plant growth. There are five different hormones produced in plants—auxins, gibberellins, cytokinins, abscisic acid, and ethylene.

Naturally occurring auxins in the terminal bud inhibit the development of lower lateral buds. For years, nursery managers have used rooting compounds to reduce the rooting time of cuttings. Gibberellins stimulate stem growth even more dramatically than auxins. They also promote cell elongation, stimulate cell division, and control enzyme release. Over 70 different Gibberellins have been discovered in various plant tissues. Commercial supplies are obtained from the fungus *Gibberella fujikuroi*. Gibberellic acid is commonly used in the grape industry to thin blossom clusters and dramatically increase fruit size. Cytokinins affect the enlargement and division of cells, tissue differentiation, dormancy, leaf aging, and leaf drop. Cytokinins produced in the plant root tips and developing seeds will only move upward toward the shoots. They slow the process of senescence by preventing the breakdown of chlorophyll in the leaves. Abscisic acid is the only natural plant-made growth inhibitor. Scientists believe it prevents seeds from germinating while they are still in the fruit. It also plays an important role in maintaining water supplies within the plant tissue. Abscisic acid has the net effect of shutting down the metabolic processes in plant cells.

The application of growth regulators to greenhouse plants can help to control their size. Cycocel applied to potted azaleas reduces stem growth, making a more compact plant, while increasing the number of flower buds formed. Decaying flowers, wilting plant tissue, and ripening fruits produce ethylene gas. Ethylene gas can cause flowers to drop their petals.

Cutless and TGR are applied on golf course putting greens to suppress annual bluegrass growth. For best results, always water-in Cutless after application. It reduces vegetative growth of both cool and warm season turfgrass, but it is not effective on seedhead reduction.

Anti-gibberellins counteract the effect of naturally occurring gibberellins in plant tissue. They reduce the rate of growth instead of increasing it. Clipper, a commercial anti-gibberellin, reduces internode elongation and slows the growth rate of shade trees.

When used properly, plant growth regulators will regulate and control many different growth patterns. Always read and follow all label directions when using any plant growth regulator.

QUESTIONS

Answer the following questions using correct spelling and complete sentences.

1. What are the two major groups of plant growth regulators?

2. How is abscisic acid different from other naturally occurring plant growth regulators?

3. What plant growth processes will cytokinins control?

4. Approximately how many different forms of Gibberellins exist?

5. How can a greenhouse grower keep plants from becoming leggy?

6. Why do florists apply naphthaleneacetic acid to the ends of cuttings?

7. How can a florist reduce the amount of ethylene in the flower cooler?

8. What is the complete name of the growth regulator called MH?

9. What do lawn care companies expect to do less of after applying Primo?

10. What is the trade name for the growth regulator Paclobutrazol?

11. Why apply Maintain A after pruning trees?

12. How can the reduction of tree and shrub flowers help people with allergy problems?

13. Why apply Off-Shoot-O to hedge plants?

14. How do anti-gibberellins differ from gibberellins?

EXPLORING

1. Visit a local florist and write a list of all the things florists do to reduce ethylene gas in the flower cooler.

2. Use rooting compounds to improve the rooting of cuttings in your school greenhouse.

3. Apply tree wound paint containing Maintain A after pruning trees on the school grounds, and observe the reduction in sucker production.

4. Visit a local golf course and list all the different plant growth regulators applied to the turf.

8

PEST MANAGEMENT

Plants are subject to damage by pests. Plant pests reduce plant quality and production. Healthy plants can be attacked and lost. They also spread diseases and affect plant appearance. Most pests can be controlled and losses reduced or eliminated.

Controlling and preventing pests is important in horticulture. There are a number of methods for controlling and preventing pests. Some methods involve using chemicals. When working with chemicals, it is important that safety practices be used in selecting, applying, storing, and disposing of chemicals.

8-1. Chemicals must be carefully used to prevent injury to people, domestic animals, desirable kinds of wildlife, and the environment. (Courtesy, Jasper S. Lee)

OBJECTIVES

Understanding pest management and safety practices is important in horticulture. This chapter has the following objectives:

1   Identify five major categories of pests

2   Explain best management practices while maintaining environmental integrity

3   Describe complete and incomplete metamorphosis of insects

4   Explain the difference between selective and nonselective herbicides

5   Discuss alternative pest control techniques

6   Describe safety precautions necessary when handling, applying, and storing chemicals

7   Explain integrated pest management (IPM)

TERMS

algae
action threshold
bacteria
bactericide
best management
  practices (BMPs)
biological pest control
biotechnology
chemical pest control
complete
  metamorphosis
cultural pest control
economic or aesthetic
  injury level
fungi
fungicide
genetic pest control

herbicide
host
incomplete
  metamorphosis
insect
insecticide
instar
integrated pest
  management (IPM)
larvae
mechanical pest
  control
molecular
  biotechnology
moss
nematicide
nematode

nymph
organismic
  biotechnology
pathogen
pest
pesticide
plant disease
postemergence
preemergence
pupae
scouting
surfactant
transgenic organism
viruses
weed

WHAT IS A PEST?

A *pest* is anything that causes injury or loss to a plant. Most pests are living organisms. They damage plants by making them less productive, by affecting reproduction, or by destroying them. Pests can be put into five major categories—insects, nematodes, weeds, diseases, and rodents and other animals.

A *host* plant is one that provides a pest with food. Conditions must be right for pests to damage plants. If a pest doesn't exist where plants are grown, it can't cause a problem. Some plants are more likely to be attacked than others. Various management practices can be used to affect the environment and lessen or eliminate damage by pests.

INSECTS AND RELATED PESTS

Insects are animals with three distinct body parts (head, thorax, and abdomen), three pairs of legs, and either two, one, or no pairs of wings. Any deviation from this definition is an insect-related pest. Such close relatives include spiders and mites (four pairs of legs and two body sections), centipedes (one pair of legs for each body section), millipedes (two pairs of legs for each body section), sowbugs and pillbugs (seven pairs of legs), and snails, crayfish, and slugs—a snail without a shell. There are over 800,000 species of insects in the world and approximately 100,000 are found in the United States. Less than 1,000 are pests of plants and people. Of those, less than 100 are pests of

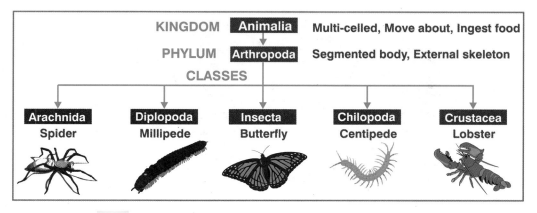

8-2. Scientific classification of insects and insect-related pests.

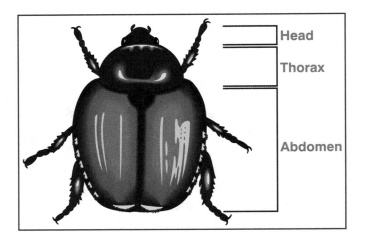

8-3. Parts of an adult insect.

ornamental plants and turf. For further classification, insect orders are included in the Appendix.

The insect body is cylindrical and segmented. It is made up of an external skeleton (body wall); internal muscles and organs; a respiratory system with openings in the sides of the body wall; and a nervous system consisting of the brain, a nerve cord running the length of the body, a network of connecting nerve cells, and sensory nerves in the antennae, eyes, mouth, and feet.

How an insect feeds is determined by the structure of its mouth. Mandibles are for chewing. Elongated beaks with an injecting organ are for piercing

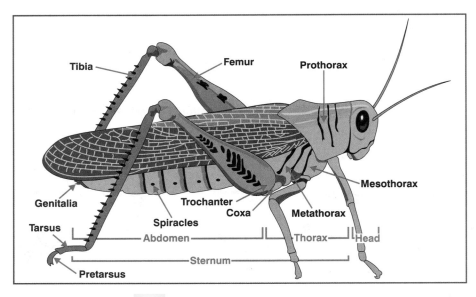

8-4. Major external parts of an insect.

8-5. Mouth-parts of a chewing and a piercing-sucking insect.

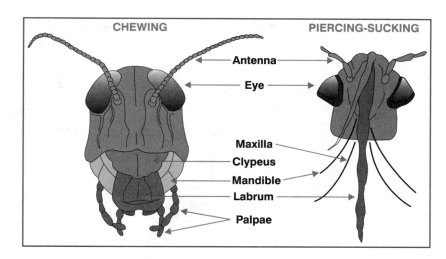

and sucking. Chewing insects include grubs, beetles, and caterpillars. Sucking insects include aphids, leafhoppers, and mosquitoes. Variations to these basic styles include siphoning parts (moths and butterflies) and rasping (thrips). How a plant is damaged is related to the mouth structure of an insect. The location of feeding is important too, such as above ground, at the surface, or below ground level in the soil.

LIFE CYCLE

The development of an insect is described as its metamorphosis. It can be gradual (simple) or complete. In simple or ***incomplete metamorphosis,*** an insect's life cycle changes from the egg through the nymph to the adult. The ***nymph*** looks similar to the adult, only differing in appearance by size and color. Examples include aphids, leafhoppers, mole crickets, and chinch bugs. An insect whose life cycle undergoes four distinct stages (egg, larvae, pupae, and adult) has ***complete metamorphosis.*** The ***larvae*** stage looks nothing like the adult stage. The ***pupae*** stage is a transformation stage and occurs in a series of instars. Examples include—caterpillars to moths or butterflies, grubs to beetles, and maggots to flies. Some insects are harmful in both the larvae and adult stages, while others are only damaging in one stage.

Insect growth is not by gradual increase in size, but by shedding of the external skeleton in four to five stages ***(instars)*** before pupation. They increase in size each time they shed. Adult insects usually do not change in appearance, size, or form, but some adults do change in color. For example, black ataenius beetles turn from red to black.

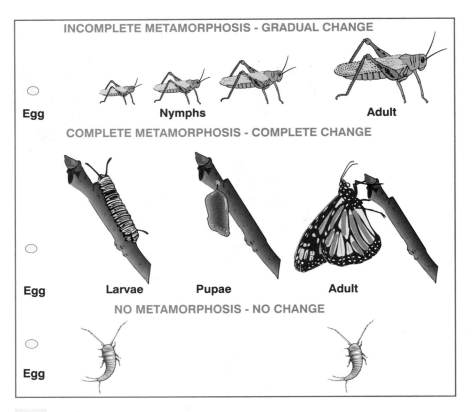

8-6. The life cycle of most insects is through incomplete or complete metamorphosis. Silverfish and springtails do not go through change.

Reproduction is sexual in most insects, but asexual in some, such as aphids. The number of generations per year will vary from several per year to several years to complete one generation. Aphids may have 10 or more generations per year. Periodical cicadas take 13, 17, or 18 years to complete one generation.

BENEFITS, DAMAGE, AND DETECTION

The stage of the insect is important from both feeding and control points of view. Some insects are beneficial, some harmful, and others simply a nuisance. Beneficial insects help plants to grow. They may improve the soil, help pollinate plants, or destroy other harmful insect pests. Examples of beneficial insects that destroy insect pests include the lady beetle, the praying mantis, and the common green lacewing. The degree to which certain insects are

| Lady Beetle | Praying Mantis | Green Lacewing |

8-7. Some beneficial insects attack and kill harmful insects—lady beetle, praying mantis, and common green lacewing.

harmful may vary from one part of the life cycle to another (larvae to adult). For example, in the larvae stage (grubs) the Japanese beetle may feed on grass roots, while in the adult stage it may feed on ornamentals.

The feeding of insect pests on plants usually reduces the quality and vigor of the host plant. However, the mere presence of other insects may be detrimental. Burrowing, nesting, and mound building damage plants. Nuisance insects may sting (bees), bite (fire ant), or simply be present (sowbug, pillbug, or earwig).

An insect control program should include these steps:

- identify the insect and population/ monitor

- determine the potential for damage/ economic threshold

- assess potential environmental issues/ hazards

- decide on integrated control measures or tactics/action threshold

- use control measures

- evaluate the results

- assess resulting environmental issues/ problems

8-8. An entomologist prepares to observe a moth's feeding pattern with night-vision goggles. (Courtesy, Agricultural Research Service, USDA)

Many ornamental and turf insects are nocturnal (night) feeders, while others feed during the day. Basic insect detection strategies can be very effective. Watch birds and their activity to detect grubs, caterpillars, etc. Use a piece of black cardboard in the vertical position to detect mites and fleas. Use a bucket (open at both ends) inserted into the soil profile and filled with soapy water to detect mole crickets and chinch bugs. Cut a rectangular surface in a lawn, for ease of roll-back, and check for possible grub activity. Place a can, with bait in the bottom, in a hole. Place a screen just above the bait to trap beetles. Use a net to catch flying insects. The best time to look for insects is at sunrise, just before they go into daily hiding.

NEMATODES

Nematodes are appendageless, nonsegmented, worm-like invertebrates. They have a body cavity and complete digestive tract, including mouth, alimentary canal, and anus. They do not have a specialized respiratory or circulatory system, but have a well-developed nervous system, an excretory system, and a set of longitudinal muscles. There are approximately 500,000 species of nematodes in the world, including 60 general and 2,200 species that are plant parasitic. Some are harmful, while others are beneficial. Some beneficial effects include maintaining a natural balance in the soil, parasitizing economically important insect pests, and biologically controlling the mole cricket, especially in Florida.

The female nematode uses a stylet mouth structure to pierce the host to feed. Sizes of plant parasitic nematodes vary from $1/75$ to $1/10$ inches long. These slender, largely transparent, unsegmented roundworms are also called eelworms or nemas. However, root-knot and cyst bodies are swollen and sac-like. Nematodes are found everywhere. Most plant parasitic nematodes live in soil. Approximately 20,000 individuals of 10 species may be in 1 pint of soil. Three billion can be found in the top 6 inches of an acre of soil.

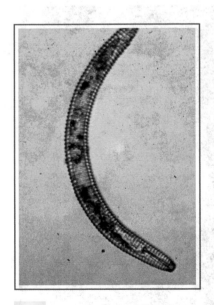

8-9. This thread-like worm is a nematode, greatly enlarged with a microscope. They invade the roots of plants and reduce the host plant's vigor. (Courtesy, Frank Killebrew, Plant Pathologist, Mississippi State University)

Nematodes' bodies are differentiated for feeding, digestion, locomotion, and reproduction. Anatomical features can be used in identification, including the size and shape of the adult female, the size and shape of stylet and tail, the shape of the esophagus and reproductive organs, and the cuticular patterns. Nematodes feed by penetrating root cells with a hollow stylet mouth structure and injecting enzymes into cells. These enzymes digest the cellular contents. Nematodes ingest the partially digested contents. Fungi and bacteria commonly enter plants through these wounds. Most nematode species are harmless and feed on decaying organic matter. Passive movement occurs by water, soil, and infected plant parts. They are also disseminated by wind, animals, tools, and equipment. Their speed of movement in the soil is approximately 12 to 30 inches per year.

DETECTION AND PREVENTION

Diagnosis of nematodes should not be made solely on symptoms being observed. The symptoms may mimic problems, such as low or unbalanced fertility, sun scald or frost damage, poor drainage, drought damage, insect or mite injury, wilt or root-rot fungi, or herbicide damage. Accurate diagnosis requires laboratory testing. Preventive measures include—using disease-free planting materials, proper site preparation, and following cultural practices to ensure steady, vigorous growth. The three categories of nematodes are ecto-parasitic (attached outside of host); endo-parasitic (feed externally on and internally within roots); and semi-endo-parasitic (partially embedded).

The nematode count, in 100 cc of soil, that merits chemical control varies with the types of nematode and host. The number at which control is recommended is called the nematode action threshold. The plant parasitic nematodes and their thresholds are sting nematodes (10), lance nematodes (60), ring nematodes (500, except in centipedegrass, which is 150), stubby-root nematodes (100), and root-knot nematodes (80). Less than these counts indicates that the kinds and numbers of nematodes detected are not expected to cause sufficient damage to justify nematicide application costs. Therefore, monitor nematodes in late spring and summer when populations peak. Correct other problems, such as pH, fertility, insects, and drainage. Mixed populations of two or more nematode types may cause damage at lower population levels.

WEEDS

The primary objective in a landscaped or turf area is to establish and maintain a uniform mulched-surface or living plant material for recreational use or aesthetic enjoyment. Uniformity of color, leaf texture, and density of the desired plant are of the highest importance for consistent ornamentation, playability, and enjoyment. To achieve uniformity in plant selection, maintain plants that are compatible and growing at similar rates for each particular zone of the landscape. Any other plant that might grow and persist would deviate from the color, texture, or density of the desired plant. These plants are weeds. They detract from the overall appearance. They compete for light, water, nutrients, and space. They reduce the vigor of the landscape planting or turf.

CLASSIFICATION AND LIFE CYCLE

A **weed** is a plant growing out of place or an unwanted plant. They may be classed as grassy (monocots), broadleaf (dicots), or other (sedges, rushes, wild onion, and wild garlic). The life cycle of the weed plant is very important in selecting control measures to defeat it. This cycle includes the stages of development, including vegetative, reproductive, and senescence (death) in annuals or dormancy in perennials. The categories include annuals (winter and summer), biennials, and perennials (seeded and creeping). Winter annuals germinate in the fall, grow during the winter, and die in the spring. Summer annuals germinate in the spring, grow during the summer, and die in the fall. Biennials grow vegetatively the first year and reproductively the second year. Perennials grow indefinitely (greater than a year). Seeded perennials reseed themselves each year. Creeping perennials will undergo a period of dormancy in most climates.

8-10. A weed is a plant growing out of place or any unwanted plant.

DAMAGE

Weeds impact desired plants by offering direct competition for water, light, space, and nutrients. Weeds can invade a damaged turf area faster than a turf will reestablish. Weeds can invade a landscape bed that has little or no mulch cover. Long standing weeds are good indicators of unsuitable conditions. Knotweed is found in compacted areas or along paths and sidewalk expansion joints. Certain weeds are indicators of varying soil or environmental conditions, such as moisture, pH, and fertility. Weeds favored under high

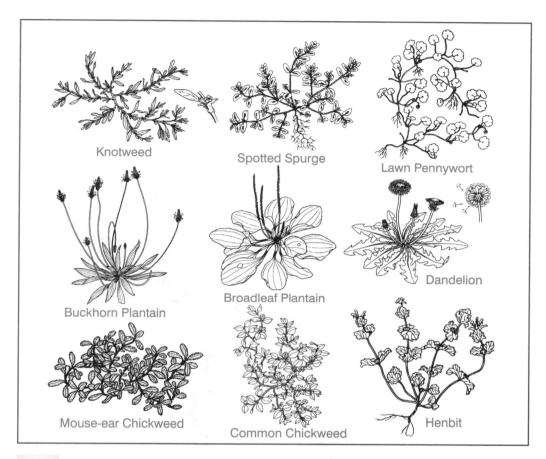

Knotweed

Spotted Spurge

Lawn Pennywort

Buckhorn Plantain

Broadleaf Plantain

Dandelion

Mouse-ear Chickweed

Common Chickweed

Henbit

8-11. **Nine common broadleaf weeds (knotweed, spotted spurge, lawn pennywort, buckhorn plantain, broadleaf plantain, dandelion, mouse-ear chickweed, common chickweed, and henbit). (Courtesy, Amchen Products, Inc.)**

moisture conditions are sedges, mosses, rushes, and annual bluegrass. Weeds favored under low moisture conditions are prostrate spurge, poorjoe, annual lespedeza, and prostrate knotweed. Weeds preferring low pH are red sorrel and broomsedge. Weeds preferring low nitrogen are clovers, legumes, and mosses. A weed preferring high phosphorus is annual bluegrass. Common chickweed prefers high nitrogen. Weeds preferring compacted soils are annual bluegrass, goosegrass, prostrate knotweed, and annual lespedeza. Weeds that are nematode indicators are prostrate spurge, Florida pusley, and prostrate knotweed.

Weeds can be divided into seasonal categories depending upon their growth cycle. These categories include weeds that germinate in the spring, in the late spring and summer, in the spring and fall, and in the fall.

Weeds must be accurately identified. Control measures should be carefully selected. Each state provides recommendations through the land-grant university.

Moss and Algae

Moss is tangled green mats composed of a branched, thread-like growth over the soil surface. It is found in closely mowed or thatched turfs or in neglected turf areas low in fertility, with poor soil aeration and drainage. Highly acidic, excessively shaded, improperly watered, and compacted conditions also provide an environment for moss. Control is achieved by maintaining vigorous plants, removing thatch, and adjusting the above factors.

8-12. Moss growing on this area is a sign of low pH and poor soil.

Algae are a group of small, primitive, filamentous, green plants that manufacture their own food. They are found under very wet conditions, forming a thick, slimy, greenish to brownish scum on bare soil. Algae is encouraged by high fertility and weak plants. It occurs in low, shaded, heavily tracked, or compacted areas. Algal scum can dry to form a tough black crust that later cracks and peels. Algae may induce iron-deficiency chlorosis in plants. To prevent such activity, establish a high density stand of plants, provide for good surface and subsurface drainage, avoid frequent watering and over-watering, aerify compacted areas, eliminate heavily trafficked areas by paths, increase light penetration and air movement, and adjust pH to recommended level.

DISEASES

Plant diseases are abnormal conditions in plants that interfere with their normal appearance, growth, structure, or function. Disease is expressed by characteristic symptoms or signs. Some diseases attack the entire plant, while others attack only part of the plant. Some diseases are foliar in activity, while others attack the roots. There are two principal groups of diseases—abiotic and biotic. Abiotic diseases are noninfectious or disorders. Biotic diseases are caused by parasites or *pathogens* that are infectious and transmissible. Most pathogenic turf diseases are caused by fungi. Others are caused by bacteria, viruses, or other infectious agents. For a disease to occur, a susceptible host, a causal agent, and a favorable environment must all be involved.

Plant parasitic diseases are contagious and can be spread between plants. Categories are identified by the temperature ranges (in Fahrenheit [F]) of the attacks. Such categories include cold weather diseases (32 to 45°F), cool weather diseases (45 to 60°F), warm weather diseases (60 to 75°F), and hot weather diseases (over 75°F). Often, the general name is used in combination with the common name to be more precise as to the causal agent.

Three basic methods of control include—(1) increasing the host's resistance; (2) altering the environment to hinder the pathogen; and (3) keeping the pathogen away from susceptible hosts.

8-13. **Pink snow mold disease on turfgrass. (Note the keys in the photograph for size comparison.)**

IDENTIFICATION

Diseases do not just happen; they result from a cause. Many things can cause disease. Careful laboratory study is often needed to accurately identify the disease.

Diseases are visible by the kind of damage seen. Symptoms of disease include the following—(1) rotting plant parts, particularly the fruit; (2) leaves turning yellow or having an unnatural color; (3) plants wilting; (4) plants having twisted leaves or stems; (5) buds, flowers, or fruit not developing or falling off; and (6) dead plants. Various combinations of these may also be found.

TYPES OF DISEASE

Diseases are of two major types—environmental and parasitic. Environmental diseases are caused by elements in the plant's environment that are not right for the plant. Since plants vary in needs, their tolerance of environments varies. Some plants can live with a condition that will cause others to die.

Environmental diseases include the following—(1) nutrient deficiencies, such as not enough potassium in the soil; (2) damage to plant parts, such as improper cultivation; (3) chemical injuries, such as using chemicals incorrectly; (4) pollution injuries, such as smoke from a factory; (5) weather, such as storms or excessive moisture; and (6) naturally-occurring genetic abnormalities. Tissue analysis and soil testing may be needed to help identify environmental diseases.

Parasitic diseases are caused by microorganisms. These organisms live in the plant and disrupt its normal life. A microscope is often needed to see the organisms. Laboratory study is used to make accurate identification.

8-14. The leaves on this shrub are discolored and have spots because of a viral disease.

ORGANISMS THAT CAUSE DISEASES

Fungi

Fungi cause more plant diseases than any other parasite. They are small, one-celled, usually filamentous, spore-bearing organisms that grow on and in plants. Fungi cause plant mildew, rusts, and smuts. Fruit rot is often caused by fungi. The Dutch elm disease is caused by fungi, and has killed many trees. Fungi are spread by wind, water, insects, and in other ways. For example, diseases in ornamental shrubs can be spread by trimming the shrubs with shears that have been infected by a shrub with disease. However, just as there are beneficial insects, there are some beneficial fungi.

Bacteria

Bacteria are small, one-celled organisms that have a primitive nucleus. Some bacteria are beneficial, and others cause plant disease. Bacteria often get into plants through cuts or breaks in the bark or epidermis. Some enter through flowers and natural openings in the stem and leaves. Apples and pears may get a bacterial disease known as fireblight. Rot in fruit and vegetables is often due to bacteria.

Viruses

Viruses are infective living agents of microorganisms that do not have an organized nucleus. They can multiply only in connection with living organisms. They are very small and are visible only with powerful electron microscopes. They cannot be seen with regular microscopes. A number of diseases are caused by viruses. Examples include cucumber mosaic, citrus tristeza, and tomato ring spot. Viruses are spread by insects, equipment, and by vegetative propagation.

8-15. Virus-resistant plants have been created using antisense technology, a form of genetic engineering in which cells are made to do the opposite of their genetic instructions. (Courtesy, Agricultural Research Service, USDA)

PLANT DISEASES IN SOILLESS CULTURE

Plant species vary in their sensitivity to soilless cultivation. Cucumbers are extremely sensitive to high salts, low light, cool temperatures, and low humidity. Tomatoes are also very sensitive to low light. Disease resistance is especially critical when selecting tomato cultivars for hydroponic production.

Plant diseases may be caused by either unfavorable environmental conditions or microorganisms. Common causes of noninfectious diseases include low or high temperatures, chemical injury, air impurities, nutrient deficiency, nutrient oversupply, and oxygen deprivation.

Four types of microorganisms can cause diseases in soilless culture crops—bacteria, fungi, viruses, and nematodes. Bacteria can deplete the oxygen supply in the nutrient solution. Fungi usually feed upon injured plant tissue, resulting in spot or rot symptom. Viruses are of particular concern since they are responsible for a wide range of plant disease symptoms. Virus injury usually results in lower crop yield and poorer quality fruit. Viruses are transmitted from one plant to another by insects, humans, and other means.

RODENTS AND OTHER ANIMALS

Small and large animal pests are also a problem. Animal pests eat the leaves, stems, fruit, and roots of plants. Birds, skunks, rodents, and armadillos disrupt the surface in search of grubs. Deer, bear, farm animals, and other wildlife cause undulations on the soil surface, as well as plant part removal. Individual growers may have large losses.

Preventing and controlling animal pests involves destroying animal habitat and getting rid of the animals. Cutting weeds destroys habitat. Other animal pest control may require the use of pesticides. However, some animals are protected by law. For example, it may be illegal to kill certain species of birds that are causing damage.

8-16. Some animal pests dig tunnels in the soil which damage land and plants.

PESTS CAUSE DAMAGE

Pests damage plants by interfering with their growth and reproduction. When this happens, the products that plants provide are damaged and/or the amount produced is less.

HOW PESTS CAUSE DAMAGE

Plants and crops produced by plants are damaged by pests. The kind of damage varies with the kind of pest and the plant. Some plants aren't bothered by the presence of pests, while other plants may be severely damaged by the same pests.

Some pests make holes in plants and their products. Insects may eat holes in the leaves of plants. This makes the plant less productive. If the plant is a leafy food crop, such as cabbage, the plant is of little value if the cabbage heads are full of holes. This kind of damage is easy to see.

Pests sometimes attack the vascular system of a plant. This kind of damage can go unseen for a long time. For example, borers can attack the vascular systems of trees. They may bore a tunnel around the tree trunk and cut the vascular system apart. When this happens, the tree may be unable to conduct water to the leaves and nutrients to the root system.

Some insects and animal pests attack the fruit of plants. This may damage the fruit or stop it from developing. Not only does this reduce fruit production, it also reduces the seed and reproduction of the plant.

Products from plants infested with pests are contaminated. Seeds may be infested with a disease. Wherever they are planted, the disease will develop. Buds and scions used for vegetative propagation carry any disease from their parent with them. The new plant will also have the disease.

Sucking pests take the nutrients out of plants. Food for the plant is reduced. The plant will not grow as well. Fruit will be smaller and less will be produced.

8-17. A fire ant mound has damaged this shrub.

Some pests dig holes and build mounds. These damage land and the plants that grow on it. Rodents may dig tunnels under the ground. Crawfish and ants may build mounds above the ground. Mounds damage plants and equipment. The fire ant found in the southern United States and parts of Central America also attacks people and animals.

How Damage Affects The Grower

People who produce plants are affected when pests strike their crops. This damage reduces the income to growers. Costs of production are increased, and the quantity produced is decreased.

Pests cause plants to produce less. When insects eat the leaves of a plant, the plant cannot carry on photosynthesis for food production. This will affect the plant growth and fruit production. Plants damaged by pests are not of the same quality as undamaged plants.

Pests increase the cost of producing plants. Controlling them is costly. Pesticides may have to be purchased and applied to the plants. Equipment to apply the pesticides and fuel to run the equipment are needed. In addition, labor will be required to do all of the activities.

CONTROLLING PESTS

Many pest problems can be prevented. Good management practices will help to reduce the problems. If pests get into plants, they must be controlled. The method selected must be right for the plant and the pest. Laws may also regulate the practices that can be used.

Pests are controlled in the following ways—cultural practices, biological methods, mechanical methods, chemical methods, genetic methods, or by a combination of two or more methods.

Cultural Pest Control

Cultural pest control uses management techniques to control pests. Cultural control includes proper maintenance programs, sanitation, and resistant varieties. Primary and secondary cultural practices must be considered as part of pest management. Such activities include mowing, irrigation, fertilization, pruning, aerification, mulching, etc. These activities will impact the presence and severity of the pest problem. Such consideration will support a more integrated approach to management. Three questions should be addressed early in the evaluation process. What is wrong? What is the source of the problem? What should be done about it?

Such problems may be labeled in the traditional pest categories of weeds, insects, nematodes, or diseases. Or they may be caused by such sources as chlorosis, chemical burns, buried debris (organic and inorganic), dog injury (liquid and solid), vandalism, moss, algae, summer drought or desiccation, winter drying out or desiccation, heaving (freezing and thawing), ice injury, late spring frosts, air pollution, salt injury, or chemical spills.

Interplanting of crops between rows of different crops can be used to confuse insects. Sex attractants can also be used to attract insects to an area away from the crop or to an insect trap. These attractants are substances secreted by insects or artificially developed in the laboratory.

8-18. An insect trap in a vegetable field. (Courtesy, Jasper S. Lee)

Air pollutants include ozone, peroxyacetyl nitrates (PAN), sulfur dioxide, nitrogen dioxide, fluorides, chlorine, hydrogen chloride, ethylene, and toxic dusts. Symptoms include a bleached appearance, followed by chlorosis, then, ultimately, necrosis. Be aware of these rising problems and be prepared to react accordingly. Most common problem pests include weeds, insects, nematodes, and diseases. With most weed and insect problems, the causal agent is identified then steps are initiated to control the problem. With diseases, the symptoms of the causal agent are seen before initiating steps to identify the causal agent. Employ an integrated approach, including biological, cultural, mechanical, and chemical measures.

Some examples of cultural pest management strategies include:

- Improve the soil.
- Select pest-resistant plants.
- Purchase quality seeds and healthy plants.
- Use sterile growing media when potting.
- Plant at recommended times.

- Design with diversity.
- Mulch plants.
- Rotate vegetable crops.
- Remove dead and diseased plant foliage.
- Water at the proper time.
- Use insect traps.
- Use proper pruning techniques.
- Set up physical barriers.

BIOLOGICAL PEST CONTROL

Biological pest control uses living organisms that are predators to control pests. Biological control of insects includes nematodes feeding on mole crickets, using bacteria and fungi, and altering the reproductive processes of the pests. Biological control of weeds includes maintaining a high density stand of plants (turf or groundcover) to choke-out weeds.

Many insects are beneficial in controlling other insects. Pests have natural enemies in the environment. Lady beetles are notorious for their roles in controlling a range of insect pests. Predatory mites have been used to control other kinds of mites on apples. Of course, many gardeners like to have a few toad frogs around to eat insects! Caution: Insecticides should not be used on plants to control insects if predatory insects are present. The insecticide will kill the beneficial insects.

8-19. Biological pest control uses living organisms to control pests. This lady beetle is feeding on aphids. (Courtesy, Agricultural Research Service, USDA)

Forms of bacteria and fungi have been developed for release into the environment. These organisms attack the pests and destroy them. A good example is the bacterium *Bacillus thuringinensis*. When released in fields, the bacteria attack and kill various species of worms.

Several major insect pests have been controlled by altering the reproductive process. Since some insects mate only once, a female that mates with a sterile (no sperm) male lays eggs that won't hatch. The male insects are raised in laboratories and sterilized. They are released in areas with insect problems. The males from the laboratory mate with the females. The eggs produced are infertile (won't hatch). This method has been used to control some pests. Increased use will be made of biological methods in the future. Research is underway to develop new ways of using plants and animals to control pests.

MECHANICAL PEST CONTROL

Mechanical pest control includes using tools or equipment for control. The pests are destroyed or removed. Here are examples of mechanical methods:

■ **Plowing**—Plowing destroys some pests, particularly weeds. The pests are cut off or dug up. Conventional cultivation with hoes, harrows, and other equipment destroys weeds. This works best in hot, dry weather. Plants can be "set out" and grow again if a rain immediately follows plowing. They aren't dead!

■ **Mowing**—Mowing cuts off weeds. Not only does it partially destroy the weeds, but it also destroys places where other pests can hide. Most weeds can sprout after being mowed. Mowing may be needed several times in a growing season. Of course, mowing cannot be used to control pests in some crops because it would destroy the crop.

8-20. Pine needle mulch used with newly planted pansies.

■ **Mulching**—Covering the ground with a layer of plastic, sawdust or other material prevents weed growth. Other pests won't have the weeds as hiding places. Mulching is used with ornamental plants and certain vegetable crops.

CHEMICAL PEST CONTROL

A *pesticide* is a chemical used to control pests. *Chemical pest control* includes using pesticides, such as herbicides on weeds, insecticides on insects, nematicides on nematodes, and fungicides on fungi. They can be further classified by how they control the insect, disease, or weed. Most of these chemicals are deadly—they are made to kill the pest. They can also cause problems for other life.

PESTICIDE	PEST CONTROLLED
Insecticide	Insects
Miticide	Mites
Acaricide	Ticks and Spiders
Molluscicide	Snails and Slugs
Fungicide	Fungi
Avicide	Birds
Rodenticide	Rodents
Nematicide	Nematodes
Bactericide	Bacteria
Herbicide	Weeds
Piscicide	Fishes
Predacide	Predatory Animals

8-21. The pesticide and the pest it is designed to control.

Most chemicals are complex compounds. They are developed in laboratories and tested on trial plots. Their use must be approved by various government agencies, such as the Environmental Protection Agency. Chemicals can be hazardous to humans.

Chemicals are often mixed with other materials when they are used. Many chemicals are mixed with water. In some cases, substances to help the solution "work" are used. One example is a surfactant. A *surfactant* is a material that helps the dispersing, spreading, wetting, or emulsifying of a pesticide formulation. Most surfactants are added to help solutions spread evenly over the waxy surfaces of leaves.

Insect Chemicals

Chemicals that control insects are known as *insecticides.* Some of the substances used are very poisonous. The material that actually does the killing is known as the active ingredient. Labels on containers describe the amount of active ingredient.

Insecticides are made in several forms—dusts, granules, powders, and solutions. Each works best under certain conditions. Most of the powders and solutions are mixed with water for spraying on the affected plants. The equipment must be properly calibrated before application.

Insecticides are classified by how they get into the insect's body.

8-22. Insecticides are poisons and must be handled with care.

- **Stomach poisons**—Stomach poisons are eaten by the insects. Layers of the insecticide must be on the plant leaves or other materials the pest eats. Stomach poisons work best on chewing insects, such as leafworms, armyworms, and bagworms.

- **Contact poisons**—These poisons are absorbed through the insect's skin. To work, the poison must come into contact with the insect. This means that the poison must be sprayed on the insect or on places where the insect will go. Contact poisons are best for controlling sucking insects, such as aphids.

- **Systemic poisons**—Systemic poisons are inside the plant. The poison is applied to the soil and taken up by the roots. Some systemic poisons are sprayed on the leaves and stems and absorbed into the plant. The vascular system moves the poison all over the plant. When an insect bites the plant or sucks its juice, it gets poison.

- **Fumigants**—Fumigants are insecticides in gas form. The poison enters the insect's body through the respiratory system. Fumigants can be used only in enclosed places. People and other animals must stay out during the treatment. Fumigants are good for controlling insects in seed or plants in greenhouses. Sometimes they are used to treat the soil. When this is done, the ground is covered with a layer of plastic to hold the gas in contact with the soil.

Care must be used in handling and applying insecticides. They are dangerous! Humans should wear protective clothing and masks. Always wash skin and clothes after using insecticides.

Nematicide Chemicals

Chemicals used to control nematodes are known as **nematicides.** Since nematodes are in the soil, treatments must be made that reach into the soil.

Fumigants are sometimes used. The soil is usually plowed and covered with a plastic sheet. The nematicide fumigant is sprayed under the plastic. In some cases, the nematicide is injected into the soil.

Tear gas (chloropicrin) is commonly used for fumigation. It kills nematodes, fungi, and weed seed.

Weed Chemicals

Chemicals used to control weeds are known as **herbicides.** Many kinds are available. Most herbicides in use today are derived from organic compounds. Herbicides may be classified by one or more of the following characteristics—type of action, chemical composition, method of application, and species of plants affected. For example, selective herbicides are effective in controlling a limited number of plant species, while nonselective herbicides destroy all vegetation. In addition, contact herbicides kill only the portions of plants that they contact, while systemic or translocated herbicides are absorbed into the plant's vascular and root system and destroy the entire plant.

Herbicides may be applied to the soil either before planting (preplant) or after planting but before crop emergence *(preemergence).* Some herbicides are applied after crop emergence *(postemergence).* Producers sometimes use both soil-applied and foliar-applied herbicides for weed control.

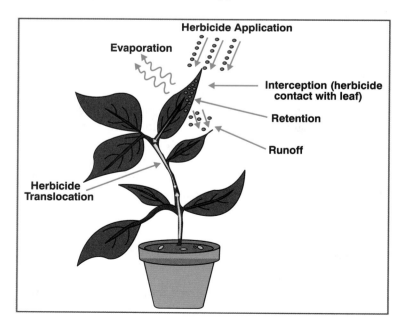

8-23. Herbicide uptake by plants.

| Cocklebur | Crab grass | Henbit | Dandelion |

8-24. Cocklebur and crab grass are summer annuals, while henbit is a winter annual, and dandelion is a perennial.

The performance of herbicides depends upon temperature, rainfall, humidity, maturity of crop and weeds, soil characteristics, and chemical concentration. Higher temperatures elevate metabolic rates in plants, thus speeding up the injurious effects of the herbicide. However, temperatures above 85°F also result in greater volatilization and decomposition of chemicals due to sunlight. Higher humidity also increases herbicide uptake and action. Young plants are more susceptible to herbicide injury.

Disease Chemicals

Many kinds of chemicals are used to control plant diseases. The kind used depends on the disease problem. Accurate diagnosis of the disease is a must. If this is not done, the wrong chemical might be used and the disease would not be controlled. When used on plants that are not diseased, they are known as protective treatments. When used on plants with disease, they are known as therapeutic chemicals or therapeutants.

Chemicals used to control disease caused by fungi are known as *fungi-*

8-25. Samples from floral crops are screened for new viruses. (Courtesy, Agricultural Research Service, USDA)

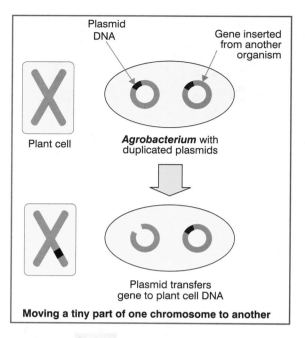

Plasmid DNA

Gene inserted from another organism

Plant cell

Agrobacterium with duplicated plasmids

Plasmid transfers gene to plant cell DNA

Moving a tiny part of one chromosome to another

8-26. Recombinant DNA.

8-27. Leaf tissue from blueberry plants is collected for DNA analysis and gene mapping. (Courtesy, Agricultural Research Service, USDA)

cides. The best fungicides are systemic in action. They get into the vascular system of a plant and reach all parts. Fungi are killed all over the plant. Fungicides that kill by contact must be sprayed directly on the growing fungi.

Bactericides are used to control bacteria. The chemicals are also known as germicides. These solutions are frequently sprayed on plants. Caution should be used to avoid damaging the plant.

Genetic Pest Control

Biotechnology is the management of biological systems for the benefit of humanity. ***Genetic pest control*** utilizes biotechnology by gene transfer or genetic manipulation to make plants resistant to specific pests. This means that the plants have some trait that repels the pest. ***Organismic biotechnology*** deals with intact or complete organisms. ***Molecular biotechnology*** involves changing the structure and parts of cells.

Gene mapping is used to identify genes and their locations along the chromosomes that make up the species. Once genes have been mapped, they can be isolated and transferred from one organism to another. Current recombinant DNA technology allows scientists to splice, clone, and insert genes to change the genetic makeup of an organism.

A ***transgenic organism*** carries a foreign gene that was inserted by laboratory techniques in all its cells. A transgenic organism is produced by placing genes from another organism into embryos. Each transferred gene is expressed in all tissues of the resulting organism. The most successful method for producing transgenic organisms involves microinjection of cloned genes into the pronucleus of a fertilized ovum.

The first transgenic squash, Freedom II, has been approved by the Food and Drug Administration. Scientists at the Asgrow Seed Company in Kalamazoo, Michigan, designed the transgenic yellow crookneck squash to resist viruses that frequently damage these gourds. Freedom II carries an inserted gene for a protein found in the watermelon mosaic virus 2 and the zucchini yellow mosaic virus. These viruses won't attack plants that contain the protein.

SAFETY PRACTICES IN PEST CONTROL

Many of the methods used to control pests are dangerous. They can injure people and other animals. They can pollute the environment and contaminate water and food. Government laws have been passed to regulate the use of pesticides. Also, pesticides are approved only for certain uses. Here are a few safety rules to follow in pest control:

UNDERSTANDING A PESTICIDE LABEL

The label is a license to sell for the manufacturer. It is a way to control the distribution, storage, sale, use, and disposal of the product to the state or federal government. To the user, the label is a source of facts on how to use the pesticide correctly and legally. To physicians, the label is a source of information on proper treatment for injury. The pesticide label is the law.

Every manufacturer has a Brand, Trade, or Product Name for its pesticide. Different manufacturers use different names for products containing the same active ingredient. The Common Name of a pesticide is a shortened version of the complex chemical name. It is included in the Ingredient Statement on the label. Only Common Names which are approved by the U.S. Environmental Protection Agency (EPA) may be used in the Ingredient Statement. The Chemical Name is the complex name which identifies the chemical components and structure of the pesticide. The Chemical Name is listed in the Ingredient Statement along with the percentage amount of each ingredient. The type of pesticide and what the pesticide will successfully control is also listed, as well as the net contents. The contents are expressed in

⬆chipco® Sevin® 80WSP

brand Carbaryl Insecticide

rP RHÔNE-POULENC

FOR COMMERICAL USE ONLY ... NOT FOR USE BY HOMEOWNERS

ACTIVE INGREDIENT:
Carbaryl (1-naphthyl N-methylcarbamate) ....80% by wt.
INERT INGREDIENTS: ....20% by wt.

E.P.A. Reg. No. 264-526 | E.P.A. Est. No.

KEEP OUT OF REACH OF CHILDREN
WARNING   AVISO

Si usted no entiende la etiqueta, busque a alguien para que se la explique a usted en detalle.
(If you do not understand the label, find someone to explain it to you in detail.)

For MEDICAL And TRANSPORTATION Emergencies ONLY Call 24 Hours A Day 1-800-334-7577
For PRODUCT USE Information Call 1-800-334-9745

STATEMENT OF PRACTICAL TREATMENT

Carbaryl is an N-Methyl Carbamate Insecticide.

IF SWALLOWED: Never give anything by mouth to an unconscious or convulsing person. If conscious and not convulsing, drink 1 to 2 glasses of water and induce vomiting by touching the back of the throat with finger.

IF IN EYES: Flush eyes with plenty of water. Get medical attention if irritation persists.

IF ON SKIN: Wash thoroughly with soap and water.

IF INHALED: Move from contaminated atmosphere and call a physician.

GENERAL
Contact a physician immediately in all cases of suspected poisoning. Transport to a physician or hospital immediately and SHOW A COPY OF THIS LABEL TO THE PHYSICIAN.

ANTIDOTE STATEMENT
ATROPINE SULFATE IS HIGHLY EFFECTIVE AS AN ANTIDOTE. See NOTE TO PHYSICIAN.

NOTE TO PHYSICIAN
Carbaryl is a carbamate insecticide, which is a cholinesterase inhibitor. Overexposure to this substance may cause toxic signs and symptoms due to stimulation of the cholinergic nervous system. These effects of overexposure are spontaneously and rapidly reversible.

Specific treatment consists of parenteral atropine sulfate. Caution should be maintained to prevent over atropinization. Mild cases may be given 1 to 2 mg intramuscularly every 10 minutes until full atropinization has been achieved and repeated thereafter whenever symptoms reappear. Severe cases should be given 2 to 4 mg intravenously every 10 minutes until fully atropinized, then intramuscularly every 30 to 60 minutes to maintain the effect for at least 12 hours. Dosages for children should be appropriately reduced. Complete recovery from overexposure is to be expected within 24 hours.

Narcotics and other sedatives should not be used. Further, drugs like 2-PAM (pyridine-2-aldoxime methiodide) are NOT recommended.

To aid in confirmation of a diagnosis, urine samples should be obtained within 24 hours of exposure and immediately frozen. Analyses will be arranged by Rhône-Poulenc Ag Company.

Consultation on therapy can be obtained at all hours by calling Rhône-Poulenc emergency number 1-800-334-7577.

PRECAUTIONARY STATEMENTS

HAZARDS TO HUMANS (& DOMESTIC ANIMALS)
WARNING

MAY BE FATAL IF SWALLOWED. Avoid breathing of dust or spray mist. Do not take internally. Avoid contact with eyes, skin or clothing.

PERSONAL PROTECTIVE EQUIPMENT:

Applicators and other handlers must wear long-sleeved shirt and long pants, waterproof gloves, shoes plus socks and chemical-resistant headgear for overhead exposure.

Discard clothing and other absorbent materials that have been drenched or heavily contaminated with this product's concentrate. Do not reuse them. Follow manufacturer's instructions for cleaning and maintaining Personal Protective Equipment (PPE). If no such instructions for washables, use detergent and hot water. Keep and wash PPE separately from other laundry.

When handlers use closed systems, enclosed cabs, or aircraft in a manner that meets the requirements listed in the Worker Protection Standard (WPS) for agricultural pesticides [40 CFR 170.240 (d) (4-6)], the handler PPE requirements may be reduced or modified as specified in the WPS.

8-28. A sample pesticide label.

ounces or pounds for dry formulations, and as gallons, quarts, or pints for liquids. Liquid formulations may also list the pounds of active ingredient per gallon of product. Pesticides are available in a variety of formulations and each is used in its own way. A wettable powder is not the same as a dust.

Signal Words and Symbols are used on the label as a clue in recognizing how potentially dangerous the product is to humans. The Signal Word must appear on the front in large letters immediately following "Keep Out of Reach of Children" which must appear on every pesticide label.

Signal words include:

- **DANGER**—indicates that the pesticide is highly toxic. Any product which is highly toxic orally, dermally, or through inhalation or causes severe eye and skin burning will be labeled "DANGER." In addition, all pesticides which are highly toxic orally, dermally, or through inhalation will carry the word **POISON** printed in red and will have the skull and crossbones symbol.

- **WARNING**—indicates that the pesticide is moderately toxic. Any pesticide which is moderately toxic orally, dermally, or through inhalation or causes moderate eye and skin irritation will be labeled "WARNING."

- **CAUTION**—signals that the product is slightly toxic. Any pesticide that is slightly toxic orally, dermally, or through inhalation or causes slight eye and skin irritation will be labeled "CAUTION."

Precautionary Statements are included on pesticide labels to help users decide the proper steps to take to protect the user, others, and domesticated animals who may be exposed. These statements are sometimes listed under the heading "Hazards to Humans and Domestic Animals."

Route of Entry Statements immediately follow the signal word, either on the front or side of the pesticide label. They indicate which route or routes of entry (mouth, skin, lungs) that must be particularly protected. Many pesticides are hazardous by more than one route. A single pesticide label may have several precautions listed.

Specific Action Statements follow the Route of Entry Statements. They tell the user the specific action that should be taken to prevent poisoning accidents. The statements are related to the toxicity of the pesticide (Signal Word) and the Route of Entry Statements. "Do not breathe vapors or mist" and "avoid contact with skin or clothing" are common Specific Action Statements.

Examples of Specific Action Statements:

■ **Protective Clothing and Equipment Statements**—listed on many pesticide labels. Long-sleeved shirts, long-legged pants, and gloves should be worn when applying all pesticides.

■ **Statement of Practical Treatment**—gives the recommended first aid treatment in case of accidental poisoning. All DANGER and some WARNING and CAUTION labels contain a note to physicians describing the appropriate medical procedures for poisoning. They may identify an antidote.

■ **Environmental Hazards Statements**—warns of potential hazards to the environment. Special warning statements should be read very closely.

■ **Special Toxicity Statements**—warn of potential hazards to wildlife, insects, or aquatic organisms. These statements help choose the safest pesticide for a particular situation.

■ **General Environmental Statements**—appear on almost every pesticide label. They are reminders to use common sense to avoid contaminating the environment.

■ **Physical or Chemical Hazards Statements**—indicate special fire, explosion, or chemical hazards the pesticide may pose.

■ **Classification Statements**—indicate whether the Environmental Protection Agency (EPA) has classified the pesticide as "general" or "restricted use." This does not mean that the product has a low hazard level. Use the Signal Words and Precautionary Statements to judge the toxicity hazard of the pesticide.

■ **Reentry Statements**—included on pesticide labels that have DANGER or WARNING signals. They tell how much time must pass before people can reenter a treated area without appropriate protective clothing and equipment. If no Reentry Statement appears on a label, sprays must be dry or dusts must be settled before reentering a treated area without protective clothing. This is the minimum legal reentry interval.

■ **Storage and Disposal Instructions**—included for a pesticide and its container on all pesticide labels.

■ **Directions for Use**—tell how to use the pesticide. It gives information about the pest the product claims to control; the crop, animal, or site it is intended to protect; the form in which the product should be applied; the proper equipment needed; how much to use; mixing directions; compatibility with other products; as well as where and when the product should be applied. Additional information may include the minimum number of days between application and the harvest of crops.

APPLYING PESTICIDES SAFELY

Understanding the proper use of pesticides is imperative to their effectiveness and to the safety of people. When applying pesticides, the user should wear the proper protective clothing and equipment the pesticide label recommends. The application equipment should be checked for leaking hoses or connections and plugged or dripping nozzles. All people, pets, or livestock should be cleared from the area before application. Pesticides should be applied only on days with no breezes to minimize drift, preferably in the morning.

Some safety guidelines when applying pesticides include:

■ Use only approved pesticides—Government regulations allow the use of certain pesticides and prohibit the use of others. Follow the law!

■ Read the label before application—review specific Warnings and Precautions and other instructions before you begin.

■ Use the pesticide with the lowest toxicity—toxicity refers to how poisonous the pesticide is.

■ Use the right equipment—the same sprayer should not be used for insecticides and herbicides since a residue can be left in the tank or other parts.

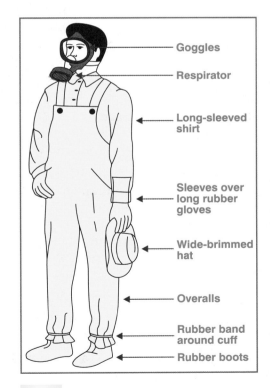

■ Mix according to the directions on the pesticide label—if mixing is required, use the recommended proportions, measure carefully, and mix thoroughly in a well-ventilated area.

■ Apply evenly—the pesticide should be sprayed uniformly with a minimum of overlap.

■ Avoid vapor damage—vaporization is the evaporation of an active ingredient during or after application and can occur when it is windy or when the temperature following application will reach 85°F.

■ Clean up—thoroughly clean all equipment inside and out following the pes-

8-29. Wear protective clothing and equipment when applying insecticides.

Goggles

Respirator

Long-sleeved shirt

Sleeves over long rubber gloves

Wide-brimmed hat

Overalls

Rubber band around cuff

Rubber boots

ticide label instructions; including any excess pesticide mixed which cannot be used.

■ Store properly—pesticides should be stored in their original containers and be protected from temperature extremes.

■ Know the correct emergency measures—read the Statement of Practical Treatment before you begin. It is a good idea to have emergency telephone numbers handy.

STORING PESTICIDES SAFELY

Proper storage of pesticides is important. Some reasons include protecting human health, preserving the environment, and maintaining chemical effectiveness. One should buy only the amount of pesticide that is needed for a particular job or for the current growing season. Smaller containers may be more expensive. However, they may be the best buy because they eliminate waste and the need for storage space.

If you need to store pesticides, always read the pesticide label for specific storage requirements. The chemical and the container must be maintained in good condition and disposed of properly.

When designing or designating a pesticide storage area, there are several considerations. The area should be easy to lock, well-ventilated, properly lighted, dry, and protected from extreme heat and freezing. The area should allow enough space so that the various types of chemicals can be separated. It should be enclosed in such a manner that leaks or spills can be easily contained and easily cleaned up. Storage areas must be designed so that there is no danger of chemicals being washed into local water supplies.

Use approved management techniques for storing pesticides safely. Some techniques include:

■ Locate your storage area where clean up materials such as absorbents and water are close at hand.

■ Keep pesticides in their original containers with pesticide labels in place.

■ Never store pesticides near food, medicine, or cleaning supplies.

■ Do not store flammable materials with pesticides.

■ Organize the materials so they are accessible and visible.

■ Mark each container with date of purchase.

■ Routinely check containers for damage or leaks.

■ Dispose of unwanted or outdated materials and containers according to the pesticide label recommendations.

INTEGRATED PEST MANAGEMENT (IPM)

Integrated pest management (IPM) is a pest management strategy that uses a combination of best management practices (BMPs) to reduce pest damage with the least disruption to the environment. Research has shown that no single control measure works consistently over a long period of time. One reason is the pest can develop a resistance to pesticides. A resurgence of the pest problem can occur. The goal of IPM is to keep pest populations below the *economic* or *aesthetic injury level,* the point at which plant losses due to the pests are equal to the cost of control. IPM provides protection against hazards to humans, domestic animals, plants, and the environment. This optimizes pest control within the overall constraints of economic, social, and environmental conditions.

Integrated Pest Management is an ecologically based pest control strategy that relies heavily on natural factors, such as natural enemies and weather, to control pest populations. IPM users seek out control strategies that disrupt these factors as little as possible. IPM also strives for effective use of naturally occurring inhibitory and mortality elements of the pest's environment

8-30. Growers strive to hold plant losses and pest populations at or below the economic or aesthetic injury level.

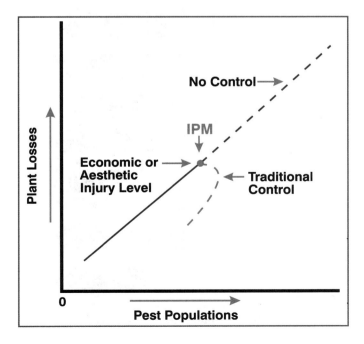

including weather, pest diseases, predators, and parasites. It employs a wide variety of techniques to take advantage of these natural forces.

PHASES OF INTEGRATED PEST MANAGEMENT (IPM)

Integrated pest management (IPM) is a composite of best management practices (BMPs) to control pests. Phase One involves pest identification, monitoring, and action thresholds. *Action threshold* is the predetermined level at which pest control is needed.

Phase Two is to evaluate all possible control measures. If the action threshold has been reached, all possible control options are evaluated, and a

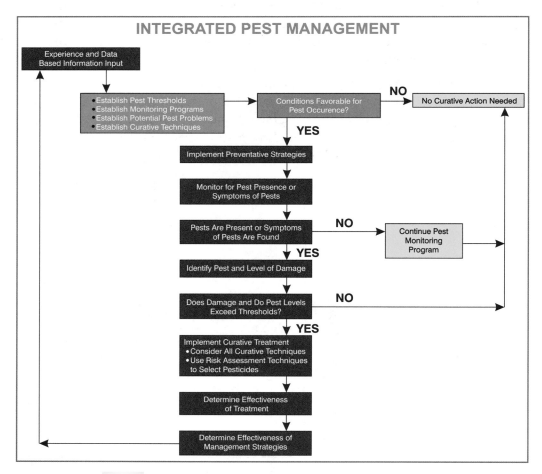

8-31. Integrated pest management (IPM) decision flowchart.

selected option is implemented. If the action threshold has not been reached, continue the monitoring program. Control options may include:

- Chemical
- Biological
- Mechanical

BEST MANAGEMENT PRACTICES (BMPs) EXAMINED

Through the use of best management practices (BMPs), pest management can coexist in harmony with a natural setting. ***Best management practices (BMPs)*** are those practices that combine scientific research with practical knowledge to optimize yields and increase crop quality while maintaining environmental integrity.

The following is a list of some of the best management practices (BMPs) used in horticulture situations.

- Management of surface and subsurface water runoff
- Erosion control
- Cultural control of pests
- Soil testing
- Timing and placement of fertilizers
- Controlled release fertilizers
- Irrigation management
- Biological control of pests
- Pesticide selection
- Correct pesticide use

8-32. Development of additional biological control measures is a top priority for agricultural researchers. (Courtesy, Agricultural Research Service, USDA)

BMPs can effectively eliminate the risk of unwanted materials reaching environmentally sensitive areas.

*B*ASIC ELEMENTS OF AN IPM PROGRAM

Basic elements of an integrated pest management (IPM) program include:

- People—system devisers and pest managers

- Knowledge and information needed to devise the system and make sound decisions

- Program for monitoring the ecosystem elements

- Pest densities at which control methods are put into action

- Techniques used to manipulate pest populations

- Agents and materials

Certain pests commonly cause major damage in any production system. It is very important to correctly identify the pest and understand its life cycle. Many control techniques involve disruption of the pest's life cycle at its weakest point. Individuals must monitor plants regularly to determine current levels of pest activity, known as ***scouting.*** Scouts check to identify the presence of a pest, the stage of development, and the amount of damage done.

IPM incorporates the changing or amending of any or all parts of the plant ecosystem in order to lower pest populations. The ecosystem of a plant includes the biotic factors, such as the living plants and animals, and the abiotic factors, such as soil and water.

8-33. Scouting is a basic element of an integrated pest management program.

REVIEWING

MAIN IDEAS

The presence of pests in our environment impacts the quality of the plants produced. The five major kinds of pests are—(1) insects, (2) nematodes, (3) weeds, (4) diseases, and (5) rodents and other animals. Controlling pests can be very expensive. It may cause other problems, such as damage to the environment. It is far better to prevent pests than it is to control them.

Many pests are not harmful to the plants, but there are several that can destroy entire communities of plants. Pests are found growing actively both above and below ground.

Weeds destroy the overall appearance of the site and compete with the host for space, nutrition, light, moisture, and air. Insects can be very damaging to plants. Understanding their life cycles improves the chances for better management. Nematodes can be a problem in growing plants. Diseases impact the quality of plants. An integrated pest management (IPM) approach is the proper strategy that uses a combination of measures to control pests and involves biological, cultural, mechanical, genetic, and chemical controls.

Always follow safety procedures when selecting, applying, storing, and disposing of chemicals. People should protect themselves and the environment from damage.

QUESTIONS

Answer the following questions using correct spelling and complete sentences.

1. Define integrated pest management (IPM).

2. What is a weed?

3. What are the five major categories of pests?

4. What is a pesticide and what are four different types?

5. What is a plant disease and what are the causes?

6. Explain the difference between organismic and molecular biotechnology.

7. What is complete metamorphosis?

8. What are important safety practices when using pesticides?

9. Name three beneficial insects.

10. How does pest damage affect the grower?

11. What is included in cultural pest control management techniques?

12. What is the meaning of the term "economic injury level"?

13. What is the difference between a selective and nonselective herbicide?

14. Name three insect detection strategies.

15. What is the purpose of a surfactant?

EXPLORING

1. Take a field trip to a park or school yard to locate specific weeds. Make a weed collection and label them by name and life cycle. Use references from the land-grant university in your state.

2. Locate specific insects in your area. Make an insect collection and label them by name, where found, and classify them by insect order. Get the assistance of your biology teacher, pest consultant, or other individual with this activity.

3. Make a survey of the landscape and grounds at your school or home. Describe the injury symptoms observed in various plants. Make a list of these symptoms and analyze plant parts to determine causal agent.

4. Make an inventory of all pesticides at your school or home. List the pest controlled, active ingredient, and toxicity level, along with the pesticide name on a sheet of paper. Discuss with your class what was found and the safety measures to be followed in using the pesticide.

9

HORTICULTURAL BUSINESS MANAGEMENT

People who own, manage, or create new businesses can be quite successful. Some businesses become wealthy. Others are not quite so successful and they lose money. Success requires a basic understanding of business management practices. Having a good understanding of how a business operates is vital to your success.

Are you creative? Could you create a business in horticulture that serves the unmet needs of people? You are probably beginning to dream about what might be possible! Good information is essential! Many opportunities exist in horticulture to start a new business.

9-1. A modern structure with window displays helps this business develop a positive image in the community and gives potential customers a good first impression. (Courtesy, Kuhn Flowers, Jacksonville, Florida)

OBJECTIVES

This chapter provides basic information on business management as related to horticulture. The following objectives are included:

1 Explain entrepreneurship

2 Describe three ways of doing business and relate these to the economic system

3 Explain how to prepare a business plan

4 Describe the five functions of management

5 Identify uses of computers in business

6 Describe labeling, pricing, displaying, and advertising in relation to marketing

7 Demonstrate knowledge of safety practices in the workplace

TERMS

advertising
balance sheet
business plan
cash-flow budget
cash-flow statement
consumer
controlling
corporation
cost
directing
direct mail
direct markup pricing
display
economic system
entrepreneurship
fixed costs

free enterprise
hardware
income and expense
    summary
management
manager
managing
marketing
marketing mix
marketing plan
mission statement
organizing
overhead
partnership
planning
pricing strategy

production schedule
profit
profit margin
profit margin pricing
promotion
publicity
risk
safety
safety policy
selling price
software
sole proprietorship
staffing
variable costs
way of doing business

ENTREPRENEURSHIP

Entrepreneurship is creating goods and services to meet the unique demands of consumers. There may be new ways of using horticulture to satisfy consumer needs and wants. You can probably think of new things that people will buy or services they need!

An entrepreneur is a person who practices entrepreneurship. The individual is most likely the owner of a business that provides goods or services. There is more to becoming an entrepreneur than owning a business. Creativity is involved. Finding new areas of need and developing new products is the creative part. Entrepreneurs need to be good managers and organizers of resources. Regardless of their good intentions, they face risk.

Risk is the possibility of losing what has been invested. Various things create risk. Not enough demand can result in too few sales to cover the costs of providing the product or service. Good planning and decision-making skills help reduce risk.

9-2. Many entrepreneurship opportunities exist in horticulture, such as in landscape maintenance.

Many kinds of businesses involve horticulture. These range from greenhouses to interior plantscape maintenance services, floral shops, and landscape design and maintenance services. Just think about all of those in the area where you live! Opportunities for new businesses exist. Creativity is needed to identify them. Good management skills are needed to be successful with them.

HOW BUSINESS IS DONE

It is possible to do business because of the economic system. The *economic system* is how people go about doing business. It is provided for by the form of government in a nation. The United States uses a form of capitalism known as free enterprise.

Free enterprise is an economic system that allows people to do business with a minimum of government interference. Some government regulations are needed. These are to assure smooth transaction of business and to protect consumers and entrepreneurs.

With free enterprise, people own the businesses. They make choices about how the businesses are operated and what will be produced. People are allowed to run their businesses as they choose. They cannot, however, damage the environment or otherwise carry out illegal activities. People need to follow government regulations. The regulations are to protect people and assure ease of economic activity.

WAYS OF DOING BUSINESS

The economic system provides ways for people to do business. A *way of doing business* is how we organize and carry out free enterprise. Businesses are of three types:

■ *Sole proprietorship*—A sole proprietorship is a business owned by one person, known as a proprietor. For example, a landscape maintenance service could be a sole proprietorship. The proprietor has responsibility for the success of the business. Some sole proprietorships are quite small; others are large. Most begin small and increase in size. Many are not designed to become large businesses. The proprietor assumes responsibility for the success

9-3. A business owned and operated by one person is known as a sole proprietorship. (Courtesy, Jasper S. Lee)

of the business. The individual must have the money and resources to start it. Records must be kept. Reports must be prepared.

- *Partnership*—A partnership is a business owned by two or more individuals. For example, two people could open a landscape maintenance service. Each would share in the business as a co-owner. In a partnership, the partners are usually bound by a written and legal contract. The role of partners often varies. What they provide to the business may vary. In some cases, partners have a silent role. They are not actively involved on a daily basis. A silent partner may provide resources, such as money. Partners share risks as well as income in the proportion to their investment in the business. Good understanding is essential.

- *Corporation*—A corporation is a way for people to do business by creating an artificial entity. The name of the business is usually followed by the letters "Inc." A corporation is viewed much as an individual in doing business, but it is not an individual. The people who form a corporation must get a charter from the appropriate agency of state government. They elect officers and have a board of directors. People are hired to run the corporation. The corporation will issue stock. People buy stock as a way of sharing in the business. People who buy stock are known as stockholders. Individuals can have losses no greater than the amount they have invested in stock. Owners of stock can receive dividends. A dividend is part of the profit that is paid to stockholders for their investment in the business.

  A cooperative is a special type of corporation. Cooperatives are intended to provide services to the people who form them. Cooperatives are often used in agriculture.

CONSUMERS

Businesses must produce goods and services that will be consumed. This means that the goods and services are used to satisfy needs or wants. A *consumer* is a person who uses goods and services. Every time you buy bedding plants, for example, you are a consumer. In some cases, consumers are viewed as businesses because they use goods and services in their operation.

Consumers make choices. This creates demand for what is needed. They can decide to plant flowers or put in a garden. When they do so, they select and buy equipment and supplies. They choose a site for their activity. The same is true with most areas of life. The economic system goes about providing these goods and services for consumers. Businesses use the system to create and deliver "things" that meet consumer demand.

THE BUSINESS PLAN

Developing a well-planned, written business plan will be the most important thing you do to help your business idea get off the ground. A *business plan* is a written document that guides the operation of a business. A plan includes the goals and objectives, ways and means, deadline dates, and methods for assessing progress. New businesses will need plans to get them started. Operating businesses also need good, up-to-date business plans.

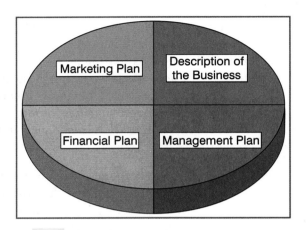

9-4. The four main parts of a business plan.

A business plan includes:

■ Description of the Business
■ Marketing Plan
■ Management Plan
■ Financial Management Plan

DESCRIPTION OF THE BUSINESS

This section includes a detailed description of the business idea, the products and services you will be providing, the location of your business, and why this location is desirable. What are your products and services? Who are your customers? What makes your business unique? The business' *mission statement* should be included. It is a brief (25 to 30 words) written description about the purpose of your business and the customers it serves. Conclude with a brief, written description of where you want the business to be at some point in the future.

THE MARKETING PLAN

Marketing plays a vital role in a successful business. How well you market your business will ultimately determine your degree of success or failure.

The ***marketing plan*** is a written, detailed plan that outlines your specific strategies and goals to get customers to purchase your products or services. Having a good product or service and not marketing it is like not having a business at all. Many business owners operate under the mistaken concept that the business will promote itself. Marketing is the lifeline of a business and should be treated as such.

THE MANAGEMENT PLAN

Managing a business requires dedication and persistence. It requires the ability to make decisions and manage employees, yourself, and finances. The management plan sets the foundation for and facilitates the success of your business.

Management is all of the activities needed to move a business toward its goals. A number of activities are included. These occur in all businesses— from the smallest to the largest. In small businesses, the owner may be the manager as well as the labor force.

Managers may need to get the help of trained professionals in the early stages. They may need an attorney and an accountant. An attorney prepares the necessary legal documents for a business. An accountant handles the details of financial status and taxes. A source of finances will be needed. This may involve using money that has been saved or getting a loan from an individual or a bank.

THE FINANCIAL MANAGEMENT PLAN

Having a sound financial management plan is one of the best ways for your business to remain profitable. Even so, this will be one of the most difficult aspects to develop. How well you manage the finances of your business is the cornerstone of every successful business venture. Each year, thousands of potentially successful businesses fail because of poor financial management. As a manager, you will need to identify and implement policies that will lead to and ensure that the business' financial goals are met.

A financial management plan includes several basic financial forecasts. Many decisions will be based on what is presented here. The basic financial statements to be included are cash-flow analysis, source and use of funds, profit and loss statement, and balance sheet. Each of these statements allows you to take a past, future, and present look at your business.

9-5. Many businesses use computer software programs for accounting, word processing, and Internet communication. (Courtesy, Ron Biondo, Illinois)

DEVELOPING A BUSINESS PLAN

Most communities have economic development offices that can provide help in developing a business pan. If not, a lending agency or accountant can likely provide information on where to get help. Questions to ask in developing a business plan include the following:

1. The idea—What is to be developed or done?

2. Competition—Is anyone else doing it, or can it be done better or cheaper?

3. Resources—What material, human, capital, and technological resources are needed to operate the business?

4. Regulations—What are the federal, state, and local regulations that will affect this business?

5. Marketing plan—Who will buy the product or service? Will it be necessary to advertise? Will sales be seasonal or constant throughout the year?

6. Location—Where will the business be located? Is the location convenient for customers?

7. Employees—Are additional employees needed? What qualifications should they have? What will be the cost of wages and benefits?

8. Management—Who will manage the business?

9. Financial plan—What finances will be needed? What records will be kept?

MANAGING A BUSINESS

All businesses have managers. A *manager* is the person who is responsible for the operation of a business. Managers direct activities in the business. In a small business, the manager and owner may be the same person. Success or failure of a business is related to the skills of the manager.

Managers are responsible for the operation of the business. Successfully operating a horticultural business involves managing the different areas of the business. *Managing* is conducting or directing the operations of a business. It requires careful planning and organization and an efficient record-keeping system. Managerial duties include giving attention to all details.

Planning is the basis upon which all other management activities take place. It involves setting goals and objectives. Organizing is the system for efficiently carrying out the plans. Directing is leading and supervising. It may involve motivating, delegating responsibilities, and evaluating. Coordinating involves unifying the different areas of the business. Controlling is the managerial responsibility of monitoring plans and processes. A manager uses knowledge and skills to perform duties. In addition, financial records are used to evaluate the business finances.

9-6. Planning is a continuing step in management.

BUSINESS MANAGEMENT PRACTICES

Management activities are in five main areas. These are often known as management duties or functions. Managers know these as the functions needed for a business to be successful.

Planning—Planning is the process of deciding how a business will operate. This includes what it will produce. Many possibilities are available. Individuals must choose the best from among the possibilities. Objectives are identified. Strategies for achieving the objectives are developed. Some plans are short-term; others are long-term. Plans should be regularly assessed to see if the business is making progress. If the business is not progressing, the plans need to be more realistic.

Organizing—Organizing is the process of setting up a system for efficiently carrying out business plans. It deals with getting the work done. Organization makes it possible to produce products or services. Consumers buy what is produced to satisfy their needs.

Directing—Directing is the process of leading and guiding employees to achieve the objectives of the business. Two major areas are involved. They are leadership and motivation. The manager will need to provide leadership so the employees perform well. Employees will also need to be motivated to do their work. Directing may also involve delegating responsibilities and evaluating employees.

Staffing—Staffing is concerned with the recruitment, selection, and training of employees. Having good, productive employees is essential for success. Employees need to be trained in their work. They will also need to be evaluated and receive suggestions on how to improve.

Controlling—Controlling is assessing goals and objectives to see if the business is making progress. It involves seeing if the performance of the business has measured up to what was expected. If not, changes should be made to improve the performance.

RECORD KEEPING

Accurate records are necessary for managing and operating a horticultural business. Information that is useful in operating the business and certain required government information is gathered and organized into records or statements. Records may help the manager evaluate the financial health and operating efficiency of the business or determine when additional employees are needed. Records are often used to compare the present to past performance. Records are used as a basis for making management decisions and providing direction for the business. They are also necessary for government tax reporting purposes.

Types of Records

Records can be classified into two types—financial and physical. Examples of financial records include the balance sheet, income and expense summary, and the cash-flow statement. Credit applications are another financial record a business may have if it extends credit to customers. Examples of physical records include information concerning the culture and production of horticultural crops, inventory records, machinery and equipment records, and labor records.

The **balance sheet** is a financial statement that shows assets (what a business owns), liabilities (what a business owes), and net worth (assets less liabilities) of a business for a specific date.

The **income and expense summary** is a listing of all revenues and expenses for a specific period. It is sometimes called the profit and loss statement because it shows the net profit (or net loss) of a business. It is a very useful tool used to analyze profitability and to determine the operating efficiency of the business.

The **cash-flow statement** shows the sources of money into and out of a business for a specific period. It summarizes the sources and uses of funds. A **cash-flow budget** is used to predict cash-flow through the business for a set period. The cash-flow statement can be compared to the cash-flow budget to help the manager effectively plan for surpluses or deficits in future periods. A business must have positive cash-flow to continue in operation. However, it is not uncommon for a horticultural business that has certain set periods where revenues are greater to have a cash deficit in other periods where additional capital (cash) may be needed.

Production records are kept for horticultural crops. They contain information concerning culture and production. They supply information used in determining the cost of a crop and production schedules. A **production schedule** is the time required to produce a crop of plants to marketable size. Production records also provide historical data for planning similar crops in the future.

Inventory records are used to keep an accurate count of materials, supplies, and plants on-hand. They let the manager know what is available for sale and when materials or other supplies may need to be reordered.

Machinery and equipment records are kept for each major machine or piece of equipment. They contain the original cost and date of purchase. They also are used to keep a listing of any repairs or maintenance that was done and when. Scheduled preventive maintenance may also be listed here.

Cash-Flow Statement

Month _____, Year ____

I. **Cash Receipts:**

Beginning cash balance . . . . $_____
Cash received . . . . $_____
Sales. . . . . $_____
Non business income . . . . $_____

     Total Cash Inflows   **$** _____

II. **Cash Disbursements (Expenses):**

Ending cash balance . . . . $_____
Operating expenses . . . . $_____
Capital expenses . . . . $_____
Debt payments (principal) . . . . $_____
Nonbusiness expenses. . . . . $_____
Income tax and social security tax . . . . $_____
Family and living expenses. . . . . $_____

     Total Cash Outflows **$** _____

III. **Ending Cash Balance**   **$** _____

9-7. An example of a cash-flow statement.

Labor records are required for reports the business must send to the government, such as reporting wages and tax withholdings. They are also used when determining the cost of labor for completing certain tasks.

CALCULATING COSTS FOR PRODUCTS AND SERVICES

Determining costs is an important responsibility for the horticultural business manager. Financial and physical records are used when calculating the costs of products and services. Managers involved in plant production must calculate the costs for producing marketable plants. Those involved in providing horticultural services, such as landscape design, landscape maintenance, or floral design, must determine the costs of the services they provide.

In order to establish a selling price, an accurately calculated cost of the product or service must be determined and recorded. *Selling price* is what the customer pays for a product or service. *Cost* is the amount of money spent by the business in producing the product or providing the service. With products, the cost is the actual cost per unit, or wholesale cost. No two busi-

Landscape Construction Estimate Sheet

Job Name_____ Date_____

Job Location _____ Designer _____

Job Description _____

Description	Quantity	Material Unit Cost	Total Material Cost	Total Labor Cost ($15.00/hr)	Total Labor and Material Costs
Plant List					
Acer saccharum, 'Green Mountain,' 3" B&B	3	275.00	825.00	90.00	915.00
Fothergilla gardenii, 3 gallon container	12	16.00	192.00	60.00	252.00
Sod	235 yds	1.20	285.50	81.25 (.25/yard)	438.75
Hedera helix, 'Thorndale,' 3" pot	300	0.58	174.00	45.00	219.00
Construction Materials					
Finish grading of site				525.00	525.00
Brick pavers, 6 cm	1230	1.61	1,980.30		1,980.30
Sand, construction grade, 2" deep	3 tons	5.00	15.00		15.00
Gravel, grade 8, 3" deep	4 tons	6.00	24.00		24.00
Landscape fabric, 3' × 5' roll	3 rolls	9.95	29.85		29.85
Patio installation	300 sq. ft.		369.50	900.00 ($3/sq. ft.)	900.00
Florida cypress mulch, 4" deep	10 yds	36.95		90.00	459.50
Subtotal Costs (total labor and material costs)					$4,758.40
Overhead Costs (subtotal material and labor costs × 20%)					$1,151.68
Cost Summation (subtotal material and labor costs + overhead)					$5,910.08
Profit (cost summation × 20%)					$1,182.00
TOTAL (cost summation + profit)					$7,092.08

9-8. An example of a landscape construction estimate sheet using fixed percentages to calculate overhead costs and profit.

nesses have the exact same type of costs. However, businesses usually record their costs as fixed costs or variable costs.

Fixed costs are the general operating expenses of running a business. Examples include: mortgage or lease payments, legal fees, utilities, insurance, office expenses, salaries, and maintenance of equipment. These are expenses that are not directly attributable to producing a specific product or providing a service. *Variable costs* are those directly associated with producing a product or providing a service. These costs increase or decrease according to the level of production. Examples of variable costs include materials and supplies used in plant production, such as seeds, media, containers, and fertilizer. Variable costs are calculated on a per unit basis.

To determine the costs for horticultural services, managers must determine what costs are involved in providing the service. These costs include labor and materials. Overhead costs and profit are also included. *Overhead* costs are the general fixed costs of running a business. *Profit* is the amount of money the business receives after deducting all of the costs. Some service-oriented businesses use a fixed percentage to calculate their overhead costs and profit when establishing a selling price for their service.

COMPUTER TECHNOLOGY
IN BUSINESS

Computer technology is all of the processes that can be performed with computers. It includes many applications found in horticultural businesses. Most of the computers used today are personal computers or microcomput-

9-9. Computers can be used to control greenhouse irrigation systems (left) or for landscape design and video imaging (right), as well as to keep records and process reports.

ers. The physical equipment that makes up a computer is known as *hardware*. Hardware includes devices such as disk drives, microprocessor, keyboard, mouse, monitor, CD-ROM, and printer. In order for the hardware to function, it must have instructions. Instructions are in the form of computer software. *Software* is the computer program instructions that run the hardware to get the worked performed as desired. Software includes both the operating system, such as Windows®, and the application program, such as Microsoft Word®, Corel WordPerfect®, Microsoft Excel®, and many others.

Computers have a wide range of uses in a horticultural business. They are best known for keeping records and processing reports. Managers often use them to develop their business plan, maintain their financial records and create the necessary reports, generate written communication or exchange messages, and as a tool to calculate costs for products and services. Computers can also be used to control or monitor processes in the business or as a tool to aid in the actual service the business provides. Examples include monitoring and controlling a greenhouse irrigation system or, with the use of a software program, designing a landscape for a client. Portable, laptop computers with presentation software can also be used as a selling tool when calling on customers.

Computers can also be linked to other computers to search and provide information or as a communication tool. The Internet (World Wide Web) is now the most common way to link with other computers. Businesses gain access to the Internet's information highway through a local Internet service provider. Browsers are the software that allow access to Internet information. Common browsers include Internet Explorer® and Netscape Navigator®. Browsers allow you to access information on the World Wide Web through a URL (Universal Resource Locator) address. An example of a URL address is http://www.interstatepublishers.com. If you do not know an address or want to search for specific information on a specific topic, browsers have search features. There are also sites on the Internet that provide search capability, such as http://www.yahoo.com or http://www.lycos.com. Careful assessment of information obtained on the World Wide Web is essential since it is not regulated. Business decisions should only be made with information retrieved from reputable sources.

Internet service providers also make available to their customers electronic mailboxes which have a unique address. These are known as email addresses. This allows businesses to quickly communicate and exchange information worldwide. Some businesses also develop a homepage and post it on the Internet so potential customers can find out about their business.

MARKETING

Marketing is providing the horticultural products and services people want. A combination of four variables is used to reach the target customers—product, place, price, and promotion. These are called the four "Ps", or *marketing mix*. Product refers to the ability of the product or service to satisfy the needs of customers. Place refers to making the product or service convenient for people to purchase. Price is the amount of money that buyers pay for the product or service. Promotion is the communication between the source of the product or service and potential customers.

9-10. Managers and employees must often communicate in planning marketing strategies.

LABELING

Labels help the customer locate and identify products and supplies. They also provide information designed to assist customers in the buying decision. They should be easy to read and uniform in style. Truthful labeling is essential. The U.S. Department of Agriculture, Federal Trade Commission, Food and Drug Administration, and the Consumer Product Safety Commission have regulations on labeling. States also have specific requirements, such as plant inspection and seed purity.

Labels should provide essential information about the product. Product labels should give quantity, intended use, any special instructions, and the name of the manufacturer. In addition, labels provide information unique to the product. Plant labels should include the common and botanical names.

They should also include cultural requirements, features, and uses of the plant. Labels may be printed directly on containers or packaging, or attached in a secure manner.

PRICING

Plants, supplies, and services must be properly priced for the business to stay in operation. Effective pricing is necessary to create enough revenue to pay expenses and make a profit. Pricing strategy is a marketing technique that can be used to improve overall competitiveness. *Pricing strategy* is determining how you are going to price your products and services.

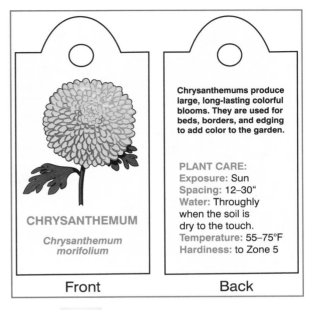

Chrysanthemums produce large, long-lasting colorful blooms. They are used for beds, borders, and edging to add color to the garden.

PLANT CARE:
Exposure: Sun
Spacing: 12–30"
Water: Throughly when the soil is dry to the touch.
Temperature: 55–75°F
Hardiness: to Zone 5

CHRYSANTHEMUM

Chrysanthemum morifolium

Front   Back

9-11. An example of a plant label.

Pricing methods based on cost include profit margin pricing and direct markup pricing. *Profit margin* is the financial return (income) in excess of direct costs and expenses. It is sometimes referred to as gross profit. The selling price for *profit margin pricing* is calculated by adding the actual cost per unit to the desired per unit percentage of profit. *Direct markup pricing* uses the actual cost per unit multiplied by a constant factor to establish the selling price.

PROFIT MARGIN PRICING		DIRECT MARKUP PRICING	
Actual Cost	$10.00	Actual Cost	$10.00
Profit Percentage	50%	Multiplier	X 2
Selling Price	$15.00	Selling Price	$20.00
— — — — — — PROFIT — — — — — —			
Actual Cost	$10.00	Selling Price	$20.00
plus PROFIT	5.00	less Actual Cost	10.00
equals		equals	
Selling Price	$15.00	PROFIT	10.00

9-12. Two examples of pricing methods based on actual unit cost.

Market demand pricing methods include pricing based on competition (either above or below) and pricing based on what the customer expects to pay. These methods do not ensure that the selling price is greater than the cost. Multiple pricing uses different pricing strategies for different times. Many businesses do this. The key to success is to develop a well-planned strategy, establish prices, and constantly monitor costs to ensure a profit.

DISPLAYING

Horticultural businesses use displays in a variety of ways to attract customers to purchase products or services. A ***display*** is an exhibit of merchandise—products or services. Some of the major types of displays include window displays, point-of-purchase displays, wall and ceiling displays, and shelf displays.

Displays help customers easily locate items in the business. They also help establish a pattern of movement throughout the business. A window display assists the business in developing a positive image in the community and gives potential customers their first impression of the business. Displays set up near check-out areas, on counters or in aisles, and other open areas are referred to as point-of-purchase displays. Usually three walls are available for displays, plus any additional temporary walls that are added to enhance display areas. Ceiling displays are used for hanging signs, for seasonal decorations, and to call attention to specific products. Shelf displays are used throughout the store to stock merchandise.

9-13. This business is using aisle, shelf, and wall displays.

Guidelines for creating successful displays include:

- Keep displays small and simple.
- Label displays clearly.
- Keep displays neat and clean.
- Group items of similar types or uses.
- Use promotional displays for new and sale items.
- Change displays frequently.

ADVERTISING

Promotion is the coordination of all seller-initiated efforts to communicate with potential customers. *Advertising* is any message communicated to potential customers about a product or service that uses mass media. The business must pay for mass media advertising. Examples of mass media include newspapers, television, radio, World Wide Web, or direct mail. *Direct mail* is printed ad materials sent directly to consumers. The purpose of advertising is to inform potential customers about a product or service and to get them to connect with the business selling the product or service. Many businesses now advertise their products and services on the Internet with a World Wide Web homepage.

Advertisements must attract and hold a potential customer's attention if they are to be successful. They must have an effective design and simple message. An ad must include the name of the business, the products or services being offered, and any special message that would interest customers. It

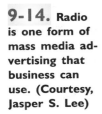

9-14. Radio is one form of mass media advertising that business can use. (Courtesy, Jasper S. Lee)

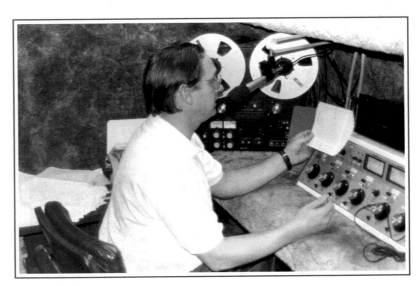

may also include the business' address, phone and fax numbers, and email address. Through advertising, consumers become more knowledgeable of the products and services available from the business.

Publicity is different from advertising. *Publicity* is mass media news coverage that includes the name of the business and product or service at no cost to the business. Businesses develop plans and strategies to promote positive publicity that will build customer confidence.

WORKPLACE SAFETY

Safety is preventing injury or loss. Safety in the workplace is an important consideration for both managers and employees. Establishing and maintaining a workplace that is as safe as possible should be a high priority. This is for everyone's well-being. A horticultural business that provides a safe work environment is more likely to have lower insurance costs. It will also reduce the chance of employee injuries and lost production time.

Horticultural businesses often use equipment and chemicals. Employees should be trained in how to properly use the equipment. Owner's manual instructions should be followed. Material Safety Data Sheets (MSDS) are now supplied with many chemicals. Information about the safe use of a chemical is included on the MSDS. Government laws have been passed to regulate the use of pesticides. Also, pesticides are approved only for certain uses. Pesticides should be applied properly using the proper protective clothing and equipment. Good environmental stewardship should be practiced when storing or disposing of chemicals.

9-15. This sign reminds employees and managers that a health hazard exists. (Courtesy, Ron Biondo, Illinois)

The success of a business' safety program depends on the business' safety policy and work rules. A *safety policy* is a brief, written definition of management's philosophy toward safety. Work rules include specific safety measures employees should follow while on the job. New employees should receive a copy of the safety policy and work rules at the time they are hired. Accident prevention regulations and guidelines of federal, state, and local governments, including OSHA, must be followed. OSHA is the Occupational Safety and Health Administration which regulates standards of compliance for various segments of business in order that every worker be provided with a safe and healthful working environment.

9-16. Chemicals must have a Material Safety Data Sheet (MSDS).

REVIEWING

*M*AIN IDEAS

Successfully operating a horticultural business involves managing the many different areas that make up the business. It requires careful planning and organization, and an efficient record-keeping system. Records are classified as financial or physical. They are used to evaluate the financial health of the business, complete government reports, calculate product costs, and have accurate lists of products and equipment on-hand.

Determining costs of production or services is an important responsibility for managers. Costs must be calculated and recorded. They are used to establish a selling price for the product or service.

Computers are used in horticultural businesses for keeping records and processing reports. They can also be used to control or monitor processes in the business or as a tool to aid in the actual service the business provides. Many businesses also use computers to communicate and search and retrieve information through the Internet.

Marketing is providing the products and services people want. Four variables are used to reach customers—product, place, price, and promotion. Labeling, pricing, and displaying are important when selling products. Labels help the customer locate and identify products and supplies. They also provide information designed to assist customers in the buying decision. Effective pricing is necessary to create enough revenue to pay expenses and make a profit. Horticultural businesses use displays in a variety of ways to attract customers to purchase products or services. Advertising is one way people are informed of products or services the business has to offer.

Safety in the workplace is an important consideration for both managers and employees. Establishing and maintaining a workplace that is as safe as possible should be a high priority for everyone. The success of a business' safety program depends on the business' safety policy and work rules.

QUESTIONS

Answer the following questions using complete sentences and correct spelling.

1. What is entrepreneurship? An entrepreneur?
2. What is an economic system?
3. What are the three ways of doing business? Briefly describe each.
4. Why are consumers important?
5. What is management?
6. What is a business plan? What are the four major parts of a business plan?
7. What are the five functions of management? Briefly explain each.
8. List four different types of financial records a business may have.
9. Explain fixed cost and variable cost. Give examples of each.
10. How can computers be used in a horticultural business?
11. What types of information are on product labels?
12. What two methods of pricing are based on cost?
13. What is the difference between advertising and publicity?
14. What determines the success of a business' safety program?

EXPLORING

1. Interview the manager of a business about the nature of the work. Determine the greatest problems in management and how these problems are solved. Investigate trends in the area and how these trends will impact the business on a long-term basis. Prepare a written report on your findings.

10

GREENHOUSE STRUCTURES

Greenhouses, hothouses, glasshouses, stove-pipe houses, hotbeds, and cold frames are all terms that have been used to describe structures designed for growing and maintaining plants. Early Egyptian and Roman civilizations used greenhouse-like structures to protect and grow tender crops, such as fruits and vegetables, during winter months.

Today, greenhouses vary in sizes, as well as forms, and are used for several purposes. Homes, offices, hospitals, and public gardens use greenhouse structures to maintain and protect plants for our enjoyment. Research institutions use greenhouse structures to conduct experiments on various crops so they can better control the environment. Commercial wholesale nurseries and greenhouse operations use greenhouse structures to propagate and grow flowering plants, foliage plants, bedding plants, and vegetables to sell to garden centers, flower shops, and grocery stores. In some cases, greenhouses are used to start crops, such as vegetable plants and tobacco, to be transplanted later into the field.

10-1. A modern greenhouse may have a roof that opens to provide ventilation and to increase light levels. (Courtesy, Ron Biondo, Illinois)

OBJECTIVES

This chapter covers the different types of greenhouse structures and has the following objectives:

1 Define greenhouse structure

2 Explain the various design styles of greenhouses

3 Illustrate the structural parts of a greenhouse

4 Explain the differences between the transparent coverings used for greenhouses

5 Describe the factors involved in locating a greenhouse range

6 Name the different types of benches and bench arrangements used in greenhouses

7 Explain the different methods used in controlling greenhouse temperature

TERMS

anchor support posts
attached even-span
attached greenhouse
bar caps
barrel vault
connected
    greenhouses
cross-benching
    arrangement
fan and pad cooling
    system
fog evaporative
    cooling system
forced-air heaters

freestanding even-
    span
freestanding
    greenhouse
gothic arch
greenhouse
greenhouse range
gutter-connected
headhouse
infrared radiant
    heaters
lean-to
longitudinal
    arrangement

peninsular
    arrangement
purlins
Quonset
ridge
sash bars
sawtooth
sidewalls
trusses
uneven-span
venlo
ventilators
window-mounted
    greenhouse

GREENHOUSE STRUCTURE TYPES

Greenhouses are structures that are covered with a transparent material that allows sufficient sunlight to enter for the purpose of growing and maintaining plants. These structures are usually artificially heated and cooled so as to manipulate the environment to produce high quality plants. In commercial operations, two or more greenhouses located together are usually referred to as a *greenhouse range.* Often times, these greenhouses are oriented or attached to a central building that is used for offices, storage, and work space. This type of building is called a *headhouse.*

Greenhouse structures are built or manufactured in many design types and sizes. These types are dependent upon the amount of space available, the type of plants to be grown, the geographical location, and the cost of construction materials. The three basic design types are attached, free-standing or detached, and connected greenhouses.

10-2. Greenhouses are used to provide a controlled environment for growing plants.

ATTACHED GREENHOUSES

An *attached greenhouse* is connected to a building. Few of these types of greenhouses are used for commercial plant production, but are used for displaying plants in retail flowershops, garden centers, offices, and homes. The main advantage of this type of greenhouse is that less construction material is needed. However, the disadvantage is that the existing building can shade

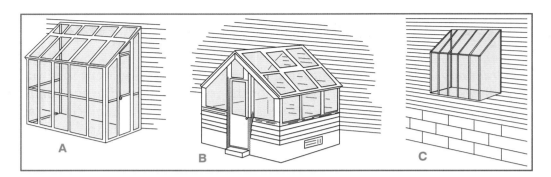

10-3. Attached greenhouses are excellent for displaying plants. Types include: (A) lean-to, (B) attached even-span, and (C) window-mounted.

and limit the amount of light that enters the greenhouse. Also, ventilation and temperature can be difficult to control. The three basic styles of attached greenhouses are:

- *Lean-to* greenhouse—This type is constructed against an existing building. The ridge of the roof is attached to the side of the building and the roof slopes away from the building. A lean-to can be as long as the building in which it is attached. The south side of a building is generally the best place for a lean-to greenhouse.

- *Attached even-span* greenhouse—This style has rafters of equal length and the end wall is attached to an existing building. More space is available in an attached even-span greenhouse than in a lean-to. Because of the more available space, this style is generally more expensive to build and heat than a lean-to.

- *Window-mounted greenhouse*—The window-mounted unit is a prefabricated style that is available to fit many standard size windows. It has very limited space and environmental conditions are difficult to control.

*F*REESTANDING GREENHOUSES

Freestanding greenhouses are separate from other buildings or greenhouses. They usually consist of sidewalls, end walls, and a roof. Since they are not attached or connected to any other structure, the maximum amount of sunlight is received by the plants within. Freestanding greenhouses allow the grower to regulate many different types of environmental conditions for various greenhouse crops.

There are a few disadvantages of freestanding houses. One is that more land is needed for the greenhouse range. A general rule for the spacing between freestanding houses should be one-half the width of the greenhouse. For exam-

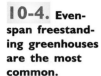 **10-4.** Even-span freestanding greenhouses are the most common.

ple, if the greenhouses are 30 feet wide, then a minimum of 15 feet is needed between each greenhouse. This will reduce the chances of the greenhouses shading each other. A second disadvantage is freestanding greenhouses have more exposed surface area, therefore require more heat, especially at night.

There are many different styles of freestanding greenhouses. Some of the more common ones are:

■ *Freestanding even-span* **greenhouses**—These styles have rafters of equal length and are the most common style of freestanding or detached greenhouses. Today, most even-span greenhouses are clear spans with truss sup-

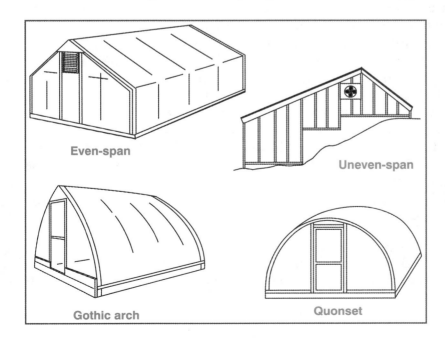

10-5. Various styles of freestanding greenhouses.

Even-span

Uneven-span

Gothic arch

Quonset

ports. This allows growers to move about the greenhouse with ease and allows for more automation of equipment.

■ *Uneven-span* **greenhouses**—Rafters of unequal lengths are used in this unit. This style of greenhouse is suitable for placing on hillsides to allow for the maximum amount of sunlight to enter. A disadvantage of the uneven-span greenhouse is that the slope can cause difficult working conditions.

■ *Quonset* **greenhouses**—They consist of curved roofs with or without side walls. Quonset style greenhouses were introduced after World War II when growers purchased surplus materials from the military.

■ *Gothic arch* **greenhouses**—Styled in the shape of a pointed arch, this greenhouse was developed at Virginia Polytechnic Institute and State University. Truss supports were eliminated so as to add a double layer of polyethylene covering.

CONNECTED GREENHOUSES

Several greenhouses joined together are called **connected greenhouses.** Generally, the sidewalls within the structures are absent. There are several advantages of connected greenhouses. With the absence of interior walls, less construction material is needed and less land is needed since the houses are attached to each other. Connected greenhouses require less heat than freestanding houses because there is less exposed surface area. Overall, the cost of these types are less for the same amount of growing area in freestanding houses. In addition, greenhouse operators find connected greenhouses more efficient because workers and equipment can move throughout the greenhouse with ease.

A major disadvantage of connected greenhouses is growing crops that require different environmental conditions, such as temperature, humidity, and light. In areas of the country in which snowfall is great, connected greenhouses can collapse unless precautions are taken. These houses should be constructed to withstand the extra weight of the snow. Also, heating cables must be installed between the ridges to melt fallen snow so that it can be removed successfully. In large connected houses, ventilation can sometimes become difficult. Most of the time these houses are equipped with large fans mounted on an exterior side wall to circulate fresh air throughout.

Many different styles of connected greenhouses are available. Some of the more common styles are:

■ *Gutter-connected* **greenhouses**—Several even-span greenhouses attached together would define this type. Sometimes this style is referred to as *ridge-and-furrow*. In a gutter-connected greenhouse the eave is replaced with a gutter which aids in the removal of rainfall and snow. Generally the interior walls are replaced by support posts.

10-6. Ridge-and-furrow greenhouses are gutter-connected. (Courtesy, Ron Biondo, Illinois)

- *Venlo* **greenhouses or Dutch houses**—These differ from gutter-connected houses in that they are normally wider and have twice as many ridges between supports. This style of glass greenhouse has been noted for its excellent percentage of light transmission. Venlo greenhouses are used in the Netherlands for the production of vegetables and flowers.

- *Barrel vault* **greenhouses**—These are Quonset style structures with sidewalls joined together. These type of greenhouses are usually covered with a flexible plastic film like polyethylene.

- *Sawtooth* **greenhouses**—These are lean-to greenhouses joined together. They are usually constructed in warmer climates like California, Florida, and Texas. The upper portions of the roof are vented in order to take advantage of natural ventilation.

GREENHOUSE CONSTRUCTION

Small greenhouse structures can be easily built by plans obtained from books or state Cooperative Extension Service agencies. Large commercial greenhouse operators find that prefabricated and preassembled greenhouses ordered from a greenhouse manufacturing company are more suitable for their needs. These preassembled greenhouse packages often provide all the parts and equipment that a grower may need for a particular area. However, one should inquire as to what these packages do and do not include. Many times, these packages do not include the greenhouse end walls, which then have to be constructed by the grower.

10-7. Glass greenhouse construction.

Before a greenhouse is constructed, it must have a foundation to provide the structural support. The foundation can be continuous, but often the support is provided by concrete footings. The footings are located below the frost line and are spaced at intervals according to the spacing of the support anchor post. Water and electrical lines should be located and placed before the foundation is constructed. It is important to correctly square and level the concrete footings for the prefabricated structure members to fit properly.

Some greenhouses have *sidewalls* or curtain walls. The sidewall of a greenhouse is located between the concrete footings and support anchor post; however, they do not provide any structural support. Sidewalls can be a few inches to several feet in height and extend around the perimeter of the greenhouse. Wood, brick, cinder or concrete blocks are often used.

On many of the polyethylene-covered houses, sidewalls consist of 1 inch by 6 inch boards to which the plastic is attached and held tight over the frame. Some of the newer polyethylene-covered houses have a lock or snap aluminum system, which has replaced the wood sidewalls. This system will hold polyethylene, shade cloth, screening, or even a tarp in place. The main advantage of this system is that it will not rot and it is easier to replace the polyethylene covering.

*G*REENHOUSE FRAMEWORK

The type of transparent covering used on a greenhouse will largely determine which type of greenhouse framework is used. At one time, wood was the primary construction material used for the greenhouse framework. Now, wood is used for small hobby-type greenhouses or construction of the greenhouse end walls. If lumber is used, it should be pressure-treated and painted white to reflect sunlight. Most commercial greenhouse frames are constructed of aluminum or galvanized steel.

Since glass greenhouses require more support for their covering, they are composed of more structural components. However, the components common to all greenhouses are:

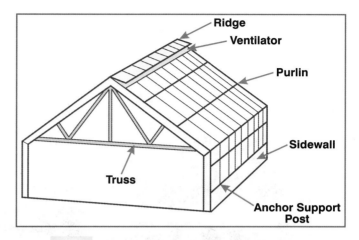

10-8. Structural components of a greenhouse.

- ■ *Ridge* is the top of the greenhouse.
- ■ *Anchor support posts,* or side posts, provide the main structural support for the greenhouse. They are spaced at regular intervals along the greenhouse and are set in the concrete footings.
- ■ *Trusses* are added to the framework for structural strength. They are available in different shapes, but are composed of rafters, chords, and struts.
- ■ *Purlins* run the length of the greenhouse and are bolted to each truss, which adds more structural strength.
- ■ *Ventilators* are moveable units of the greenhouse to allow for natural ventilation. They may be attached to the sides or the ridge.

GREENHOUSE COVERINGS

The most important function of a greenhouse covering is to allow the maximum amount of light into the greenhouse for the growth and development of the plants. Several types of transparent coverings or glazing materials for greenhouses are available, but each varies as to its light transmission.

Glass

At one time, only glass was used to cover greenhouses because it provided the best light transmission. Glass-covered greenhouses have been known to

10-9. Glass-covered greenhouses may be used for research.

last over a lifetime. However, the rise in the cost of glass and fuel for heating has made other glazing materials more appealing to growers. Glass is available in different grades, weights, and sizes. For greenhouse glazing purposes, a double strength, grade B is usually used. In some extreme-cold areas, double layers of glass are used to reduce the heat loss. This type of glass is called thermopane, which consists of two sheets of glass with a dead air space of about ¼ to ½ inch that acts as an insulation layer.

Glass-covered greenhouses are available in many styles. Today, most glass-covered greenhouses are constructed of aluminum or steel. Glass greenhouses require much more structural support than other types. Glass panes are attached to **sash bars** with a glazing compound and then **bar caps** are attached to the outside to hold the glass in place. These structural members can cast shadows during the winter months, which can reduce the growth of the plants. To avoid this, greenhouse glass panes have become larger in size than the original 16-inch by 18-inch pane. Now glass panes for greenhouses can range up to sizes of 32-inch by 36-inch. The Venlo greenhouses have glass panes that are 28-inch by 65-inch, which allow more light to enter.

*P*olyethylene

Several types of flexible-film plastic for greenhouse coverings are available, but polyethylene is the most widely used. Polyethylene became popular in the 1950s when growers realized the savings of polyethylene over glass. The initial cost of installation is cheaper, as well as fuel cost, which is 40 percent lower than glass or fiberglass-reinforced plastic. The biggest disadvantage of polyethylene is its short life span. Ultraviolet (UV) rays from the sunlight break down

10-10. Greenhouses covered with polyethylene.

the polyethylene causing it to become dark and brittle. Without UV inhibitors, polyethylene would only last one winter season, but greenhouse polyethylene with UV inhibitors can last an average of three years.

Greenhouse polyethylene is available from greenhouse supply companies in various widths and lengths. Polyethylene is also available in various thicknesses, but for greenhouse coverings a 4 mil or 6 mil thick (0.1 or 0.15 mm) polyethylene is used. Now, most polyethylene-covered greenhouses use a double layer, with the outer layer being 6 mil, and the inner layer being 4 mil.

Polyethylene coverings can often produce condensation on the inside surface. This is caused by warm, humid air coming in contact with the cold polyethylene covering. The condensation then beads up and drops down onto the plants, wetting their foliage. The wet foliage can increase chances of disease occurrence. Also, water droplets on the inside of the polyethylene can reduce the amount of sunlight entering the greenhouse. Products are available, such as Sun Clear®, that can be sprayed onto the inside surface to reduce condensation. Also, polyethylene is available with anti-condensation chemicals manufactured into the film.

Fiberglass-Reinforced Plastic

At one time, corrugated fiberglass-reinforced plastic panels were used more to cover greenhouses than they are today. Fiberglass-reinforced plastic panels are lower in cost than glass, easy to install, strong, and have a longer life span than polyethylene. If properly maintained, fiberglass-reinforced plastic can last up to 15 years before replacement is needed. The corrugated panels are available in 51-inch width and various lengths.

One of the biggest advantages of fiberglass-reinforced plastic is it allows almost the same amount of sunlight through as glass. However, fiberglass-reinforced plastic is very susceptible to ultraviolet, dust, and pollution degradation. Only clear, fiberglass-reinforced panels, labeled for greenhouse coverings, should be used. Also, fiberglass-reinforced panels are highly flammable; therefore, insurance rates may be higher than with the use of another type of covering.

10-11. Endwalls covered with corrugated polycarbonate.

Acrylic and Polycarbonate

Rigid, double layer acrylic and polycarbonate panels have been introduced as an effective greenhouse covering and have almost replaced the use of fiberglass-reinforced plastic. Acrylic and polycarbonate panels are lightweight and easy to install. Both have excellent heat insulating ability and good light transmission. Eight-millimeter-thick polycarbonate panels have a useful life of about 15 years, while acrylic panels—though more expensive—last 20 years or more. A disadvantage of the acrylic panel is that it is flammable. Both can be purchased in various widths and lengths. Many growers are using the double layer panels on the end and side walls of the greenhouse and covering the roof with polyethylene. Also, polycarbonate is available in a single layer, which is clear and corrugated. Some growers are using the clear, corrugated polycarbonate panels to replace fiberglass-reinforced plastic.

LOCATING A GREENHOUSE

After investigating all the different types and styles of greenhouses, one must determine the most suitable location for the greenhouse or greenhouse range. If buying an existing range, then one needs to weigh the cost of buying

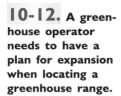

10-12. A greenhouse operator needs to have a plan for expansion when locating a greenhouse range.

versus building. Several factors must be considered when determining the location. Some of these are:

- **Market**—Before investing any money in a greenhouse business, one should determine if there is a market or if anyone will buy the plants that are to be grown.

- **Accessibility**—It is best to locate the greenhouse near a major highway so supplies can be easily received and crops can be shipped out. Most greenhouse crops are shipped by trucks. If the greenhouse operation is to be a retail outlet, then it needs to be highly visible to the general public.

- **Climatic Conditions**—The type of crops to be grown should always be kept in mind when choosing a location. The available sunlight, average minimum temperature, snowfall, and areas with high winds are some of the climatic conditions that will dictate the most suitable location for that particular greenhouse crop.

- **Topography and Drainage**—The location of the greenhouse should be reasonably level with sufficient surface and subsurface drainage. Labor and material movement are easier on a more level site.

- **Water and Other Utilities**—A reliable source of good quality water is needed in a greenhouse operation. Also, other utilities, such as electricity, fuel, and waste disposal cost, should be taken into consideration.

- **Zoning Regulations**—In many populated areas, greenhouses are prohibited. One needs to check the zoning regulations for the particular site before the property is purchased.

- **Labor Supply**—A dependable source of skilled and unskilled labor is needed to operate a successful greenhouse business.

■ **Expansion**—A greenhouse operator needs to plan for future expansion of the greenhouse range. Also, the operator needs to consider expansion in supply storage, as well as work area.

LAYOUT OF A GREENHOUSE RANGE

Most buildings are constructed perpendicular or parallel to the road. Greenhouses are different. The range layout and the individual greenhouse floor plan needs to take into consideration environmental influences on the site, as well as the circulation of people, plants, and supplies.

RANGE ORIENTATION

The range should be oriented to prevent the least amount of shadows and achieve the maximum light intensity. The orientation of the greenhouse range is much more important in northern latitudes of North America during the winter months. Freestanding single greenhouses above the 40° north latitude should be built with the ridge running east to west; whereas, those below the 40° north latitude should run in a north to south direction. Connected greenhouses should be oriented with the ridges running north to south in all latitudes. This will provide the least amount of shadows cast created from the gutters.

Also, the direction of the prevailing winds is important in the orientation of the greenhouse range. Prevailing winter winds will influence the amount of heat loss, while the prevailing cool summer breeze may aid in keeping the greenhouses cooler.

When laying out a greenhouse range, consideration should be given to entrances and drives so that they are accessible by large trucks, while providing sufficient parking for customers and employees. In addition, the location of the headhouse and storage facilities should be easily accessible for unloading supplies and loading plants to be shipped.

GREENHOUSE FLOOR PLAN

The type of crops to be grown, layout of outbuildings, and planned future expansions will often dictate the floor plan of the greenhouse. Selection of the type and arrangement of benches or beds that the plants are grown on will influence how efficiently the crops are produced.

Bench and Bed Arrangements

In-ground beds are generally used for fresh flower crops, like roses, carnations, and chrysanthemums, that may grow tall; while raised benches are used for pot crops, like florist mums, azaleas, and poinsettias. Whatever the crop that is to be produced, the grower should try to arrange the benches or beds to achieve the maximum amount of growing area while considering the movement of employees in and around the plants. The most common type of bench arrangements found in greenhouses are:

- **Longitudinal**—The *longitudinal arrangement* runs the entire length of the greenhouse. This type is commonly used for beds in which fresh flowers are grown. It allows for more mechanization, but often is difficult for employees to move across the greenhouse.

- **Cross-benching**—The *cross-benching arrangement* runs the width of the greenhouse. Usually, aisles are located along the sidewalls as well as between the benches. This type allows for easy accessibility to all benches but does not maximize growing space.

- **Peninsular**—The *peninsular arrangement* is similar to the cross-benching arrangement, except a central aisle runs the length of the greenhouse and the benches extend to the sidewalls. Accessibility by employees is increased with this type arrangement, while still maximizing the growing area.

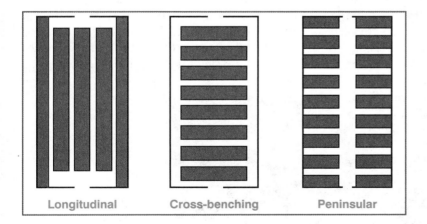

10-13. Three common styles of bench arrangements.

Longitudinal   Cross-benching   Peninsular

Types of Greenhouse Beds and Benches

Beds are commonly constructed from concrete. They are built so as to isolate the growing medium which reduce the possibility of soil-borne diseases. In-ground beds should be constructed with drain tiles in the bottom to allow for drainage of excess water and permit steam to be introduced for medium pasteurization.

Raised benches are normally built 24–30 inches in height for potted plants. Frames or supports for raised benches are constructed from wood, pipe, or cinder blocks. Wood that is resistant to decay, such as redwood or cedar, should be used. The widths of the benches vary depending on the arrangement in the greenhouse. If they are located against a wall, they should not be wider than 3 feet and if they are accessible from both sides, they should not be wider than 6 feet. Some of the more common types of raised bench tops are:

- **Wood Bench Fabric**—This type of bench top is sometimes called lath or snow fencing. It is treated strips of wood connected together by wire and is rolled out over the bench frame. Wood bench fabric is good to use when growing large pot plants or flats, but not suitable for small, 3-inch pot plants.

- **Welded Wire**—This type of bench top consists of 14-gauge galvanized wire, which may be coated with plastic and has a mesh size of 1 to 2 inches and is good for growing a variety of pot plant sizes.

- **Prefabricated Plastic**—This is a lightweight, but strong, vinyl material that is sold by the square foot and can be customized to fit the greenhouse bench arrangement. Prefabricated plastic is best suited for growing large pot plants or flats.

- **Movable Benches**—Movable benches allow for optimum use of greenhouse space. The tops are made of 14-gauge galvanized expanded wire on frames with steel rollers. This type of bench is sometimes called the aisle-eliminator or floating aisle because the rollers allow the entire bench top to move either to the right or left. Thus, only one aisle is needed for the whole greenhouse.

10-14. Movable benches maximize greenhouse space. (Courtesy, Ron Biondo, Illinois)

10-15. Plants placed directly on the greenhouse floor.

Some growers grow pot plants directly on the floor. The area used is covered with gravel or a porous concrete that allows water to percolate through but inhibits weed development. Bedding plants and some foliage plants can be grown on the greenhouse floor in this manner because they require less handling by employees. This type of floor plan greatly increases the use of greenhouse space.

Aisle and Walks

If aisle or walks are heavily traveled with equipment, plants, or people, they should be constructed of concrete. Less traveled walks need only gravel. The minimum width for walks is 18 inches. However, one should consider the amount of traffic and the type of equipment, such as wheelbarrows and pesticide sprayers, that will be used in the greenhouse. Normally, walks and aisles are 3 to 4 feet wide.

CONTROLLING GREENHOUSE TEMPERATURE

For most greenhouse crops, a night temperature of 55 to 65°F, and a day temperature 10 to 15°F higher than the night, must be maintained so that the high quality crops are ready to sell at a certain date. Temperatures can be controlled in the greenhouse with the use of heaters, ventilation and evaporative cooling systems, and shading materials.

Two types of sensing devices are used to monitor greenhouse temperature. A thermometer is a sensing device that registers the temperature. A thermostat is an automatic sensing device used to regulate heating and cooling equipment.

GREENHOUSE HEATERS

Providing heat in the greenhouse is one of the greatest expenses of a greenhouse operation. Most heat is lost by conduction through the greenhouse covering. However, it is also lost through radiation and air infiltration through the covering, wall, door, and other greenhouse parts. The grower must provide heat and evenly distribute that heat in order to maintain a desired temperature.

Several types of greenhouse heating systems are available. When selecting a greenhouse heating system, a grower should not only look at the initial cost of the system, but also investigate the type of fuel best suited, the ease of automated operation, and the maintenance. To maintain a true reading of the temperature near the plants, greenhouse thermostatic controls need to be shielded from the sun and placed in the center of the greenhouse at plant height. The grower should also consider emergency heaters and back-up generators. Some of the types of heating systems commonly used in greenhouse operations are:

■ **Steam Heating Systems**—Pressurized steam is produced in a boiler, which is then circulated to the greenhouses. The steam heat is distributed through

10-16. Unit heaters provide the necessary heat in this greenhouse. (Courtesy, Ron Biondo, Illinois)

pipes located along the sidewalls and overhead. As the steam cools and gives off heat, it condenses into water. The water is pumped back into the boiler where it is re-heated and turned back into steam. The steam produced in the boiler can also be used to pasteurize growing media.

- **Hot Water Heating Systems**—Hot water heating systems use the heat given off by hot water. This system consists of a boiler and heating pipes. The heating pipes are located throughout the greenhouse—along the walls and underneath benches and growing tables.

- **Forced-air Heaters**—*Forced-air heaters* are localized heater units that force hot air directly into the greenhouse or through a perforated plastic tube system, which is placed overhead down the length of the greenhouse. This type of heater is commonly used in commercial greenhouse ranges.

- **Infrared Radiant Heaters**—*Infrared radiant heaters* are individual unit heaters that produce infrared radiation that travels through the air and directly warms plants and other objects in the greenhouse. These type of heaters conserve energy but may be difficult for the grower to monitor because the air temperature is, on the average, 7 degrees cooler than conventionally heated greenhouses.

VENTILATION AND COOLING SYSTEMS

Greenhouses need to be ventilated and cooled throughout the year. They collect and retain solar heat energy. During the summer and even in the winter months, the air temperature can become too high for proper plant growth. In addition, fresh air needs to be introduced into the greenhouse to renew the supply of carbon dioxide, which is needed for photosynthesis. Also, sufficient air circulation reduces the chance of plant disease. As with greenhouse heating systems, thermostatic controls for ventilation and cooling need to be placed with the thermostatic controls for heating. Some types of ventilation and cooling systems are:

- **Natural Ventilation System**—This system operates on the principle that hot air rises and cooler air settles. Air can be exchanged through open ridge and side vents controlled by thermostats. This is the oldest method of cooling. However, for winter cooling, this form may cause injury to plants located near the vents.

- **Fan Ventilation System**—Fans blows cooler, dryer outside air into the greenhouse. When the system starts, the exhaust fans automatically turn on. The exhaust fans pull the humid air out of the greenhouse.

- **Fan-tube or Fan-jet Ventilation System**—Fans bring in cooler, dryer outside air and mix it with the warm air. The air is distributed through a perforated plastic tube that runs the length of the greenhouse. The exhaust fans pull the

10-17. Self-contained arctic cooling units and side vents are used to cool the greenhouse.

mixed air through and out of the greenhouse. This system is often used in conjunction with the heating system.

■ *Fan and Pad Cooling System*—This wet-pad system is used for summer cooling. Large exhaust fans draw air through a moistened cellulose pad mounted on the opposite end of the greenhouse. The air is cooled through the evaporation process as it moves through the pad. This system works best in areas with low humidity.

■ *Fog Evaporative Cooling System*—This mist system is also used for summer cooling. Fog is generated with water sprays from nozzles or atomizers above the plants. Most of the fog evaporates reducing the greenhouse temperature before reaching the level of the plants. However, some of the mist settles on the plant where it also helps to reduce leaf temperature.

SHADING MATERIALS

Greenhouse temperatures can be reduced by using shading materials on the greenhouse covering. Shading helps prevent excessive heat buildup in the greenhouse. Liquid shading compounds are often applied to glass or polyethylene covered greenhouses to reduce the light intensity, as well as the temperature. These compounds can be purchased from greenhouse supply companies and are usually applied around March, when the light intensity and temperature begin to increase.

Black polypropylene shade fabric can be pulled over the greenhouse to reduce light intensity and temperature. These shade fabrics come in various percentages of shade and can be manufactured to a specific size.

10-18. Black polypropylene shade fabric is used to reduce the light intensity and temperature of the greenhouse.

REVIEWING

*M*AIN IDEAS

Greenhouses are transparent structures that artificially control the environment to grow many horticultural and some agricultural crops. Greenhouses are fun and entertaining for the gardener and often used for therapy in hospitals, prisons, and nursing homes. However, they are considered a major piece of equipment for making money by the commercial wholesale plant grower.

There are basically three types of greenhouse design types. These are the attached, freestanding, and connected greenhouses. Many variations of these exist. No matter what type of greenhouse, almost all have sidewalls, a ridge, trusses, purlins, anchor supports, ventilators, and a method to heat and cool the greenhouse during the different seasons. All greenhouses have a transparent covering that should allow the maximum amount of light in for the growth and development of plants. The common types of coverings are glass, polyethylene, fiberglass-reinforced plastic, and acrylic or polycarbonate.

Before building a greenhouse, one should select the best possible location. The market, accessibility, climatic conditions, topography, water, zoning regulations, labor supply, and future expansion will all influence the most suitable location. After selecting a location, the layout of the greenhouse range is important. The environmental influences, as well as the movement of people, plants, and supplies, will affect the efficiency of the growing operation.

Heating and cooling is very important in maintaining or growing high quality plants. For most plants, a minimum night temperature of 55 to 65°F should be maintained, with a day temperature of 10 to 15°F higher. Temperature can be adequately controlled with the use of heaters, ventilators, cooling units, and shading materials.

QUESTIONS

Answer the following questions. Use complete sentences and correct spelling.

1. What is a greenhouse? What are the three basic design types?
2. What are the three basic styles of attached greenhouses? Briefly describe each.
3. How do freestanding greenhouses differ from connected greenhouses?
4. What are four styles of freestanding greenhouses? Briefly describe each.
5. What are four styles of connected greenhouses? Briefly describe each.
6. How can polyethylene be held tight against the greenhouse frame?
7. What are the common components in the frame of a greenhouse?
8. What are four types of greenhouse coverings? Which is the most popular? Which has the longest life span?
9. What are the factors to consider when trying to locate a greenhouse?
10. How should a greenhouse be oriented? What is your location in relation to the 40° north latitude?
11. What are the most common types of bench arrangements found in greenhouses? Sketch the bench arrangements.
12. What are four types of bench tops? Briefly describe each.
13. Name four types of heating systems used in greenhouses. Briefly describe each.
14. How is the fan and pad cooling system different from the fog evaporative cooling system?

EXPLORING

1. Visit a local greenhouse range or plant nursery with greenhouses. Study the construction, the type of greenhouses, and how they are heated and cooled.

2. Construct models or architecturally draw the various styles of greenhouses.

3. Working in cooperative groups, research the best possible location for a greenhouse range in the community. Identify the crops that should be grown.

4. Design an actual greenhouse range and discuss in class the factors involved with building a greenhouse range.

11

GROWING CROPS IN THE GREENHOUSE

Growing ornamental crops in the greenhouse is becoming more technical and specialized as consumers' demands for quantity and quality increase. Producing greenhouse crops is enjoyable and challenging. Different greenhouse crops are needed at various times of the year. Growers must have the knowledge and skills to produce crops on a year round basis and deliver them to their customers on schedule. These skills include selecting containers, mixing and preparing the growing medium, providing adequate moisture and nutrients for the plants, controlling growth factors, and preventing or eliminating greenhouse pests.

Experienced growers know how to regulate environmental factors to produce greenhouse crops. The control or management of light, temperature, air, and water to promote healthy growth is often referred to as cultural practices. Understanding basic cultural requirements of greenhouse crops is of great importance to all whose careers include producing crops in the greenhouse.

11-1. Growing greenhouse crops is enjoyable and challenging. These Georgia students are learning about horticulture in the greenhouse at their school. (Courtesy, Jasper S. Lee)

245

OBJECTIVES

This chapter focuses on general greenhouse practices used in the production of ornamental plants. It has the following objectives:

1 Explain the importance of environmental conditions on greenhouse crops

2 Explain the various types of containers used for growing greenhouse crops

3 Describe the functions of a good growing media and the components of soilless media

4 Describe the different types of irrigation systems used in greenhouse production

5 Explain the difference between organic and inorganic fertilizers used in greenhouse production

6 Describe the various methods of applying fertilizers to greenhouse crops

7 Describe the different methods of pinching greenhouse crops

8 Explain the use of plant growth regulators on greenhouse crops

9 Describe the different practices of greenhouse integrated pest management

TERMS

azalea pot
bark-based mixes
bulb pan
capillary mat system
carbon dioxide
   fertilization
chemical pinching
   agents
DIF
disbudding
fertilizer proportioner
   or injector

fluoridation
forcing
hard pinch
headspace
inorganic fertilizers
organic fertilizers
peat-lite mixes
peat moss
peat pellets
peat pot
perlite

photoperiodism
pinching
soft pinch
spaghetti tube
   irrigation
standard pot
thermoperiodic
thermotropism
vermiculite

GREENHOUSE ENVIRONMENT

Greenhouses provide sunlight for growth and protect crops from adverse weather. Many times, the grower has to also manipulate the environment or plant to achieve the desired amount and form of growth. Plant height and flowering are two commonly controlled growth responses.

TEMPERATURE

Temperature has a major effect on plant growth. Temperature effects photosynthesis and respiration. Temperature is a key factor in a number of plant responses. A night temperature of 55 to 65°F and a day temperature 10 to 15°F higher than the night must be maintained for most greenhouse crops. **Thermotropism** is plant growth response to temperature, and **thermoperiodic** describes plant response to changes in day and night temperature. Poinsettias and holiday cactus are examples of thermoperiodic plants. A period of cool temperature along with short days causes them to initiate flowering.

11-2. Poinsettias require a period of cool temperature along with short days to initiate flowering. (Courtesy, Paul Ecke Poinsettias)

For some greenhouse crops, a period of cold temperature is required for flowering. This physiological process is known as vernalization. Bulb crops including tulips, narcissus, and Easter lilies must undergo a period of cold temperature for flowering.

LIGHT

Light is necessary for photosynthesis to occur. The quality, quantity, and duration of sunlight affects plant growth. Light quality is the wavelength or color of light. Growers are not generally concerned with the manipulation of the quality or color of light since sunlight provides the proper spectrum of radiation. They are most concerned with light intensity and duration.

11-3. Proper spacing is important to maximize the light intensity.

Light Intensity

The intensity or brightness of light describes light quantity. Without light, photosynthesis and plant growth cannot occur. The more light a plant receives, the better capability it has to produce more food reserves. Every plant species has a favorable light intensity for maximum growth. Some plants prefer bright, indirect light. Others prefer to grow in full or partial shade.

Maximum light intensity can be achieved in several ways. It should begin by properly designing and planning the greenhouse or range layout. Growers should routinely clean and maintain greenhouse coverings so that the crop does not suffer from low levels of light. In the case of polyethylene covering, cleaning not only increases the light intensity, but also extends the life of the cover. Proper plant spacing on the bench or in the bed is important so the plants do not shade each other, reducing the light intensity. In northern latitudes of North America during the winter months, the light intensity drops to very low levels. Some growers have to add supplemental lighting to the crops to raise the intensity so that the quality is not affected.

The development of high energy discharge lighting systems has now given growers a more efficient and effective artificial lighting source. Supplemental light sources include incandescent lights, fluorescent lights, metal halide lamps, and low and high pressure sodium lamps.

11-4. High energy discharge lighting makes the best supplemental lighting system. (Courtesy, Ron Biondo, Illinois)

Reducing light intensity is primarily achieved by applying shade compounds or installing shade fabric. Shading compounds are opaque materials applied directly to the panes of a greenhouse. Shade materials are usually applied in the spring, around March, and removed in September. Shading compounds do not generally need to be removed because they will naturally dissolve.

*L*ight Duration

Light duration involves the length of exposure to light in a 24-hour period. Plants have an internal mechanism for keeping track of light exposure they receive each day. The length of days can have a direct effect on the growth of many plants. Researchers have found that it is length of darkness that influences plant growth. This is called *photoperiodism.* Greenhouse crops like chrysanthemums and poinsettias will begin to flower naturally in the fall when the day length is shorter. They are said to be short-day plants. These plants grow vegetatively when the days are long.

11-5. Poinsettias will begin to flower naturally in the fall when the day length is shorter.

In order to produce enough vegetative growth on the plants, growers have to extend the day by artificially lighting the plants at night. Incandescent lights are turned on from 10 p.m. until 2 a.m. This is usually done with the use of time clocks. Flower initiation, or development of the flower's buds, begins when the days are short. If the days are naturally long, growers shorten the days by pulling a light-proof cover over the plants. The covers are pulled automatically or by hand around 5 p.m. and removed around 8 o'clock the next morning. This shortens the day length and induces flowering.

Shasta daisies and coneflower plants flower as days lengthen in the summer. They are termed long-day plants. A third group of plants is unaffected by day length. They are classified as day-neutral plants. Examples of these include many foliage plants, African violets, and carnations.

Growers control the length of days in a greenhouse to bring on one of two plant responses. The plants will be kept in a vegetative stage of growth or encouraged to flower.

*A*IR

Air has carbon dioxide and oxygen that are critical for photosynthesis and respiration. Exchange of oxygen and carbon dioxide through stomata in the leaves keeps photosynthesis operating at peak efficiency. The ability of air to move in and out of the soil is important in providing oxygen for healthy root growth. Root cells must have oxygen to undergo the chemical process of respiration. Air quality is also an ingredient to producing healthy plants in the greenhouse.

I I-6. CO_2 burners release carbon dioxide gas into the greenhouse atmosphere. (Courtesy, Ron Biondo, Illinois)

*C*arbon Dioxide Levels

Plants use carbon dioxide in the photosynthesis process. The supply of carbon dioxide is usually adequate during the time the greenhouse is ventilated. Ventilation of the greenhouse helps to replenish CO_2 in the greenhouse atmosphere. When greenhouse vents are closed, CO_2 levels can drop significantly. Carbon dioxide becomes a limiting factor and photosynthesis is slowed. Plant growth is increased significantly by injecting the greenhouse air with carbon dioxide. This is accomplished with special burners for the purpose of increasing the rate of photosynthesis. This process is called ***carbon dioxide fertilization.*** Supplemental CO_2 is most effective when used to give young plants a boost. Photosynthesis uses CO_2 to produce simple sugars during the dark phase of its chemical process. The result is higher quality crops and a shorter production time. Carbon dioxide is usually added during the night between 9:00 p.m. and 6:00 a.m.

Humidity

Humidity, which is water vapor in the air, affects plant growth. The growth rate of plants increases under conditions of high humidity. High humidity reduces water stress of a plant so photosynthesis can function smoothly. If the humidity is low, the dryness of the air can put stress on the plant. This is especially true if soil moisture is inadequate and wilting occurs. One drawback of excessive humidity in the greenhouse is the increase of leaf and flower diseases.

SELECTING CONTAINERS FOR CROPS

Many greenhouse crops are grown and sold in containers. The type of container a grower selects will influence the cultural practices, marketing

11-7. Many greenhouse crops are sold in plastic cell packs. (Courtesy, Ron Biondo, Illinois)

plans, and ultimately, the profit made. Many years ago, most greenhouse crops were grown in clay pots. Today, the majority of the greenhouse crops are produced in plastic containers.

Growers have many choices of containers. Size is one of the first considerations when selecting containers. The type market, whether a retail flower shop or mass market, will often dictate the size of container a grower selects. For instance, mass market outlets generally sell more smaller size poinsettias during the Christmas holidays than retail flower shops, which prefer larger size poinsettias. Other factors that influence growers selection are drainage, durability, ease of handling for mechanization, and color. Color of the containers is important in the marketing of the plant. The color of the containers should enhance and compliment the plants.

Containers are grouped according to the type of crop grown. There are certain containers for rooting, bedding plants, green and flowering plants, and hanging baskets.

ROOTING CONTAINERS

Some rooting containers are manufactured from organic compounds such as peat moss. **Peat pellets** are compressed peat moss that expands when moistened to become small peat-filled pots. These are used for seeding or rooting cuttings. Similar to these are the **peat pots**, which are compressed peat moss pots, that the grower fills with growing medium. Others are made from inorganic compounds, such as plastic or metal. Some growers use plastic foam cubes or blocks to root cuttings, while others prefer to root cuttings in open plastic or metal flats.

11-8. Peat pellets and peat pots are good for growing vegetable plants.

BEDDING PLANT CONTAINERS

Producing bedding plants as plug seedlings allows growers to increase the number of plants they

11-9. Producing bedding plants allows growers to increase the number of plants they produce. Plugs can be automatically transplanted to cell packs or pots. (Courtesy, Ron Biondo, Illinois)

produce. These seedlings are produced in cell pack containers and, when they reach marketable size, are ready for transplanting with their root systems intact and undisturbed. They do not overcrowd as quickly as seedlings in open trays or flats. Plug production also allows growers to keep a more accurate inventory of seedlings available for transplanting.

The majority of bedding plants are grown in plastic cell packs. These cell packs are designed to fit into 11" × 21" plastic flats. Many numbered combinations are available. The most popular are 72 cells per flat, 48 cells per flat, and 36 cells per flat. Each cell has a small hole in the bottom to allow for drainage. Many growers use automated equipment to fill the cell packs in flats with growing media and automated seeders to plant each individual cell. Plug seedlings maximize greenhouse space, reduce production time, and allow for mechanization. Some growers also produce larger size bedding plants in 4 inch plastic containers for the landscaper and late-season gardener.

GREEN AND FLOWERING PLANT CONTAINERS

Most green and flowering plants are grown in plastic containers. However, a few crops are still produced in clay containers. Clay containers are used because they are porous, allowing more aeration for the roots, and heavier, providing more stability for top-heavy plants.

Plastic containers are available in square or round shapes—round being the most popular. The type of pot used depends on the growth habits of the plants. *Standard pots* are containers equal in width and height. *Azalea pots* are slightly shorter than the standards. This gives the plants more stability and a more pleasing appearance. Azalea pots tend to be the most popular. **Bulb pans** are half as high as their widths. These shallow containers are designed for bulb crops, such as tulips and hyacinths.

Hanging Baskets

A wide variety of greenhouse crops can be grown in hanging baskets. The majority of hanging baskets are made of plastic. However, wire with preformed wood fiber liners and clay are available. The most popular sizes are 8 and 10 inch diameter baskets. Some have snap-on saucers, while others have self-contained saucers.

GROWING MEDIA

Production of greenhouse crops has some specific demands that are often different from the production of outside plants. Rapid growth is required of greenhouse crops. Growers produce a crop quickly, ship it, and bring in a new crop. Therefore, the growing medium must provide conditions that encourage plant growth. Another requirement of greenhouse crops is that they be uniform. A crop in which all the plants have the same rate of growth and development is highly valued. To ensure that the plants grow at the same rate, the growing medium used must be uniform.

Selection of the growing medium for a greenhouse crop is one of the most important decisions involved in the production of the crop. Medium influences not only the cultural practices used on the crop while in the greenhouse, but also influences shipping cost and customer satisfaction after the plant is purchased. The growing medium serves several important functions in the growth and development of the plants, such as:

- **Moisture holding capacity and aeration**—The growing medium should provide adequate moisture for plant growth, while retaining sufficient air space for root development. If plant roots are not provided sufficient air space for exchange of gases, growth will be slowed and, thus, the target selling date will be missed.

- **Nutrient holding capacity**—The growing medium should provide available nutrients for plant growth. The pH of the medium needs to remain fairly stable throughout the life of the crop so that nutrients are available to the plants. The ideal medium pH for most greenhouse crops is 5.5 to 6.5. Also, the medium needs to be able to hold the nutrients against leaching so plants do not have to be fertilized as often.

■ **Support for the plant**—The growing medium needs to have sufficient bulk density to support the top portion of the plant. Lightweight growing medium may not provide sufficient support. The plant may easily fall over on the bench due to improper support or the roots may become damaged during shipment because the top flops about. In hydroponic culture, or plants grown in nutrient solution, gravel or some other means of artificial support is provided.

■ **Reproducible and available**— Many growers want to have their own secret growing mix that produces monster plants. However, the ingredients need to be readily

11-10. The growing media provides plant support, water, nutrients, and air. (Courtesy, *Greenhouse Product News*).

available for the next crop and they should be mixed in the same manner each time so the cultural practices do not have to change.

COMPONENTS OF GROWING MEDIA

Many growers have shifted away from soil-based medium in greenhouse pot plants. Soil, when mixed with other ingredients, can provide an excellent growing medium, but good quality topsoil is not consistently available and the shipping cost increases because of the increased bulk density.

Soilless medium is a medium that does not contain topsoil. Numerous organic and inorganic materials are available to form different soilless media. Some of the more common components are:

■ **Peat moss**—*Peat moss* is a moss plant that grows in heath bogs. Peat moss may vary depending on source. Sphagnum peat moss is the best. It has a high moisture and nutrient holding capacity.

■ **Bark**—It is best to use aged hardwood or pine bark. Bark may be used as a substitute for peat moss or in combination with peat.

■ **Sand**—Sand is often used in combination with organic matter, such as peat moss, to increase the aeration and drainage. It has a low nutrient holding capacity. A clean, coarse grade of sand should be used.

11-11. Soilless medium components should be readily available. (Shown: top left, bark; top right, sand; bottom left, perlite; bottom right, vermiculite)

- **Perlite**—*Perlite* is heat-treated lava rock. It is a light-weight product with low nutrient and moisture holding capacity. It increases aeration to the medium.

- **Vermiculite**—*Vermiculite* is heat-treated mica that has high nutrient and moisture holding capacity. It is light-weight and should be used only with organic matter.

These components, along with fertilizer amendments, are formulated into mixes. They can be mixed by the grower or commercially prepared. A common formulation of soilless media is the ***peat-lite mixes*** developed at Cornell University in the 1960s. Mix A of the peat-lite mixes is composed of one part sphagnum peat moss to one part vermiculite, which is used for starting bedding plant seeds. Mix B substitutes the vermiculite for perlite and is used mostly for pot-plant crops.

Many commercially-prepared mixes that use the peat-lite formulas are available from greenhouse supply companies. These mixes have had the pH adjusted to the proper range. Nutrients and wetting agents have been added so the mix is ready to use. Commercially-prepared mixes are shipped dry in bags or compressed bales and need to be moistened prior to planting.

Some growing media formulas have substituted aged pine bark for the sphagnum peat moss. These mixes are known as ***bark-based mixes.*** Com-

11-12. Commercially prepared medium is being placed in a machine that automatically fills flats. (Courtesy, Ron Biondo, Illinois)

mercially-prepared, bark-based mixes are also available from greenhouse supply companies.

When mixing peat-lite or bark-based media, growers will always add at least three fertilizer amendments and a wetting agent. First, the pH level must be adjusted. In most cases, dolomitic limestone is added at the rate of 10 pounds per cubic yard to adjust the pH into the desired range. If the pH must be lowered, then sulfur is added. Phosphorus is incorporated into the mix by adding super-phosphate or triple super-phosphate. The third amendment is a micronutrient fertilizer, which supplies all the minor nutrients needed for sufficient plant growth. Sometimes growers will add additional nitrogen and potassium when mixing.

Soil sterilization may be necessary for controlling insects and diseases in growing media. Some sterilization equipment uses steam as the heat source. Other soil sterilization equipment contains electrical heating elements to heat the growing media to a set temperature.

IRRIGATION PRACTICES

A big responsibility of growers is the daily task of watering, which effects the overall quality of the greenhouse crop. Often times, irrigation of plants is taken too lightly. Every employee of a greenhouse operation should realize that water is needed by the plants more than any other substance.

11-13. Employees need to be trained in proper irrigation techniques. (Courtesy, Ron Biondo, Illinois)

Several factors affect how often greenhouse crops need watering. Some containers have porous sides which allows for faster evaporation. Plants located near the outside of a greenhouse bench or near heaters and fans dry out faster. Plants use more water on hot, bright summer days than they do on cloudy or winter days. The water holding capacity of the growing medium has a direct effect. The type, stage of growth, and size of plants must be considered. For example, cacti require less water than tropical foliage plants. Young, rapid growing plants require more frequent watering than older, mature plants that grow slower. Plants with more leaf surface also lose more water through transpiration than do smaller-leaved plants.

The person responsible for watering should be trained in the proper techniques for watering plants and know the effects of watering on plants. Some of the common watering practices are:

■ **Drainage**—Growing medium should be selected so that it provides sufficient air space while retaining moisture. Poorly drained medium will not allow for good root development and the quality of crop will be affected. Growers may remove a plant from its container to check the bottom of the container for moisture. If the container is wet or if water can be squeezed from the plant's root ball, water is not needed.

■ **Timing**—Watering should be done just before the plant shows signs of water deficiency. This is often difficult to determine. The season of the year, type of crop, stage of development, type of growing medium, and air circulation patterns will all influence when to irrigate. One must observe the plants and growing medium carefully each day to determine the irrigation needs. Growers often use moisture meters as a guide to determine when watering may be necessary. Another method uses a dipstick inserted into the growing medium. If the stick is dry and does not have medium stuck to it when removed, the plant probably needs watering.

■ **Amount**—Plants should be thoroughly watered each time they are irrigated. The entire medium surrounding the root system should be moistened. Water breakers should be used on hose ends to control the flow of water. When crops are planted, sufficient space should be left at the top of the container so water can be filled in this space and allowed to flow into the growing medium. This ¾ to 1 inch space is sometimes referred to as the container's *headspace*. To wet all the medium in the container, apply water until it runs out of the bottom.

■ **Water carefully**—Water should be applied to the growing medium, and not to the foliage or flowers. Water allowed

11-14. Space should be left at the top of the container to allow for proper watering.

to stand on foliage and flowers increases the chances of disease. Also, it is best to irrigate in the morning so water that has splashed onto foliage has time to dry before dark. To prevent contamination by disease organisms, hooks should be used to hold the hose ends or nozzles off the greenhouse floor.

Plants express similar symptoms from underwatering as overwatering. At one time, underwatering was a method of controlling crop heights and to induce flowering. When insufficient amounts of water are applied to crops, growth is slowed and crop production time is increased. Smaller leaves and shorter internodes usually result from underwatering. When water is held back too long from the plant, leaf margins will burn, leaves will drop, and the plant may wilt without recovering. Growers now use growth regulators and retardants to control growth instead of restricting the water supply to the plant.

On the reverse side, overwatering can reduce plant quality by producing soft, succulent growth. Plants in this situation will wilt easily and do not have a long shelf life once they reach the market. If an excessive amount of water is applied, the supply of oxygen in the growing medium is depleted and the root system becomes damaged. This causes wilting, nutrient deficiency, and stunted growth.

TYPES OF IRRIGATION SYSTEMS

Many types of irrigation systems exist. They range from manual hose watering to automatic irrigation systems that irrigate entire greenhouse ranges. Systems differ depending on the type of greenhouse crop. Irrigation systems can be connected to timers for more automation. Some of the common types of systems are:

- **Hand Watering**—Manually watering plants is time consuming. It may become boring to the employee doing the watering, allowing for more error. However, every greenhouse needs hand watering for a select few plants that become dry before others on the bench. For hand watering, a water breaker is usually attached to the end of the water hose.

- **Spaghetti Tube Irrigation**—*Spaghetti Tube irrigation* involves small tubes connected to a main line. Water is carried to each pot from the ¾ inch polyethylene or PVC pipe main line by a thin polyethylene micro-tube placed in each pot. The micro-tube is weighted or anchored in each pot. When operating, water dribbles through the tubes, watering all the pots on the bench at the same time. Foliage remains dry with this system.

- **Drip Irrigation**—Drip irrigation is similar to spaghetti tube irrigation except the tube provides a steady drip with little runoff. Drip irrigation is widely used with hanging baskets.

- **Ebb and Flood**— Ebb and flood is a method where water is pumped into the system at regular intervals, filling the benches. The con-

11-15. **Spaghetti tube irrigation on foliage plants.**

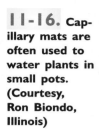
11-16. Capillary mats are often used to water plants in small pots. (Courtesy, Ron Biondo, Illinois)

tainers sit in the water. The medium in the containers slowly absorbs water. After a certain period, the water is drains from the bench. Ebb and flood benches conserve water, provide accurate nutrient levels, and allow for optimum spacing of plants.

■ **Capillary Mat System**—*Capillary mat system* is a form of sub-irrigation in which potted plants are set on a moist, porous synthetic mat. Water moves upward through the drainage holes into the growing medium by wick action. This system allows for different size containers to be used. However, algal buildup on the mats may become a problem.

■ **Overhead Irrigation**—Water is applied over the canopy of the plants with spray nozzles mounted on risers. This system is commonly used on propagation benches and bedding plants. Many growers are hesitant to use overhead irrigation on other crops because the wet foliage increases chances of disease. Overhead irrigation is simple to install and easy to regulate.

■ **Perimeter Irrigation**—This system is used primarily on fresh flowers growing in beds. A black plastic or PVC pipe is fitted around the perimeter of the bed with emitters that spray water over the growing medium. The emitters, or nozzles, are located below the foliage. Perimeter irrigation systems are used mostly in the production of carnation and rose crops.

■ **Soaker Hose System**—Various types of soaker hoses are available. These hoses are used in fresh flower beds and drip or ooze water from the length of the hose. Water is applied slowly to the growing medium allowing thorough saturation of the medium. The hoses can be easily installed and can be removed between crops.

WATER QUALITY

Water quality describes the quantities and types of materials dissolved in water. The quality of the water source should be checked. Water quality varies from region to region. Many greenhouse crops are sensitive to certain salts or minerals present in the water. Soluble salt content and pH are two factors that affect water quality. Soluble salt buildup in growing medium prevents plant roots from absorbing water properly, especially young plants. Many greenhouses use groundwater as the water source. Most groundwater contains mineral impurities. Calcium, sulfur, and iron are examples of mineral impurities found in groundwater.

Chlorinated water from city water systems generally does not harm plants, but fluoridation can cause injury to some greenhouse crops. *Fluoridation* is the addition of low concentrations of fluoride into a municipal water system for the prevention of tooth decay. Some plants, like the Easter lily, are highly sensitive to these low concentrations of fluoride. Many green plants, like the spider plant, corn plant, and prayer plant, are also affected by fluoride in the water. Common symptoms of fluoride toxicity of these plants

11-17. Watering plants until water drains from the bottom of the container helps reduce soluble salt buildup.

11-18. Tip burn on a spider plant can be caused by fluoride in the water.

are leaf tip burn and leaf margin necrosis. This can be corrected by avoiding fluoridated water or slightly raising the pH of the growing medium. Growers frequently check the soluble salt content and pH of irrigation water to determine water quality.

FERTILIZATION

The pH of the growing medium is adjusted, usually with dolomitic limestone, before the greenhouse crop is planted. The pH of the growing medium effects the nutrient availability to the plant. Soilless growing medium has very little, if any, available plant nutrients. Some fertilizer amendments are added during mixing of the medium. These fertilizers supply nutrients for a short period of time. Thereafter, additional fertilizers are needed for adequate growth of the crop to continue.

There are several types of fertilizers from which to choose. The type fertilizer used will determine the method in which they are applied. The two general categories of fertilizer are organic and inorganic fertilizers.

ORGANIC FERTILIZERS

Organic fertilizers are naturally occurring nutrient materials that are derived from plants or animals. They release nutrients slowly as they decompose in the medium. Large commercial greenhouse growers do not normally use organic fertilizers. One reason is that the organic fertilizer analysis is low; therefore, large amounts are needed to obtain the desired results. Second, the nutrients are released as the compound decomposes. Sometimes this decomposition rate is variable and dependant on environmental conditions. Last, some organic fertilizers have undesirable effects, such as weed seeds or offensive odors.

Some specialty plant growers or hobbyists use organic fertilizers with favor-

11-19. Some common organic fertilizers.

able results. Some of the fertilizers that they incorporate into the medium or topdress are—bloodmeal; bonemeal; cotton seed meal; fish emulsion; manures, which include cow, chicken, sheep or bat; mushroom compost; and sewage sludge.

I I-20. A complete controlled-release fertilizer.

INORGANIC FERTILIZERS

Inorganic fertilizers are synthetic nutrient compounds that are derived from mineral salts. Single-element fertilizer formulations are available, such as ammonium nitrate (NH_4NO_3), superphosphate (P_2O_5), or potassium nitrate (KNO_3). Complete fertilizers are also available, such as 10-10-10, which contain nitrogen, phosphorus, and potassium.

Three formulations of inorganic fertilizers are commonly used in greenhouse operations. Dry or granular formulations are used mostly as starter fertilizers because of their low analysis. Soluble liquid fertilizer formulations normally have a high analysis and are applied through the water system. Controlled release fertilizers are synthetically designed to release nutrients slowly over a specific time. Osmocote® and Sierra® are trade names of controlled release fertilizers. A variety of analyses of these types are available.

Methods of Fertilizer Application

Two methods of adding dry or controlled release granular fertilizers to the medium are—incorporation or topdress. Incorporation is normally done when the fertilizer does not leach readily through the medium or an initial amount is needed for starting the crop. Starter fertilizer, like 5-10-10; slowly soluble fertilizer, like limestone; micronutrient, like Perk®; and controlled release fertilizers, like Osmocote® may be incorporated into the growing me-

11-21. A fertilizer proportioner introduces concentrated fertilizer solution into the irrigation line.

dium. Growing medium that is incorporated with fertilizers and then moistened should be used quickly so the nutrients do not leach to the bottom.

Topdressing, or applying fertilizers to the top of the medium in the container, is done on long term crops. This is usually restricted to controlled release formulations and used as a supplement to other methods of applications.

Soluble liquid fertilizers are best dissolved in warm water in a storage tank that is connected to the irrigation line. As the crop is irrigated, a specific concentration of fertilizer is introduced into the irrigation system. This practice is known as fertigation. Devices that are used to introduce and meter the concentration of soluble liquid fertilizer into an irrigation system are called a ***fertilizer proportioner or injector***. The proportioner or injector continuously mixes a small quantity of the concentrated fertilizer solution into the irrigation system. Proportioners or injectors are available with dilution ratios that range from 1:12 to 1:200. The soluble liquid fertilizer usually contains an indicator dye of blue or green so that the grower can determine if the fertilizer is being applied. Most growers apply soluble liquid fertilizer every time the crop is irrigated, while some fertilize only on a weekly basis.

*F*ertilizer Rates

Dry formulations of fertilizer that are incorporated into the growing medium are measured in pounds per cubic yard of growing medium. For smaller volumes of medium, a grower may use ounces or grams per cubic foot. If the fertilizer is topdressed, then it is measured in ounces or tablespoons per diameter of pot size. For the most accurate measurement of dry formulation, it is always best to measure by weight instead of volume of fertilizer.

Soluble liquid fertilizers are measured and applied in parts per million (ppm). This is the measure of the concentration of the fertilizer stock solution. One part per million is equal to one milligram per liter. However, growers convert this to ounces per gallon or pounds per 100 gallons. The rate of many greenhouse crops that are continually fertilized at every watering will range from 100 to 250 ppm of nitrogen.

Monitoring Crop Nutrient Levels

Three ways in which growers monitor the nutrient levels of greenhouse crop are by visual observations, media testing, and foliar analysis. Visual observations of the plants' nutrient levels should not be the only method utilized. In many cases, when only visual observations of nutrient levels are used, the damage that occurs may not be reversible and the crop becomes a total loss. Many large growers use medium or soil testing, which tells what nutrients in the medium are available to the plant. Some also include measurements of the pH level and the level of soluble salts in the medium. Media that is to be tested is usually sent to state or private laboratories. However, many growers monitor the soluble salt levels and the pH levels of the media by using soluble salt meters and pH meters. Foliar analysis involves leaf tissue testing.

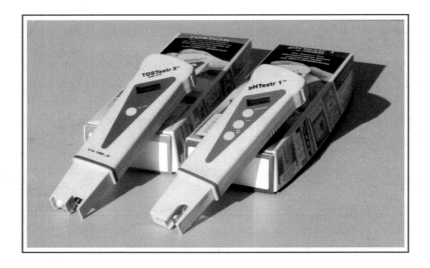

11-22. Soluble salt meter and pH meter are handy instruments to monitor the crop's nutrient level.

CONTROLLING PLANT GROWTH

Disbudding is a practice used for improving the size and quality of flowers. **Disbudding** is the removal of all lateral buds and only the terminal bud is allowed to develop. It is often used on roses and carnations. The removal of the lateral flower buds allows for the terminal flower to become larger. Center bud removal (CBR) will improve the spray formation of chrysanthemums. CBR will encourage flower stems to branch and produce a better mum for use in flower shops.

PINCHING

Pinching is the simplest form of pruning and is used to encourage lateral branching or to remove unwanted growth. Pinching is done with the thumb and forefinger. No other special tools are needed. The more plant material that is removed in pinching, the longer it takes for the plant to recover. There are three basic types of pinches:

- **Roll-out pinch**—A roll out pinch is the removal of just the meristematic tip of the stem. The roll out pinch is recommended in winter when growth is slowed.

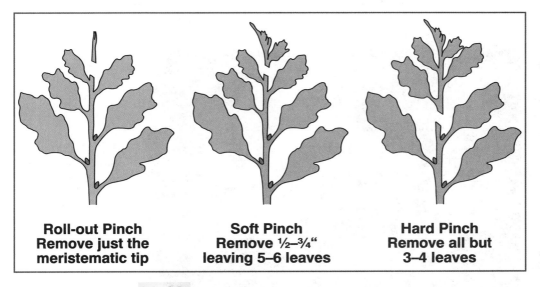

Roll-out Pinch
Remove just the
meristematic tip

Soft Pinch
Remove ½–¾"
leaving 5–6 leaves

Hard Pinch
Remove all but
3–4 leaves

11-23. There are three basic types of pinches.

- **Soft pinch**—A *soft pinch* is the removal of the terminal bud and ½ inch to ¾ inch of the stem. Five to six leaves are left on the cutting. Branches or shoots should be given soft pinches whenever possible, because the smallest amount of growth is removed. Poinsettias are given a soft pinch to produce branched plants.

- **Hard pinch**—A *hard pinch* is the removal of the terminal bud and everything but three to four leaves. Hard pinches may be used in summer with tall cuttings. However, hard pinches are not very often used.

GREENHOUSE GROWTH REGULATORS

A variety of chemicals, both natural and synthetic, are available to modify and control growth on greenhouse crops. These chemicals are usually called growth regulators and they produce a number of different growth responses. Some of the chemical growth regulators used on greenhouse crops are:

- **Auxins**—These chemicals are used to promote the rooting of cuttings. Plants naturally produce auxins, but growers usually apply a synthetic auxin, such as indole-3-butyric acid (IBA), to hasten root initiation and development. The auxins are used in very low concentrations and are often diluted with talc or alcohol.

- **Gibberellic acid (GA)**—Sprays of gibberellic acid (GA) are applied to florist azalea to overcome the cold requirement needed for flowering. Also,

11-24. Cuttings are dipped into an auxin solution to promote rooting.

gibberellic acid is used to increase the flower size of geraniums and accelerate flowering of cyclamen.

■ **Ethephon**—Ethephon is used to initiate flowers on certain plants, like bromeliads. Once applied to the crop, the chemical undergoes changes and releases ethylene gas, which induces the flower.

■ **Height retardants**—Height retardants are available that shorten the stem internodes resulting in a nice compact plant. These are applied as a drench or spray. Some of the chemicals are Cycocel®, B-Nine SP®, and A-Rest®. All of these are labeled for different crops. As with any chemical, always read and follow the label directions.

■ **Chemical pinching agents**—*Chemical pinching agents* promote branching by either causing death to the terminal bud or temporarily stopping shoot elongation. Off-Shoot-O® and Atrimmec® are two chemicals that are used as chemical pinching agents.

GREENHOUSE INTEGRATED PEST MANAGEMENT

Growers use a variety of practices to control unwanted pests in the greenhouse. These integrated pest management practices include prevention, careful observation, and corrective measures. The greenhouse environment

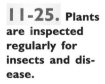

11-25. Plants are inspected regularly for insects and disease.

11-26. A sticky trap helps monitor the population of insects.

is favorable for the development of weeds, insects, and diseases that may be detrimental to the crop. To help prevent pests in the greenhouse, a number of practices are often used. These are:

■ **Pest entry prevention**—All plants that enter the greenhouse should be carefully inspected for insects and diseases. Any signs of pests on incoming plants should be isolated and treated.

■ **Weed control**—Weeds can compete with crops for moisture and nutrients, but most importantly, they can harbor insects and diseases. Weeds should be eradicated from inside, as well as outside, the greenhouse.

■ **Sanitation Practices**—Many preventive measures can be taken before the crop is planted to reduce the chances of pests. Growing medium should be pasteurized and containers, tools, and benches should be disinfected. Dead leaves, flowers, and stems should be removed from the greenhouse.

■ **Crop inspection**—Crops need to be inspected regularly. Plants especially need to be checked closely in the buds and under the leaves.

■ **Environmental manipulation**— If favorable environmental conditions for pest development are known, then sometimes the environment may be slightly adjusted to prevent their development. For example, black root rot disease can be prevented by slightly lowering the growing medium pH to between 4.5 and 5.0.

■ **Pest eradication**—The pest needs to be positively identified before eradication can begin. Biological control measures and pesticides are forms of pest eradication.

Good preventive practices and observations can reduce the amount of pesticides that have to be applied to the crops. Pesticide use can also be reduced by becoming more knowledgeable of the types of greenhouse pests, the crops that they prefer, their life cycles, and the type environment that encourages their development.

SCHEDULING PLANT PRODUCTION

Greenhouse growers produce a variety of flowering and foliage plants. Flowers help express and enhance the mood and spirit of various holidays and seasons of the year. Bedding plants enhance the spring and the fall. Easter lilies and poinsettias are examples of important holiday crops.

Many hours are spent planning greenhouse crop production in order to have the proper crop ready for sale during a certain season or specific holiday. When determining the production schedules, growers need to know how many weeks the crops will need to grow to marketable size and, if flowering, be in bloom. Growers also schedule to produce continuous crops in the greenhouse or more than one crop at a time.

CULTURAL REQUIREMENTS FOR SELECTED CROPS

Growers need to know the cultural requirements of the crops they produce. These include temperature and lighting requirements and plant spacing necessary for proper development. They also must know when to water and fertilize.

Scheduling varies among growers in different regions of the country. This is due to the difference in natural environmental conditions—temperature and sunlight intensity and duration. Important greenhouse crops include chrysanthemums, poinsettias, Easter lilies, bedding plants, and foliage plants.

Chrysanthemum (Dendranthema grandiflora)

Chrysanthemum varieties are very diverse. Chrysanthemums come in many colors and the flower forms differ. There are also physiological differences that influence production schedules. Mums are classified based on these differences:

- Group—One grouping is based on the number of weeks it takes them to flower from the time they begin receiving short days. For instance, a nine-

11-27. Chrysanthemums have different flower forms.

week variety takes nine weeks to flower once it is given short-day treatments. The response groups of mums range from 6 weeks to 15 weeks. Garden mums include six-, seven-, and eight-week varieties. The majority of potted chrysanthemums have a 9- or 10-week response.

■ Height—A second grouping involves the height of the plants. Chrysanthemum varieties are identified as being tall, medium, or short.

■ Flower Form—Mums are also grouped by their flower forms, which include standard, spider, spray, anemone, decorative, pompon, and spoon.

Today, a few specialized companies provide over 95 percent of all the chrysanthemum cuttings. They are grown as single stem or pinched plants. They prefer a growing medium with a pH of 6.0 to 6.5. Azalea pots are the pots of choice for most growers. Typically, one cutting is used in a 4½-inch pot. Plant cuttings shallow and at a 45-degree angle with the tops extending over the lip of the pot. After planting, the grower waters the plants with a balanced and complete fertilizer solution.

There are four main periods of growth and development in producing a potted chrysanthemum crop. The first phase or vegetative phase involves the promotion of root and leaf growth. The vegetative phase lasts three to five weeks depending on the variety and the time of the year. Water when the growing medium approaches drying. Provide 65 to 70°F night temperatures, 75 to 80°F day temperatures, and high humidity. Long days must be provided to keep chrysanthemums in a vegetative stage of growth. Interrupt the night darkness from December through February by lighting the plants from 10:00 p.m. to 2:00 a.m. From September through November and March through May, light for three hours. In the summer months, light for one to two hours to be sure plants stay vegetative.

Grow for 7 to 14 days or until roots appear at the bottom of the pot before pinching the plants. Pinching is the removal of the growing point of the stem. Pinching encourages branching of the plants. The result of pinching includes better-shaped plants that appear fuller and produce more flowers.

11-28. A quality chrysanthemum in bloom.

After pinching, the grower provides the mums with short-day conditions. Turn off night lighting to provide short-day treatment for the initiation of flower buds. Cover the crop with black cloth if necessary to extend the period of light duration. The desirable temperature range for flower bud initiation is 62 to 65°F nights and a day temperature not exceeding 80°F. Also, lower the humidity levels in the greenhouse. Reduce fertilizer rates. In the summer months, temperatures can build up under black cloth resulting in heat delay. The result is a delay of flowering by several days to weeks. Use of a chemical growth retardant effectively controls the height of tall varieties two weeks af-

Table 11-1. Simplified Growing Schedule for Chrysanthemums				
Tall varieties	Vegetative 1 week	Short days 1 week	Pinch	
Medium varieties	Vegetative 1–2 weeks	Pinch	Short days	
Short varieties	Vegetative 2 weeks	Pinch	Vegetative 1 week	Short days

ter pinching. As the plants grow and develop, the grower increases the space between the containers.

The grower keeps the plants under short-day conditions until flower buds develop and begin to show color. As the flower buds develop, they may be removed to improve the overall quality of the plants. This practice is known as disbudding. One method of disbudding is center bud removal, pinching off the most terminal flower bud. This allows the lateral flower buds to develop. Center bud removal is performed with decorative-type mums. Another disbudding practice involves the removal of all the lateral buds. The plant responds by sending all its energy to the remaining bud. The result is larger showier flowers.

Chrysanthemums are ready for market when 50 percent of the buds are in bloom. To finish production, drop temperatures to 55 to 58°F at night to increase color intensity and quality. Growth of the plant at this point has stopped. Stop fertilization and allow the growing medium to dry a bit more than before. A quality final product will be symmetrical and have all of its flowers opening at the same time.

*P*oinsettia (Euphorbia pulcherrima)

Poinsettias are usually produced for marketing during the winter holiday season. Therefore, growers plan to have their poinsettias in bloom by early to mid-December. Growers can buy stock plants and take their own stem cuttings or purchase 2¼ inch rooted cuttings.

11-29. Poinsettias are pinched in mid-September. (Courtesy, Ron Biondo, Illinois)

Poinsettias require a well-drained and well-aerated growing medium with a pH between 5.0 and 6.5. The recommended pot for most production is the azalea pot. It is important for all of the pots to have the same amount of medium. This helps with watering practices and in uniform development of the entire crop. Plant the cuttings in the center of the pot and at the same depth. Shallow placement promotes healthy root growth. After planting, drench the pots with fungicide to control root and stem rot.

As short-day plants, poinsettias initiate flower buds as days get shorter. Flower bud initiation begins when nights reach about 11 hours and 50 minutes of darkness. Lowering night temperatures in the greenhouse to 62 to 64° F during flower bud initiation produces a favorable response. Poinsettia production can be broken down into four stages of development. A growing schedule for 6 to 6½-inch multi-flowered poinsettia crop in soilless medium follows:

- Late August (Vegetative Stage)—Pot poinsettias in a finishing pot. Drench with a fungicide solution and place the containers side by side on the greenhouse bench. Promote vegetative growth by lighting plants from 10 p.m. to 2 a.m. Fertilize and provide 68 to 70° F night temperatures and 70 to 80° F day temperatures.

- September 10 (Vegetative Stage)—Pinch the plant leaving four to six leaves. Maintain high humidity to encourage breaks. Note, that the longer the period between pinching and flower bud initiation, the larger the plant. Raise fertilizer rate and maintain night temperatures of 68 to 70°F.

11-30. Poinsettias are associated with the holiday season.

- September 20 to 25 (Flower Bud Initiation Stage)—Turn lights off and give plants short days. If nearby lights interfere with the natural darkness, cover the crop with black cloth from 5:00 p.m. until 8:00 a.m. Drop night temperatures to 62 to 64°F and day temperatures to 70 to 72°F. Initiation of flower buds occurs over a period of 8 to 10 days. Space plants 15 inches by 15 inches to avoid stretching.

- October 10 (Flower Bud Development Stage)—Flower buds begin to develop. Stop black cloth treatment. Adjust temperatures to 64 to 66°F at night and 70 to 75°F during the day. Apply a chemical growth regulator to reduce internode length of stems and to keep the plants compact. Drench with a fungicide to control root rot.

- November 15 (Flowering Stage)—Begin to finish plants. Drop temperature to 58 to 62°F nights to deepen bract color. Reduce fertilizer rate. Growers discontinue fertilization about a week before the sale date in order to prevent bract burn.

Easter Lily (Lilium longeflorum)

The Easter lily crop is difficult to grow because Easter varies year to year. Easter is the first Sunday following a full moon after March 21. The earliest Easter can be is March 21. The latest date for Easter is around April 20. Therefore, growers must schedule their crops differently each year.

Easter lilies are potted in 6-inch lily pots as soon as the bulbs arrive in late October. The growing medium should have a high bulk density and a pH between 6.5 and 7.0. Place the bulbs near the bottom of the pot. The bulb is then covered with several inches of growing medium. This potting practice promotes the formation of stem roots. The stem roots help to stabilize the plant and serve as a backup in case the bottom roots die. *Forcing* is a term used to describe the practices that get bulbs to grow and produce flowers. The simple schedule for forcing naturally cooled bulbs is to pot the bulbs, keep them at 50 to 60°F for two to three weeks, cool, and force. Precooled or case cooled bulbs are potted, given one to two weeks at 50 to 60°F, and then forced. With six weeks of cooling, expect 110 to 115 days to force the plants.

Temperature affects the speed of forcing. It can be adjusted to speed or slow growth. If the lilies are too advanced, they can be given cooler temperatures to slow their growth. If the lilies are behind schedule, temperatures are raised. Generally, if Easter is early, bulbs are forced at 63 to 65°F nights. If Easter is late, the bulbs are forced at 60°F nights. When Easter is somewhere in the middle, they are forced at 62°F nights. Flower buds should be visible by about 40 days before sale. Long-day treatments can be used as a substitute for insufficient cooling or to induce earlier flower formation. Lighting plants from 10:00 p.m. to 2:00 a.m., beginning when shoots emerge and continuing for 10 to 14 days, speeds development. Each four-hour per day lighting substitutes one full day of cooling.

Lilies should be watered sparingly at first. After the shoots have emerged, allow the medium to dry between waterings. Overhead watering is most commonly used with Easter lilies. Root rot can be a problem with wet medium, so capillary mats and ebb and flow systems are avoided. Fertilization keeps plants healthy and growing. Slow-release fertilizers are often incorporated in soilless mixes before potting. Early fertilization produces vigorous stem growth and leaf development.

11-31. Easter lilies are grown as a potted flowering crop for the Easter season.

The height of Easter lilies is influenced by a number of factors. Light, temperature, humidity and watering practices affect plant height. Lower light levels result in taller plants. The more water or fertilizer the plants receive, the taller the plant. Therefore, high temperatures that increase the frequency of water and fertilizer applications, result in taller lilies. High humidity also produces taller plants.

DIF and growth retardants are employed to control the height of lilies. *DIF* is the mathematical difference between daytime and nighttime temperatures. A positive DIF is a lower nighttime temperature. When the nighttime temperature is higher than daytime, a negative DIF occurs. DIF is a very easy way to maintain the height of lilies. A zero or slightly negative DIF is the most affective. Large negative DIF causes the leaves to droop and curve down. Growth regulators can be used as a drench or as a spray when shoots are 6 to 8 inches tall. It is applied twice with a seven day interval between applications.

The practice of leaf counting is the best way to monitor the progress of the crop and to determine whether growth needs to be speeded or slowed. It also allows the timing of the crop to begin on January 15, leaving plenty of time for adjustments. It involves calculating the number of leaves unfolding per day by dividing the number of new leaves unfolded by the days. (Example: Five newly unfolded leaves divided by four days = 1.25 leaves unfolding per day.)

Easter lilies are ready for market when the largest floral bud has reached the puffy white stage. If necessary, plants that open early can be held in a cooler with temperatures between 35 and 40°F. They can tolerate cold storage for about two weeks. The grower removes the plants from the cooler a few days before they are to be sold. Warm them slowly when they are removed from cold storage. This allows them to adjust to the outside temperature. If flower buds have opened, the anthers should be removed. Removal of the anthers prolongs the life of the flower.

11-32. Easter lilies with puffy white buds are ready for market. (Courtesy, Ron Biondo, Illinois)

REVIEWING

*M*AIN IDEAS

This chapter focused on the general greenhouse practices used in the production of ornamental crops and the variety of choices the greenhouse grower has to produce the crops on schedule.

Many times, growth of greenhouse crops needs to be regulated. Manipulation of light on some crops, like chrysanthemums and poinsettias, is used to initiate flowering. In areas with harsh winters, CO_2 fertilization, and supplemental lighting are used to increase growth. Pinching by hand is still used today to increase branching of the crop. Chemical growth regulators are also used to promote a variety of growth responses, such as plant height reduction, root initiation, flowering, and branching.

Growers have to select containers suitable for the production and marketing of the crop. The growing medium that is used in the containers will ultimately influence the quality of the crop. Several types of growing media are available. Growers can mix their own or they can purchase commercially prepared mixes.

The quality of the water should be checked before selecting the type of crop to be grown. Several types of irrigation systems are available for greenhouse use. Manual hand watering is not efficient, but each greenhouse needs access to hose watering for watering plants in areas that may dry out before others. Most growers fertilize plants through the irrigation system with fertilizer proportioners or injectors. Other types of fertilization programs are available.

Growers are practicing a variety of methods of pest management instead of relying solely on the use of pesticides. These practices not only reduce costs for the grower, but help protect the environment.

QUESTIONS

Answer the following questions. Use complete sentences and correct spelling.

1. What are two organic containers?

2. What are three popular plastic containers? standard, azelia, bulb pans

3. What are four important considerations in selecting a good growing medium? moisture, air, nutrient holding capacity, stability

4. Describe the components used in a peat-lite mix. What ratios of these components are used? peat moss, perlite, vermiculite

5. What are some crops that are harmed by fluoridation of water systems? What are some typical symptoms of fluoride toxicity of these crops?

6. How are fertilizers applied to greenhouse crops?

7. List and describe three ways in which growers monitor the nutrient levels of greenhouse crops. visual, media testing, tissue analysis

8. How do light intensity and duration affect greenhouse crops?

9. What is the practice that growers use to increase the day length on crops? How do they decrease it? *artificial lighting*

10. What is the purpose of pinching greenhouse crops? Name three types of pinches. *encourages lateral branching*

11. What are the five types of growth regulators used on greenhouse crops? Explain the use of each. *auxin, roots, GA, ethophon*

12. What are six methods of greenhouse integrated pest management? Describe each.

EXPLORING

1. Visit a local greenhouse range and observe the materials used and their practices. Prepare a report of your findings.

2. Using a video camera, prepare a film to present to the class of the various practices used in greenhouse operations.

3. Grow a greenhouse crop in a small greenhouse. Keep detailed records of the types of plants grown, medium used, type of container, fertilization program, irrigation practices, growth regulators used, and pest management practices.

4. Obtain several plants of one kind. Use different fertilizers, various application methods, and rates. Record your observations and growth results.

5. Obtain different types of greenhouse crop plants. Practice the various techniques of pinching. Compile your results in a report.

6. Arrange for a professional to speak to your class on integrated pest management practices. Summarize his/her presentation.

12

NURSERY FACILITIES

Have you ever visited a nursery to by a tree or a shrub? If so, you probably noticed several different looking structures and pieces of equipment. Did you wonder what that plastic covered, arched building was or about that big truck with a tree in some type of jaws on the back?

A nursery can be an exciting adventure. Learning about what it takes to establish and operate a nursery is important. Most important is the selection of a proper location. Careful consideration must be given to all aspects of nursery design and development.

One goal of all nursery owners is to maximize profits. It is a business. A comprehensive plan will allow a grower to utilize labor efficiently and maximize production. Good planning will also result in good environmental stewardship.

12-1. Nurseries can be very profitable and aesthetically pleasing places to work.

OBJECTIVES

This chapter gives information on the design and structures needed for a production nursery. It has the following objectives:

1   Describe the importance of proper nursery site selection

2   Discuss the environmental factors considered in nursery site selection

3   Identify economic factors involved in selecting a nursery site

4   Explain the layout of container and field nurseries

5   Describe the different types of nursery structures

TERMS

air pollution
coldframe
container nursery
drainage
field nursery
hotbed
lathhouse

overwintering structure
plant hardiness map
plant heat-zone map
shadehouse
temperature
topography

NURSERY SITE SELECTION

Having land and special nursery structures are necessary for growing nursery plants. Structures protect plants from harsh weather and other environmental conditions. Proper equipment is also essential to nursery plant production. However, the selection of a proper nursery site is the most important part of establishing a new production facility.

The selection of a proper site is critical for the overall success of a nursery. A good site will be more economical over the years. It can be expensive to attempt to correct the problems of a poor site. Environmental and economic factors must be considered.

Environmental factors to consider when selecting a site include climate, soil and topography, and available sources of water. Economic factors include available capital and labor, market potential of target sales area, type of plant material to be grown, and production method to be used. A field nursery grows plant in fields and a container nursery grows plants in containers. A container nursery will require less land to produce the same amount of plants as a field nursery. It is important to visit each site being considered. Each factor should be evaluated as to the effect it will have on the overall operation and success of the nursery.

ENVIRONMENTAL FACTORS

Knowing the crop to be grown will help identify if the site chosen for a nursery is suitable for production. Factors to consider include winter and

12-2. Climate, soil and topography, and available sources of water must be considered when selecting a site for a nursery.

summer temperatures, wind, soil type, topography of the land, natural drainage, rainfall, water availability and quantity, air pollution and plant pests. A check-off list with various environmental factors under each of the above categories will help narrow the number of potential sites suitable for nursery production in your area.

Temperature

Temperature refers to how hot or cold something is and is used as a measure of the heat energy of molecules as they move about. The type of nursery crop being grown will dictate the importance of temperature in site selection. Young plants are sensitive to rapid changes in temperature. Due to milder winter conditions and a longer growing season, many producers of container-grown plants are located in the southeastern United States or in California. Overwintering of container-grown plants is not a major consideration in southern California due to radiational cooling, but Arctic air sweeping south with cold fronts in the eastern United States makes overwintering practices very important.

Many field growers are located near large bodies of water, such as lakes and oceans, which have a moderating effect on temperatures. All nursery growers should be familiar with the USDA *Plant Hardiness Map* of the United States. It identifies 11 zones by average annual minimum temperatures for each zone. Another useful map is the American Horticultural Society *Plant Heat-Zone Map* which identifies 12 different zones in the United States based on the average number of days above 86°F. This map is important since many plants that can tolerate cold temperatures do poorly in areas with extended summers and high temperatures.

Wind

Severe wind erosion can occur, especially on sandy soils. High winds carrying soil particles can damage the foliage and stems of plants, especially seedlings. It may also interfere with irrigation application and uniformity. In container nurseries, plants blowing over can be a big problem.

Windbreaks can be used to protect exposed sites from excessive wind damage. The benefits of windbreaks can include better plant growth, decreased soil erosion, reduced transpiration from plants, reduced evaporation from soils, and more efficient spraying. The purpose of a windbreak is to cause a filtering effect. The ideal windbreak is not solid, but is 40 to 50 per-

12-3. **Plants which have too much top growth in relation to container size often blow over in windy conditions.**

cent wind permeable. A windbreak can have a sheltering effect up to 20 times its height. However, maximal protection occurs at a distance of 3 to 6 times the height of the windbreak. Artificial windbreaks are generally limited due to the effective height to which they can be constructed. Natural windbreaks are considered to be the best long-term investment. Problems associated with using natural windbreaks can be unwanted shading, competition for nutrients and water, and windbreaks may serve as hosts for certain pests.

Soil and Topography

Although soils in general contain the same components, they differ in their properties and characteristics. Soils differ in texture, structure, color, consistency, and native fertility. Field nurseries grow their trees and shrubs in the nursery's fields. To grow these plants successfully, the grower must know what components make up the soil and in what quantities they exist. Chapter 6 discussed soils and fertility in depth.

The type of production method used will depend on the type of soil present at the site. If balled and burlapped (B&B) stock is to be produced, the soil should hold together when plants are dug. Seedlings or bare root (BR) plants are best grown in sandy or loam soils. These soils do not tend to crust over after rain or irrigation. They can be worked soon after a rain and plants can be removed with less damage to their root system.

Soils which are chosen for the production of field grown nursery stock should have good physical and chemical qualities. The soil must be properly

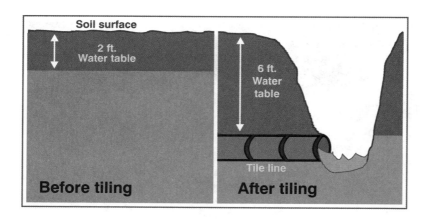

Soil surface

2 ft.
Water table

6 ft.
Water
table

Tile line

Before tiling   **After tiling**

12-4. Tile drainage is used to lower the water table.

managed to meet crop requirements. Soils vary greatly in organic matter content, pH, native fertility, and other factors. The organic matter content of soils can range from nearly 100 percent in the peat soils of southern Florida to almost zero in sands. The suggested range of organic matter content for production soils is 2 to 5 percent.

Natural drainage is an important consideration when deciding on a nursery site. A number of advantages can be gained from proper water management and land drainage. *Drainage* describes the removal of surface or ground water or the manner in which it is removed. Benefits of such a program include warmer soil temperatures in the spring, uniformity of soil moisture, decreased levels of soil pathogens, decreased loss of soil nitrogen from denitrification, reduced surface erosion, and increased operation efficiency.

Topography also plays an important part in the efficient operation of a nursery. *Topography* is the surface features of an area, including human-made changes. Land should be relatively level with a slope of 1 to 2 percent to provide for adequate drainage. While rolling land can be terraced, steep slopes should be avoided. Variations in topography can cause water to accumulate in low lying areas which can be subject to increased chances of freezing temperatures. Steep or irregular land makes the efficient use of nursery machinery difficult and complicates the installation of roads and irrigation systems.

Rainfall

Areas with high rainfall during critical times for nursery operations should be avoided. During the winter months, occasional heavy rainfall can have disastrous effects on field operations. Common problems encountered when the soil becomes saturated are an inability to get equipment into the field, delays in lifting of trees and seedlings, damage to soil structure, flooding, and possibly erosion. Summer thundershowers often create problems with seedbed washout, flooding, and erosion.

Heavy rains can kill or stunt seedlings by washing them out of beds, covering them with soil, or by causing physical damage. Other problems associated with excessive rainfall include the increased occurrence of certain pathogens, decreased efficiency of pesticides, stimulation of weed growth, and excessive leaching of soil nutrients. Hail is another problem associated with thundershowers. It will generally do the most damage to young seedlings. However, large plants, machinery, and buildings may also be damaged.

Water

Selection of a nursery site is often dependent upon the availability of quality irrigation water. In most areas, water rights must be obtained and water usage is regulated by state and local water boards. The importance of water usage has become extremely important in recent years as seen by the regulation of irrigation hours. Water needs for current nursery practices and any future expansion must be considered.

The following are some questions which should be answered before selecting on a site:

1. Are there restrictions on flow and periods of application?

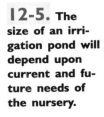

12-5. The size of an irrigation pond will depend upon current and future needs of the nursery.

2. Is the water source reliable, especially during drought?

3. Are back-up water sources available?

4. Is the delivery system reliable?

5. What are expected repair time periods and costs?

6. Is domestic water available? If so, what is the cost and are there agricultural rates?

7. What is the water quality?

Normally, adequate rainfall does not always occur throughout the year on a timely basis. A nursery needs to have access to a dependable supply of water. Possible sources of irrigation water include municipal sources, wells, lakes, ponds, streams, and rivers. The cost of municipal water is often quite high and may be regulated.

Wells are the most common and best sources of irrigation water for most nurseries. Well water is often of high quality. Draw-down and pumping capacity must be checked to determine if this system will meet the irrigation needs of the nursery. Lakes and ponds are also common sources for irrigation water. They will need to be checked regularly for contaminants. Streams and rivers are not commonly used sources of irrigation water.

In many areas of the U.S., high quality water is not available. Ideal irrigation water should have a slightly acidic pH and the soluble salts level should be under 500 ppm. Water which comes from open sources, such as lakes and ponds, may contain particulate matter and microorganisms which will need to be filtered out before use.

Air Pollution

Air quality can have profound effects on the quality of plants being grown. *Air pollution* is when harmful or degrading materials get into the air. In many areas, major pollutants are produced from automobile emissions and industry activity. Some of the compounds known to cause problems include chlorides, fluorides, ozone, and sulfur dioxide. The direction of the prevailing wind should be taken into consideration when selecting a nursery site.

Plant Pests

A variety of pests exist which can have detrimental effects on the health and vigor of field-grown plants. A thorough survey should be conducted to

12-6. Pests can be a problem for a number of ornamental plants.

determine which pests are present. Numerous insects may cause damage. Insect pests can be placed into the following categories—twig, stem, and root boring; defoliators of conifers; defoliators of hardwoods; root-feeding; and sucking insects. Mites are not insects, but are arachnids. Mites can be a severe problem during hot, dry weather throughout the spring and summer months or year round in the southern United States. Deer often cause problems in nurseries by rubbing tree trunks with their antlers or eating tender new foliage.

Depending on previous land usage and cropping, root rot disease, such as *Phytophthora*, may be present. A variety of other leaf, stem, and twig diseases may be present. Their impact will depend on the crop to be grown, as well as other cultural and environmental factors. Chapter 8 has additional information on pests and pest management.

*E*CONOMIC FACTORS

The cost of land and its proximity to urban areas can determine if a nursery can be established on a given site. Garden centers in urban areas can generally afford to pay more for land compared to field nurseries assuming other factors do not limit growth and sales. Labor has become a major issue in the nursery industry with many having to rely on migrant workers. Be certain to check the availability of reliable transportation for shipping. Having several wholesale nurseries in the same area can be beneficial for sales, marketing and purchasing.

*L*and Cost and Availability

In many instances, the best nursery sites will also be high priced farmland. For many, it is not considered economical to produce field-grown crops on high priced land. It is quite often more economical to buy a good site than it is to try and grow plants on one less suitable.

The initial cost of the land will often determine if it should be considered as a potential nursery site. Undeveloped land may initially be cheaper. However, costs associated with developing the site may be more expensive. The total cost of the developed site should be considered when making purchasing decisions.

*L*abor

The nursery industry can be characterized as being a labor intensive industry with seasonal demands. A dependable source of labor located close to the nursery is essential. Due to the seasonal nature of field nurseries, seasons of intensive labor requirements are followed by seasons where little help is needed. A field nursery will need one permanent employee for every seven to eight acres of production. A container nursery will need one permanent employee for every acre of production. Adherence to migrant labor laws and the Worker Protection Standard are extremely important for nurseries.

*T*ransportation and Markets

Good transportation facilities are an essential part of any nursery. The market which a nursery will ship to will determine its needs. For many nurs-

12-7. Large nurseries may ship several truckloads of plants daily, facilitating the need for large shipping areas.

eries, a location near an interstate highway is beneficial. It will expedite the shipment of plants, especially out-of-state.

Nursery stock can be shipped by truck, rail, bus, or plane. Nurseries may have their own fleet of trucks or may hire a trucking company to provide this service. Easy access to a nursery is a must. The location of a nursery may be very important if the primary market area is a large metropolitan area. The location may not be as important if plants are being shipped long distances.

Utilities and Services

The availability of telephone and electrical power services is essential to the everyday functioning of a business. If a nursery is located in a rural or remote area, initial hook-up costs may be quite costly. The availability of prompt repair and maintenance should also be a consideration.

Competition

Competition from other nurseries may or may not be a problem. If several growers are located in the same area and serve a limited number of customers, location will be more important. A nursery should consider the need for its product and the availability of a profitable market share before entering into business where several nurseries already exist.

If the market area is expanded, the benefits of selecting a site near other nurseries will often overcome any competition. Numerous advantages exist for growers to purchase supplies cooperatively at reduced costs when bulk orders are placed. Discount prices are often available on supplies, such as chemicals and fertilizers.

NURSERY DEVELOPMENT

In general, the production of container-grown plants requires two to three times less land per plant compared to field-grown plants. Only 60 to 70 percent of available land is usable for production. The remainder is used for support areas, such as office space, sales display areas, potting areas, and roads. Planning is very important, and many mistakes can be avoided.

Nursery Layout

There are two different types of nursery production facility layouts. One is designed to produce nursery crops in containers. The other grows nursery crops in fields. They have different needs and requirements when the nursery is designed.

Container Nurseries

A ***container nursery*** grows nursery crops to marketable size in containers. A chart of production practices in container nurseries is useful for developing a nursery layout. First, an overall plan which includes irrigation systems, roads, ponds and service areas should be developed. Areas for propagation, potting, container production, and shade and overwintering structures should be developed next. Finally, development of office and shipping areas can be undertaken. It is important that plants and people flow from one area of the nursery to the next. Therefore, nursery structures and

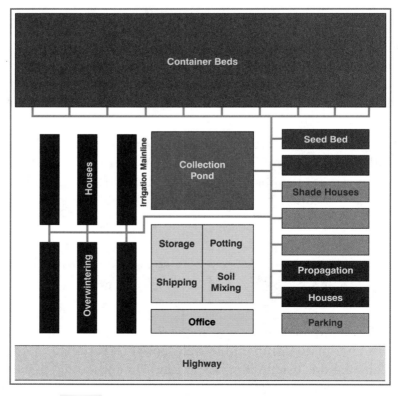

12-8. **General layout of a small container nursery.**

growing areas are generally arranged such that each step from propagation to shipping allows for maximum productivity.

Field Nurseries

A *field nursery* grows nursery crops to marketable size in fields. A chart of production practices for field nurseries is extremely valuable when designing a functional layout. If seedlings or planting stock are being produced, a field operation may require a storage building for storing bare root (BR) plants. If finished plants are produced, an acclimation area consisting of shade and overhead mist systems will be required if summer digging is to occur. In general, fields are laid out in blocks surrounded by dirt roads. Soil characteristics such as drainage and ability of the soil to hold together when plants are dug are important considerations for field nurseries. Field nurseries also require considerably more land to establish a production facility compared with container nurseries.

12-9. Holly plants being produced in a field nursery operation.

LAND PREPARATION

Once the nursery has been planned on paper land preparation is the next step. Be sure to check local and state regulations regarding water runoff from nursery facilities. Holding ponds to catch runoff are a wise investment. Roads, drainage areas, and irrigation systems all need to be properly sized for current production and future expansion.

*L*and Leveling

Once the nursery site has been selected, the natural features should be examined so that a design can be created. Topographical maps, soil survey maps, and aerial photographs will be helpful for this process. For the initial design, field areas, roads, and irrigation systems should be laid out.

The next step will be to clear the site. All major objects, such as trees, stumps, rocks, and other debris, should be removed. The natural contours of the site must be taken into consideration. If the site is reasonably level, with a slope of 2 to 3 percent, the field can probably be leveled using a land plane. If there is considerable variation in the existing topography, top soil may need to be removed, stockpiled, and returned to sites which have since been leveled.

12-10. Laser leveling a field.

*R*oad Design and Construction

A road system should be designed to provide ample space, to allow for the continuous movement of nursery stock, equipment, and personnel. It should not have any turns or obstacles which would make maneuvering difficult.

Different areas of a nursery require different road widths and surfaces. Roads which serve as main entrances, and those which lead to shipping, loading, and storage areas, need to be wide enough and stable enough to handle large trucks and equipment. Depending on the type of soil, it may be necessary to pave such roads with concrete or asphalt. Secondary roads throughout the nursery need to be designed to handle trucks and tractors with trailers. Many of these roads can be left as soil or covered with a suitable material, such as gravel or sea shell material.

Irrigation

The most important aspects of an irrigation system are a reliable year-round source of water; sufficient pumping and pressurizing capacity, well-designed and appropriately sized main and lateral irrigation lines, and an application system which allows for uniform distribution of water. Most container nurseries use overhead sprinkler irrigation systems. Drip irrigation has become popular for irrigation of larger containers and field nursery operations.

For container nurseries, daily water requirements may exceed 30,000 gallons per acre of nursery stock. If applied through an overhead irrigation system, 80 to 90 percent of the applied water may be lost to runoff. Plans should be made to contain all runoff in holding ponds on the nursery site.

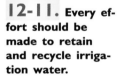

12-11. Every effort should be made to retain and recycle irrigation water.

Drainage

Once fields have been leveled and the roads installed, various areas throughout the nursery should be checked for proper drainage. If fields and roads are properly sloped and designed, then all runoff water should be channeled to one or two locations within the nursery. Collection ponds will often be included in the design of the nursery to collect all runoff water. These collection ponds are normally located at low spots on the property.

Free-standing water in and along roadways is often a problem in many nurseries. This is often the result of continuous compaction of moist soil by heavy machinery and equipment. In some instances, it may be necessary to construct roads which are crowned in the middle in order to facilitate drainage. Water should flow so that it does not pool up alongside the road as this can lead to potential disease problems.

NURSERY FACILITIES AND EQUIPMENT

A variety of structures are needed for producing plants. Not all plants can be propagated outdoors. It is necessary to provide structures where environmental conditions can be regulated. Office space, work areas, and storage are also needed. Structures used in the production of nursery plants include greenhouses, coldframes, hotbeds, shadehouses, and lathhouses. Some considerations when choosing structures include type of structure, cost, type of covering, and heating and cooling requirements.

Equipment is essential to nursery plant production. Nursery equipment ranges in size from hand and power tools to large tractors. If used properly, tools and equipment help make many tasks much easier and safer.

NURSERY STRUCTURES AND AREAS

Greenhouses—A greenhouse can be described as an enclosed structure designed to provide ideal conditions of light, temperature, and humidity for

12-12. A propagation house covered with clear polyethylene and shade cloth.

the growth of plants. In general, propagation greenhouses should be no larger than 25 feet wide by 100 feet long in order to provide optimal conditions for propagation.

Several greenhouse coverings are available. Different types of plastics are now being used instead of glass. Fiberglass has been a popular greenhouse covering because of its low cost and ability to withstand environmental abuse.

Coldframes—A coldframe is the simplest and most economical outside propagation structure. A *coldframe* consists of a wooden or concrete block frame with heat supplied by solar radiation through a glass or other transparent covering. Coldframes should have an east-west orientation with the low side towards the south to maximize absorption of solar radiation. Coldframes are generally suited for germinating seeds, rooting cuttings and protecting tender plants during winter months. Close attention must be given to monitoring the environmental conditions within this structure.

Hotbeds—*Hotbeds* are similar to coldframes except that electric or hot water heating is used. Hotbeds are efficient propagating structures during cold weather because thermostatically-controlled heating can speed the germination of seedlings or rooting of plants.

12-13. Heated beds are sometimes used to start plants early in the season.

Headhouse—A headhouse gets its name from the fact that it is usually located on the northern end, or at the "head" of a greenhouse. Types of facilities often found in a headhouse include work benches, storage bins for potting media, storage areas for chemicals, sinks and hose bibs, and good environmental conditions for workers. Tasks often performed in a headhouse include sowing seed, transplanting, potting, grafting, and cutting preparation.

Shadehouses—**Shadehouses** are permanent structures used to protect plants from environmental factors, such as wind, temperature, hail, heavy rain, and solar radiation. Shading may be provided by using lath strips on a frame structure (**lathhouse**) or with shade cloth. They can also be used for propagation.

The lath strips on a lathhouse are usually made of wood or aluminum. They are placed on the structure with a north-south orientation. One width apart is a general rule when spacing pieces which creates 50 percent shading. Shade cloth is now commonly used in place of lath due to lower initial costs and uniformity of shading. Shade cloth has a useful life of approximately seven years. Percentages of shade, ranging from 30 to 73 percent are commonly used in the nursery industry.

12-14. Shadehouses can be used to produce plants which do not grow well in full sunlight.

Overwintering Houses—Polyhouses are popular since they may be constructed with lower investment costs compared to traditional greenhouses. A polyhut is a smaller structure which many growers use to propagate cuttings and overwinter them in raised beds. While the frame is considered permanent, the covering is usually replaced yearly. Usually, 4 to 6 mil polyethylene sheets are used to cover polyhouses. Polyhouses may also be used as overwintering structures. **Overwintering structures** have a permanent framework and are covered annually with polyethylene to prevent winter damage to nursery crops. The coverings are usually taken off during the summer months to prevent excessive heat build-up inside the structures.

12-15. Over-wintering structures covered with white polyethylene to prevent winter damage.

Cold Storage—For field nurseries which dig bare root (BR) plant material, a cold storage building is essential. Cold storage buildings are usually large, well insulated structures which may rely on cold temperatures or refrigeration units to maintain interior temperatures. Temperatures between 34 to 40°F and high relative humidity are ideal for the storage of bare root (BR) plants. Bare root (BR) plants are generally harvested in the fall and graded according to size in the cold storage facility. After grading, the plants can be stored in large bins which stack on top of each other, shipped, or potted and overwintered. Care should be taken not to let the plants dry out during storage.

Potting Areas—Potting areas are usually centrally located in nurseries to facilitate movement of plants throughout the nursery. Many nurseries have large, covered potting sheds which allow for potting operations during wet

12-16. A soil mixer and mixing area in a large nursery operation.

weather. All potting media should be mixed or stored on a hard surface, such as concrete, that is well drained to help prevent disease. Potting areas should also be kept free of weeds.

Proper mixing of potting media is critical. A non-uniform batch of potting medium will result in poor or uneven plant growth. Many smaller nurseries use a front-end loader on a tractor to mix potting media. Larger nurseries use cement-mixers or commercial soil mixers in their operations.

Shipping—Many nurseries have a centralized holding shelter where plant material is stored until it is ready to be loaded for shipment. If large trucks are being used, a raised loading dock makes loading the trucks easier. Plenty of space should be provided to allow for easy delivery and turning of large vehicles.

Offices—A well landscaped office area can be an important sales tool. Specimen plants can be used to show potential customers how they can be used in the landscape and what they will look like. Office space is required for sales staff, managers, and accounting personnel.

The entry area into the nursery can also be utilized as a well-landscaped sales tool. A large, visible sign which is easy to see from the highway is essential. Parking areas for customers should be kept separate from employee parking.

Pesticide Storage and Mixing Areas—All chemicals should be stored in buildings approved for the storage of pesticides. Nursery managers are responsible for following guidelines established by the Environmental Protection Agency for safe storage and handling of chemicals. All critical information is contained on the pesticide label.

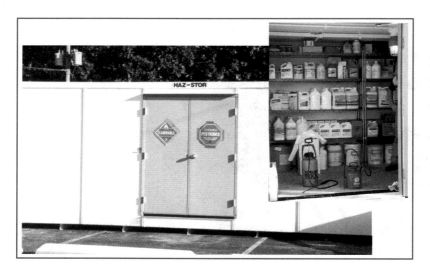

12-17. The latest design in self-contained storage facilities for chemicals.

Pesticide storage buildings should be constructed. Concrete floors which will contain spills should be used. Chemical storage areas should be well ventilated. All chemicals should be stored at temperatures above freezing.

Chemicals should be mixed in areas where spills can be contained. An absorbent material should be available to clean up any pesticide contamination. Clean, fresh water should be available for employees to wash after handling pesticides. All protective clothing should be stored in a separate room away from chemicals.

Storage Buildings—Storage buildings should be available to keep equipment from being exposed to harsh weather conditions. Many larger nurseries also have maintenance shops where equipment can be maintained and repaired. Dry storage areas are required for tools, fertilizers, pots, labels, and other nursery supplies.

REVIEWING

MAIN IDEAS

The proper selection of a nursery site will be of paramount importance to the success of such an enterprise. Site selection should not be determined by one individual, but, rather, a teamwork-approach should be taken. A small group of specialists, including the nursery manager and specialists trained in soils, entomology, pathology, and civil engineering, will be most valuable in making recommendations for the usefulness of a potential site. Such initial planning will prevent future problems from being encountered which were not expected, thereby further ensuring success of the nursery.

One of the most important environmental factors to consider is temperature, both high and low. The USDA Plant Hardiness Map and the AHS Heat-Zone Map should be consulted. Wind direction and speed need to be considered. Soil types, topography and drainage are all important for successful nursery production. If rainfall is not reliable, quality and quantity of available water become extremely important. Proximity to large cities or industrial areas may influence air quality. Plant pests and prior crop history of the land should be considered when developing a nursery site.

Economic factors to consider include land cost and availability, labor resources, availability of transportation and proximity to markets, utilities and services, and competition or cooperation with other nurseries in the area. The two main types of production nurseries are container and field. Field nurseries grow crops to marketable size in the ground. Labor requirements are greater for container nurseries compared to field operations.

Considerations for land preparation include leveling, road design and construction, irrigation design and drainage and collection of runoff. Structures typically seen at nurseries include greenhouses, coldframes, hotbeds, shadehouses, lathhouses, overwintering houses, and cold storage buildings. Other areas found in the nursery may include propagation, offices, potting, shipping, dry storage and pesticide storage and mixing.

QUESTIONS

Answer the following questions using correct spelling and complete sentences.

1. Why is proper nursery site selection important?
2. Explain the usefulness of the USDA Plant Hardiness Map or AHS Heat-Zone Map.
3. Why is seasonal rainfall important for nursery production?
4. Describe the importance of windbreaks.
5. Explain the differences in preferred soil types for balled and burlapped (B&B) stock versus bare root (BR) plants.
6. Why is drainage important in nursery site selection?
7. What are some considerations in determining water needs for a nursery?
8. Name two common sources of irrigation water for nurseries.
9. List several general categories of nursery pests.
10. Describe the labor needs of container and field nurseries.
11. What transportation criteria need to be considered before choosing a nursery site?
12. Why is proximity to competition important?
13. Describe differences in layout between container and field nurseries.
14. Describe the different facilities often found in most nurseries.

EXPLORING

1. Visit a real estate office and choose a potential nursery site from actual properties. Present a report on why the site was chosen.

2. Visit several nurseries and describe the differences in layout of the different nurseries. Note any similarities and differences.

3. Visit different city and county offices to determine which permits are needed to start a nursery business in your community.

4. Review different greenhouse and nursery supply catalogs to become familiar with different structures and equipment that are available.

13

PRODUCING NURSERY CROPS

The horticultural industry is the fastest growing segment of agriculture. Landscape plants are a big part of that growth. Producing nursery crops involves the production of landscape plants including trees, shrubs, and groundcovers. It differs in several ways from growing ornamental crops in the greenhouse. Often, nursery crops take several months to several years to reach marketable size.

Growers produce nursery crops in containers of growing medium or in fields in the soil. They need skills in propagation, growing media preparation, planting, and harvesting techniques. Growers and employees must know the cultural requirements of the crops they produce.

13-1. Aerial view of a California nursery operation.

OBJECTIVES

This chapter covers container and field-grown nursery crop production. It has the following objectives:

1 Describe the importance of nursery production in the United States

2 Why are marketing and product mix important to nurseries

3 Discuss the differences between container and field-grown plants

4 Describe the importance of nutrition and monitoring in producing container-grown plants

5 Explain why temperature limits the production of plants

TERMS

aeration
balled and burlapped (B&B)
bare root (BR)
broadleaf evergreen
container
container bed
controlled release fertilizer
dolomitic limestone
evaporation
foliar analysis
gro-bag
heeling-in

liner
liquid feed system
liquid fertilizer
mechanically-harvested
narrowleaf evergreen
nursery stock
pot-in-pot
pour-through method
root circling
root pruning
stubbing
water holding capacity

NURSERY PRODUCTION IN THE UNITED STATES

Nursery production is the growing of nursery stock in fields or in containers. ***Nursery stock*** are nursery plants produced for resale. They include landscape plants, such as trees, shrubs, and groundcovers, and specialty areas, such as fruit trees and vines.

Container nursery production is considered to be relatively new, having its origins in California in the 1950s. Many nurseries in the northeastern United States have been producing field-grown plants since the 1800s. According to the latest USDA statistics the top ten states for greenhouse/nursery production in terms of cash receipts in the United States were:

1. California
2. Florida
3. North Carolina
4. Texas
5. Ohio
6. Oregon
7. Michigan
8. Pennsylvania
9. Georgia
10. Oklahoma

13-2. A well-organized container nursery operation.

CONTAINER NURSERY PRODUCTION

Product mix will often determine the propagation needs of a nursery. In Georgia, container nurseries grow approximately 75 percent broadleaf evergreens, 12 percent narrowleaf evergreens, and 6 percent deciduous shrubs and ornamental trees. In general, growers should try to produce approximately 70 percent of their plants that are staples in the industry and 30 percent of the plants should be new crops which have market potential.

Another option is to become a specialty nursery. A grower could specialize in a particular type of plant, such as rhododendrons or native plants. A grower needs to have a market before planning production. The market is just as important as producing quality plants.

13-3. A container nursery may specialize in particular types of plants.

PROPAGATION

Most nurseries propagate plants from cuttings or seeds. For cuttings, plant material is usually gathered from a stock plant area. This area may be an actual blocks of plants grown in a field and used exclusively as parent material or a production area from which plant material is taken. Since field-grown stock plant areas may be costly to maintain, many nurseries use cuttings taken from container production areas. Stock plant areas should be physically separated from the propagation area and all plants should be maintained in excellent condition for optimal propagation results.

A variety of growing media is available. A good medium should firmly hold the cuttings in place, be free of insects and disease, have good water holding capacity, and provide excellent aeration and drainage. *Aeration* is the exchange of air in the soil with the air in the atmosphere. Some commonly used propagation media components are sand, perlite, vermiculite, peat moss, and bark. Sand, perlite, and bark all provide excellent drainage.

Vermiculite and peat moss increase the water holding capacity of a medium. The *water holding capacity* is the medium's total pore space less air space after drainage. An acceptable air space for container media after drainage is between 20 and 30 percent. Air space greater than 30 percent indicates low water holding capacity. This medium will need frequent irrigation. Air space below 15 percent can cause damage due to low oxygen concentrations in the medium. Poor

13-4. Cuttings of narrowleaf evergreens being grown in outdoor beds.

aeration in container media is a common cause of poor root growth. A water holding capacity between 30 and 50 percent is recommended for container media. It may be as high as 85 percent for some container media.

13-5. Growers need to have bulk growing media components readily available.

When growers propagate plants from seed, there are several factors to be considered. These include the propagation medium, container type, type of seed, and the environment in which the seedlings will be grown.

A common medium used for seed germination is a 1:1 mixture of perlite and peat moss on a volume basis. Any number of commercially available media will work for germinating seeds as long as the medium has good physical and chemical properties. Proper sanitation is important. Carelessness can result in poor germination and seedling growth. Disease problems are often the cause of poor seedling performance. Flats are the most common container for sowing seeds. Trays or packs may be used with automated seeding machines or seeds may be sown directly into individual containers.

As a general rule, seeds should not be sown any deeper than twice the diameter of the seed. Very fine seeds should not be covered but firmly pressed in contact with the propagation medium. Large seeded plants can be sown directly into individual pots or sown in outdoor beds. Depending on the operation, seeds can be sown in rows or broadcast over an area. Many ornamental trees and shrubs are grown in outdoor beds.

CONTAINER-GROWN PRODUCTION

The *container* provides protection for the plant's root system and facilitates transplanting and survival. They should be easy to handle and ship. Criteria to consider when selecting a container are:

- adequate drainage
- ability to hold sufficient volume of media
- lightweight
- easy to handle
- durable
- free of toxic substances
- prevent root circling

Rooting containers are often used for propagation instead of ground beds because transplanting losses are minimized along with handling costs. Examples of rooting containers used include flats (wooden, plastic, and metal), clay pots, plastic pots, compressed fiber pots, peat pots, cell packs, root cubes, paper pots, and poly bags. Each grower should select a rooting container which works best for their system and produces the highest quality plant.

13-6. The trees at this nursery are being grown in wooden containers.

When a young plant of suitable size is ready to be planted into a larger container for growing into a larger plant, it is called a *liner*. A liner can be a seedling, either bare root or in a rooting container, a rooted cutting, or a tissue culture plantlet. It is important that liners have well-branched root systems for optimal development.

Container Selection and Pruning

The criteria for selecting a growing container are similar to those for the selection of a rooting container. An added factor to consider is consumer ap-

13-7. A traditional black plastic container (left) and a container made from recycled newspaper (right).

13-8. Preventing root circling of weeping willow roots in containers using copper hydroxide in latex paint on left; plant on right was not treated.

peal. Plants will often be sold in the container they were grown in. Most container-grown plants are grown in black plastic containers because of their longevity (resistance to breakdown from ultraviolet light) and their consumer appeal.

One problem with growing plants in containers is ***root circling*** (pot-bound) and girdling. Plants become pot-bound when their root systems become too long for their containers. Different methods have been devised to help prevent root circling. Currently, air root pruning, bottomless containers and the use of copper compounds, such as SpinOut™ (Griffin L.L.C., Valdosta, Georgia), are the most popular.

Pruning container plants may also be necessary to induce growth and develop the plant canopies into desirable forms. It is necessary to remove undesirable growth and damaged or dead stems. Some plants may also require staking to support them in an upright position.

FERTILIZATION OF CONTAINER-GROWN PLANTS

Most components of container media contain few nutrients that contribute to plant growth. Because container media differ from field soils, nutrition of container-grown plants can be more complex and intensive. Almost all container nurseries use pre-plant amendments in their media. Common approaches to regular feeding are controlled release fertilizers, quick-release mixed fertilizers, and liquid feed systems.

Dolomitic limestone and micronutrients are examples of pre-plant amendments. *Dolomitic limestone* is used to provide calcium and magnesium for plant growth, as well as neutralizing the acidity of certain media components, such as pine bark and peat moss. Five to eight pounds of dolomitic limestone per cubic yard of potting media is sufficient for most plants. Lower rates should be used for acid-loving plants, such as azaleas and rhododendrons. The amount of dolomitic limestone to be added should be adjusted based on observations of water quality, plant growth, and the pH of the growing medium. The ideal pH for most plants is between 5.0 to 6.0, but certain species, such as boxwood and nandina, grow best near a neutral pH of 7.0. The optimal pH of container growing media is lower than that of field soils. Availability of micronutrients in container growing media is the most important consideration.

Micronutrients are required in small quantities, but are essential for plant growth. Always follow label recommendations. Micronutrients commonly found in commercial preparations include iron, manganese, zinc, copper, boron, molybdenum, and sulfur. Proper use of micronutrients can help prevent chlorosis and stunting of plant growth. Most micronutrient applications are good for one year.

Superphosphate has been recommended in the past as a pre-plant amendment for growing media. This practice is no longer recommended since superphosphate is highly soluble and readily leaches from the container media in a short period of time. Adequate phosphorus is supplied by other fertilizers applied during the growing season.

Nitrogen, phosphorus, and potassium are supplied by regular fertilizer applications. A ratio of 3-1-2 of N-P-K is ideal for the growth of most woody nursery plants. With current concerns about nutrient pollution and increased labor costs, controlled release fertilizers are a viable alternative. *Controlled release fertilizer* is one in which one or more of the nutrients have limited solubility so they become available to the growing plant over an extended period. With controlled release fertilizers, one

13-9. Controlled release fertilizer has been added to the surface of the growing medium in this container.

or more applications can result in a full season of adequate nutrition, depending upon formulation. Different controlled release fertilizers have different release mechanisms and fertilizer components. The duration of release for most controlled release fertilizers is based on a growing medium temperature between 70 and 80°F.

If the growing medium is going to be used in a short time period, incorporation of a controlled release fertilizer into the medium is recommended. The fertilizer can also be applied to the surface of the growing medium after potting. The disadvantage of applying fertilizer to the surface of the medium is that it may be washed out of the container or may spill during handling. Containers can also blow over. Most plants grow equally well whether the fertilizer is incorporated or applied to the surface of the growing medium. The longevity of these products depends on time of application, irrigation rates, and temperature of the growing medium throughout the season. A fertilizer containing nitrogen, phosphorus, and potassium applied at the rate of 3 to 4 pounds of nitrogen per cubic yard of growing medium should supply sufficient nutrients for nine to twelve months.

Quick-release mixed fertilizers, such as a 10-10-10 formulation, have been popular due to their lower cost. However, since the fertilizer is readily soluble, applications must be made regularly at short intervals, increasing labor costs. Quick-release fertilizers must be applied when the foliage is dry and removed from the foliage before irrigation to prevent foliar burn. This type of fertilization program is not generally recommended.

Liquid feed programs are used by growers who have injection systems to pump fertilizer solutions through the irrigation system. Once the sprinkler

13-10. The concentrated fertilizer solution in these tanks is injected into the irrigation system and the plants are fertilized at the same time they are watered.

system and injectors are in place, fertilizer costs may be one-fourth the price of controlled release fertilizers. With environmental concerns, many growers are using liquid feed systems to supplement controlled release fertilizer programs.

Liquid fertilizer is a fluid in which the plant nutrients are in true solution. With *liquid feed systems,* concentrated fertilizers are mixed in a holding tank before being diluted to the appropriate amount and injected into an irrigation system. This is known as fertigation. Plants can be fertilized daily using 75–100 ppm nitrogen. More concentrated applications may be made less frequently. During the fall and winter months, frequency or concentration of fertilizer is reduced. Growers should adjust the concentrations and ratios of nutrients in their liquid feed programs based on plant growth and regular monitoring of the nutritional status of the crop.

Monitoring the Nutrition of Container-Grown Plants

Producers of nursery and greenhouse crops should make every attempt to regularly monitor the nutritional status of their growing media. This allows the grower to make immediate decisions regarding fertility and leaching requirements, and reduce the occurrence of nutritional problems. Periodic monitoring is required since crop growth may be decreased without the development of visual symptoms. High nutrient concentrations can occur due to improper watering and fertilization methods. Excessive rainfall or irriga-

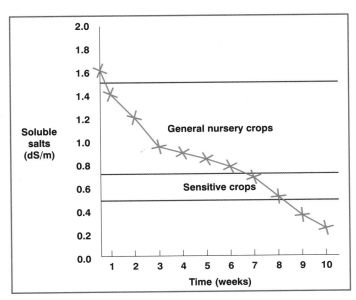

13-11. Nutrition of container-grown plants can be monitored by graphically charting test results.

tion can also leach nutrients from the medium which results in poor plant growth.

Few growers monitor the nutritional status of their plants. Two simple tests that can be performed at the nursery are the measurement of pH and soluble salts. Reliable instruments are available that any grower can afford. While several methods of media testing exist, a simple method for use in nursery operations is the *pour-through method* developed at Virginia Tech. The advantages of the pour-through method over other procedures are: (1) a short time period for sample extraction, (2) soluble salt and pH analysis can be conducted in the field, (3) no container medium is handled, (4) no specialized equipment is required for sample extraction, and (5) there is no chance of rupturing controlled release fertilizer particles, which could result in false readings. The pour-through method should be compared to other testing procedures before full scale implementation.

Factors to consider in interpretation of media analysis include different nutritional requirements of different species, stage of crop growth, time of year, fertilization program, growing medium, and other environmental and cultural factors. Media should be tested every two weeks to determine if adequate nutrient levels are being maintained. Readings should be plotted graphically so trends can be seen. Since the soluble salt level gives an indication of the concentration of total salts and not individual elements, nursery personnel should have individual elements checked every four to six weeks. A growth medium that tests in the low range will not have sufficient nutrients for good growth. Soluble salt readings of 3.0 dS/m or greater will generally result in decreased plant quality or injury to sensitive species or seedlings.

If soluble salt levels are high, plants should be leached with an application of water equal to two container capacities. Container capacity refers to the amount of water a container with growing media will hold after being saturated and allowed to drain. If soluble salt levels are low, it may be time to reapply fertilizer based on the particular needs of the crop. Time of year will often dictate when and how much fertilizer should be applied.

When the pH of the growing medium is too high, injection of acids into the irrigation system is the quickest way to resolve the problem. Sulfuric acid is most commonly used to decrease the pH of growing media. Iron sulphate is also effective as a topdress or liquid drench application. Ammonium-nitrogen fertilizers will also decrease the pH of growing media. When the pH is too low, dolomitic limestone is incorporated into the growing media. If magnesium is not needed, calcium carbonate can be used. Hydrated lime can be used on established plants, but should be used with caution since it is very caustic.

Foliar analysis is a leaf tissue test used to diagnose nutrient deficiency symptoms in plants. Plants of the same species that have been grown together under similar conditions can be grouped together for sampling. Plants of different species, or the same species grown under different conditions, should be sampled separately. Twenty to thirty of the uppermost, fully expanded leaves should be randomly sampled just before a new flush of growth begins. For diagnostic purposes, samples of tissue from healthy and abnormal plants should be sampled. Samples should be placed in paper bags with proper identification before mailing to a state or private laboratory for analysis. Once the results are returned, the range of nutrient concentrations, as well as the ratio of different nutrients to each other, should be considered. Accurate record keeping will aid in the development of fertility management programs.

WATERING CONTAINER-GROWN PLANTS

Proper watering of container-grown plants is important. Irrigation should be based on the demand of the crop. This can be determined by weighing the pots, feeling the growing medium, or using indicator plants that readily show moisture stress. Soil moisture sensors are not as reliable in coarse organic media as they are in field soils. Plants should be grouped according to water requirements and similar container sizes.

The evaporation of moisture is influenced by temperature. *Evaporation* is the changing of water from a liquid to a gaseous state. To conserve water, irrigation applications should take place during the early morning hours to minimize evaporative losses. In general, the volume of water leached from the container should not exceed 25 percent of the water reaching the surface of the

13-12. For larger container nursery operations, drip irrigation can be used to conserve water and still meet the needs of the plants.

growing medium. Plants with dense canopies or overgrown plants may have lower percentages of applied irrigation reaching the surface of the container media. With overhead sprinklers, a general rule is that only 15 percent of the applied water reaches the surface of the media. Irrigation systems should be checked regularly for uniformity of application. Cyclic irrigation, or applying the daily requirement of water in two or more applications throughout the early part of the day, has been shown to conserve water, compared to applying all the water at once.

For larger containers, low volume irrigation is a viable alternative. Drip irrigation can be used to apply small amounts of water to specific areas as needed, resulting in increased irrigation efficiency. In general, low volume irrigation systems can use up to 75 percent less water than conventional overhead sprinkler systems. Economically, drip systems are best used on container sizes of #5 and above. Common problems associated with low volume irrigation systems include clogging of emitters, the need for excessive water filtration, and large variances in emitter application rates.

13-13. Container spacing affects plant growth and quality.

CONTAINER BEDS

Container beds stand alone and only contain container plants. When designing container beds, factors to consider include—the size of the plant being produced, container spacing requirements, production practices, such as pruning and fertilization, weight of containers, and irrigation design and coverage.

Container spacing is important in container production. It affects plant growth and quality. Containers are spaced tight during the initial growing period. As the plants grow, the spacing between the containers is increased. The increase in spacing allows sunlight to reach the lower leaves on the plants.

Container beds should be designed so similar sized plant material with common cultural requirements can be produced in the same area. If plants are grown in large containers or in heavy media, the distance that employees have to carry plant material becomes important. Shade and overwintering structures need to be taken into consideration when designing container beds.

Irrigation systems also influence container bed design. If overhead sprinklers are used, a row of sprinklers can be run down the middle of a bed or sprinklers can be located on the outer edges of the bed. Sprinkler system design generally limits bed width due to sprinkler coverage patterns. The use of drip or low volume emitters generally limits bed length due to changes in water pressure.

Container beds can be covered with a variety of materials. The most common materials are black polyethylene, ground cover cloth, gravel, and crushed sea or clam shells. Black polyethylene is readily available and commonly used. Two limitations to the use of black polyethylene are its short period of usefulness (usually less than 1 year) and disposal of old plastic. Ground cover cloths are becoming more popular since they have a longer useful period (3–5 years) and work well when placed over black polyethylene. Gravel and crushed shell material are generally cheapest to install and have a long life, but problems exist with plants rooting into the container beds and weeds are more of a problem.

Two general container bed designs are used. Traditionally, container beds have been graded such that the center of the bed is raised and the irrigation water drains to either side of the bed. The disadvantage of this design is that employees and equipment have to pass through muddy, wet areas before

13-14. **Plants growing on woven black ground cloth.**

stepping onto the container bed. The other design is to have the container bed sloped towards the middle so that irrigation water flows down the center of the bed and drains off in only one area. This eliminates soft, wet areas around the entire perimeter of the container bed.

OVERWINTERING CONTAINER-GROWN PLANTS

Nurseries in most parts of the country need to protect plants from low temperatures during the winter months. The roots of plants are not as cold hardy as shoots. Container-grown plants are susceptible to winter damage since their root systems are not insulated by the soil like plants growing in the soil. The difference between the ambient air temperature and the temperature inside a container is usually less than 5°F. Because of this slight difference, a plant which may be killed at 10°F in the landscape may be killed if root-zone temperatures reach 25°F in the container. Plants in containers are also subject to desiccation when the roots are frozen. For this reason, containers should not be allowed to dry out during the winter months.

To begin to prepare plants for winter, fertilizer programs should be reduced and all pruning finished in ample time for the plants to acclimate. In areas with mild winters, the most common form of winter protection is to place the containers close together and surround the outer edges with containers filled with media or to use cloth or plastic wraps. Shade cloth or other porous materials can also be used to cover plants. These materials protect the plants by trapping heat released from the soil.

13-15. Placing plants close together and wrapping them is a method of cold protection.

For sensitive crops or in areas where regular winter protection is required, covering plants with white polyethylene or thermal blankets can be used. Overwintering houses covered with polyethylene are also commonly used by many nurseries. If required, heaters can also be used inside overwintering houses to help maintain acceptable temperatures. White polyethylene coverings are preferred to clear polyethylene since they prevent the build up of high temperatures on sunny days. All covered structures should be regularly monitored and vented if temperatures become excessive.

PEST CONTROL

Nursery plants are regularly attack by a number of insect, mite, disease, and weed pests. All nurseries should have a crop pest management plan. Many nurseries have a person, known as a scout, who is trained in the identification of plant pests. The job of the scout is to regularly monitor and record pest populations in the nursery and assist in making pest control recommendations. Regular monitoring and using integrated pest management techniques can result in substantial savings through the use of fewer pesticides and more effective applications. When pesticides are used, the safest chemical should be chosen. County agents or state extension specialists are good sources of information for recommendations on pest control.

Weed control is a major problem in container nurseries. Many nurseries use preemergent herbicides for improved weed control. Examples of common weed problems in containers are spotted spurge, eclipta, bittercress, common groundsel, and oxalis. Most preemergent herbicides can be applied soon after potting to prevent weed growth from occurring. When preemergent herbicides are used, the cost for weed control in nurseries can be one-third that of hand weeding alone. Having weed-free containers makes shipping plants easier and also eliminates alternate hosts for other pest organisms. For all pests, sanitation in all areas of the nursery is important.

FIELD NURSERY PRODUCTION

Most field nurseries produce shade trees, flowering trees, and a variety of evergreen and deciduous shrubs. Some nurseries specialize in producing seedlings, liners, and whips, which are used by other field and container growers. According to a recent survey, the most popular shade tree in the

```
┌─────────────────────────────────────────────────┐
│        ┌──────────────────────────────┐          │
│        │     MOST POPULAR TREES       │          │
│        └──────────────────────────────┘          │
│                                                   │
│   Shade                Flowering                  │
│                                                   │
│   Red Maple            Crab Apple                 │
│   Pin Oak              Callery Pear               │
│   Honey Locust         Flowering Dogwood          │
│   Sugar Maple          Eastern Redbud             │
│   Norway Maple         Washington Hawthorn        │
│   Green Ash            Kousa Dogwood              │
│   Littleleaf Linden    Serviceberry               │
│   White Ash            Oriental Cherry            │
│   Red Oak              Flowering Plum             │
│   Willow Oak           Saucer Magnolia            │
│                                                   │
└─────────────────────────────────────────────────┘
```

13-16. Most popular shade and flowering trees in the United States.

United States was the red maple. The most popular flowering tree was the crab apple.

Plants selected for production should be based on a thorough market analysis. Sales of large field-grown plants will not begin until the third to fourth year of production. The breakeven point for field operations often occurs between the sixth and eighth year. Since this type of production is a long-term investment, proper selection of plant material is critical.

FIELD PRODUCTION

Fields should be checked to determine nutritional requirements and the need for soil amendments. Organic matter content and pH levels should be determined. Proper adjustments can be made by incorporating necessary materials into the soil before planting. Cover crops can be a good source of organic matter and will serve to prevent erosion problems.

Soil fungi, insects, and weeds should also need to be controlled. Depending on the crop to be grown, it may be necessary to fumigate the soil. This is often a recommended procedure if seedling beds are to be established. For certain weeds and insects, mechanical control may be the best choice. A number of insecticides and herbicides are available to control insect and weed problems.

Plant Spacing

Distance between rows of plants will depend on the final size of the plant being grown and the type of equipment required for harvest. For shade and flowering trees, a general rule of thumb for plant spacing is three feet for every inch of stem diameter at harvest. For instance, if you plan to harvest 2-inch diameter dogwood trees, they should be spaced six feet apart within a row. Smaller shrubs or trees to be harvested bare root can be planted closer together. Long-term crops, which are widely spaced, can be inter-planted with short-term crops.

Access to the field on a year-round basis is critical. Development of sod aisles between rows of trees helps accessibility by supporting equipment even under wet conditions. Sod aisles should be mowed regularly whereas tree rows should be kept weed-free.

13-17. Tractor-pulled planter allowing workers to plant five rows of bare root (BR) seedlings into a nursery bed.

Liner Selection and Planting

Just as important as deciding which plants to grow is the selection of quality liners. Many growers do not have the expertise or facilities for the production of liners. Initially, they purchase liners from other growers.)

Seedlings of certain species are readily available. Seedlings are sold either by stem diameter size or at a fixed cost per tree in ungraded lots. Seedling liners can range in height from a few inches up to three feet or more depending upon species. Topping of seedling trees increases caliper size and uniformity.

13-18. A red maple that has been stubbed and allowed to develop a new trunk.

13-19. Nursery plants can be marketed in containers, bare root, and balled and burlapped.

Tree seedlings that have been topped are preferred as lining out stock for budding, bench grafting, and stubbing. *Stubbing* is the process of cutting a seedling down to stub and allowing the plant to regenerate a new top. Un-topped seedlings are preferred by field growers who will not be doing any further propagation with the seedlings.

Seed source is an important factor to consider when purchasing seed. Selection of seed from sources that are adapted to the area in which the seedlings will be grown will result in higher quality plants. Many tree species from southern seed sources are not winter hardy in the northern part of their natural ranges. Whereas, seedlings from northern areas often perform poorly in southern latitudes.

Liners can also be purchased as vegetatively propagated plants. Examples would be plants grown from cuttings, grafted plants, or plants that have been tissue cultured. There is a great deal of interest in plants grown on their own roots from vegetative cuttings. Certain cultivars of red maple have been found to have a bud union incompatibility problem, which results in the loss of trees due to stem breakage at the bud union site. This problem can be overcome by growing red maples on their own roots.

Several species are now available as tissue cultured liners. These plants may have some of the same advantages as those plants grown on their own roots. However, there has been little research on the growth and performance of tissue cultured plants compared to traditional methods of propagation. Mountain laurel is an example of a plant that is now readily available as a tissue cultured liner.

Liners can be purchased as bare root (BR) plants or container-grown. ***Bare root (BR)*** describes trees grown in a nursery or field and dug without taking soil. Many broadleaf tree species can now be purchased in containers. While container-grown liners may be more expensive, there may be less transplant shock and increased survival in the field compared to bare root plants. Only top quality liners should be purchased and planted. It is important that liners be pest free with good branch structure and a well developed root system. Always beware of container-grown liners that are potbound and have circling roots. An attempt to save a few cents per plant in the beginning could cost a grower in the future.

Most liners can be planted with a transplanter or larger plants can be placed in individual planting holes. Many species respond well to fall transplanting. In colder areas, late summer or early spring transplanting may be best. It is important that the root system of liners does not dry out during the transplanting process. Newly planted liners will perform best if they receive supplemental irrigation.

13-20. A plant that was kept in a container too long. Notice the deformed roots.

NURSERY PRODUCTION SYSTEMS AND CARE

Traditionally, field-grown nursery stock has been dug by hand with a spade. The root balls were covered with burlap and sold as balled and burlapped (B&B) plants. ***Balled and burlapped (B&B)*** describes trees grown in a field, dug keeping a ball of soil surrounding the root system, and covered with burlap material. There are several advantages to digging plants B&B. Since balled and burlapped plants can be dug and held for limited periods of time, the digging and transplanting season can be extended. When compared

13-21. **Balled and burlapped (B&B) plants ready for shipment.**

to plants dug bare root, difficult to transplant species often react more favorably when dug balled and burlapped. Since the plant has an intact root ball with native soil, transplanting shock is also minimized.

Most larger nursery stock is **mechanically-harvested** (gathered with equipment), such as with a mechanized tree spade. Plants greater than 1¼ inch in stem diameter or that require an 18-inch diameter root ball are often mechanically harvested. Mechanical tree spades are available in a variety of sizes, which will dig root balls from 15 inches to greater than 60 inches. The American Nursery and Landscape Association provides standards for root ball diameter in relation to trunk diameter for trees and other standards for all types of nursery stock. Mechanically-harvested trees are often put into burlap-lined wire baskets for transport to holding areas before they are sold.

Trees dug by hand or with a tree spade have similar disadvantages. A portion of the root system is removed when the plant is dug, decreasing the

13-22. **A truck-mounted hydraulic tree spade mechanically harvesting a tree.**

chance for quick establishment in the landscape. Time of digging is often dictated by soil moisture conditions. Soils that are too wet or too dry limit digging of field-grown plants. The root balls of plants dug balled and burlapped or mechanically are awkward and very heavy. This makes handling more difficult and shipping expensive. Trees dug B&B require a skilled labor force familiar with proper harvesting techniques.

Smaller trees and shrubs can be harvested bare root. The advantages to producing bare root (BR) trees for harvest are the plants are lightweight, which makes shipping more economical. The initial cost per plant to the grower is lower and this cost savings can be passed on to the consumer. Major disadvantages to digging bare root trees are handling problems and field survival. Short periods of exposure to the heat and sun during digging can damage the roots. Since little or no root activity occurs until bud break for most bare root plants, the timing of digging and transplanting can be limited. In general, bare root trees should not be transplanted in the fall. However, certain species can be dug in the fall and successfully overwintered in storage facilities.

Proper storage and holding of nursery stock dug from the field is necessary to keep the plants' root system from being damaged and drying out. Growers temporarily store nursery stock by heeling it in. **Heeling-in** involves temporarily placing a plant that has been dug in a field in a bed with its root system covered. It will remain there until it is transplanted or sold. The material used to cover the root system, such as soil, saw dust, or peat moss, should be kept moist.

Two modifications of field-production are the use of gro-bags and the pot-in-pot production system. A **gro-bag** is a cylinder with a porous fabric side and a plastic bottom. It prevents the formation of difficult to harvest taproots. The porous fabric side allows the movement of water to the root system and keeps the plant's roots within the bag. The idea behind the success of the gro-bag is that 80 percent or more of the roots would be harvested, increasing survival and growth when transplanted.

Gro-bags offer some advantages when it comes to harvesting. The possibility of digging trees in the summer is possible because plants are harvested with a greater percentage of intact roots. If plants are harvested during the growing season, this should be done when the plants are between flushes of growth. A hardening-off period will be required for plants harvested during the growing season. Special attention should be given to reducing evaporative demand and keeping the root system protected. With gro-bags, expensive digging equipment is not required and a crew can harvest gro-bags in about one-

13-23. A gro-bag keeps the plant's root system within the bag.

fourth the time required for balled and burlapped plants. Fabric bags treated with SpinOut are now available.

13-24. Plants being grown using a pot-in-pot production system.

Some problems associated with the production of large plants in containers are they blow over and the roots are killed by cold temperatures in the winter and hot temperatures in the summer. The *pot-in-pot* system solves these problems by keeping the plants upright while protecting the root system. A holder pot is placed in the ground with the lip of the pot remaining above grade. A planted container is then placed into the holder pot for the growing season. This system is a good match of field and container production systems. Special attention must be paid to drainage (heavy soils drain poorly) and problems associated with harvesting plants. These include plants that have rooted into the ground and exposure of the root system after harvest to high temperatures from solar radiation striking the sides of the planted container.

FERTILIZATION

Steps in a good fertilization program include determining soil type(s), performing soil tests, interpreting results, and making a corrective application of fertilizer before planting. The key to successful soil sampling is to get a representative sample from a given field. A composite sample of 15 to 20 soil cores (4 to 8 inches deep) should be taken for each field to be analyzed. When taking soil samples, avoid areas near roads, field borders, old fertilizer bands, eroded areas, and places where organic materials, such as mulch, have been stored.

Nutrients, such as phosphorus, potassium, calcium and magnesium, are incorporated into the soil before planting. If liming is required to raise the pH of the soil, dolomitic limestone is often used since it also supplies magnesium. Recommendations for adjusting pH of the soil and for the nutrient requirements of the crop are available from a state extension specialist or county agent.

Tissue analysis of nursery stock can be used to monitor plant nutritional status. Fall sampling is particularly useful since the spring flush of growth will depend upon nutrient accumulation during the previous growing season. Uniform sampling of mature leaves is important for meaningful test results. The uppermost leaves of woody ornamental plants should be used for tissue analysis.

Foliar Tissue Levels	
Nitrogen	2.0-2.5%
Phosphorus	0.2-0.4%
Potassium	1.5-2.0%
Calcium	0.5-1.0%
Magnesium	0.3-0.8%

13-25. Foliar tissue levels for ornamental plants on a percent dry weight basis.

Current recommendations for nitrogen fertilization of nursery crops ranges from 100 to 300 pounds of nitrogen per acre per year. Deciduous trees generally require the highest rate of nitrogen (250 pounds N/year/acre) followed by narrowleaf evergreens (200 pounds N/year/acre). *Narrowleaf evergreens* are those that retain their needle-like leaves through the winter. Broadleaf evergreens are often fertilized at the rate of 100 pounds N/year/acre. *Broadleaf evergreens* are those that do not have needle-like leaves but hold their leaves throughout the winter. Nitrogen fertilizers should be put on in split applications, which can range from two to six times per season. Slow release forms of nitrogen should be used to increase uniformity of feeding and decrease nitrogen losses. Timing of fertilizer applications will be dependant upon the climate and crop grown.

PRUNING

Pruning helps control the growth and shape of nursery stock. It is necessary to develop plants with desirable forms and compact growth. The plant appearance is improved. Dead, diseased, and damaged stems are removed. Pruning also serves to control plant size and remove any weak areas in the branching pattern. Proper pruning helps develop a desirable, natural form and strong branches.

Another type of pruning needed for field-grown plants is root pruning. *Root pruning* is the process of removing outward root tips which encourages the plant's root system to develop within a small area near the base of the plant. Root pruning helps prepare a plant for later transplanting. Field nurseries root prune every two to three years. It is performed in the fall when the plant's growth has slowed or is inactive.

IRRIGATION

Supplemental irrigation is important during the transplanting and establishment phases of field production. It has been shown to affect the survival and growth rate of field-grown plants. Properly irrigated nursery plants will be more vigorous and larger in a shorter period of time, reducing production time. Irrigation needs can be determined with a tensiometer or can be based on environmental conditions, such as daily net evaporation. Irrigation requirements based on net evaporation can only be calculated if plant size and accurate correction factors are known. Determining when to water is an acquired skill. The goal of applying irrigation is to prevent soil moisture stress, which limits plant growth and development.

Several methods may be used to irrigate nursery stock in the field. Flooding the field or allowing water to run down furrows along the rows of plants are two. Overhead sprinklers are more common. Drip irrigation may be a viable alternative. Sprinklers or flooding may waste water by irrigating a larger area than the plant's root zone. With drip irrigation, water passes through a plastic pipe and drips through tiny holes in tubes at the base of each plant. It slowly provides water to each plant's root zone. It has been estimated that drip irrigation can reduce water consumption by 75 percent and water runoff by up to 90 percent. A 75 percent reduction in water and pumping costs would greatly benefit nursery growers.

PEST CONTROL

Regular scouting and appropriate corrective measures are important. Individual insect and disease pests will vary by crop. Weed control is a major concern for producers of field-grown nursery stock. While the control of weeds had traditionally been handled with physical or cultural means, increased labor costs and advances in herbicide technology now make the use of herbicides a viable alternative. All nurseries should have a weed control program that combines the benefits of both herbicides and traditional weed control methods, including the use of integrated pest management techniques. Worker Protection Standards must be observed when chemicals are used.

REVIEWING

MAIN IDEAS

Sales of landscape plants and garden items have been on the rise in recent years. Homeowners and realtors are acknowledging the value of a well landscaped home. With population growth and increased housing, there is an increased demand for nursery plants. Both container and field-grown nursery operations are viable agricultural alternatives. With proper planning and attention to production and marketing details, profitable production of nursery crops is possible.

Container nursery production is on the rise throughout the United States. Marketing and production must be considered before starting a nursery. Some nurseries specialize in propagation. Media and container selection are very important production considerations. Different crops may require different media and fertilizers. Controlled release fertilizers and liquid fertilizers are commonly used in container nurseries. Regular monitoring of the nutritional status of container grown plants is essential for success. Overhead watering for small containers and drip irrigation for larger containers is commonly used. Overwintering protection of container grown plants is critical for many parts of the United States.

Considerations for field production include fertility, pests in the soil, plant spacing, and year-round access to the field. Liner selection can be critical since the plants will often be in the field for several years. Liners can be from seed, cuttings, or tissue culture. Field grown plants can be containerized, sold bare root or as balled and burlapped plants. Large trees and shrubs are often dug with mechanical harvesters or tree spades. New field production methods include the use of grobags or the pot-in-pot production system. Fertility, irrigation, and pest control will vary with the crop produced.

QUESTIONS

Answer the following questions using correct spelling and complete sentences.

1. Describe the current economic status of the nursery industry in the United States.
2. Name the top 10 nursery producing states.
3. List four general categories of nursery-grown plants.
4. List several characteristics of a good propagation medium.
5. Name several criteria used to select a container.
6. List the four functions of a good growing medium.
7. What are the physical factors that influence growing media?
8. Describe the differences between controlled-release fertilizers, quick-release mixed fertilizers, and liquid-feed systems.
9. Why is monitoring the nutrition of container-grown plants important?
10. Describe the pour-through method and its uses.
11. Discuss how to correct pH and soluble salt problems in containers.
12. Why is winter protection important for container-grown plants?
13. Describe the importance of proper liner selection for field nurseries.
14. Compare and contrast balled & burlapped, mechanically harvested, and bare-root production systems.
15. Discuss fertilization of field-grown plants.

EXPLORING

1. Use local soil maps and visit a Soil Conservation Service office to determine good potential field nursery sites in your area.

2. Attend a nursery trade show to familiarize yourself with the industry.

3. Produce a container-grown crop for sale. Select an evergreen hardwood crop, semi-hardwood crop, softwood crop, and deciduous hardwood crop common to your area.

4. Visit nurseries accompanied by an extension agent to learn about nursery problems.

14

FLORAL DESIGN

Everyone finds flowers appealing. Attractive, colorful flowers add meaning and express emotion. Flowers must be properly used if they are to have the desired effect. In addition, proper care ensures that top quality is maintained longer.

The ways flowers are arranged add special appeal. Understanding color and floral design and using the proper tools helps in making floral arrangements. Knowledge of the floral materials is a big step in creating a visually pleasing arrangement. Having a beautiful arrangement requires little more than one with less appeal.

14-1. Learning basic design skills is important for the beginning floral designer. (Courtesy, Dianne Noland, Illinois)

OBJECTIVES

This chapter covers the fundamentals of arranging flowers, including design, conditioning, and construction. It has the following objectives:

1 List ways flowers were used in past civilizations

2 Describe the different types of permanent flowers

3 Discuss techniques to increase flower vase life

4 Discuss the rules that control correct color use

5 List the art principles used in floral design

6 Demonstrate four major wiring techniques

TERMS

analogous colors
asymmetrical balance
balance
central axis
clutch method
complementary colors
corsage
design elements
dominance
everlasting flowers
filler flowers
floral design
floral foam
floral garland
floral tape
florist shears

flower preservatives
focal point
form
form flowers
gauge
greening
greening pins
hook method
hue
Ikenobo design
line
pattern
pierce method
primary colors
principles of design
proportion

rhythm
scale
secondary colors
shade
shape
silk flower
splinting method
symmetrical balance
texture
tint
vase life
Victorian design style
water tubes
Williamsburg design
  style

HISTORY OF FLORAL DESIGN

Flowers have played an important role in all civilizations since the beginning of recorded history. Written records show that the Egyptian civilization dates back to the year 2800 BC. The Greek and Roman cultures followed the Egyptian's and flourished between 600 BC and 300 AD. Each of these civilizations incorporated the use of flowers into their culture.

CLASSICAL FLOWER ARRANGING

Egyptians used floral arrangements to beautify their homes. They also utilized flowers in ceremonies as offerings to their gods and to their dead. Egyptians designed flower arrangements in large vases, jars, and bowls. They created arrangements that had rows of the same colored flowers. A simple design style describes their floral arrangements. Sometimes the Egyptians placed flowers in a container so that none of the flower stems were visible.

Egyptian designers used many kinds of flowers in their arrangements: roses, narcissus, water lilies, bachelor buttons, and violets. Their arrangements also contained a variety of foliage, such as ivy, olive, or palm leaves.

14-2. Flowers have played an important role in all civilizations.

The Greeks adopted much of the Egyptian culture around 600 BC. However, their style of flower arranging did not follow the Egyptian principles. They did not arrange flowers in vases or jars as the Egyptians did. Greeks started the practice of scattering flower petals on the ground during festivals or making floral wreaths to be worn or carried.

The Romans conquered Greece and incorporated many Greek traditions into the Roman way of life. Romans continued the Greek tradition of floral wreaths, but added the floral garland type arrangement to their culture. A *floral garland* is similar to a wreath but it is not made into a circular form. These long floral streamers were displayed by draping them on tables or wearing them around their necks.

14-3. A floral garland (left) is similar to a wreath (right) but not made into a circular form.

EUROPEAN FLORAL DESIGN

After the fall of the Roman Empire, about 300 AD, many different countries emerged in this region that is now called Europe. Italy, France, Holland, Germany, and England all developed their individual societies and floral customs.

During the next 1,000 years each of these countries developed different design styles. The following names are associated with the different design periods: Renaissance, Baroque, Flemish, and Victorian. These floral designs all have their roots in the traditional Greek and Roman cultures.

ORIENTAL FLORAL DESIGN

Oriental cultures developed while the Greek and Roman cultures did, but in another part of the world. Life in oriental societies evolves around their religious beliefs and practices. Buddhist priests in India started using flowers in religious ceremonies by placing flowering shrub branches on their altars. However, their religious beliefs prevented them from cutting live tree branches, but it did allow using any branches broken off during storms.

Around the 1st century AD, the Chinese adopted the Buddhist religion from India. However, the Chinese priests thought it highly improper to clutter the altar with broken flowering tree branches. Instead, they developed the practice of placing the branches in large pottery urns.

Buddhism was introduced into the Japanese culture around 600 AD from China. The Japanese culture created a style of floral design called the linear asymmetrical balance. A very famous Japanese Buddhist priest named

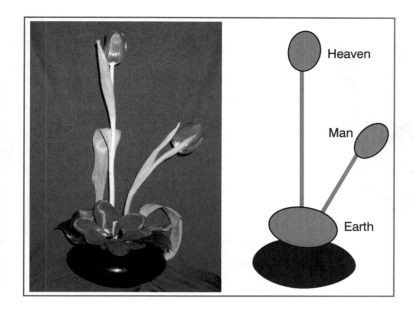

14-4. **The three main placements in a Japanese design. (Courtesy, Dianne Noland, Illinois)**

Ikenobo (Ike-no-bo) spent his life meditating on religious themes and arranging flowers for religious ceremonies.

During the early 11th century, the Ikenobo school of floral design started and specific written rules were established. Today these same rules still apply to all Japanese floral arrangements created in the Ikenobo tradition. With *Ikenobo design,* each flower has a specific meaning and an exact location in the arrangement. Presently Ikenobo floral design schools in Japan and the United States teach these traditional design principles and rules.

AMERICAN FLORAL DESIGN

Early settlers in this country brought to America their Euro-

14-5. **A Williamsburg design.**

14-6. A Carnation corsage.

pean cultures. The *Victorian design style* became the pattern for early American floral arrangements. The term *Williamsburg design style* describes the American version of the Victorian design style. Large round or globe shaped floral arrangements made by using different kinds and colors of flowers define this Colonial Williamsburg floral tradition.

During the 1920s, a small bouquet for a woman to wear, known as a *corsage,* became a popular form of floral expression. This trend continues today. The strong European influence dominated the American floral designs until after World War II. Modern contemporary flower arrangements in this country combine the European mass design style with the formal line design of the Orient.

CARE AND HANDLING OF FRESH FLOWERS AND FOLIAGES

Today the majority of cut flowers are imported from South American countries. By the time the florist receives these flowers, they have spent several days in transit on airplanes and trucks. During this entire time the flowers are not in water. The florist must process and condition cut flowers when they arrive at the flower shop.

When re-cutting stems, bubbles of air enter the end of the flower stem and block water movement up the xylem tissue. Make a fresh stem cut and remove ½ to 1 inch of the old stem. Many florists follow the practice of re-cutting stems under water to prevent air bubbles from entering the stem. Immediately after re-cutting the stems, place the flowers in a bucket of water.

Fill clean flower buckets with warm tap water before putting flowers in them. The correct water temperature is 100 to 105°F. Remove all leaves that will be below the water in the flower buckets. Bacteria will quickly decompose the leaves and reduce the quality of fresh flowers. Carefully remove rose

14-7. Cutting stems before putting in bucket of water and floral preservative.

thorns to prevent damage to other flowers. Use a floral preservative solution mixed at the proper concentration. Never put cut flowers in metal buckets, always use plastic. Flower preservatives can chemically react with the metal and produce substances that are toxic to flowers. Let the flowers remain at room temperature in a well-lit area for 2 to 3 hours to allow for the uptake of the floral preservative solution. Then place buckets in a 34 to 38°F flower cooler to complete the conditioning process.

14-8. Excess foliage being cut from the stem of a fresh flower for an arrangement.

STEPS FOR CONDITIONING FLOWERS AND FOLIAGE

The following steps for conditioning flowers will increase both the longevity of flowers and the efficiency of the business:

1. Unpack and inspect (for proper amount and quality) the flowers immediately upon receiving them. Report any missing or poor-quality flowers.

2. Prioritize the order of processing the flowers—condition wilt-prone (gerberas, bouvardia, baby's breath) and expensive (lilies, orchids, roses) ones first.

3. Remove sleeves, ties, and any foliage that will be below the water level to prevent rotting and bacterial formation.

4. Re-cut all stems; remove ½ to 1 inch (under warm water, especially for roses).

5. Use specific treatment solutions as needed, such as a hydration solution or an ethylene inhibitor solution.

6. Place in a floral preservative solution mixed at the proper concentration.

7. Let the flowers remain at room temperature in a well-lit area for 2 to 3 hours to allow for the uptake of the floral preservative solution.

8. Place the flowers in a cooler set at 34 to 38°F, with high humidity (80 to 90 percent), and constant lighting.

9. The flowers are ready to be used in a design!

FLOWER PRESERVATIVES

Using floral preservatives can almost double the vase life of most fresh flowers. Several companies manufacture and sell flower preservatives. *Flower preservatives* are chemicals which are added to water to increase the vase life of fresh cut flowers. *Vase life* is the period of time after cutting flowers that they can be held and still look good. Flower preservatives help prevent bacterial growth in the buckets. Bacterial growth can cause water to have an odor and shorten the flower's vase life. Read and follow label directions when mixing these preservatives and water.

Effective floral preservatives have three main ingredients. Each of these ingredients has an important function in keeping flowers and foliages fresh and long lasting. Either powder or liquid formulations should be used in high-quality water.

THE PRINCIPLES OF FLORAL DESIGN

Floral design is the art of organizing the design elements inherent in plant materials, container, and accessories according to the principles of design. The *principles of design* are rules and guidelines to help a floral designer create a beautiful composition. The major principles of design are proportion and scale, balance, rhythm, and dominance. Other minor design principles include radiation, repetition, transition, variation, contrast, and focal point. Proper use of the art principles is essential in creating any kind of art, including arranging plant materials.

PROPORTION AND SCALE

Proportion is the principle of art that is the foundation of all the other principles. Good *proportion* means the pleasing relationship in size and shape among objects or parts of objects. *Scale* is a part of proportion, dealing with relative size only among things, not shapes. The finished floral design must not be too large or too small for its container. Most American florists use the Japanese Ikenobo rules of proportion when establishing the height of

14-9. The height of an arrangement should be at least 1½ times the container's greatest dimension.

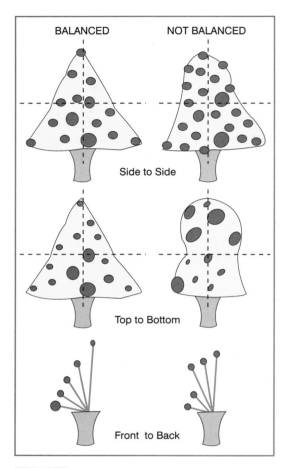

BALANCED   NOT BALANCED

Side to Side

Top to Bottom

Front to Back

14-10. Visual balance refers to balance from side to side, top to bottom, and front to back on both sides of the central axis.

the arrangement. Floral arrangements made in tall containers should be one and one-half to twice as tall as the container height. Arrangements made in low containers should be one and one-half to twice as tall as the container is wide.

BALANCE

Balance is a key part of the beauty of an arrangement. **Balance** is the physical or visual stability of a floral design. Balance refers to the arrangement's equilibrium or equality in weight, both physical and visual. Physical balance is the actual stability of plant materials within the container. A design with physical balance can stand on its own in a stable manner and not fall over.

Visual balance refers to the perception of an arrangement being in balance or being equal in weight on both sides of the central axis. The **central axis** is an imaginary central line running through the center of a floral arrangement. A floral design lacking balance is visually unsettling like a crooked picture or a shirt buttoned the wrong way. Poor visual balance in the floral design will overshadow other attractive aspects of the design, such as proper proportion or an effective center of interest.

Physical and visual balance must both occur for a design to be successful. An arrangement may be physically secure through good mechanics, yet lack visual balance. A beautiful visually balanced design may even topple over because the physical balance is poor.

Floral designs may be symmetrically or asymmetrically balanced. Visually locate the central axis of the arrangement. If each side of the central axis is equal in size and shape, then the arrangement has **symmetrical bal-**

14-11. Symmetrical balance is formal and dignified. (Courtesy, Dianne Noland, Illinois)

14-12. Asymmetrical balance is informal, creative, and dynamic. (Courtesy, Dianne Noland, Illinois)

ance. They seem to have the same physical weight. However, if the two sides are not equal, the arrangement has *asymmetrical balance.* Asymmetrical balance is informal and had its roots in Japanese and Chinese flower arranging. Floral designers often vary the use of both symmetrical and asymmetrical balance to create variety when creating different floral arrangements.

RHYTHM

Rhythm is related, orderly organization of the design elements to create a dominant visual pathway. The related aspects of the design (colors, shapes) tie the design together and gives it flow. Rhythm is "frozen motion" and suggests movement. A floral design with pleasing rhythm has continuity that invites the eye to view the entire design.

DOMINANCE

The effective use of dominance or emphasis lets the viewer know what is most important in the design. *Dominance* in floral design means that one

14-13. A focal point visually ties the entire arrangement together.

design element or characteristic is more prevalent or noticeable and other elements are subordinate to the main feature. Interest and attention are captured and held if one feature dominates and others are secondary in importance.

A center of interest is an important part of many design styles. A *focal point* visually ties the entire arrangement together. The location of the center of interest is usually centered in the lower part of the arrangement just above the container rim.

To create the most effective center of interest, always plan this dominant area before beginning the arrangement. In general, the greatest visual weights are concentrated at the center of interest. Some of the following methods of developing a center of interest or focal point—large flowers, special-form flowers, dark shades, use of framing foliage nearby, strong color contrasts, and accessories.

Some contemporary designs will actually have several focal points. Although many design styles have a focal point, some floral designs do not. Centerpieces and vase arrangements contain equal flower placements throughout the arrangement with no single area of focus.

THE DESIGN ELEMENTS

The *design elements* are the physical characteristics of the plant materials that a designer uses in a floral design. Thorough knowledge of the design elements of line and form (shape), space, texture, pattern, and color is so important because floral designers select and organize these "tools," the design elements, to create beautiful floral designs.

LINE AND FORM

Line is the visual movement between two points within a design. The four types of line are vertical, horizontal, curvilinear, and diagonal. All of these lines should appear to radiate from a central part of the design. Lines may be forceful or subtle, continuous, interrupted or implied.

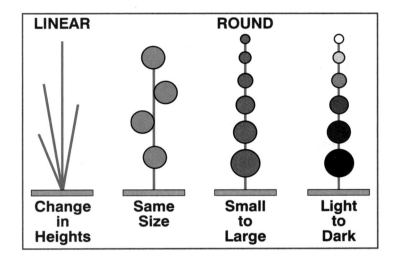

14-14. A line can be created with both linear and round plant material.

Line can be created not only with linear plant materials, but also with round flowers. Place round plant material in a line and in a progression from upward flower facings at the top to outward flower facings near the container. The round flowers can be all the same size and color or can be varied from light to dark, small to large, or dull to bright.

Line and form are closely related because appropriate lines are needed to create designs with beautiful form. *Form* is the three-dimensional shape of the outline of a floral design. *Shape* is the two-dimensional term for form. Floral arrangements are created using the two common geometric shapes, the circle and triangle. Floral design forms or shapes can be either one-sided or all-around (also called free-standing). Triangular floral designs will look like a symmetrical or asymmetrical triangle. Circular floral arrangements look like a mound made from flowers.

Most plant materials can be divided into four general categories based upon shape—line materials, form materials, mass materials, or filler flowers. The combination of these three types of plant shapes or forms lends variety to a design.

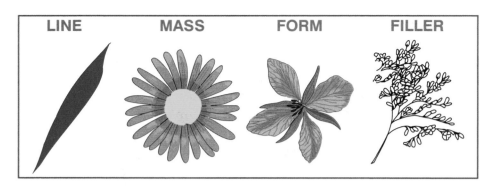

14-15. Plant material can be classified in three main shapes or forms—line, mass (round or form flowers), and filler flowers.

Line plant materials, such as snapdragons, larkspur, liatris, gladioli, and delphinium, heather, or foliages, such as eucalyptus, Scotch broom, or horsetail, are ideal to form the outline of an arrangement. *Greening* describes adding green foliage to a floral design. Linear plant materials are adaptable; line foliages can be curved, tied, wired, and trimmed to suit the arrangement; line flowers can be cut into smaller sections to use as fillers or mass flowers.

Form flowers, also called special form flowers, are flowers with distinctive shapes, such alstroemerias, freesia, iris, lilies, orchids, open roses, and most tropical flowers, such as bird of paradise, anthuriums, and proteas. Form flowers and materials create a striking center of interest and add uniqueness to a design.

Mass plant materials can be classified as round or solid flowers. Although mass types are usually flowers, some dried pods and cones also fit this category. Mass plant materials fill out the shape of a design. Examples of mass flowers are allium, chrysanthemums, carnations, daisies, and lotus pods.

Filler plant materials, *filler flowers* (also called "fillers"), are small flowers that are used to add texture, color, and depth as well as fill space between the mass and line flowers in a floral design. Examples are sea lavender, baby's breath, statice, lady's mantle, and golden aster.

SPACE

Another mark of an experienced floral designer is the generous incorporation of effective space within an arrangement. Space is an important design element in every arrangement, adding vitality and life. By creating meaningful open spaces, the solids are sharply defined; the beauty of separate plant

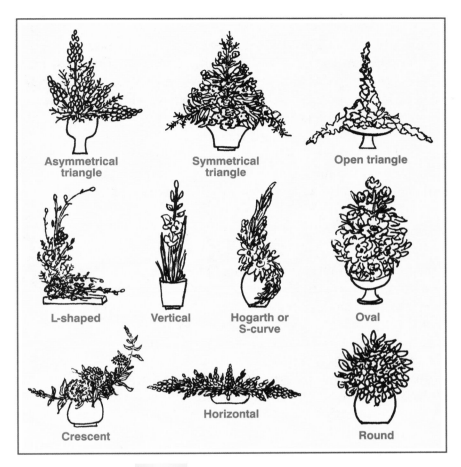

14-16. Basic floral design shapes.

materials can be fully appreciated. Beautiful flowers should be seen clearly without crowding.

Open space is inherent within many types of plant materials, such as ferns, iris, and clustered flowers like alstroemerias. Create space by incorporating branches and linear or unique shapes. Allow plenty of space near these materials for enjoyment of their interesting shapes and spaces. Most floral designs have more spaciousness at the top with progressively less space toward the center of interest.

TEXTURE

Texture is the design element that is directly related to the sense of touch. In plant materials, **texture** is determined by both the surface quality and the structure or placement of the plant parts in the design. Used wisely, interesting textures enhance and increase interest in a design. Plant materials that have texture appealing to the sense of touch, can be described as silky, satiny, feathery, smooth, rough, hairy, furry, woolly, velvety, or prickly. Plants also have many variations in their structure. This visual value can be described as airy or dense, delicate or coarse, lacy or solid.

The textures of the flowers and the container should be repeated to enhance each other. A glossy vase can be distracting if the flowers do not repeat the polished, shiny, or glossy texture. The background, fabrics, and accessories should also harmonize with the texture of the arrangement.

PATTERN

Pattern is determined by the physical characteristics of the plant materials. The arrangement of the leaves and petals create many structural patterns. A fern may have a repeated pattern, expressing precision. Other foliages, such as a saddleleaf philodendron, have an irregular leaf pattern and may seem informal or casual. Many plants have color patterns on their leaves. Pattern is closely tied to texture because leaves with color patterns appear textured or leathery when the surface is actually smooth. Color patterns in both foliage and flowers add interest to an arrangement. Color patterns are often described as mottled or variegated. Plant materials with color patterns should be used in small amounts as accents.

COLOR

Color is the most important element in the visual arts. Proper use of color, when designing arrangements, can help overcome minor problems of balance, form, and proportion. All colors can be created from the three *primary colors*—red, yellow, and blue.

Combining two primary colors creates the three *secondary colors*—green (yellow–blue), violet (blue–red), and orange (red–yellow). Colors give off a feeling of either warmth or cold. Red and yellow, the colors of fire, advance to the viewer and are called warm colors. Blue and green, the color of ice, move away from the viewer and are known as cool colors.

Hue is the name of a color. *Hue* refers to the pure color without adding white or black. White is added to a hue to make a *tint*. For example, adding white to red makes a pink tint. Black is added to a hue to make a *shade*. For example, adding black to red creates a burgundy shade.

Hues next to each other on the color wheel are called *analogous colors.* Analogous hues go well together. Combining three touching hues creates an analogous color scheme. Yellow, orange, and red hues create a beautiful analogous fall color scheme. Hues directly opposite each other, one primary the other secondary, on the color wheel are known as *complementary colors.* Complementary color schemes are always dramatic. The Christmas hues of red and green represent a vivid example of a complementary color scheme.

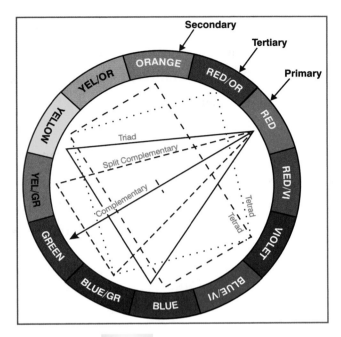

14-17. The color wheel.

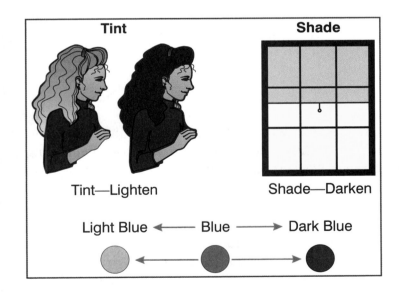

14-18. Tints are light values or a white added. Shades are dark values or a hue with black added..

TEN CARDINAL DESIGN RULES

1. Create the arrangement height with proper placement of the first flower

2. Work down from the top of the arrangement to the container

3. Place small flowers or buds at top of the arrangement

4. Place large flowers (open) near the container

5. Use light color flowers near the top

6. Use dark color flowers near the container

7. Place flowers at the top of arrangements farther apart

8. Place flowers near the container closer together

9. Start flowers at back of arrangement and work toward the front

10. Establish the focal point at the top edge of the container

FLORAL DESIGN EQUIPMENT AND TECHNIQUES

Knowing and using the correct tools and supplies is very important to the floral designer. Since new products and supplies are constantly being developed, a designer needs to stay current and look for new items that may make the job even easier.

CUTTING TOOLS

With experience, a floral designer will learn that a knife is the fastest and most efficient cutting tool to use. Any type of knife can be chosen depending upon the designer's preference. Pockets knives are convenient because they can be folded; knives that resemble paring knives are another option. Always keep the knife sharp for the best results.

Florist shears are a cutting tool with short, serrated blades for cutting thick or woody stems. Some tropical flowers and small branches can be easily cut with shears rather that a knife or scissors. Florist shears are preferable over ordinary scissors for cutting plant material because the blades will not pinch the stem. Pruning shears will cut very thick or tough branches with a smooth cut. Often, florist shears may not be able to cleanly cut through a big branch. When selecting pruning shears, be sure to select a type with a sharp cutting blade; always avoid the anvil-type, which cuts the stem by pinching it.

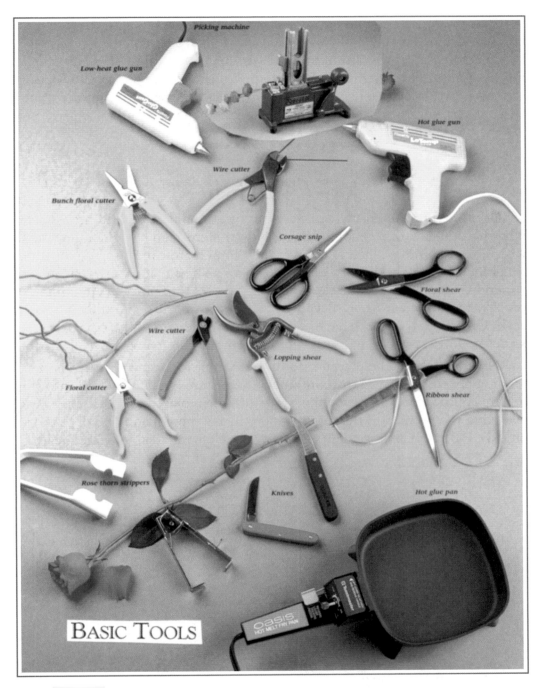

Picking machine

Low-heat glue gun

Hot glue gun

Wire cutter

Bunch floral cutter

Corsage snip

Floral shear

Wire cutter

Lopping shear

Floral cutter

Ribbon shear

Rose thorn strippers

Knives

Hot glue pan

BASIC TOOLS

oasis
HOT MELT FRY PAN

14-19. Basic tools for the designer. (Courtesy, Big Sky Pictures, Troy, Michigan)

Both ribbon scissors and utility scissors will be useful for a designer. Ribbon scissors are cutting tools with long, sharp blades designed for cutting ribbon, netting, or fabric. To maintain their sharpness, these scissors should not be used to cut anything else, including plant materials. Utility scissors are scissors that can be used for trimming leaves, paper, or other materials. Having a pair of utility scissors will allow the ribbon scissors to be used only for their intended purpose.

Supplies

Floral foam is a porous material designed to hold water and provide stability for stems within a floral design. Flower stems can be held at any angle when inserted into a block of floral foam. Floral foam comes in two formulations for both fresh and dry plant materials. The fresh foam is green and commonly available as a brick or a block. A brick of foam is approximately the size of a fireplace brick. Because of its porous nature, floral foam can be easily cut to any size or shape, depending on the container size. A florist knife or a heavy-gauge wire can be used to cut the foam.

The foam should be placed into the container with some of the foam extending above the rim of the container. Some stems may be placed into this top area of the foam. If the foam is too short, stem placements will look awkward and too upward; if the foam is too tall, the top portion will dry out too quickly. For small arrangements, leave a ½-inch foam area; for larger designs,

14-20. Wire mesh can provide support in vases and as a covering over foam.

one or two inches of foam can extend above the container. For foam placed into glass containers, the green foam can be camouflaged with silver foil or the appropriate color of cellophane. Corsage bags or other plastic can also be used to downplay the green color showing through the crystal vase. Always provide a small area to add water to the arrangement.

Other supplies that may be needed in design work are greening pins, water tubes, and wooden picks or hyacinth stakes. *Greening pins* look like hairpins with an "S" or flat top and are used to secure moss or foliage to a design. Short pieces of wire can also be bent in half to secure moss if greening pins are not available.

Water tubes are small, rubber-capped, plastic tubes used for holding water and a single flower or a small cluster of filler flowers. A flower stem is placed through the small opening in the rubber cap and into the small reservoir of the water tube. Water tubes come in various sizes, in green or clear colors, and are rounded or pointed. The rounded tubes may be used for boxed or packaged flowers that may be out of water for a while; the pointed water tubes are ideal for placing into floral foam. To extend the stem length of a flower in a design, a pointed water tube may be used alone to add a small amount of additional stem length. For greater stem length needs in a tall arrangement, the water tube can be wired to a wooden pick or attached to a hyacinth stake. Some water tubes are made to fit over a hyacinth stake; other water tubes can be wired, taped, or both to a hyacinth stake.

Floral tape is made of paraffin-coated paper that is used to cover wires and stems in a unobtrusive way. Floral tape does not feel sticky to human hands; only when the floral tape is stretched and pulled tightly will it adhere to itself. A beginning floral designer may find that pre-stretching the tape will help the taping to be more secure. Floral tape comes in two widths—½ inch and 1 inch. The narrower width is commonly used for corsage work; the wider width may be used for bouquets, wreaths, or other design work. Floral tape is offered in a wide array of colors, including light and dark green, tan, white, gray, pink, yellow, black, and red.

Florist wire is sold in 18 inch lengths by the gauge. The *gauge* of a wire describes its thickness or thinness. The smaller the gauge number, the thicker the wire. Florist wire comes in #18 to #30 gauge wire. Commonly used gauges are #26 for bows, #22 for medium-weight flowers, such as mini carnations, spray chrysanthemums, or daisies, #20 for heavier flowers, such as standard carnations, roses, or a fully open lily. Filler flowers may be wired with #28 wire.

Ribbon has its own terminology and differing widths. Ribbon is typically sold as #1, #1½, #3, #5, #9, #40. A thinner pixie ribbon is also offered.

14-21. A variety of vases and containers are suitable for basic design styles.

Use #1, #1½, and #3 ribbon when making corsage bows. Ribbons for flowering potted plants are #5 and #9, while #40 is used for bows in funeral sprays. Regular and wired-edge ribbon may come in satin, cotton, silk, sheer, paper, or burlap. Satin ribbon will have a shiny side and a matte or dull side. Some of the ribbons, including printed cotton, will have a definite front side and underside.

Learning the basic design styles will help build the skill and confidence for a floral designer to tackle more difficult styles. Some of the basic design styles include bud vases, vase arrangements, basic geometric shapes, and wrapping a potted plant. A variety of vases and containers can be used to create basic designs. Glass, pottery, plastic, ceramic, metals, and baskets can all provide an interesting foundation for basic designs. Bud vases are quite slender; vases for vase arrangements are taller and wider; basic geometric designs can be created in a multitude of sizes and shapes.

*F*LOWER WIRING TECHNIQUES

Wiring flowers is a very important entry level job in the flower shop. The beginner must master these techniques before moving up to more complex jobs in the flower shop.

Florist wire comes in 18 inch long pieces that vary in gauge from #18 to #30. The smaller the wire gauge number the larger the wire diameter will be. Larger gauge wire sizes, #18, #20, and #22, are usually used to add support to flower stems for funeral bouquets and other fresh flower arrangements. Smaller gauge wire sizes, #24–#30, are used for corsage and wedding bouquets.

Stronger flower stems today require less wiring than those stems of a few years ago. Wrapping wire around a stem can provide the extra support necessary to hold the flower in the proper place in an arrangement. A floral tape is used to cover the unattractive wire stems. Depending on the kind of flower and how it will be used determines which of the four major wiring techniques to use—pierce, hook, clutch, or splinting. Wiring is very important when constructing a corsage, boutonniere, or wedding bouquet.

Pierce Method

The pierce method involves inserting a wire through the calyx of a flower. It is effective for flowers that have a thick calyx just below the flower head, such as a rose or carnation.

■ Use a #20 or #22 wire

■ Cut the stem ½ inch below the calyx

■ Hold the wire close to the end and push it through the side of the calyx. Continue pushing the wire until half the wire is through the calyx

■ Bend the wire ends down and create an artificial stem, then cover the wire with floral tape

14-22. Pierce wiring method using a carnation.

Daisy Hook Method

The daisy hook method (or hook method) is a technique of inserting a wire through a flower stem and the flower itself and making a hook at the end of the wire and pulling it down into the flower. This technique will securely wire flowers that have a hard central disc without a prominent calyx, such as daisies

14-23. Hook wiring method using daisy mums.

and chrysanthemums. Other individual florets of delphinium, dendrobium orchids, and hyacinths can also be wired in this manner, using thinner gauge wire (#26 gauge). When constructing a corsage or boutonniere made of daisies, the hook wiring method is used.

- Use a #22 or #24 wire
- Cut the stem 1 inch below the calyx
- Insert the wire up the stem until two inches of wire are above the flower petals
- Bend the top end of the wire over a knife to form a hook and gently pull this hook into the flower
- Start wrapping floral tape just under the flower petals and continue down the stem and wire

14-24. Wrap-around wiring technique using baby's breath.

Wrap-Around Method

This technique works best to make a small cluster of filler flowers for use in corsages and wedding bouquets. The wire encircles the stems and is bent downward to form a new stem. Fine textured foliages, such as plumosa fern, tree fern or ming fern, can also be held together effectively with this technique.

- Use #26 or #28 wire
- Cut several filler flower stems 1 to 1½ inches long
- Hold the flowers in one hand and wrap the wire around the cluster of flower stems with the other hand
- Finish by covering the stems and wire with floral tape

Splinting Method

Splinting helps to support weak flower stems when designing a fresh floral arrangement. This technique is also used when making wedding bouquets. Rose and carnation flowers usually need the extra support provided by this splinting method of wiring.

- Use #18 or #20 wire

- Insert ½ inch of the wire into the flower calyx and spirally wrap the wire down the stem

- At the end of the stem turn the wire and wrap spirally up the stem

- For wedding bouquets cover the wire with floral tape first or secure with ribbon

- Hand-held bouquets have the flowers inserted directly into the oasis of the bouquet holder

EVERLASTING FLOWERS

Today, the word flower doesn't necessarily mean fresh, living plant material. During the early 1950s, companies manufactured imitation flowers made from polyethylene plastic. Plastic flowers quickly became known as cheap imitations of the real thing and their use declined. In the 1970s, florists began to create floral arrangements using artificial flowers made from polyester fabric. The name *silk flower* describes these flowers, although they are not made from silk fabric. The term permanent flower became popular during the 1980s to describe silk flowers and the new and improved plastic flowers.

While naturally dried flowers and weeds were used by early settlers, it was not until the 1980s that they were rediscovered and came back into fashion. Most florists refer to this group as *everlasting flowers*.

Designers of dried or silk arrangements must employ the same principles and elements of design that are used with fresh flowers. However, some of the mechanics and specific techniques are slightly different. Designing with dried and silk flowers gives much flexibility to the designer. Stems can be

14-25. Designing with everlasting flowers gives the designer much flexibility. Many styles and containers (including those that do not hold water) are possible with dried flowers. (Courtesy, Dianne Noland, Illinois)

lengthened or manipulated into various shapes. Materials may be glued, wired, and taped in a wide variety of ways that are not possible with fresh materials. A diverse range of containers can be used, including those that do not hold water. In many ways, designing with dried or silk flowers is much easier than with fresh flowers; because, wilting is no longer a concern.

FLORAL DESIGNS FOR SPECIAL OCCASIONS

Floral arrangements are an important part of many holidays and special occasions. They may set the mood or theme for a holiday, or provide a way

14-26. Examples of floral designs for special occasions.

for people to express their thoughts and feelings. Valentine's Day, Easter, Mother's Day, Thanksgiving, and Christmas are major holidays when flowers are often used. Weddings, funerals, anniversaries, birthdays, school proms and homecomings, and banquets are special occasions where marketing opportunities exist for floral shops.

REVIEWING

*M*AIN IDEAS

Flowers have been used by people throughout the world from the beginning of recorded history. The Egyptians used flowers to beautify their home, as did the Greeks over 5,000 years ago. The Greeks, however, did not put flowers in containers but spread flower petals on the ground or made floral wreaths to wear or carry.

The oriental cultures developed entirely different design techniques from the Egyptians and Greeks. Oriental designs use a lot of linear asymmetrical balance in their flower arrangements. The Ikenobo school of floral design was started hundreds of years ago in Japan and even today Ikenobo design schools teach the same principles of floral design made famous years ago.

While many people like fresh living flowers, others enjoy permanent flowers made from silk, plastic, or naturally dried flowers. These permanent flowers allow the floral designer greater latitude and allow the use of unique containers and unusual designs.

Fresh flowers must be properly conditioned to increase their vase life. Always make a fresh cut on a stem before putting the flowers in warm water. Using a floral preservative in the flower water will also increase the vase life of the flowers.

Well designed floral arrangements can be created by following the basic art principles of design.

Four major flower wiring techniques are described. The pierce, daisy hook, wrap-around, and the splinting wiring methods are used to provide extra support necessary when using flowers.

*Q*UESTIONS

Answer the following questions. Use complete sentences and correct spelling.

1. What civilization started using flowers?

2. What civilization started using the flower garland?

3. What country developed the Ikenobo floral design style?

 4. How do oriental and American designs differ?

 5. When did plastic flowers first become popular in the United States?

 6. How hot should water be in flower buckets?

 7. Why should flower buckets be kept clean?

 8. What are analogous and complementary colors?

 9. Why should flower preservatives be used when conditioning flowers?

10. Where in a floral arrangement should light-colored flowers be placed?

11. What different wire sizes are used by a florist?

12. What is the difference between circular and freestanding flower arrangements?

13. Florist usually create a symmetrical triangular style arrangement for what occasions?

14. What flowers need the pierce wiring technique?

15. What flowers need the wrap-around wiring method?

EXPLORING

1. Visit a flower shop in your area. Talk with the florist about job requirements.

2. Practice identifying different wire sizes by feel.

3. Practice wiring flowers using the pierce, hook, wrap-around and splint methods.

15

INTERIORSCAPING

Plants are used inside of buildings to make the environment more appealing. Open areas in lobbies and along hallways may have large masses of carefully selected plants. Areas in offices and near entrances may have single plants or small clusters. Choosing the plant material is an important first step.

Species of plants are selected to achieve the desired effect and on the basis of suitability. Some plants grow well in interior environments. Knowing the growth and environment needs of plants helps in making selections. Providing proper care is essential in assuring that the plants are attractive and long-lasting.

15-1. Plants are selected based on their ability to withstand the growing conditions. (Courtesy, Society of American Florists)

OBJECTIVES

This chapter presents important principles of interiorscaping, including plant environments, plant care, and species characteristics. It has the following objectives:

1 Describe good indoor plant growing conditions

2 Differentiate between plant growth and plant maintenance

3 List foliage plants by light tolerance levels

4 List insect pests found on foliage plants

5 Describe why soil amendments are added to a growing medium

6 Explain plant acclimation

7 Identify cultural and biological pest control strategies

8 Describe growing plants in a terrarium

TERMS

acclimated
acute plant problem
atrium
chilling injury
chronic plant problem

foot candles (FC)
humidity
interiorscaper
plant growth
plant station

INTERIORSCAPE

Interiorscaping uses foliage plants inside buildings to create the feeling of outdoors. An ***interiorscaper*** is a person who designs and creates pleasing and comfortable areas inside buildings using plants.

Interiorscaping began in this country in the early 1970s by using plants to create a pleasing environment for office workers or mall shoppers. Plants help define space in buildings and offices. They have become a status symbol for many people. People relate growing plants to a clean, good environment. Plants keep people in touch with the environment.

Plants develop living screens to separate space in office buildings. They can filter the air and reduce noise in an office area. Aesthetic plant qualities should not be underestimated, plants are beautiful and people appreciate this beauty. Skilled designers create outstanding indoor gardens using hundreds of different kinds of plants. Designers use the term ***plant station*** to define an area where a plant will look better than any other object.

Today many architects are designing buildings with special rooms designed just for growing plants called ***atriums.*** True indoor gardens can be created in these "green rooms." Shoppers in the winter consider the shopping mall atrium as a "breath of spring" which relaxes them.

15-2. Foliage plants are used to create a pleasing environment inside buildings.

PLANT ENVIRONMENT

TEMPERATURE

Most foliage plants belong to the tropical plant group. This group of plants will die if exposed to cold or freezing temperatures. *Chilling injury* can damage tropical plants in the 35 to 45°F temperature range, well above freezing.

Acute plant problems occur suddenly and cause immediate damage. Chilling injury from cold drafts, burning of plants with smoking materials, or improper application of chemicals are all examples of acute plant stress. However, *chronic plant problems* develop over a long time and the plant slowly declines in vigor. Low light intensity, over-watering, and nutrient deficiencies are all examples of chronic plant problems.

Plants placed close to a door or window can be damaged from cold drafts in the winter. The remedy may be in selecting hardier plants or in moving the plants away from the draft. Air conditioner vents directed on plants can sometimes damage plants by directing very cold air on sensitive plants. Use a recording thermometer to help isolate this condition. Slow plant growth or no growth may suggest chronic cold injury. Sometimes powdery mildew is associated with cold temperature damage.

High temperature problems, while rare, can exist. Plants placed too close to fireplaces, windows, or heat vents are subject to high temperature injury. Moving the plant or replacing it with a hardier variety usually solves this

15-3. An Anthurium before (left) and after (right) exposure to cold temperature. (Courtesy, Jasper S. Lee)

problem. Interiorscape plants grow best when the day temperature ranges between 70 and 80°F and a night temperature of 60–65°F.

HUMIDITY

Humidity is the amount of moisture in the air. In the rain forests where tropical plants grow naturally, humidity levels rarely drop below 80 percent. However, in air conditioned environments the humidity level usually runs from 40 to 65 percent. During the heating season, 15 and 20 percent relative humidity levels are very common. People and plants both find this humidity range to be uncomfortable.

Plants respond to low humidity levels by not producing any new leaves or only forming very small leaves. These leaves may become brittle or have burnt leaf edges. Meters are used to measure the relative humidity in the air around the plants. The goal is to maintain a humidity level of 60 percent or higher. Good watering practices will also help reduce plant damage from low humidity. Reducing air movement around plants can also help reduce damage.

LIGHT

Light determines the rate of plant growth and controls many other related plant processes. Light is the most important environmental condition related to plant growth. Photosynthesis combines water and carbon dioxide in the presence of light to produce sugar (plant food). The production of chlorophyll, the substance that makes leaves green, requires light. Chlorophyll also plays an important role in photosynthesis.

On a bright sunny summer day 10,000 foot candles (FC) of light reach the surface of the

15-4. Plants grown under low light conditions have long, thin stems.

15-5. Chinese evergreen, *Aglaonema* (left), and Prayer plant, *Maranta* species (right), both tolerate low light conditions.

earth. A ***Foot candle (FC)*** is the amount of light cast by a single candle on a square foot of surface. Inside office buildings, shopping malls, and homes, light levels will range from 50 to 150 foot candles of light. Plant growth requires high light intensity. Many foliage plants are native to the tropical rain forests of South America. These plants in their native environment receive around 1,000 foot candles of light. Foliage plants with colored foliage require higher light intensities than plants with only green leaves.

   Plant growth can be defined as plants increasing in size by producing new leaves and stems. Plants may exist for one or two years at much lower light intensities, but they do not grow. New leaves and branches are not produced but the plant exists without growing or dying. This condition is called

15-6. *Dracaena deremenisis* tolerates low light while *Dracaena fragrans* and *Dracaena marginata* tolerate medium light. Three different types of *Dracaenas* are shown in the foreground.

Table 15-1. Growing Guidelines for Selected Foliage Plants

	Light	Water	Humidity	Problems
Asparagus fern *Asparagus sprengeri*	Medium 2,500–4,500 FC	Let dry slightly	Medium	Spider mites
Baby's tears *Helxine soleirolii*	Medium 2,000 to 4,000 FC	Keep moist	High	
Begonias *Begonia* species	Medium 2,500 to 5,000 FC	Let dry slightly	Medium	Mites
Brassaia (Schefflera) *Brassaia actinophylla*	Medium 3,000 to 5,000 FC	Let dry slightly	Medium	Mites, Mealybugs, Scale, Thrips
Bromeliad *Neoregelia* species	Medium 2,500 to 4,000 FC	Let dry slightly	Medium	
Cacti Many species	High 4,000 to 8,000 FC	Let dry	Low	
Cast iron plant *Aspidistra elatior*	Low 1,500 to 3,000 FC	Let dry slightly	Medium	
Chinese evergreen *Aglaonema* species	Low 1,000 to 2,500 FC	Keep moist	Medium	Mealybugs, Scale
Croton *Codiaeum* species	High 3,000 to 8,000 FC	Let dry slightly	Medium	Mites, Mealybugs, Scale
Dracaena *Dracaena* species	Medium 2,000 to 4,000 FC	Let dry slightly	Medium	Mites, Mealybugs, Thrips
Dumbcane *Dieffenbachia* species	Low 1,500 to 2,500 FC	Let dry slightly	Medium	Mites, Mealybugs, Scale
English ivy *Hedera helix*	Low 1,500 to 2,500 FC	Let dry slightly	Medium	Mites, Scale
False aralia *Dizygotheca elegantissima*	Medium 2,000 to 4,000 FC	Let dry slightly	Medium	Mites, Mealybugs, Scale
Ferns Many species	Low 1,500 to 3,000 FC	Keep moist	High	Scale, Caterpillars, Mealybugs
Figs *Ficus* species	High 4,000 to 8,000 FC	Let dry slightly	Medium	Mite, Mealybugs, Scale, White fly
Fittonia *Fittonia verschaffeltii*	Low 1,000 to 2,500 FC	Let dry slightly	High	Mealybugs
Grape ivy *Cissus rhombifolia*	Low 1,500 to 2,500 FC	Let dry slightly	Medium	Mites
Jade plant *Crassula argentea*	High 6,000 to 8,000 FC	Let dry	Low	Mealybugs, Scale
Mother-in-law's tongue *Sansevieria* species	Adaptable 1,500 to 6,000 FC	Let dry	Low	

(Continued)

Table 15-1 (Continued)			
	Light	**Water**	**Humidity**	**Problems**
Nephthytis *Syngonium podophyllum*	Low 1,500 to 3,000 FC	Let dry slightly	Medium	Mites, Mealybugs
Norfolk Island pine *Araucaria heterophylla*	High 4,000 to 8,000 FC	Let dry slightly	Medium	Mites
Palms Various genus	Low-medium 3,000 to 6,000 FC	Keep moist	Medium	Mites, Mealybugs, Scale
Peace lily *Spathiphyllum* species	Low 1,500 to 2,500 FC	Let dry slightly	Medium	Mealybugs, Scale
Peperomia *Peperomia* species	Low 1,500 to 3,000 FC	Let dry slightly	Medium	
Philodendron *Philodendron* species	Low 1,500 to 3,500 FC	Let dry slightly	Medium	Mites, Mealybugs, Scale, Thrips
Pilea Many species	Low 1,500 to 2,500 FC	Keep moist	Medium	
Pothos *Epipremnum* species	Low 1,500 to 3,000 FC	Let dry slightly	Medium	Mealybugs, Thrips
Prayer plant *Maranta* species	Low 1,500 to 3,500 FC	Keep moist	High	Mites, Mealybugs
Purple passion vine *Gynura sarmentosa*	Medium 3,000 to 5,000 FC	Let dry slightly	Medium	
Schefflera *Schefflera arbicola*	Low 1,500 to 3,000 FC	Let dry slightly	Medium	Mites, Mealybugs, Scale, Thrips
Spider plant *Chlorophytum comosum*	Low 1,000 to 2,500 FC	Let dry slightly	Low	
Strawberry begonia *Saxifraga sarmentosa*	Medium 3,000 to 5,000 FC	Let dry slightly	Medium	
Swedish Ivy *Plectranthus australis*	Medium 3,000 to 5,000 FC	Keep moist	Medium	Mealybugs
Ti plant *Cordyline terminalis*	Low 1,500 to 3,500 FC	Keep Moist	Medium	
Wandering Jew *Zebrina pendula*	Medium 2,500 to 4,000 FC	Let dry slightly	Medium	
Wax plant *Hoya carnosa*	Low 1,500 to 3,000 FC	Let dry	Medium	Mealybugs
Yucca *Yucca elephantipes*	Medium 3,000 to 5,000 FC	Let dry	Low	
Zebra plant *Aphelandra squarrosa*	Low 1,000 to 1,500 FC	Keep moist	Medium	Mites

Source: R. J. Biondo and D. A. Noland. *Floriculture: From Greenhouse Production to Floral Design.* Danville, Illinois: Interstate Publishers, Inc. 2000.

plant maintenance. Light intensities of 50 to 100 FC are adequate for plant maintenance.

Natural sun light will only provide usable light for a maximum of 15 feet from a window. Artificial light sources must be used when little or no natural light is available. Light intensity is measured with a meter that reads light in foot candles (FC). Light intensity cannot be determined by walking through a building or office. Taking light intensity measurements provides the interior plantscaper important information about existing light conditions.

ARTIFICIAL LIGHT SOURCES

Light bulbs, called incandescent lights, emit light in the red spectrum. Plants growing under incandescent lights usually develop small spindly leaves. Light bulbs used by themselves do not provide a good light source for growing plants. Incandescent lights may also produce large amount of heat that can burn plant leaves.

Fluorescent lights are cooler, have a wider range of light wavelengths, and last longer. Fluorescent lights provide the quality of light necessary for good plant growth. Although several different kinds of fluorescent lamps are available today, the cool white lamps will provide the best artificial light source. Light intensity diminishes rapidly as the distance between the lights and plant foliage increases. Plants need to be kept close to the light source.

GROWING MEDIA

Plant roots are out of sight and easily forgotten. Unless a plant has a quality root system it cannot do well, even in a perfect environment. A good media helps produce a quality root system. Today, most plants are not grown in soil but in an artificial material called a soilless growing medium.

The ideal growing medium should have the following qualities:

- Be well-drained
- Have good moisture holding capacity
- Have good aeration
- Provide support for the plant in the container

Growing media contains both solid particles and pore spaces (air holes between media particles). Growing media pore spaces fill with water the same way a sponge soaks up water. By volume, solid particles should take up

15-7. A commercially prepared growing medium is being used for these plants. (Courtesy, Ron Biondo, Illinois)

half the media space. The other half of the media should be pore spaces. These pore spaces can be filled with air when the soil is dry or filled with water when the soil is wet.

Ideal growing media will have half the pore spaces filled with water and the other half still dry after watering. Good soil will not drain well and provide enough air filled pore spaces for good plant growth. Mixing soil amendments with soil can increase the soil's ability to drain. Soil amendments are materials added to soil to improve its composition. Many growers use a growing media that contains no soil only artificial amendments.

Peat moss is a very popular organic amendment added to growing media. There are many kinds of peat moss, but only the sphagnum-type peat moss should be used for growing foliage plants. Peat moss can hold large quantities of water and also provide the plant's root system good aeration necessary for growth.

Inorganic soil amendments include perlite and vermiculite. Perlite is a heat-treated lava rock. It is light-weight with low nutrient and moisture holding capacity. It increases aeration to the medium. Perlite is dusty and can float to the surface of the growing medium making a mess after watering plants. Vermiculite is a heat-mica that has high nutrient and moisture holding capacity. It also provides good pore space for air in the growing medium. It should only be used with organic matter. Several different grades of vermiculite are available.

Growers mix several soil amendments together to create their own growing medium or purchase a commercially available premixed medium. Following are a few of the commercial premixed brands on the market—Jiffy mix, Promix, and Metro Mix. While each is different, they all provide a suitable artificial growing media for plants.

FERTILIZATION

Indoor plants do not need large quantities of fertilizer. They are not actively growing, but existing. They should not be over stimulated with fertilizer. A liquid fertilizer solution should be applied when normally watering indoor plants. Fertilize plants in the spring and summer, not in the winter. The fertilizer should contain nitrogen, phosphorus, and potassium elements in equal parts, such as: 15-15-15 or 20-20-20 fertilizer analysis.

WATERING

Over-watering of indoor plants is common, both by the homeowner and the professional. Under-watering, sometimes called neglect, is less likely to occur. Over-watering usually occurs when small plants are being grown in large containers. The growing medium simply does not have time to dry between watering which can cause root diseases to develop.

Most root diseases develop when the growing medium remains wet. These diseases can injure the root system and

15-8. Inserting a fertilizer spike into the growing medium of this plant provides nutrients for several weeks.

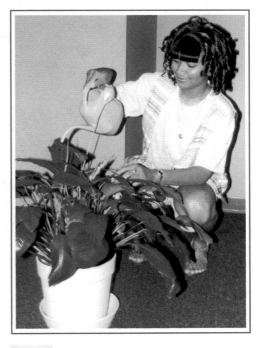

15-9. Indoor plants need more water in the spring and summer and less in the winter. (Courtesy, Jasper S. Lee)

cause plant death. A moisture meter with probe should be used to determine when plants need watering. Indoor plants need more water in the spring and summer and less in the winter. Plants should be checked every three to four days and watered as needed.

ACCLIMATION

In Florida, growers aggressively produce foliage plants under high light conditions and constant fertilization. These plants need to become accustomed or **acclimated** to lower light levels found in homes or commercial office buildings. Plant leaves must adapt to the new environmental conditions. This change may take several weeks to several months to complete. Heavy leaf drop may occur during this acclimation period. Slowly changing the plant environment from high light levels to low light levels will acclimate plants. Growers use large warehouses or shaded greenhouses to provide the conditions necessary for plant acclimation.

15-10. Growers acclimate plants in the greenhouse before they are sold to customers. (Courtesy, Cheryl Schroeder, Illinois)

INTEGRATED PEST MANAGEMENT (IPM)

Integrated Pest Management (IPM) uses cultural, biological, and chemical strategies to control plant pests. In the past, the weekly application of pesticide chemicals controlled most plant insect and disease problems. To-

Table 15-2. Disease Problems on Selected Foliage Plants	
Disease	**Plants Usually Attacked**
Leaf spot	Dracaena Palm
Crown and root rot	African violet Norfolk Island pine Philodendron Pothos Schefflera
Virus	Dumb cane Piggy back plant Schefflera
Powdery mildew	African violet Begonia Jade plant

day, interiorscape managers employ cultural and biological strategies along with pesticides to control pests.

Growing plants in artificial media can reduce plant root disease problems. Prevent root diseases by using a moisture meter to decide watering schedules. Improving air circulation around the plants will also prevent damage from leaf diseases. Remove and destroy insect and disease infested plants from the interiorscape. Using IPM can reduce much of the routine chemical use of the past.

Several destructive plant insects can be controlled by beneficial insects. Good bugs eat the bad bugs. Plant breeders have also developed foliage plants that have a natural immunity from many diseases. Biological control means using one living organism to control another living organism (the pest). These biological strategies can control many interiorscape plant problems.

Pesticide chemicals still have a place in controlling pests on foliage plants. Apply fungicides to prevent or control plant diseases caused by fungi and insecticides to control insect pests on plants. They should only be applied after other strategies have failed to control the pest. All pesticides can be dangerous to humans and other animals. Read and follow all label directions when using pesticides. Wear all required protective clothing shown on the pesticide label.

POPULAR INDOOR PLANTS

15-11. African violet.

AFRICAN VIOLET

Botanical name: *Saintpaulia ionantha*

Description: Popular flowering foliage plant grown for its variety of beautiful white, pink, blue, purple, or red flowers. The thick, dark green leaves usually are covered with hair and grow in a rosette pattern from the center of the plant.

Light: Requires 300 FC of light, but not direct sunlight.

Water: Keep foliage dry when watering. Allow plant to dry between watering.

Pests: Aphids, crown rot, botrytis, and powdery mildew.

Other: To keep African violets growing, do not water with cold water and only fertilize when in bloom.

15-12. Amaryllis.

AMARYLLIS

Botanical name: *Hippeastrum* sp.

Description: Beautiful lily-like flowers in red, pink, orange, white, or striped colors. Buy dormant bulbs in mid winter and pot at once. Correct pot size will allow 1 inch of space between pot and bulb. Plant bulb in pot with top half of bulb out of the soil.

Light: Place pot in moderate sun until leaves develop, then move into full sun and turn occasionally to keep plant from leaning toward light.

Water: Keep soil on the dry side until leaves develop. Increase the amount of water as the leaves and flowers develop.

Pests: No major insect or disease pests.

Other: In fall, place pot in a cool dry location and stop watering until January. Bring into warm room to reflower the amaryllis each year.

ASPARAGUS FERN

Botanical name: *Asparagus densiflorus* 'Sprengeri'

Description: A relative of the edible asparagus but not a true fern. Needle-like foliage with sharp spines on the stems. Good in hanging baskets.

Light: Bright, indirect light.

Water: Allow to dry between watering.

Pests: Spider mites, occasionally aphids.

Other: Leaf yellowing and loss can occur from growing in low light conditions or too much or too little water.

15-13. **Asparagus fern.**

BABY RUBBER PLANT

Botanical name: *Peperomia obtusifolia*

Description: Waxy dark green leaves that are 2 to 3 inches in diameter and almost round. Good plant for dish gardens and terrariums. Over 100 different cultivars of Peperomia, some with white leaf margins.

Light: Bright, indirect light.

Water: Keep dry.

Pests: Spider mites and mealybugs.

Other: Leggy growth will occur under low light conditions.

15-14. **Baby rubber plant.**

BANANA

Botanical name: *Musa acuminata*

Description: Large herbaceous perennial with very large dark green leaves. Stem will produce banana in seven years, then stem dies. New stems develop at base of old stem.

Light: Full sun.

Water: Stores water in large fleshy root and stem. Do not over water.

Pests: Spider mites and scale insects.

Other: Plant outdoors in the summer around pools, take inside in the fall.

15-15. **Banana.**

15-16. **Bird of paradise.**

BIRD OF PARADISE

Botanical name: *Strelitzia reginae*

Description: The bird-of-paradise is one of the most exotic looking flowers. The large flowers are brilliant orange and red with tongues of vivid blue. In warm climates it is gown outdoors as a landscape plant. Very easy to grow indoors.

Light: Needs full sun, place in front of south window.

Water: Pot in well drained soil mix and keep wet.

Pests: No major insect or disease pests.

Other: Both the flowers and the foliage of "birds" are used by florists to create exotic floral arrangements.

15-17. **Boston fern.**

BOSTON FERN

Botanical name: *Nephrolepis exaltata*

Description: Ferns are usually sold in hanging pots which show the delicate cascading fern fronds. Spore bearing capsules appear as brown spots on underside of leaf at certain times of the year.

Light: Filtered light, never direct sun. Minimum of 75 FC required. Gradual yellowing of foliage suggests too low of light conditions.

Water: Keep soil moist, but do not over water.

Pests: Spider mites.

Other: Fronds turn brown when handled by people. Do not touch! Do not over fertilize.

15-18. **Bougainvillea.**

BOUGAINVILLEA

Botanical name: *Bougainvillea* sp.

Description: This tropical flowering vine or bush makes an excellent house plant. The true flowers are inconspicuous, but they are surrounded with spectacular red, pink, purple, or orange petal-like bracts. Blooms develop in late winter and continue for many weeks.

Light: Full sun in the winter.

Water: Keep thoroughly moist and avoid moisture stress.

Pests: Mealybugs.

Other: Move pot outdoors in the summer into a mostly sunny location. Place plant in a cool dry location in the fall and let the plant go dormant. After Christmas, bring into a sunny window and start watering.

CACTUS

Botanical name: *Cercus* sp.

Description: Grows as a column with 6 or 8 prominent scalloped ribs. White funnel-shaped flowers only open at night.

Light: Full sun.

Water: Keep dry.

Pests: Scale insects.

Other: Pot in sandy soil and maintain cool night temperature (55°F). Over watering will induce root rot.

15-19. **Several cactus species.**

CALADIUM

Botanical name: *Caladium* sp.

Description: Arrow-shaped leaves 6 inches to 2 feet long. Various patterned with white, pink, or red leaves with small calla-like white flowers. Grown from bulb-like tubers.

Light: Moderate sun.

Water: Plant in well drained soil, moderate moisture.

Pests: Tobacco mosaic virus.

Other: Must have a rest period each year. When leaves wither, remove tuber from soil and store in a dry warm place for three or four months. Repot and grow at 70°F.

15-20. **Caladium.**

CHINESE EVERGREEN

Botanical name: *Aglaonema* sp.

Description: Oval shaped leaves, with patterns of green, gray, and cream colors. Mature height of 2 to 5 feet. Popular foliage plant, easy to grow.

Light: Will exist under very low light intensities. Can survive 10 to 15 FC light conditions.

Water: Sensitive to over watering. Let media dry between watering.

Pests: Aphids, mealybugs, spider mites, and root rot.

Other: Day temperatures between 70 and 80°F. Sensitive to cold drafts.

15-21. **Chinese evergreen.**

15-22. Christmas cactus.

15-23. Corn plant.

15-24. Croton.

CHRISTMAS CACTUS

Botanical name: *Schlumbergera bridgesii*

Description: Flattened, green, and branching stems without leaves produce food by photosynthesis. Cherry red colored flowers form at the end of each stem around Christmas each year.

Light: Full sun.

Water: Keep dry.

Pests: Spider mites and scale insects.

Other: Good hanging basket plants. Over watering can cause root rot.

CORN PLANT

Botanical name: *Dracaena fragrans*

Description: Several large canes are usually planted in a single pot. It has narrow dark green leaves, 2 inches wide and 2 feet long, that resemble corn leaves.

Light: Can survive under low light conditions less than 50 fc.

Water: Keep thoroughly moist and avoid water stress.

Pests: Spider mites.

Other: Several varieties available with different leaf coloration.

CROTON

Botanical name: *Codiaeum variegatum*

Description: Leathery leaves may be curly or flat. Many irregular shaped leaves vary from mitten-shaped to ribbon-like. Leaves have many colorful patterns of green, brown, yellow, red, and orange.

Light: Needs 150 FC or more. It will lose bright leaf color under low light conditions.

Water: Keep thoroughly moist. Avoid water stress.

Pests: Spider mite problems can cause leaf drop.

Other: Avoid placing crotons in drafts. Many different cultivated varieties.

DUMBCANE

Botanical name: *Dieffenbachia* sp.

Description: Develops a large, thick trunk called a cane. Variegated green and cream colored leaves.

Light: Filtered sunlight 100 to 150 FC needed for plant maintenance.

Water: Allow soil to dry between watering.

Pests: Mosaic virus, stem rot, and mealybug problems.

Other: Calcium oxalate in the sap causes inflammation of skin. If eaten may cause temporary numbness or inability to speak.

15-25. Dumbcane.

FALSE ARALIA

Botanical name: *Dizygotheca elegantissima*

Description: Very long, narrow, dark greenish–brown compound leaves. Usually grown as a single stem plant with no branching.

Light: Provide a minimum of 150 FC of light.

Water: False aralia sensitive to drought and wet soil conditions. Lower leaves can fall off if over watered or under watered.

Pests: Spider mites and mealybug problems.

Other: If temperature falls below 60°F, lower leaves will fall off.

15-26. False aralia.

GOLD DUST TREE

Botanical name: *Aucuba japonica 'Variegata'*

Description: Leathery leaves have a high gloss with small blotches of butter yellow. Inconspicuous purple flowers.

Light: Moderate shade.

Water: Moderate moisture.

Pests: No serious pests.

Other: Can be grown outdoors in warm climate.

15-27. Gold dust tree.

15-28. Heart-leaf philodendron.

15-29. Kentia palm.

15-30. Mother-in-law's tongue.

HEART-LEAF PHILODENDRON

Botanical name: *Philodendron scandens*

Description: A vine with dark green, 6-inch long, heart-shaped leaves.

Light: Tolerates low light conditions less than 50 fc. However, does better in 75 to 100 fc.

Water: Allow soil to dry between watering. Can grow in water, no soil.

Pests: Leaf spot, mosaic virus, tip burn, and spider mites.

Other: Very popular, easy to grow, foliage plant. Grows best if soil is kept on the dry side.

KENTIA PALM

Botanical name: *Howea fosterana*

Description: Tough, leathery leaflets on slow-growing fronds. One of the most widely available palms.

Light: Bright indirect light. Foliage will burn in direct light.

Water: Keep soil evenly moist, allow to dry between watering.

Pests: Spider mites, occasionally scale insects, and root rot.

Other: Never allow to completely dry out. Keep out of drafts.

MOTHER-IN-LAW'S TONGUE OR SNAKE PLANT

Botanical name: *Sansevieria trifasciata*

Description: Stemless plant having long sword-shaped leaves growing out of the soil. Leaves may be dark green or variegated.

Light: Will tolerate low light levels, less than 50 fc.

Water: Do not over water, but difficult to under water. Can tolerate neglect.

Pests: Rots and nematodes.

Other: Several different varieties, one has a birdnest-form instead of long leaves.

NORFOLK ISLAND PINE

Botanical name: *Araucaria heterophylla*

Description: Looks like a miniature pine tree. Very slow growing, usually only 6 to 8 feet tall when grown indoors.

Light: Needs 150 FC of light. Keep from direct sunlight.

Water: Let soil dry between watering.

Pests: Mealybugs, scale insects, and spider mites.

Other: People touching plant can injure sensitive foliage, place away from people.

POINSETTIA

Botanical name: *Euphorbia pulcherrima*

Description: Most popular Christmas flowering pot plant. Brought to the United States from Mexico by Joel Poinsett in the late 1800s. Small yellow flowers are surrounded by red, white, pink, or variegated colored leaves called bracts.

Light: Full sun.

Water: Do not over water.

Pests: Many root rot problems and white fly insects.

Other: Grown outdoors in warm climate.

POTHOS

Botanical name: *Epipremnum aureum*

Description: Popular vine, usually grown in hanging pots. Variegated green and yellow heart-shaped leaves.

Light: Can be maintained at 50 FC of light. Popular plant for very low light conditions.

Water: Allow plant to dry between watering. Lower leaves will yellow if over watered.

Pests: Spider mites, leaf spot, and root rot.

Other: Many varieties differing in leaf color.

15-31. **Norfolk Island pine.**

15-32. **Poinsettia.**

15-33. **Pothos.**

15-34. Rubber tree.

15-35. Schefflera.

15-36. Spider plant.

RUBBER TREE

Botanical name: *Ficus elastica*

Description: Large, thick, leathery leaves 12 to 15 inches long and 4 to 6 inches wide. Rubber plant has a sticky white milky sap.

Light: Medium light over 200 FC required.

Water: Requires a lot of water; can drop leaves if maintained on the dry side.

Pests: Few problems.

Other: Many varieties grown with different leaf sizes and coloration.

SCHEFFLERA

Botanical name: *Brassaia actinophylla*

Description: Dark green compound leaf resembling an umbrella. Medium to large growing indoor plant.

Light: Needs 150 FC or more, does well in indirect sunlight. Leaf drop common when grown under low light conditions.

Water: Over watering may cause older leaves to drop.

Pests: Spider mites, mealybugs, leaf spot, and root rot.

Other: Use as a specimen floor plant.

SPIDER PLANT

Botanical name: *Chlorophytum comosum*

Description: A hanging plant with long, strap-shaped, green leaves 18 inches long and ¾ inch wide. New plants develop at the end of shoot tips.

Light: Needs 150 to 400 FC of light.

Water: Keep soil thoroughly moist.

Pests: Scale insects and mealybugs.

Other: Fluoride in city water can cause leaf tip browning. Short day plant requiring less than 12 hours of light for flowering.

STAGHORN FERN

Botanical name: *Platycereum bifurcatum*

Description: An epiphytic fern (air plant) that grows on a tree trunk or a slab of wood. An unusual fern with leaves that resemble cabbage leaves.

Light: Bright indirect light.

Water: Moist, dunk plant and root system in water.

Pests: Mealybugs.

Other: Difficult to grow because usually hard to water.

15-37. Staghorn fern.

WEEPING FIG

Botanical name: *Ficus benjamina*

Description: Shiny, pointed, dark green leaves on drooping branches. Popular medium size tree used in shopping malls. May have aerial roots develop along the branches.

Light: Needs 150 FC or more for plant maintenance.

Water: Keep soil moist. Leaves can fall if under watered.

Pests: Spider mites and scale insects.

Other: This plant is very sensitive to changes in light or temperature conditions that can cause extensive leaf drop. Novelty plants are created with braided trunks, poodle-clipped, or bonsai shapes.

15-38. Weeping fig.

ZEBRA PLANT

Botanical name: *Aphelandra squarrosa*

Description: Large ovate dark green leaves with white veins that are similar to zebra stripes. Has large yellow flower spikes in the fall that will last 6 weeks. Several popular cultivars with different leaf colors and more compact growth habit.

Light: Bright indirect.

Water: Evenly moist.

Pests: Scale, mealybugs, and aphids.

Other: Difficult to grow in the home.

15-39. Zebra plant.

TERRARIUMS

You do not need a plot of ground or even a sunny window to grow a garden. A *terrarium,* or bottle garden, is a miniature landscape growing in a covered or closed glass or plastic container. The terrarium was originally invented around 1850 as a way of transporting living plants from the far parts of the world when sea voyages took months or years to complete. Today's terrariums are more decorative in nature, but these mini gardens can be maintained for long periods of time with a minimum amount of attention.

Common misconceptions about terrariums are that they require no care and that just about any kind of plant will thrive in a bottle. Plant selection is very important when starting a terrarium. Cactus and succulent type plants do not do well in the conditions found in a terrarium. Instead, pick plants that thrive under the tropical conditions of high humidity and low light found inside a glass bottle.

15-40. Terrariums are containers enclosing a garden of growing plants.

Any number of containers can be used as terrariums, as long as the material is either clear glass or plastic. Large-necked bottles and fish tanks make good terrarium containers because they are easy to reach inside when positioning the plants. Small-necked bottles will challenge the gardener to be creative in establishing plants through the small bottle neck. Bottle gardens do best in bright, but not direct, sunlight. If the sun shines directly into the bottle for more than an hour, the plants inside are likely to be cooked.

Before planting the terrarium, always clean the inside of the container. Add 1 inch of a planting medium composed of 3 parts commercial potting soil and 1 part crushed charcoal. Remove about half of the soil from the plant root ball before planting it in the terrarium. Firm the soil around the plant roots with your hand or a blunt object. After the plants are in place, apply a layer of ground fir bark to cover the soil surface. Mist the plants lightly and set the terrarium in an area that has bright, indirect light.

Covered terrariums will not require additional water, but moisture can condense on the inner glass surface and destroy the beauty of the terrarium. Most plants are killed by over watering, not under watering, them. When in doubt about whether or not to water the terrarium, do not water it. Fertilizing terrarium plants can cause them to quickly outgrow the container and thus require constant pruning.

REVIEWING

MAIN IDEAS

Today, growing large foliage plants inside office buildings and shopping malls is big business. Plants give people the feeling of the outdoors when inside. They feel better when plants surround them. Interiorscape managers must maintain foliage plants under conditions that may not be the best for plant growth.

Foliage plants are tropical and must have very warm temperature with high humidity. These plants can be easily damaged from cold chilling temperatures. Never place plants close to window or near outside doors where cold drafts can injure the plants. High temperatures created by placing plants too close to light bulbs or heat ducts can also damage foliage plants by drying out the leaf tissue. Most interiorscape plants are native to the tropics and therefore require high humidity. Spraying the foliage is one way to increase the humidity around these plants.

All plants need a good light source, either the sun or artificial lamps provide the necessary light. Fluorescent lamps provide the quality of light necessary for good plant growth when natural sunlight is not available. Several plants are listed that can survive under low or medium light situations.

Interiorscape plants must have a well mixed growing media if they are to exist under adverse conditions. Peat moss, vermiculite, and perlite are amendments that when added to the soil will improve the soil drainage and water holding capacity. Exercise care when watering indoor plants. It is very easy to over-water and kill the root system. However, under-watering can also injure the plant when too little water is available for good plant growth. Use a moisture meter to measure the available water in the soil before adding more water to the plant pot.

Most foliage plants are propagated and grown under high light conditions in Florida and Texas. Before these plants can be placed in office buildings or shopping malls, they must be acclimated to these future low light conditions. It will take several weeks for the foliage plants to acclimate to a low light environment.

Many interiorscape plant managers control insects and diseases using the integrated pest management practices. Prevention of plant pest problems by good cultural growing practices can reduce the need for routine pesticide spraying.

QUESTIONS

Answer the following questions. Use complete sentences and correct spelling.

1. What is interiorscaping? An interiorscaper?
2. What temperature will cause chilling injury to plants?
3. What is the difference between acute and chronic plant problems?
4. What is the ideal temperature range for interiorscape plants?
5. What is plant maintenance?
6. How can plants be acclimated to a new environment?
7. Name ten plants that do well under low or medium light conditions.
8. Why don't foliage plants grow well in good soil?
9. Why are soil amendments added to the growing medium?
10. Name four plant disease problems.
11. What fertilizer requirements do interiorscape plants have?
12. What is IPM and why is it important when growing plants?
13. What are the major flower colors of an African violet?
14. What is the major insect problem on the Boston fern?
15. What damage will fluoride in city water cause to spider plant leaves?

EXPLORING

1. Visit a local shopping mall and identify the different plants by name.

2. Grow several interiorscape plants, requiring low light conditions, under full sun and monitor their growth.

3. Become responsible for the care and maintenance of interiorscape plants at your school.

4. Grow several heart-leaf philodendrons under different light conditions—fluorescent lamps, incandescent lights, and natural sunlight. Compare their growth rate and type of growth under each different light condition.

16

DESIGNING LANDSCAPES

People like to see attractive outside areas. They like to see healthy green plants and beautiful flowers. Everyone admires a well-kept lawn and garden. A beautiful landscape doesn't just happen—it must be planned! Planners must know plant materials and how to use plants.

Plants fill important psychological needs. Green has a calming effect on people. Living plants promote human health and help people overcome the stress of living. Plants in a landscape make work and play more enjoyable. At our homes, tending plants provides a quality use of leisure time. At our places of work, plants make attractive surroundings. They make where we live exciting and comfortable.

16-1. A beautiful landscape makes life more enjoyable.

385

OBJECTIVES

This chapter covers the fundamentals of landscape design. It has the following objectives:

1 Explain the importance of landscaping and landscape planning

2 Describe procedures in landscape planning

3 Explain elements of design and effect

4 Describe xeriscape planning

5 Explain how plants are selected

TERMS

accent planting
balance
bush planting
corner planting
drought tolerance
element of effect
focalization
foundation planting
ground covers
landscape architect
landscape design
landscape designer
landscape planning
landscape symbol
line planting
outdoor ceiling
outdoor floor

outdoor room
outdoor walls
principles of design
private area
proportion
public area
rhythm and line
service area
shrub
simplicity
site analysis
sitescaping
specimen plant
unity
water requirement
water zone
xeriscaping

PLANNING THE LANDSCAPE

Landscaping is using plants to make outdoor areas attractive to people. The landscape plantings that surround a home or business play a major role in creating the beauty that is seen. They provide the framework to blend a home with its surroundings. In some cases, wood, rocks, statuary, pavings, and other items may be used. Without adequate plant knowledge, landscaping may fail.

16-2. An attractive landscape provides a good first impression to this business.

Bush planting is when an untrained person tries to landscape without a knowledge of plant materials. The result is a hodge-podge of plants that are incorrectly chosen and planted. Good landscaping begins with a design and includes installation and maintenance.

PURPOSES OF LANDSCAPING

Landscaping achieves many purposes. Here are several of those purposes:

- Create an aesthetically pleasing outside environment—Areas around homes, businesses, and parks are made attractive and appealing. Properly installed landscapes add value to property.

- Create pride in our homes, places of work, and communities—An attractive, neat environment helps people feel better about themselves and where they live.

16-3. Overgrown landscape plants cover up the house windows.

- Reduce noise from factories, highways, and other sources—Shrubs absorb noise and abate its interference in the lives of people.

- Help buildings and other structures blend into the terrain—Landscaped areas provide a careful blend of buildings with the environment.

- Provide privacy by blocking the view into backyards and other locations— Shrubs and other plants can be used to create private areas at homes and other places.

- Conserve natural resources and reduce pollution—Plants and other materials used in landscaping keep soil from washing or blowing away. They also help

16-4. The natural feature of the landscape site must be taken into account. (Courtesy, Church Landscapes)

prevent the run-off of water so that it soaks into the soil. Plants use carbon dioxide and give off oxygen.

■ Create places for recreation and outdoor activities—Landscaping is used to make playing fields and other areas safe, appealing, and useful.

■ Increase property value—Good landscaping increases the value of homes and commercial sites. Real estate with no landscaping or with poor landscaping is not as appealing to a buyer as property with good landscaping.

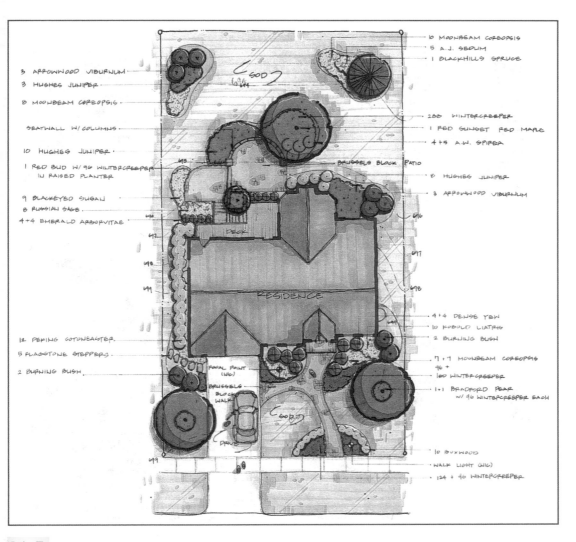

16-5. Example of a residential landscape design in color. (Courtesy, Matthew Haber, Western DuPage Landscaping, Inc.)

Landscape design is the practice of creating a plan to make the best use of available space in the most attractive way. Consideration is given to the relationships between the land, building, plants, and people. Outdoor areas are studied. Appropriate plant materials are selected to achieve the purposes of landscaping. The locations of the plants can create a desirable relationship between buildings, people, and nature.

Landscape designs are depicted on landscape drawings with supporting information, known as specifications. The drawings show the natural features of a site, how the site is to be prepared, and the kinds of plants to use

16-6. Landscape design includes landforms, pavements, constructed features, plantings, and water.

and their location. Locations of irrigation equipment, drainage systems, and other features are included.

Landscape designers are trained in the art of design and the science of growing horticultural plants. Landscape designers work primarily with residential landscape designs and small commercial sites. *Landscape architects* are trained in engineering, graphic arts, and architectural technology. The vast majority of their work involves large-scale projects, such as parks, golf courses, community planning, and large corporate complexes.

Special Uses

Landscape plans may involve projects to meet special needs. Within a landscape, sitescaping may be used. *Sitescaping* is landscaping a small part of a larger area. It is used to meet the particular needs of people. For example, only the public area of a building may be landscaped. Sitescaping is used when costs of landscaping must be spread over time or when landscaping is needed on only a portion of an area.

Xeriscaping is a form of landscaping that uses plants which require small amounts of water. Plants needing substantial water would not be used in deserts. Native cacti and other plants that require small amounts of water would be used. Xeriscaping often involves using plants that grow naturally in the environment where the landscape site is located.

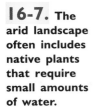

16-7. The arid landscape often includes native plants that require small amounts of water.

LANDSCAPE PLANNING

Landscape planning is preparing the details of how a site will be landscaped. The planning phase occurs before the designer develops any design for the landscape site. It begins with meeting the customer. A family inventory is often used to gather information on the customer's needs, desires, and priorities. After surveying the customer's needs, the landscape site is analyzed. This includes a base plan which is a drawing of the house on the lot. Existing physical features, such as the driveway, patio, walks, and fences are included.

16-8. The first step in landscape design is meeting with the customer. (Courtesy, Ron Biondo, Illinois)

It is important for the designer to visit and study the landscape site and to record observations. Visits at different times of the day and under different weather conditions are helpful.

PREPARING A LANDSCAPE DESIGN

Several important steps are used in preparing a landscape plan. Leaving out any step could result in a flawed plan. The steps are: determining the purpose of the landscape, analyzing the site, determining the use of the site, designing plantings, and preparing a written plan.

PURPOSE OF THE LANDSCAPE

Landscapes may serve a variety of purposes. Most are to create an attractive, useful area. Sometimes, one landscape may serve several purposes. For example, an aesthetically pleasing landscape may also abate noise, add value to property, and direct the behavior of people.

ANALYZING THE SITE

Site analysis is studying the site to identify its natural features and the appropriate treatment that would achieve the determined purpose. It includes a site analysis plan which is a piece of paper with an accurate sketch of the house and lot on which observations are made. The intended use of the area being landscaped is an important factor in planning.

Site analysis includes several important activities. Orientation of the house on the lot is a factor in locating activity areas and selecting plants. Knowing which direction is north is vital to a functional plan. Natural conditions should be assessed, such as soil type, drainage, wind movement, and air circulation. Existing vegetation should be noted. Structures on the site should be evaluated and relationships with the surrounding features determined. Study and make notes on the views in various direction from the site.

Site analysis will also include how the area is used. The nature of the users is important. A site for younger children should be free of plants with

16-9. It is important for the designer to make an on-site visit to gather information. (Courtesy, Ron Biondo, Illinois)

sharp points, such as spines on holly, and without pools or other features that are hazards. Older people like areas without steps and free of objects that could cause them to fall.

A topographic map may be used to show the different elevations and other features of the land. Slopes, rock outcrops, trees, streams, and other natural features are shown on the topographic map. Larger areas may involve the use of aerial photographs that show features of the land.

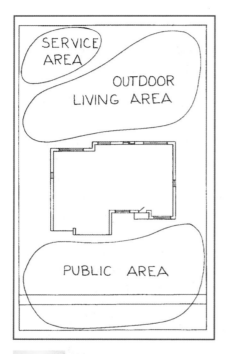

16-10. Landscape design showing the public, outdoor living (private), and service areas.

USE OF THE SITE

Landscape designs are divided into three main areas on the basis of use—public, outdoor living (private), and service. These areas are particularly used in residential landscaping.

The *public area* is the part of the landscape that will be seen from the street. It is usually in front of the building. Appearance is important. Plants should be selected for aesthetic appeal and hardiness. The public area is normally the first that is landscaped.

The *private area* or outdoor living area is out of the view of the general public. It is usually toward the rear of the building. This is where people relax and enjoy being outside. The area may include sitting and cooking areas, such as a table and barbecue grill. People often refer to the private area as the outdoor living area.

The *service area* is near the rear or the side of the house and is relatively isolated from the public and private areas. It is where the trash cans are kept and storage facilities are maintained. Some landscape plans include outdoor cooking equipment in the service area.

DESIGNING THE PLANTINGS

The outdoor living area is designed differently than the public area for several reasons. It includes all the property to the rear of the house except the service area. The outdoor living area is designed with the family's gardening

16-11. The patio and turf provide the floor and the trees provide the sense of overhead protection in this outdoor room. (Courtesy, R.S. Hursthouse & Associates, Inc.)

interests and entertainment activities taken into consideration. Designing plantings for the private area begins with the *outdoor room* concept. An outdoor room has boundaries, ground coverings, flowers, patios, and traffic areas. The canopies of larger trees may provide a ceiling. Plants are selected and planted based on what is needed to furnish the outdoor room. If shade is wanted in a particular area, a tree needs to be planted. Trees grow slowly.

Boundaries of an Outdoor Room

Outdoor rooms have floors, ceilings, and walls. The design of the landscape determines the size and nature of these boundaries.

The *outdoor floor* is the ground covering. In most cases, ground covering is turf. Sand, gravel, brick, wood, water, and other materials may be used as the outdoor floor. Establishing a good turf requires careful analysis of the soil and selecting an appropriate species.

The *outdoor ceiling* is the upper limit of the landscape. Covered patios and trees often provide the outdoor ceiling. The kind of ceiling has a big influence on the outdoor floor materials. Grasses that must have full sun may not survive under a tree canopy.

The *outdoor walls* set the boundaries of the outdoor room. Outdoor walls are made with flower beds, shrubs, fences, and walls of structures. The outdoor walls determine the area included in the outdoor room. Walls with little space between provide small outdoor rooms. People in urban areas or patio homes often have small outdoor rooms. Space simply doesn't allow for a large room.

*A*rranging Plant Materials

Four kinds of plantings are used in landscaping: corner, foundation, line, and accent. The kind of plant material must be appropriate for the kind of planting.

Corner plantings are plants placed at the corners of the house. The selection and arrangement of plant materials in a corner planting help it achieve its purpose. In general, taller plants are placed at the back and shorter plants at the front. Use plants with rounded forms and arrange them in groups. Some landscapers prefer to use an odd number of plants in a corner planting. No more than three or four different species of plants should be used in a corner planting unless the bed is especially large. The most attractive plants are used where the two lines forming the corners meet, called the in-curve. This is the center focal point. Corner plantings are in landscape beds.

16-12. Examples of corner plantings.

A *foundation planting* is the planting along the walls or foundations of buildings. It is designed to tie the building into the landscape. Taller plants are placed near corners, with lower plants below windows. Plant materials should be selected on the basis of growth. Dwarf and low-growing materials are used along foundations. A foundation planting may extend beyond the walls of the building. The intent is to focus attention on the entrance.

A *line planting* creates the walls of the outdoor room. They are often used to provide privacy and screen the area from public view. Selecting plant materials to give some variety in size and shape adds interest in a line planting. Taller shrubs are placed at the back, with shorter shrubs in the front. Line plantings are often used along property boundaries. Plant no more that three to five species in a line planting.

16-13. An example of a line planting.

An ***accent planting*** is an area of particular beauty or interest established in a landscape. Occasionally, it may be only one plant. In others, masses of plant materials of the same species may be placed in beds. Accent plantings are often used in the public and private areas to add interest and color. A general rule is to avoid placing accent plants directly in the middle of a lawn area. They are placed to the side and often used to create the illusion that the area is larger than it is. Flowering shrubs, beds of tulips, and small trees can be used to form accent plantings. Statuary and fountains may also be used as a part of the planting.

16-14. A high interest planting is placed close to this patio. (Courtesy, R.S. Hursthouse & Associates, Inc.)

PREPARING A WRITTEN PLAN

Landscape designs use line drawings, symbols, and lists of plant material specifications. A plan should be drawn to scale and accurate. Various symbols and procedures are used.

16-15. Attractive, functional landscape plans can be produced using low cost drafting equipment.

Drawing Instruments

Landscape designs may be drawn on paper with pencil or pen or designed using computer-aided landscape design program. Some people begin with a simple sketch on paper. All lines should be straight, except as used to show angular and rounded shapes. Lettering should be legible and attractive. Computer-aided design will provide neat, uniform lettering, shapes, and lines.

Plant Symbols

Landscape symbols depict different kinds of plants. These show the overhead views of shrubs, trees, and other items in the design. Symbols are

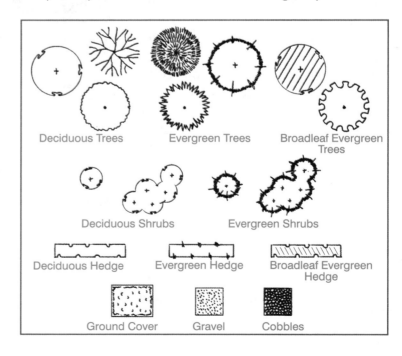

16-16. Examples of plant symbols used in landscape designs.

Deciduous Trees   Evergreen Trees   Broadleaf Evergreen Trees

Deciduous Shrubs   Evergreen Shrubs

Deciduous Hedge   Evergreen Hedge   Broadleaf Evergreen Hedge

Ground Cover   Gravel   Cobbles

also used to show different building construction features. The common landscape symbols are shown in Figure 16-16.

Skill Development

Landscape design is both an art and a science. Much can be learned from carefully observing the plans that have been prepared by other people. Look at the designs to determine the public, private, and service areas. Study the symbols to determine the kinds of plant materials that have been used. Observe the site where the plan was carried out. Analyze the site features and plant materials selected for the site.

Materials Selection

Selecting the plant materials to implement a plan requires knowledge of the different materials. In general, five types of plant materials are used in landscapes: trees, shrubs, ground covers, vines, and flowers.

A tree has one main trunk. Trees often grow quite large. Many trees are planted for shade and beauty. A few may be selected for products they produce, such as flowers on the dogwood and nuts on a pecan. Trees may be small when set out. It is easy to forget how large they may grow in a few years. Trees should be located far enough from a building so that they do not cause problems, such as from falling limbs. Trees are used in accent

16-17. Trees too close to a house.

plantings and, in some cases, line plantings. The trees in a landscape are often those that grow in the local area.

Shrubs are woody plants, but are much smaller than trees. They often have more than one main trunk. Two kinds of shrubs are used—evergreen and deciduous. The growth habit of a shrub determines how it can be used in the landscape. Low-growing shrubs are used as foundation plantings. Taller-growing shrubs are used in corner, line, and accent plantings. Examples of shrubs include hollies, forsythia, and azaleas.

Ground covers are woody or herbaceous plants that form a mat less than one foot high covering the ground. They are often used on banks or beds to direct foot traffic. Some species of ivy are suited to steep banks because of their value in controlling erosion. A vine is a plant that has a tendency to climb naturally or with support. Many ground covers can also be used as vines.

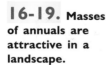 **16-18.** Ground cover is often used to control erosion and provide an attractive landscape bed covering.

Flowers are plants that have attractive blossoms. Some are seeded each year as annuals; others are perennials. Some grow from bulbs, corms, or other root structures. Examples

16-19. Masses of annuals are attractive in a landscape.

include the daffodil, iris, and tulip. People like attractive beds of flowering annuals, such as petunias, impatiens, and zinnias. Flowering plants are used in accent beds or in front of a corner, foundation, or as line plantings.

Materials Specifications

The plant materials selected in a design should be appropriate for the site, climate, use, and other needs. An attached listing of plant material specifications may be a part of the plan.

Materials should always be selected to best serve the needs of the site. Planting the wrong materials causes future problems with the landscape. Planting materials that grow tall in front of windows will cause problems in a few years. Know the different species and varieties of plants. Study their growth habits, climate needs, and disease resistance.

After the customer and the landscape designer select the preliminary plan, the designer develops a final design. This usually done on a quality vellum paper. Color may be added to the symbols. Lettering on the final plan is necessary to identify the symbols representing plant materials and structures. Lettering is also needed to supply additional notes concerning the development and installation of the design. A plant list key for the final plan is also included.

PRINCIPLES OF DESIGN

Landscape design involves following several principles. These *principles of design* help people develop good designs that are economical to install and easy to maintain. The principles of design are general guidelines for applying art to designing landscapes. The principles are sometimes known as elements of design.

SIMPLICITY

Simplicity is designing the landscape so that it falls within the range of acceptable landscaping. Simplicity involves using groups of five to seven plants. These groups are known as masses. These groups can be repeated in the landscape. Planting a row of boxwoods along a foundation is an example of simplification in design.

BALANCE

Balance implies equilibrium whether the design is formal or informal. Symmetrical or formal balance involves planting the same number of the same species on both sides of a landscape. Asymmetrical or informal balance is planting different numbers and different species. For example, five dogwoods on the left may be asymmetrically balanced with seven crape myrtles on the right of a landscape.

PROPORTION

Proportion is the relationship of the size of different species of plants. Dwarf plants should not be planted next to large plants. Heights should be aligned so that the plants and structures appear comfortable and pleasing. Taller plants are used with taller buildings, while shorter plants are used with single-story buildings.

FOCALIZATION

Focalization is the creation of focal points in a landscape. It is where the eyes of people will be attracted. Only one focal point should be in each view. If more than one is used, the landscape may appear as a hodge-podge of plant materials. A *specimen plant* is distinctive and is often used to create a focal point in the landscape. With the public area of residences, the focal point is typically on or near the front entrance. Landscaping is often planned to highlight the entrance.

Landscapes are sometimes designed so that the focal point changes with the season. For example, a flowering plant may be the focal point in the spring and a deciduous tree with colorful leaves in the fall.

16-20. A specimen plant is often used to create a focal point in the landscape. This shows the same residence in spring and winter. Notice how the focal point changes with the season.

RHYTHM AND LINE

Rhythm and line is a principle dealing with flow throughout a landscape. The shape and direction of beds and heights of plant materials affect rhythm and line. All parts of a landscape should fit together easily. The parts should not appear forced together. With rhythm and line, the landscape wall, ceiling, and floor fit together in an aesthetically pleasing manner.

UNITY

Unity is the extent to which the landscape is complete and whole. The goal of the landscape has been achieved. Appropriate plants have been selected and planted. Colors, sizes, forms, and textures all work together for a pleasing effect. The overall appearance excites the viewer. No single part of the landscape detracts.

16-21. Correct use of design principles will create a pleasing outdoor landscape when viewed from the inside.

ELEMENTS OF EFFECT

Elements of effect are artistic features in a landscape that create moods or feelings in people. Though important, they aren't emphasized as much by some landscape designers. The important elements are:

- **Line**—Line is the continuity of a landscape. Geometric shapes and curved patterns allow eyes to move about the landscape.

■ **Form**—Form is the shape of the individual plants, as well as the figures and shapes, seen from a particular view. Some forms have the shape of pyramids and move upward. Other forms have weeping effects and move downward.

■ **Texture**—Texture is the coarseness or fineness of the materials in a landscape. Plants vary in the extent of coarseness and fineness. Coarse textured materials make a large space appear smaller. Fine textured materials make a small space appear larger.

USING XERISCAPING

Xeriscaping is using a landscape design that conserves water and protects the environment. The water qualities of an area are determined, and plant materials are selected accordingly. Some locations have more moisture than others. Irrigation systems may be installed to provide supplemental water. Overall, the emphasis in a xeriscape is the use of plants that require less moisture. Xeriscaping allows for lower landscape maintenance costs and provides for the protection of the environment through the efficient use of natural resources.

Water zones are areas in a landscape based on the amount of water needed by the plants. Irrigation may be needed if natural moisture supplies are inadequate. Three zones are typically used:

■ **Very low water zone**—The very low water zone usually requires no water beyond what is naturally available, except to establish the plants. Plants that

16-22. Landscapes in arid areas often use native, water efficient plants, such as cactus.

are being established may need irrigation. Plants in the very low water zone need to be selected for low water need. Annual flowering plants are not appropriate because of their high water need. Native shrubs, cacti, and other species are often best.

■ **Low water zone**—Plants in a low water zone usually need more water than what is available naturally. Some irrigation is needed. Shrubs, ground cover materials, and limited annuals may be appropriate.

■ **Moderate water zone**—The plants in the moderate water zone typically need supplemental water. Focal points in the landscape and areas of high use near entrances are in this zone. Annuals and other succulent materials may be used.

Xeriscaping requires careful planning, design, and installation. A soil analysis is needed to determine nutrient needs and necessary soil improvements. Appropriate plants for the climate should be selected. An efficient irrigation system may need to be selected and properly installed. The amount of turfgrass is limited. Most turfgrass requires large amounts of moisture. The use of annual flowering plants is limited. Drought-tolerant perennial flowers and ornamental grasses should be selected. Mulch should be used to cool the surface, prevent erosion, discourage the growth of weeds, and conserve moisture. After installation, proper maintenance is an important part of having a successful xeriscape.

PLANT SELECTION

Appropriate plant materials must be in a landscape plan. A landscape will last indefinitely when the right plants are used. A landscape with the wrong plants will not be successful. The plants will die and will not give the desired effect.

FACTORS IN PLANT SELECTION

Factors in plant selection include water requirements, growth rate, color, hardiness, and nutrient and pH needs.

Water Requirements

Water requirement refers to the amount of water plants need to live and grow. Some plants need more water than others. Increasingly, plants that need less water are being selected in landscaping.

Drought tolerance is the ability of a plant to live and grow with low amounts of moisture. The climate in the area being landscaped is important in plant selection. Plants that need an abundance of water should be used in places with high natural moisture.

Growth Rate and Maturity

Plants grow at different rates and to different sizes. Plant selection for a landscape must consider the growth of plants. A plant that is very small when planted may grow into a large plant at maturity. Even giant trees were once small!

Trees and shrubs are classified by height and spread. Height is how tall a plant grows. It is the vertical space needs of a plant. Some dwarf shrubs grow only a few inches in height. Trees may reach a hundred feet or more. Plants should be selected and placed in a landscape on the basis of mature height.

Spread is the size and fullness of the canopy. It is the horizontal space needs of a plant. Trees may have huge canopies. Small shrubs may have only a few inches of spread. The space available to achieve the purposes of the landscape must be considered. Masses of annuals with little spread may require placing the plants close together.

Color

Color includes leaves, bark, flowers, fruit, and limb and branch structure. Some flowers are showy; others are inconspicuous. Plants should be selected for long-term or year-round color, and not just color when flowering.

Year-round color is provided by evergreen trees, shrubs, and vines. Deciduous plants provide important color as well, especially those with colorful leaves in the fall. Deciduous trees also create the need to rake leaves in many landscapes.

Hardiness and Other Adaptations

Hardiness refers to how well a plant is suited to the climate. Some plants withstand cold; others don't. Some plants require full sun; others will grow well in partial sun or shade. Low-growing plants set underneath taller plants must be able to grow in less than full sun.

Hardiness zones have been established based on plant temperature tolerance. The 11 zones are based on the average minimum temperatures plants need. The zones include temperate, subtropic, tropic, and others. When a

16-23. Landscape plantings provide texture and color throughout the year.

tree or shrub is assigned a hardiness zone rating, it should survive the winter in that zone or any zone having a higher numbered rating. See Figure 16-24.

Plants must be able to tolerate salt in some locations. Two examples include coastal areas near saltwater and along highways where salt is used to control snow and ice.

Pest resistance is important in species selection. Diseases, insects, and other pests can destroy expensive plants. Selecting species that have resis-

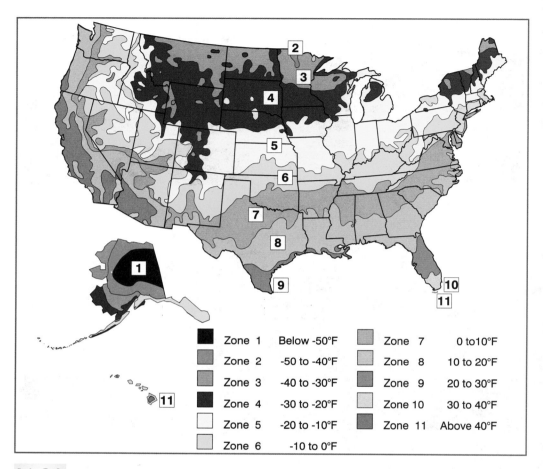

16-24. The USDA Plant Hardiness zone map defines 11 areas with average annual minimum temperatures.

tance will improve the landscape, as well as reduce costs associated with pest control.

Plants vary in nutrient and pH requirements. Soil analysis determines the nutrients available in the soil. Fertilizer can be used to increase the nutrient supply. The pH can be adjusted to meet the needs of plants. Selection should consider the needs of plants. Matching plants to the naturally available nutrients and pH reduces costs of soil modification.

SOURCES OF INFORMATION IN SELECTING PLANTS

Good information is needed to help in selecting plants. Designers need to know plant materials. The important information about plants includes:

- common and scientific names
- whether the plant is deciduous or evergreen
- height and spread of the plant
- growth, training, and trimming requirements
- characteristics of flowers and fruit, if any
- hardiness zone
- pest resistance and problems
- sun requirements (full sun, partial sun, or shade)
- water requirements
- life span (perennial, biennial, or annual)
- method of propagation
- others, such as poisonous foliage, thorns, and potential problems from roots, leaves, and flowers or fruit

Information about plants is presented in charts that summarize the major features of the species. Some common tree species and shrubs, tubers, bulbs, corms, vines, perennials, and annuals are presented in Chapter 17. **Appendix C includes information on common plants used in the landscape by classification.** Reputable local nurseries often have useful information. Agriculture offices at land-grant universities usually have information and specialists who can help with plant selection.

REVIEWING

MAIN IDEAS

Landscaping is used to create appealing outside areas. In landscaping, people try to achieve various purposes related to beauty, use, and comfort.

Landscape planning is preparing a detailed plan of how a site will be landscaped. Xeriscaping involves using plants on the basis of water needs and conservation. Careful attention must be given to the site and how it will be used. Plantings are designed using the concept of an outdoor room. Ceiling, floor, and walls are a part of the outdoor room. These are created with plant materials and other structures in the landscape.

Plant materials are arranged into corner, foundation, line, and accent plantings. Materials are selected and planted to achieve a purpose. Landscape plans are prepared using computer aided design or by drafting them on paper. Symbols are used to represent different plants and features of the site. Principles of design and effect are used in landscape design.

Careful attention must be given to matching the plant species with the needs of the landscape. A good knowledge of plant materials is essential.

QUESTIONS

Answer the following questions. Use complete sentences and correct spelling.

1. What is landscaping? How is it distinguished from bush planting?
2. Name three purposes of landscaping. Select the one that you think is most important and indicate why.
3. What is landscape planning?
4. What procedures are followed in designing a landscape?
5. What three areas are used in a residential landscape? Describe each.
6. Explain the concept of an outdoor room. Describe the boundaries of the room.
7. What are the four kinds of plantings used in landscaping?
8. Draw the symbols used on a landscape design for the following plants: evergreen with needles, deciduous shrub, deciduous tree, and broadleaf evergreen.
9. What five types of plant materials are used in landscaping?
10. What are the principles of design?
11. What is xeriscaping?
12. What are the important factors in plant selection?

EXPLORING

1. Take a field trip to a residence that has been recently landscaped. Study the site. Identify the plant materials and observe how they have been planted. What is your assessment of the landscape? Prepare a sketch that shows major features of the landscape.

2. Prepare a landscape design for a new or fictional residence. Select a scale and make careful measurements and assessments of the site. Draw a design that uses appropriate symbols and follows design principles. Display your design.

3. Visit a local nursery or garden center and study the kinds of plants that are available. Classify them as annual flowering plants, shrubs, trees, turf, and ground cover. Include both the common and scientific names. Determine the nature of the plant, including height and canopy, water requirements, hardiness, and its general characteristics, including deciduous or evergreen, and whether it produces flowers and fruit.

17

ESTABLISHING LANDSCAPES

Once a landscape has been planned, proper installation is needed. The plan must be carried out. Materials must be carefully selected and placed in the landscape. A landscape can be no better than the quality of the materials and the way they are installed.

A landscape often includes more than plants. Fences, lighting, mulch, paved areas, water, and other features enhance a landscape. Properly installing these features makes a landscape attractive and long-lasting. They add appeal at night as well as during the day.

17-1. Establishing landscapes includes installing plant material and hardscape features.

OBJECTIVES

This chapter will help develop an understanding of landscape establishment and construction. The objectives of this chapter are:

1 Explain the scope of landscape construction work

2 Name the major plant growing techniques

3 Describe steps in proper tree and shrub planting

4 Define annual and perennial flower types

5 Describe landscape surfacing materials

6 Contrast different kinds of fence material

7 Name four types of outdoor lighting techniques

TERMS

anti-transpirant
container grown
down lighting
flower bed
flower border
guying
hardscaping
Joint Utility Location and
    Information Exchange
    (J.U.L.I.E.)
landscape construction

landscape contractor
night-lighting
planting plan
silhouette lighting
soil ball
sunscald
treated fence posts
up lighting
walk lighting
winter burn

BEGINNING LANDSCAPE INSTALLATION

Once a design has been completed and the planting plan developed, work can begin on construction. **Landscape construction** is the segment of landscaping that involves the installation of materials identified in the landscape design. Construction projects range from being very simple to very elaborate. Large projects may involve moving or grading the soil, installing drainage systems, building permanent structures, and planting plants. The term *hardscaping* describes these permanent structures, such as fences, patios, walks and driveways, water features, and retaining walls.

The construction of a landscape project is performed by a **landscape contractor**. A landscape contractor is hired to install the landscape. Their job is to transfer the design provided on paper to the actual landscape. Therefore, landscape contractors must be able to read and understand the landscape plans.

The landscape designer and contractor may work for the same company. Sometimes, the contractor hires a designer or the contractor may be trained in landscape design. The important thing is that the contractor fully understands the intent of the design.

17-2. Landscape construction may involve moving soil, installing drainage systems, building permanent structures, and planting plants.

THE PLANTING PLAN

The landscape designer talked with the homeowner about their specific outdoor needs, desires, and problems. The landscape designer through the

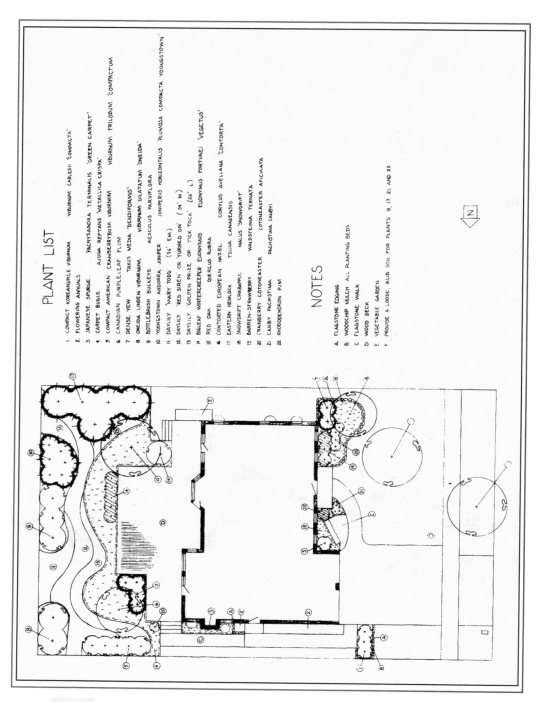

PLANT LIST

1. COMPACT KOREANSPICE VIBURNUM   VIBURNUM CARLESII 'COMPACTA'
2. FLOWERING ANNUALS
3. JAPANESE SPURGE   PACHYSANDRA TERMINALIS 'GREEN CARPET'
4. CARPET BUGLE   AJUGA REPTANS 'METALLICA CRISPA'
5. COMPACT AMERICAN CRANBERRYBUSH VIBURNUM   VIBURNUM TRILOBUM 'COMPACTUM'
6. CANADIAN PURPLELEAF PLUM
7. DENSE YEW   TAXUS MEDIA 'DENSIFORMIS'
8. ONEIDA LINDEN VIBURNUM   VIBURNUM DILATATUM 'ONEIDA'
9. BOTTLEBRUSH BUCKEYE   AESCULUS PARVIFLORA
10. YOUNGSTOWN ANDORRA JUNIPER   JUNIPERUS HORIZONTALIS PLUMOSA COMPACTA YOUNGSTOWN
11. DAYLILY 'MARY TODD'   (24" EM)
12. DAYLILY 'RED SIREN OR TURNED ON' (34" M)
13. DAYLILY 'GOLDEN PRIZE OR TICK TOCK' (24" L)
14. BIGLEAF WINTERCREEPER EUONYMUS   EUONYMUS FORTUNEI 'VEGETUS'
15. RED OAK   QUERCUS RUBRA
16. CONTORTED EUROPEAN HAZEL   CORYLUS AVELLANA 'CONTORTA'
17. EASTERN HEMLOCK   TSUGA CANADENSIS
18. SNOWDRIFT CRABAPPLE   MALUS 'SNOWDRIFT'
19. BARREN-STRAWBERRY   WALDSTEINIA TERNATA
20. CRANBERRY COTONEASTER   COTONEASTER APICULATA
21. CANBY PACHISTIMA   PACHISTIMA CANBYI
22. RHODODENDRON PJM

NOTES

A. FLAGSTONE EDGING
B. WOODCHIP MULCH ALL PLANTING BEDS
C. FLAGSTONE WALK
D. WOOD DECK
E. VEGETABLE GARDEN
F. PROVIDE A LOOSE, ACID SOIL FOR PLANTS 9, 17, 21, AND 22

17-3. An example of a planting plan to be used when installing the landscape.

17-4. Loading a tree that has been balled and burlapped at the nursery for a landscape contractor.

design process developed a ***planting plan*** which shows the exact location for plant materials, includes a plant list, and the permanent structures to be installed. Landscape contractors take the plan from the landscape designer and begin the installation process by planting the trees, shrubs, and flowers as illustrated on the landscape plan.

Nurseries sell the many varieties of trees and shrubs used by the landscape contractor. Landscape designers recommend planting balled and burlapped (B&B), bare root (BR), or container grown plants based on the time of year and the budget for each landscape job. Plants sold in the container in which they are grown are called ***container grown*** or container plants. Evergreen shrubs best survive transplanting to the landscape if grown as balled and burlapped or as container plants. Small and medium size trees and shrubs can be sold and planted using any of the three techniques. However, larger trees transplant best by the balled and burlapped method.

Hydraulic-powered tree spades are used to dig and transplant into the land-

17-5. Large trees are transplanted using a hydraulic tree spade.

scape large trees with a trunk diameter of four to twelve inches. These machines operate by hydraulic fluid pushing three or four large spades into the soil.

Guidelines for Selecting Trees

The general appearance of a tree will reveal much about its quality and potential for success when transplanted. Is the trunk straight and is the crown symmetrical? Does the tree show signs of current season growth? Signs of viability include expanding buds, new leaves, and elongated shoots. Are there any signs of disease or insect damage? Shade trees should have a strong, well-defined central leader with equally spaced branches forming a symmetrical crown. Trees having multiple leaders, crossing, rubbing, or overly-crowded branches have not been pruned properly and are less valuable.

Discolored, sunken, or swollen areas in the trunk are additional signs to look for when selecting a tree. Bark cuts and scrapes are also undesirable. Trunks with visible wood borer damage and those showing signs of sunscald or cracking should be avoided.

LANDSCAPE TOOLS AND EQUIPMENT

Landscape contractors need a variety of tools and equipment to prepare the site for planting and to install plant materials. Tools and equipment used range from hand tools to large engine-powered equipment. Proper preparation and installation techniques are essential in executing the planting plan.

When using tools or operating equipment, safe practices should be followed. Follow manufacturers' guidelines for operating equipment and power tools. Tools and equipment must be maintained and kept clean if they are to save time and function properly. Some good tool and machine operation tips include:

- Select the proper tool or machine for each job.
- Read and follow owner's manual recommendations.
- Check the condition of the tool or equipment before use.
- Perform daily maintenance procedures.
- Warm up the engine before operation.
- Operate at safe speeds.
- Check instrument readings while in operation.

Colorado Blue Spruce (*Picea pungens f. glauca*)

Eastern Redbud (*Cercis canadensis*)

Bradford Pear (*Pyrus calleryana 'Bradford'*)

Flowering Cherry (*Prunus serrulata*)

Flowering White Dogwood (*Cornus florida*)

Sugar Maple (*Acer saccharum*)

White Birch (*Betula pendula*)

Japanese Maple (*Acer palmatum*)

17-6. Some examples of trees used in the landscape.

17-7. Some examples of shrubs used in the landscape.

17-8. Some tools and equipment used in landscape construction.

SITE PREPARATION AND PLANTING TECHNIQUES

Implementing the planting plan while dealing with the unique characteristics of the site is the responsibility of the landscape contractor. Usually there will be some alteration needed at the site. This may include changing the surface grade, conditioning the soil, installing an irrigation system, or other hardscape features. Planting holes will have to be dug for plant materials. The landscape contractor must locate any underground utilities before any digging can begin at the landscape site.

Two days before starting a landscape installation, the landscape supervisor will call the *Joint Utility Location and Information Exchange (J.U.L.I.E.).* They will locate and mark all underground utilities at the landscape site. These utilities can include electricity, sewer, water, telephone, gas, cable television, and others. Most state laws require contractors to call J.U.L.I.E. before digging.

17-9. Underground utilities have been marked with flags before construction begins.

The landscape supervisor will usually arrive at the job site the day before the crew to stake out the job. Using the planting plan as a guide, the supervisor will stake all planting holes, landscape planting beds, and other landscape features. Small wooden or plastic stakes containing the plant names are driven into the ground to show the exact center of each planting hole. This will show where to dig each planting hole and the name of the plant. Besides staking plant material locations, the supervisor will also locate the patio, retaining walls, fences, or other permanent structures with stakes.

BED PREPARATION

It is necessary to provide a favorable environment for plant growth and survival. All turf should be removed from the new planting beds before planting. Land-

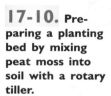

17-10. Preparing a planting bed by mixing peat moss into soil with a rotary tiller.

scape sites with poor soil conditions may require the addition of sand or peat moss to improve the soil texture. This is accomplished by covering the existing soil with a 2-inch layer of peat moss, followed with a 2-inch layer of sand. Mix the sand and peat into the top 6 inches of soil using a rotary tiller. Always plant landscape shrubs, flowers, and vegetables in a well prepared planting bed.

Planting Techniques

By following a few simple rules, the landscape contractor can guarantee successful tree growth after transplanting. A clean sharp spade will always make this planting job safer and easier. Use a file to keep the spade sharp and a putty knife to keep it clean. Different landscape contractors may use different tree planting methods. However, the following technique describes one common way to transplant trees.

The *soil ball* is the soil surrounding the root system which has been balled and burlapped. When planting a tree that has been balled and burlapped, always dig the planting hole a minimum of 12 inches larger than the soil ball to be planted. This will allow enough space between the soil ball and the edge of the planting hole for new roots to expand. The top of the soil balls should be at the top of the new planting hole. Never plant a tree deeper than the height of the soil ball. This can cause plant death from root rot. If

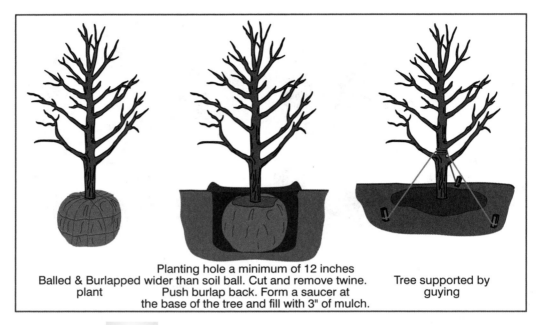

Planting hole a minimum of 12 inches
Balled & Burlapped wider than soil ball. Cut and remove twine.   Tree supported by
plant   Push burlap back. Form a saucer at   guying
the base of the tree and fill with 3" of mulch.

17-11. Planting technique for a balled and burlapped tree.

necessary, add soil to the bottom of the hole so the soil ball will be at the proper height.

Place the tree in the hole without removing any burlap or twine and check hole for proper size. After the hole is determined to be the proper depth, untie and remove all twine from the soil ball. Place the tree in the hole and pull the burlap back from the top of the soil ball. Push it down between the soil ball and the edge of the planting hole. Be sure to examine the "burlap" covering. If the "burlap" is made of plastic or other non-biodegradable material, it should be completely removed. It can be slid out from under the root ball after the plant has been placed in the hole.

Before placing a container grown plant in the planting hole, remove the container from the soil ball. Plants should be well-rooted and firmly established in their containers. The root mass should retain its shape and hold together when removed from the container. Place the soil ball in the hole and check for proper hole size. Position the plant in the hole with the best looking side toward the main viewing point.

Bare root (BR) trees should be stored in a cool, shady area with their roots protected from drying by packing or burlap. Their root system should be damp and flexible. When Bare Root trees are ready to be planted in the landscape, their root systems should continue to be protected until they are actually placed in the ground. Their exposure to the sun should be limited as much as possible.

Back Filling the Hole

Fill the space between the soil ball and the edge of the hole with soil dug from the planting hole. Break up all large soil clumps with the shovel before back filling the hole. Remove air pockets around the soil ball by using the shovel handle to tamp the back fill material. Construct a water saucer slightly larger than the soil ball with the left-over soil from the hole. The saucer will help retain rain or irrigation water for use by the plant. Fill the saucer with water immediately after back filling the hole.

17-12. Planting large trees by filling soil around ball.

Watering Practices

Newly transplanted trees and shrubs require deep and thorough watering the entire first year after planting. A good watering technique will thoroughly wet the soil 12 inches deep and provide adequate soil moisture for most plants.

Plant Staking

Trees 6 to 12 feet tall require additional support after planting. Give this support by driving two or three long wooden stakes next to the soil ball and

17-13. Bracing a tree by staking.   **17-14.** Bracing a tree by guying.

attach a wire between the stakes and the tree trunk. Protect the soft trunk tissue by covering the wire with a short piece of rubber hose. Shade trees over 15 feet tall need heavy steel cable attached between metal stakes and the tree trunk. Contractors call this trunk support method **guying.**

Fertilizer and Anti-transpirant

Most landscape contractors do not fertilize trees and shrubs at the time of planting. Over-fertilization can cause more stress to newly transplanted trees than under or no fertilization. Shade trees may require minor pruning of dead or broken branches at planting time. Evergreen shrubs should be sprayed with an anti-transpirant to reduce water loss by the plant. An **anti-transpirant** seals the leaf stomata and helps prevent leaf scorch and leaf burn.

Trunk Wrapping

Many landscape contractors wrap tree trunks with a paper tree wrap material to reduce sunscald. The term **winter burn** or **sunscald** describes the

17-15. A newly transplanted tree with its trunk wrapped and staked for support.

condition that causes the bark to blister from intense winter sunlight. While wrapping the trunk may not prevent trunk injury, it should not hurt the tree. Start wrapping the paper spirally around the trunk from the ground and continue up to the first branch. Securely attach the tree wrap paper to the tree trunk with staples or twine.

Final Planting Steps

Apply a 1½ to 2 inch layer of mulch around the base of the plant. Mulch minimizes soil moisture loss through evaporation from the soil surface. It also reduces the soil temperature around the plant's roots.

Cleaning up the planting site completes the planting operation. Landscape construction jobs may require a crew to work for several hours or many days at one job site. Each day brings a crew new challenges and opportunities to provide customers a beautiful and functional landscape.

17-16. Newly planted trees are mulched to reduce soil moisture loss and the soil temperature around the root system.

FLOWERS IN THE LANDSCAPE

Vivid flower colors contrasted with green shrubs attract attention to the landscape. Homes built before 1950 contained large flower beds in the front

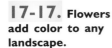
17-17. Flowers add color to any landscape.

17-18. People of all ages can enjoy planting flowers.

yard. Bright colored flowers attract the public's eye and can detract from the total home appearance. Today, designers normally place the flower beds in the back yard where they can be enjoyed close-up by family and friends. Planting flowers in the back yard also adds many colorful features to the landscape.

*F**LOWER BORDERS OR BEDS*

Landscape designers distinguish between flower borders and flower beds. Flowers planted in front of shrubs create a *flower border*, with the shrubs providing the back drop. *Flower beds* stand alone and only contain flowers. Plant tall growing flower varieties at the back or center of the flower bed and short varieties at the edge of the bed. Keep flower bed design simple. Try to plant groups of the same flowers together.

FLOWER GROUPS

Flowers are selected from the following major groups—annuals, perennials, and hardy or tender bulbs. Planting time, cost, and cultural requirements are used by the designer to select the best flowers for the job.

Annual flowers, such as petunias, marigolds, and geraniums, add seasonal color to the landscape at a reasonable cost. They must be replanted each year. Annual flowers are planted into the landscape using transplants or seeds. Transplants are planted 8 to 10 inches apart at the same depth they were originally growing. Firm the soil around each plant and water throughly. Seeds are sown in rows or scattered over the planting bed surface following the seed packet instructions. Cover the seeds lightly and water gently without splashing the seeds from the soil.

Perennial flowers usually cost more than annual flowers. However, once planted, perennials give the landscape many years of color and beauty without additional expense. Chrysanthemums, daisies, and coral bells are examples of perennial flowers.

Plant hardy flower bulbs, such as tulips, daffodils, and hyacinths, in the fall. They will give the landscape added color the following spring. In colder climates, tender bulbs, such as cannas, should be planted in the spring. They will bloom throughout the summer. In the fall, dig tender bulbs and store them through the winter for re-planting the following spring.

FLOWER PLANTING

Flowers grow best in well-prepared beds. Prepare the flower planting beds following the same basic steps as described above for shrub beds. Cutting corners during the bed preparation process will haunt the landscape contractor during the summer heat stress periods.

Some flowers require planting in full sun, while others need a shady planting location. Always know the different kinds of flowers and their individual growth requirements and plant them in the right place.

Marigold (*Tagetes* sp.)

Ageratum (*Ageratum houstonianum*)

Celosia (*Celosia argentea*)

Carnation (*Dianthus* sp.)

Pink (*Dianthus chinensis*)

Coleus (*Coleus blumeii*)

Sweet Alyssum (*Lobularia maritima*)

Impatiens (*Impatiens walleriana*)

Aster (*Callistephus chinensis*)

17-19a. Some annual flowers used in the landscape.

Snapdragon (*Antirrhinum majus*) Lobelia (*Lobelia* sp.) Salvia (*Salvia splendens*)

Geranium (*Pelargonium hortorum*) Petunia (*Petunia x hybrida*) Rudbeckia (*Rudbeckia* sp.)

Begonia (*Begonia semperflorens*) Rose Moss (*Portulaca grandiflora*) Zinnia (*Zinnia* sp.)

17-19b. Some annual flowers used in the landscape.

Painted Daisy
(*Chrysanthemum coccineum*)

Hibiscus (*Hibiscus* sp.)

Lupine (*Lupine* sp.)

Primrose (*Primula* sp.)

Oriental Poppy (*Papaver orientale*)

Globeflower (*Trollius europaeus*)

Iceland Poppy (*Papaver nudicaule*)

Yarrow (*Achillea* sp.)

Summer Phlox (*Phlox paniculata*)

17-20a. Some perennial flowers used in the landscape.

Columbine (*Aquilegia* sp.)

Astilbe (*Astilbe x arendsii*)

Aster (*Aster* sp.)

False Indigo (*Baptisia australis*)

Bellflower (*Campanula* sp.)

Delphinium (*Delphinium* sp.)

Pink (*Dianthus* sp.)

Bleeding Heart (*Dicentra spectabilis*)

Balloon Flower
(*Platycodon grandiflorus*)

17-20b. Some perennial flowers used in the landscape.

Gladiolus (*Gladiolus* sp.)   Daffodil (*Narcissus* sp.)   Morning Glory (*Ipomoea purpurea*)

Canna (*Canna* sp.)   Spring Crocus (*Crocus* sp.)   Grape Hyacinth (*Muscari* sp.)

Lily (*Lilium* sp.)   Tuberous Begonia (*Begonia x tuberhybrida*)   Tulip (*Tulipa* sp.)

17-21. Some bulbs, corms, tubers, and vines used in the landscape.

Perennial Gardens

One of the fastest growing segments of landscaping is the perennial garden. Perennials offer a vast array of colors and textures. The proper choice of perennial plant material can mean beautiful blooms from spring, through summer, and into the fall.

There are several considerations when selecting a site for a perennial garden. The amount of sun or shade the site is exposed to plays an important part in plant selection. Another key is proper drainage. The site selected should have good drainage so the plants do not sit in standing water. A good way to determine if the site has good drainage is to dig a hole at the site and fill it with water. Let the water drain and fill the hole again. If the water drains out at less than 1 inch per hour, the drainage will need to be improved. The other choice will be to select plant material that will thrive in wet conditions.

17-22. Perennial gardens are often used to add color and beauty to the private area.

After the site has been selected and the needs have been determined, the perennial bed pattern shape can now be laid out. An easy way to accomplish this is to lay a garden hose or rope on the ground in the desired shape of the garden. The shape of the bed should be in harmony with the rest of the landscape. If other landscape beds are curved, the perennial bed should be curved. If they are straight or 45 degree angles, the perennial garden should be also. For ease of maintenance, design planting beds to be no deeper than 5 feet.

The selection of plant material is the most important decision for the perennial garden. Perennials are available in a variety of colors, sizes, and flowering times. A number of groupings of plants generally look better than single plants spread throughout the garden. Taller plants should be placed towards the rear with shorter ones toward the front. There should be a plan used when selecting plant material.

Generally, perennials are planted in the spring or in the fall. Site preparation should begin some time before the actual planting. The bed should be tilled about 6 inches down and the soil mixed thoroughly. Organic material such as peat moss or leaf compost should be added. The soil should be kept loose until planting. Soil should be tested in order to determine fertilizer requirements and pH.

AFTER PLANTING CARE

Flower beds need plenty of hand watering and often require constant weeding. Without an intensive maintenance program, flower beds do not look good. A herbicide (weed control chemical), such as Preen, should be applied to prevent weed seeds from germinating in the flower bed. Always read and follow the pesticide label directions. A wood chip mulch around the individual plants is also helpful with weed control Flower beds should be fertilized at least once or twice a year by applying a liquid fertilizer when watering.

WATER FEATURES

Water features provide great visual impact in the landscape. Moving water is a great source of relaxation. Water feature are focal points that enhance the landscape.

WATER GARDENING

The most important consideration in water gardening is site selection. Most water plants need 6 to 8 hours of sun per day to thrive. The area should be level and not low where surface drainage would run into the pond. It should also be away from trees that would drop leaves into the pond in the autumn. The water garden should be located where it can be enjoyed and viewed from areas frequented in the house or landscape.

17-23. Water appeals to the sense of sight and sound in the landscape. (Courtesy, R.S. Hursthouse & Associates)

A water garden is a closed ecosystem. Guidelines for proper balance in a water garden are:

- Circulate the water volume every 1 to 2 hours.

- Floating plants should cover 65 to 75 percent of the pond surface.

- Oxygenating and marginal/bog plants should number 1 plant per 2 sq. ft. of pond surface, taking into account that the total number of plants in the pond should cover 2/3 to 3/4 of the water surface.

- There should never be more than 1 inch of fish per 5 gallons of water (for example, 6 fish 5 inches long would require a pond with at least 150 gallons of water; 6 × 5 = 30 × 5 = 150 gallons).

- There should never be more than 1 inch of snail per square foot of bottom surface area.

Water used to fill a pond may have elements that are harmful to plants and fish. City water may have chlorine, chloramine, or heavy metals. Chlorine will dissipate in 3 to 5 days. Chloramine needs to be treated and heavy metals neutralized. Country or well water may have nitrates or pesticides present. The best practice is to have the water tested before filling the pond.

Proper water circulation is important. The proper sized pump should circulate the entire water volume every 1 to 2 hours. A larger pump will be needed if the water garden has a waterfall, fountain, or other features which involves lifting water. Pumps are rated by gph (gallons per hour). The pump should be placed a minimum of 12 inches below the water surface. The out-

17-24. Some water garden installation steps include (1) excavating and installing the pond liner; (2) grading the site for proper drainage; (3) Installing power, lighting, and landscape plants; (4) planting water plants in containers or baskets; (5) acclimating fish to the pond's water temperature; and (6) enjoying the plants and fish in the water garden.

take of the pump should be placed away from the intake for proper circulation. The pump should be checked and cleaned if necessary on a regular schedule.

To determine the approximate number of gallons (water volume) a pond holds, use the following formulas based on the shape of the pond:

- Rectangular Pond: Length × Width × Depth × 7.5 = gallons
- Oval Pond: Length × Width × Depth × 6.7 = gallons
- Circular Pond: 3.14 × ½ Diameter × ½ Diameter × Depth × 7.5 = gallons

Water plants are an essential part of a water garden. When plants are fully developed, they should cover between 2/3 and 3/4 of the water surface area. Most water plants such as water lilies prefer full sun for 6 to 8 hours per day to thrive. Plants that require containers should be placed at the proper depth. They should be fertilized once a month with pond tabs from spring to fall, and twice a month during hot weather. Brown or yellow growth should be clipped off down to the crown of the plant. It is also important to note that water lilies do not like too much turbulence. They need to be placed away from waterfalls and fountains.

A proper environment is needed for fish. Fish prefer cold water because it holds more oxygen. Do not overstock the pond with too many fish. Ideally, the proper number of fish can support themselves on natural food in the pond. This doesn't mean not to feed the fish. However, they should not be fed more than once a day and only as much food as they can eat in 5 minutes. Uneaten, nutrient-rich food acts as fertilizer for algae. When the temperature reaches 40°F or below, do not feed the fish at all. Fish enter partial hibernation at 40°F.

17-25. People enjoy feeding and watching fish in a water garden.

As temperatures cool in the fall, the water garden must be prepared for the coming winter months. Depending on the geographic location of the pond, winter can be anything from icy cold to relatively balmy. The pond's surface should be lowered 6 inches to allow room for the water to expand with freezing. The pump and any filters can be removed and cleaned or left running during the winter. If you choose to leave the pump running, it is important that it be raised to just below the surface of the water. Allowing the pump to circulate the water at lower depths is stressful to fish. If the pump is removed, a heater will be needed to keep the pond from totally freezing on the surface. A horse or cattle tank heater or bird bath deicer can be purchased and used.

LANDSCAPE SURFACING MATERIALS

The landscape contractor installs many different types of surfacing materials to cover the soil and prevent soil erosion and compaction. Paving materials effectively prevent tracking of dirt into the house. Ground covers, flowers, and shrubs are planted in landscape areas with no foot traffic. Establish lawns are used in those areas receiving moderate traffic. However, heavy traffic areas require installing hard surface paving materials. Walks, driveways, and patios are examples of landscape areas that need hard surface paving.

17-26. A patio is an example of a landscape area needing hard surface materials.

MATERIAL SELECTION

Paving materials are divided into two groups—hard paving and soft paving materials. Soft paving materials lack form, such as loose aggregates. In most parts of the United States, concrete, bricks, paver blocks, and stone are the most popular types of paving material. Once installed, these provide the homeowner with a durable, long lasting, and low maintenance surface.

Concrete is chosen for ease of installation and economical installation price make concrete the first choice for most homeowners. Concrete, on the other hand, looks ordinary and does not create interesting or unusual landscape designs.

Many homeowners select bricks or paver blocks as the surfacing material to build walks, driveways, and patios. It is not easy to build brick patios nor are they cheap, but the final product is worth the extra effort. Place the bricks on top of 3 to 4 inches of compacted sand, then sift extra sand in the cracks between the bricks. Many contractors install a new kind of brick made from colored concrete instead of clay.

Pieces of natural stone laid on compact sand give a different look

17-27. A paver block walk is being installed at this landscape site.

for walks and patios. Different parts of the country provide a variety of stone for walk and patio construction projects. While the beauty of natural stone patios will impress many, the installation is expensive. High construction costs may keep stone patios out of reach for all but a few homeowners.

Loose aggregates such as crushed stone, marble chips, or wood chips can also provide a durable surface at a lower cost than concrete. These are known as soft paving materials. They easily conform and fill unusually shaped small landscape areas.

During the selection process, consider durability, installation cost, maintenance cost, and the final appearance of each material for the intended landscape use. While no surfacing material can be called the best for all situations, each does have its good points and short comings.

LANDSCAPE FENCING

Fences and walls serve to enclose a section of the landscape. A fence is an enclosure having posts, rails, and infill. Privacy, screening, property bound-

17-28. A basket weave fence can provide this privacy for outdoor activities.

aries, and security considerations dictate the selection of the type of fencing to be used. Landscape designers try to create a private area behind the house for outdoor family activities. Fences made of solid material can provide this privacy for outdoor activities.

FENCE SELECTION

Chain link fences give little privacy, but afford the homeowner the maximum in security. Several fence types allow both privacy and some security for the homeowner. Rail-type fences give no security or privacy, but show property line location. Some fence styles are installed just to beautify the landscape or give a backdrop to a planting bed.

Basket weave, board on board, and stockade type fences may be purchased as manufactured fence panels from a lumber yard or building supply store. These panels measure 6 feet high and 8 feet long. Wood fences use treated wood posts. **Treated fence posts** contain chemicals that prevent wood rot and extends the post life. Metal fences use metal posts. Install chain link fence fabric on steel pipe posts which are set 10 feet apart.

INSTALLATION PRACTICES

Stretch a nylon string between two stakes at the fence line location. Drive additional stakes along this string line to mark each post location. Dig the post hole with a hand post hole digger or powered digger. Several companies manufacture and sell gasoline or electric powered post hole diggers. Use them to increase job productivity. Level the post using a post level or a carpenter level. After leveling the fence post, back fill the hole with concrete. Allow the concrete to cure for several days before attaching the fence panels.

LIGHTING THE LANDSCAPE

For years, landscape lighting consisted of one or two spot lights attached on the back of the house. Today, modern low voltage lamps not only provide security, but also add beauty to the landscape. ***Night-lighting,*** describes the use of ornamental lighting to enhance the landscape at night. Now landscape designers can choose from hundreds of different style lamps to solve a variety of night-lighting problems.

17-29. Night-lighting is the use of ornamental lighting to enhance the landscape at night. Notice the up lighting on the tree and the small lights wrapped around the tree trunk.

LANDSCAPE LIGHTING SAFETY

Electricity and damp soil can be a hazardous combination. Many city ordinances require a licensed electrician to install all outdoor lamps. Always check the local laws before working with electricity. Install a ground fault circuit interrupter, known as a GFCI device, on all outdoor lighting equipment. Before people can be shocked, the GFCI will disconnect the electricity to a lamp when it detects a short circuit.

Low voltage landscape lighting equipment lamps (bulbs) are similar to bulbs used in automotive lighting. These systems run on 24 volts of electricity, not the 120 volts that are standard in a house. Low voltage electric lighting systems reduce the likelihood of electrical accidents that could permanently injure people.

17-30. Lights placed along a sidewalk for night-lighting.

Working at night, the landscape crew locates the lamp positions; then during the following day, they permanently place the lighting devices in the landscape. Landscape crews should always have customer safety and their own safety foremost on their minds while installing landscape lighting devices. Safety is no accident.

LIGHTING TECHNIQUES

There are four major night lighting techniques: walk lighting, up lighting, silhouette lighting, and down lighting. *Walk lighting* offers improved safety for pedestrians by illuminating walkways while providing an interesting lighting effect in the landscape. The *up lighting* technique places a lamp, which shines up, at a tree base. Walk lighting and up lighting are the most common lighting techniques in use today.

17-31. Up lighting at the base of a tree.

While the following two kinds of night lighting are less popular, they do offer a unique opportunity to show off the landscape. ***Silhouette lighting*** places a lamp behind an unusually shaped plant to accent and silhouette this plant. ***Down lighting*** positions a lamp in a tree, causing the light to shine down and casting an interesting shadow pattern on the ground.

REVIEWING

M AIN IDEAS

Landscape installation begins after developing the final landscape plan. Small trees and shrubs are usually sold as bare root or in plastic containers. Large landscape trees are dug from the nursery and sold as balled and burlapped.

Dig the planting hole for trees and shrubs larger in diameter than the soil ball to allow for root growth. Back-fill the hole with the native soil removed when digging the planting hole. Spray the foliage of evergreen shrubs with an anti-transpirant to prevent excess water loss by transpiration. Mulch over the planting hole with bark chips to reduce soil moisture and prevent weed growth.

Annual flowering bedding plants give the landscape instant color, but must be replanted each new growing season. Perennial flowers will survive winter temperatures and flower for many years. Spring flowering bulbs must be planted in the fall around Thanksgiving, and then flower the following March, April, and May. Apply Preen herbicide to flower beds to prevent weed seed germination and hand weeding in the summer.

A variety of hard surfacing materials can be used to construct a walk or a patio. Concrete is easy to install and inexpensive, but it is rather plain. Brick or flag stone will create an unusual walk or patio, but is more expensive.

Landscape designers recommend different kinds of landscape fence sections for different situations. Privacy fences need to be solid and prevent people from seeing into the property. However, security fences need to be strong, but do not have to be solid.

Today, a variety of new landscape light fixtures are available to light up the landscape. When installing outdoor lighting, always use a GFCI device for all electrical outlets and follow the manufacturer's directions. Place landscape lamps in a temporary position at night to determine their best location, then permanently install them the next day.

QUESTIONS

Answer the following questions. Use complete sentences and correct spelling.

1. Why is the landscape site staked?

2. How is a small tree correctly planted, from digging the hole to watering after planting?

3. What is J.U.L.I.E., and why call there before digging at the landscape site?

4. Why should newly transplanted trees be staked?

5. What is an anti-transpirant and why should it be used when transplanting evergreens?

6. Why wrap tree trunks after transplanting?

7. What is the difference between a flower border and a flower bed?

8. What does the term "hardscape" mean?

9. What is the normal size of most fence panels?

10. Name three annual flowers used in landscape planting beds.

11. Why does the landscape contractor install hard surfacing materials?

12. Why install GFCI on all outdoor electric outlets?

13. What is the difference between up lighting and down lighting?

EXPLORING

1. Plant landscape trees and shrubs around your school.

2. Visit a landscape nursery and identify B&B, BR, and container-grown plants.

3. Practice installing different types of landscape fence panels.

4. Plant hardy bulbs in the school planting beds.

18

LANDSCAPE MAINTENANCE

Attractive landscapes require regular maintenance. The work to be done depends on the season, kind and age of plant, and desired effect. Newly-installed landscapes must receive care to help the plants become established and grow. Installing an appealing landscape is wasted without proper maintenance.

Most landscapes need some attention on a regular basis throughout the year. Watering, trimming, fertilizing, mowing, removing leaves, and other duties help maintain a landscape. Following the basic procedures in this chapter will go a long way in the proper care of a landscape.

18-1. Well designed and maintained landscapes are beautiful.

OBJECTIVES

This chapter will explain many principles and practices necessary for maintaining beautiful home landscapes. It has the following objectives:

1   List goals for good plant pruning

2   Name the parts of a tree related to pruning

3   Describe proper tree pruning practices

4   Show proper shrub pruning

5   Explain tree and shrub fertilizing principles

6   Describe the major tree fertilization techniques

7   List the major destructive plant insects

8   Describe plant disease control techniques

9   Name weed control techniques used in landscape beds

10   List the major different mulch materials

TERMS

arboriculture

canes

canopy

crotch

dead zone

dripline

heading cuts

leader

lopping shears

pole saw

post-emergent

pre-emergent

rejuvenation pruning

renewal pruning

scaffold

sucker

thinning cuts

tree topping

trunk

watersprouts

LANDSCAPE MAINTENANCE BASICS

Landscape maintenance is the care and upkeep of the landscape materials after installation. Landscape maintenance includes tasks, such as mowing grass, fertilizing landscape plants, pruning landscape plants, applying pesticides, weeding flower beds, edging beds, removing leaves in the fall, cultivating soil, and applying mulch to planting beds. The goal of a good landscape maintenance program is to keep the landscape as attractive and functional as intended in the original landscape design.

Most landscape companies focus on landscape construction or maintenance. The specialized equipment required for each type of work differs. Landscape workers are trained to do specific types of work and are familiar with certain pieces of equipment. However, some landscape businesses do both, but usually with different crews.

18-2. Maintenance involves weed control in planting beds. (Courtesy, Ron Biondo, Illinois)

PRUNING

Good landscape design and planting practices can reduce the need for pruning in the landscape. Trying to maintain a large growing tree in a confined space, by severe pruning, does not make good sense.

Pruning can be defined as the judicious removal of certain plant parts. Several misconceptions exist about proper tree pruning techniques. However, a good basic set of ground rules will result in healthy plant growth after pruning. Correct pruning practices enhance the landscape effect and the

18-3. A chipper unit used to grind up tree limbs into a mulch.

value of plants. Always apply good pruning practices, along with common sense, while pruning landscape plants.

Pruning permanently alters the direction of plant growth. The removal of tree branches is a natural process that people must understand. Sometimes people prune trees just to have something to do. Always have specific goals in mind before using pruning equipment.

Good pruning goals include:

- Trim plants to maintain their natural beauty
- Regulate plant growth to control plant size
- Eliminate dangerous branches hanging over houses
- Remove broken or damaged branches to improve plant health
- Thin branches to increase light penetration at the center
- Trim plants to increase flower and fruit production

TREE PARTS

Before pruning trees, you should become familiar with tree part names as they relate to pruning. The major parts of a tree include crown or canopy, crotch, trunk, scaffold branches, terminal or leader, and suckers or watersprouts.

Scaffold branches establish the outline of the tree top and create the tree *canopy*. As students become familiar with different kinds of trees, they can

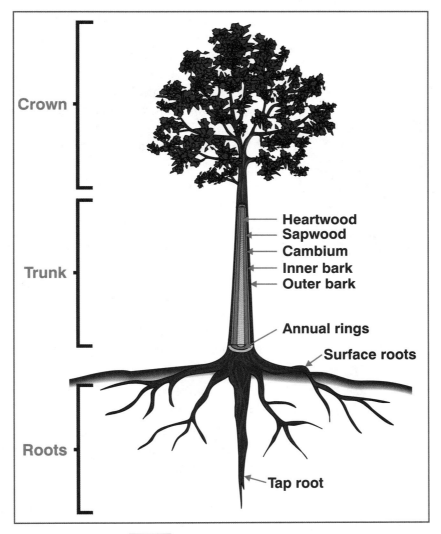

18-4. The major parts of a tree.

identify a tree from a distance by the shape of the canopy. The term *scaffold* branch defines the major tree branches extending from the trunk. Removing a scaffold branch will alter the general canopy shape. Thus, carefully consider the canopy size and shape before removing major branches.

Most trees have a central stem or *trunk.* The term *crotch* describes the junction of scaffold branches and the tree trunk. A tree's main growing point, the tip end of the trunk, is called the *leader* or terminal. This terminal bud controls the growth of the buds and branches below it. Removing the termi-

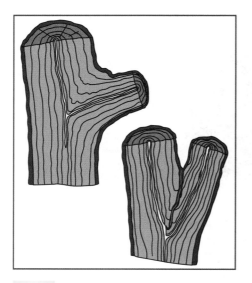

18-5. A comparison of U-shaped (left) and V-shaped (right) crotches.

nal bud will modify the shape of a tree; therefore, be careful not to damage this bud.

Most tree trimmers use the terms suckers and **watersprouts** interchangeably. **Suckers** are branches that arise from latent buds near wounded areas on a tree trunk or branch. They usually grow very fast and straight up, not horizontally as normal branches do. Remove suckers as close to the main stem as possible to prevent their regrowth.

PRUNING EQUIPMENT

An anvil-and-blade pruner has one blade that has a sharp cutting edge. It cuts through small branches up to ¾ inch thick onto the anvil-shaped surface. A by-pass pruner is similar except that both blades are sharp and the cutting edges pass one another much like scissors.

Lopping shears have a top cutting blade and a curved bottom blade to grip the wood while cutting. Their long handles allow for greater cutting and reaching power. Hedge shears have two sharp scissor-like cutting blades. Hedge shears are effective for pruning young, tender growth on shrubs. A *pole saw* consists of a saw mounted on a pole. It allows a tree trimmer to remain on the ground and still cut medium tree branches 6 to 12 feet high. Pole pruners allow the cutting of tree branches up to 1½ inches in diameter without climbing the tree.

Curved-blade and bow pruning saws have much larger teeth than a carpenter's hand saw. The large teeth will prevent the saw from binding when cutting through sap-filled tree branches. They are useful in cutting medium to large size branches. Chain saws have generally replaced the hand saws. They have cutting teeth that move around a grooved bar. They allow a person to easily cut large tree branches.

Pruning equipment is sharp and, if not safely handled, can lead to injury. Always use equipment correctly and follow all manufacturers' recommendations on use and maintenance. Federal labor laws prohibit minors from operating power saws. Eye and ear protection must be worn when operating chain

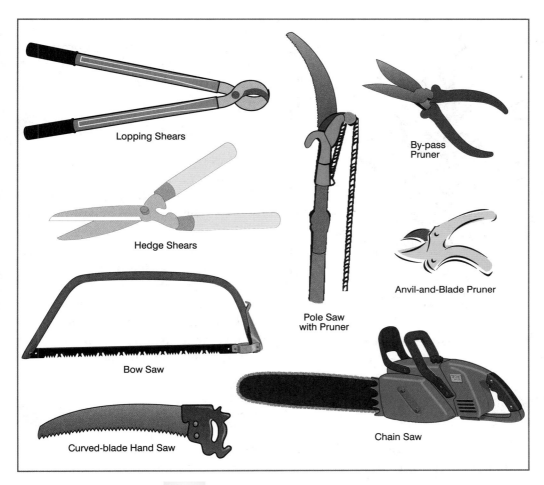

Lopping Shears

Hedge Shears

By-pass Pruner

Pole Saw with Pruner

Anvil-and-Blade Pruner

Bow Saw

Curved-blade Hand Saw

Chain Saw

18-6. **Examples of pruning equipment.**

saws. Also, wear protective clothing when cutting tree branches with power saws. Chapter 24 covers guidelines for chain saw safety in more detail.

TRIMMING DECIDUOUS TREES

Shade trees growing in the home landscape usually require little pruning. Most pruning relates to removing storm damaged branches for safety concerns. The study of trees, their growth and culture, is called *arboriculture*. An arborist takes care of trees by pruning, fertilizing, and controlling insects and diseases. Today, tree trimmer describes people that trim and maintain

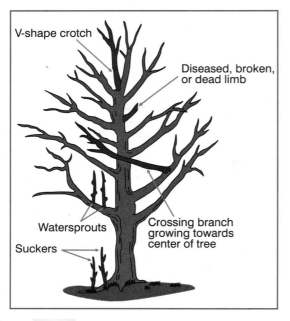

V-shape crotch

Diseased, broken, or dead limb

Watersprouts

Crossing branch growing towards center of tree

Suckers

18-7. Prune obvious faults from a tree.

trees. Some companies will only trim trees. However, landscape companies usually provide a variety of maintenance services besides tree trimming.

Tree Topping

Many homeowners still request tree companies to top their trees. The term *tree topping* refers to reducing the total tree height by $1/3$ to $1/2$ by removing upper scaffold branches. Tree topping usually shortens the life of a tree by allowing disease and insects easy entrance into the tree trunk. Tree topping severely weakens the trunk and future storm damage will usually occur. Topping also destroys the natural shape of the tree canopy.

Storm Damage

To keep trees healthy, immediately remove any broken or damaged tree branches after wind storms. When removing limbs larger than your wrist,

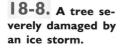

18-8. A tree severely damaged by an ice storm.

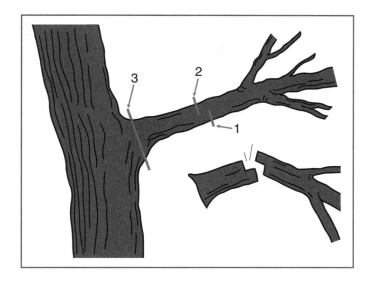

18-9. Use three cuts to remove most medium to large limbs.

make several cuts to prevent tree trunk damage by the bark ripping. Complete this pruning technique by removing the stub close to the trunk.

In the past, tree trimmers made large flush cuts when cutting large tree branches. Today, current research suggests that the flush cut may not be necessary and can initiate the tree's decline. A smaller cut has replaced the larger, flush-type cut. When removing large limbs, always cut them back to the trunk or to another branch. Never leave a stub that can rot and provide insects easy entrance into the tree tissue.

PRUNING RESPONSES AND IMPORTANT ASPECTS

The growth of the tree can be largely directed by pruning. There are several types of pruning cuts. ***Thinning cuts*** remove entire shoots. ***Heading cuts*** remove part of a shoot and direct the growth into a side limb or bud. If a limb is too low, prune leaving an upward growing shoot. If a limb is too high, prune leaving a more horizontal limb. If just buds are present and you want to direct the growth, prune just above a bud pointing in the direction you want the bud to grow. The growth of side or lateral buds is limited by high concentrations of auxins flowing downward from the shoot tip. If you prune off the shoot tip, the auxin flow is stopped and the top buds break. When limbs are in a horizontal position, auxin flow to the top of the limb is diminished and shoots on the top of the limbs break and grow. These strong vertical shoots coming off of a limb are called watersprouts. The more horizontal a limb, the more watersprouts will be produced.

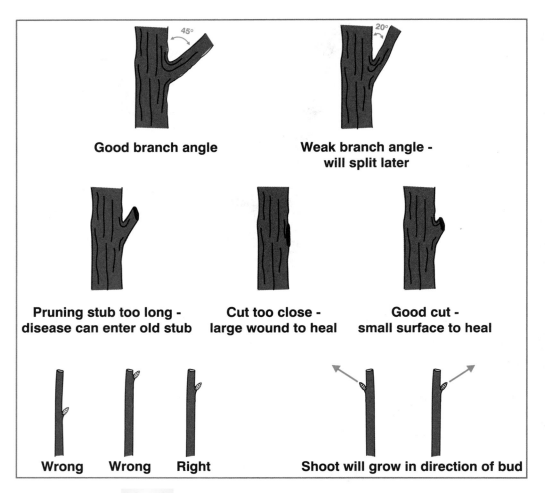

18-10. A guide on how to make proper pruning cuts.

Proper pruning cuts are very important in keeping trees healthy. If pruning stubs are left, disease may start on the dead stub and move down into the tree.

Other Types of Tree Pruning

A good tree maintenance program will provide for the trimming and removing of low growing tree branches that hang over the street, house, or patio. Always try to visualize the after pruning tree shape before removing any branches. Branches accidentally removed cannot be reattached. A good tree maintenance program will provide for minor trimming on a regular five-year

18-11. A good tree maintenance program will provide for the trimming and removing of low growing tree branches.

schedule. This type of preventive maintenance schedule can prevent minor tree problems from becoming major problems.

Tree Wound Paint

Painting the cut with a tree wound material is no longer done. Tree wound materials tend to seal disease organisms in the freshly cut surface. These organisms increase the chances of tree death. Do not paint tree cuts or wounds with any kind of material.

Time of Year to Prune

Early settlers trimmed their fruit trees in the winter because few other farming operations could be done at this time. Dormant winter pruning has been the standard for the landscape industry for years. University research now shows that trees can be trimmed at other times of the year besides the winter. Never prune trees in the spring as the new leaves are unfolding and expanding. Pruning at other times of the year does not seem to injure trees.

A few trees should be trimmed when in full leaf to prevent "bleeding," the excess loss of tree sap. Hard maple, birch, and beech trees are all bleeders. Trees bleeding sap should not be confused with the liquid given by trees infected with wet wood disease.

PRUNING FLOWERING SHRUBS

Woody landscape plants with multiple stems growing out of the ground instead of a single trunk are called shrubs. Individual stems called **canes** develop at the crown or growing point. The junction of roots and shoots defines the shrub's crown area.

Shrubs furnish the landscape with a spectacular splash of color for one or two weeks each year. Proper pruning must be done to keep them flowering year after year. Flowering shrubs without flowers look like any ordinary green plant. Shrubs develop next year's flower buds immediately after flowering each year.

Rejuvenation Pruning

Cut all the canes within 6 inches of the ground during **rejuvenation pruning.** This pruning technique can be done on shrubs that have grown too large for their planting location or for shrubs that have stopped flowering. Rejuvenation pruning will be done in late winter before the shrubs start their spring growth.

Rejuvenation pruning is best done by using loppers and chain saws. Loppers are used to cut medium size shrub branches. An entire flowering season will be lost after rejuvenation pruning of shrubs. Not all shrubs can tolerate this type of pruning.

18-12. Rejuvenation—cutting all shrub canes within 6 inches of the ground.

Renewal Pruning

The *renewal pruning* technique gradually rejuvenates a shrub over a three or four year period. Remove 1/3 of the shrub canes at ground level each year for three consecutive years. Canes less than four years old generally provide the best flowers. The renewal pruning process prevents any shrub cane from growing more than three years. All shrubs respond to renewal-type pruning. Renewal pruning also allows for better air circulation around the leaves. Improving air movement around plants can reduce a variety of leaf diseases.

Renewal pruning is done in the early spring before spring growth starts. The lopper-type pruners work best for removing the large shrub canes. Some remaining canes may be too long and will need to be shortened with heading cuts.

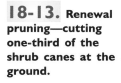

18-13. Renewal pruning—cutting one-third of the shrub canes at the ground.

PRUNING EVERGREEN SHRUBS

Small and medium size evergreens growing in a foundation planting need yearly pruning. Pruning may prevent them from growing too large for the landscape site. Landscape designers recommend planting Yew, Juniper, Arborvitae, or Boxwood evergreen varieties.

Each of these names represents a group of plants with hundreds of individual selections that differ from each other in size, shape, and leaf color. For years, foundation evergreens were sheared using a hedge trimmer. Many

18-14. Trimming a shrub with an electric trimmer. (Courtesy, Jasper S. Lee)

home owners still practice this shearing type of pruning with their electric hedge trimmers.

Shearing evergreens creates the "frustrated barber" landscape look. Mounds and mounds of green flow from one end of the house to the other. Shrubs lose their individual identity and become part of a massive green growing thing.

Today, this look is out; and the informal look is in. Evergreens should never look like they were trimmed, they should have a natural look. Careful pruning using a hand pruner, not the hedge trimmer, can produce this natural look.

Evergreens have a *dead zone* 6 to 12 inches below the green needles. Most types of evergreens cannot develop new leaf buds after pruning into the dead zone. Pruning evergreens severely can permanently disfigure the shrub. Never prune evergreens into the dead zone. To keep evergreens looking good, lightly prune them in early summer after the new growth has completed development. Moderate or heavy pruning of evergreens should be done in early spring before new growth starts to develop. Pruning medium size evergreen branches will require using lopper-type pruning equipment.

MULCHING

Mulching is the practice of spreading material over the surface of the soil. Mulching reduces water loss from surface evaporation and keeps soil temperatures uniform. Mulching also discourages weed growth and improves the general appearance of landscape planting beds. Organic and inorganic mulch are the two different types of mulch to choose from. In the case of organic much, soil tilth can also be improved as the mulch decays. Before spreading mulch over the soil, many landscapers install landscape fabric in the planting bed. The landscape fabric is helpful in reducing weed growth. Select a mulch that is easy to obtain, easy to apply, and inexpensive. Mulch should be uniformly applied to a depth of 3 to 4 inches.

ORGANIC MULCH

Organic mulches come from material that was once living–shredded tree bark, pine needles, rice hulls, peat moss, wood chips, crushed corn cobs, and cocoa bean hulls. Each region of the country produces several different kinds of organic mulch materials. Not all mulch materials are available in all parts of the country. While organic mulches have a very pleasing earthy appearance, they decay quickly in the landscape. Normally, beds require a yearly top dressing with new mulch to keep them looking good.

INORGANIC MULCH

Inorganic mulches come from materials that were never living. Inorganic mulch materials are more permanent and seldom require top dressing with new mulch. Gravel, crushed stone, river gravel, volcanic rock, sand, and brick chips furnish the landscape contractor a variety of materials to choose from.

18-15. Installing landscape fabric in a planting bed is helpful in reducing weed growth.

18-16. Pine needle mulch has been used in this flower bed.

FERTILIZATION

Chapter 6 in this book covered the many aspect of fertilization. The following information will be helpful when using fertilizer as part of a landscape maintenance program.

SHRUB FERTILIZATION

Several common fertilizer application techniques exist for fertilizing shrubs. Shrub feeding roots grow in the top 10 to 14 inches of soil. For fertilization to improve plant growth, it must move into the shrub root zone.

Fertilizer Analysis

Fertilization of shrubs may increase flower production, but it can also increase the shrub's total growth, requiring additional pruning. A wide range of fertilizer analysis can be applied. Good shrub fertilizer will contain nitrogen, phosphorus, and potassium in about equal amounts. Use one of the following fertilizer grades: 5-10-10 or 10-10-10, etc.

Plants respond well to a late winter or early spring fertilizer application. They need quickly available nutrients to start growth after the winter dormant period. Try to avoid late summer or early fall fertilizer application. Fertilization at this time of the year may delay a plant's preparation for the winter dormant period. Winter injury can result from late summer fertilizer applications.

Application Techniques

Fertilize shrubs in foundation planting beds by scattering granular fertilizer around the shrubs. Spread one hand full of fertilizer over a 3-foot by 3-foot bed area or 2–4 pounds of 5-10-10 fertilizer per 100 square feet of bed. Absorption of nutrients by the mulch can be a problem when fertilizing beds with granules.

Many homeowners fertilize landscape shrubs with liquid fertilizer solutions. Dilute concentrated fertilizer solutions with water and apply to the shrub beds. Always read and follow all label directions.

Fertilizer spikes can also be used. Drive small fertilizer spikes into the soil next to each shrub. Several different companies manufacture these fertilizer spikes. The label will describe the correct spacing for the fertilizer spikes.

TREE FERTILIZATION

The same general principles discussed when fertilizing shrubs will also apply to tree fertilization practices. Young trees benefit more from fertilization than old mature trees. Fertilize trees in the winter or early spring for best results and avoid late summer applications. Landscapers usually fertilize trees using the granular or liquid injection method. However, both homeowners and landscapers use fertilizer spikes.

Application Techniques

Drill holes 30 inches apart under the tree canopy and fill the holes with granular fertilizer. Use a large, heavy duty, ½ horse power electric drill to make the 2-inch diameter holes 12 to 15 inches deep. Be careful not to drill into any underground utility lines. Fill each hole with 1 cup of 15-5-10 granular fertilizer; then back fill the holes with the soil removed by drilling.

Inject liquid fertilizer solutions into the root zone using a fertilizer needle connected to a high pressure pump. Large power sprayers provide the 200 psi pressure necessary to inject the solution into the soil.

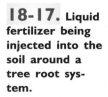

18-17. Liquid fertilizer being injected into the soil around a tree root system.

Make a standard fertilizer solution by dissolving 30 pounds of 20-20-20 fertilizer in 200 gallons of water. Insert the feeding needle 12 to 14 inches into the soil every 30 inches under the entire tree canopy. Turn on the feeding needle valve for five seconds at each injection site.

Fertilizer Spikes

During the 1980s, fertilizer spikes were developed by several companies. Manufacturing plants make fertilizer spikes by compressing granular fertilizer into a spike shaped form. A binding chemical provides the glue to hold the fertilizer particles together. Most spikes are 6 inches long and about the diameter of a shovel handle.

Drive the spikes into the soil at the tree dripline. The *dripline* is the outer edge of a tree branch system that creates an imaginary circle on the ground. Always read the directions on the box of tree spikes for exact spacing information. Golf courses and landscape companies both find this an easy tree fertilization technique.

18-18. Drive fertilizer spikes near the tree dripline.

PEST CONTROL

Pests can severely injure or kill landscape plants. However, the damage is usually more of a nuisance than life threatening to the tree. Nature provides birds and other insects to control most plant damaging bugs. Occasionally, people have interfered with this natural control process by importing insects into the United States from other parts of the world. If their natural enemies are not also imported, then an insect epidemic can develop.

During the 1920s, a boat load of elm logs containing the European bark beetle arrived in New York harbor. The bark beetle carried a fungus that caused Dutch elm disease. As the beetle eats the young elm shoots, it also transmits the fungus that can kill the tree. From 1920 to 1970 the bark beetles slowly migrated from New York to California, infecting and killing most American elm trees.

CHEWING AND SUCKING INSECTS

Insects from both the chewing and sucking groups damage trees. Chewing insects, such as caterpillars and bag worms, injure trees by eating all or part of the leaves. Under severe conditions they can eat all the leaves and thus defoliate the entire tree. Caterpillars' names can be very helpful in insect identification, as in the yellow-necked caterpillar or the fall tent caterpillar.

Sucking insects damage trees by sucking plant sap from the leaves. Aphids, scale insects, and mites represent the major sucking insects that damage trees and shrubs. Leaves damaged by sucking insects usually turn yellow. Using a magnifying lens can help in finding these very small insects.

18-19. Tent caterpillars construct silk tents and can eat all of the leaves off of trees.

Most tree insect damage occurs during the summer months. Collect and identify insects before selecting and applying insecticides. The term insecticide describes pesticide used in controlling insect pests.

PESTICIDE SAFETY

The term insecticide literally means insect killer. Insecticides can also pose a health threat to humans. Always read and follow all pesticide label directions to protect yourself and the environment. Wear protective clothing, rubber boots, rubber gloves, goggles, and respirators while mixing and applying pesticides. Check with the state Agriculture Extension office for specific information about controlling insect pests in your part of the country.

TREE AND SHRUB DISEASES

Most trees and shrubs show symptoms of minor disease damage. However, most diseases do not kill the plants. Leaf spotting, leaf yellowing, and early leaf dropping describe the major plant disease symptoms. Bacteria, fungi, and virus represent the major plant disease groups.

Disease Control

Landscapers use several different disease control methods. Planting disease resistant varieties of trees and shrubs can prevent many serious prob-

18-20. Plants growing in damp locations are more susceptible to plant disease.

lems. Damp and cloudy weather conditions provide the ideal environment for plant disease development. Plants growing in dark, damp landscape locations are more susceptible to plant diseases. Pruning trees can increase light penetration to the foliage and inhibit disease development. Pruning can also improve air circulation around shrubs and reduce plant diseases.

Plant diseases can be controlled by applying chemicals called fungicides that kill fungal organisms that attack plants. Read and follow all label directions when applying fungicides. Fungicides only provide temporary protection to the plant leaves. Reapply fungicides every two to three weeks during the growing season.

Weed Control in the Landscape

A weed describes any plant growing out of place or any unwanted plant. Weeds generally grow faster than desirable landscape plants because of a difference in their rate of photosynthesis. Weeds can destroy the beauty of the best designed landscape planting. They also rob water and necessary nutrients from landscape planting.

Control unwanted weed growth in landscape beds by using one of the following methods: landscape fabric, mulch, mechanical control, or herbicide. Each method has certain advantages and some disadvantages when compared to the other techniques.

Landscape Fabric

In the past, contractors installed solid sheets of black plastic under the mulch material to prevent weed growth. Poor plant growth can result from reduced soil moisture and soil aeration according to recent research studies. Today, a woven plastic fabric replaces the solid plastic sheets of yesterday. Few, if any, weeds penetrate this new landscape fabric. After installing the fabric, cover it with mulch materials.

Mechanical Control

This term describes weed control by hand pulling or hoeing. A few weeds will always germinate and develop in any landscape mulch materials. In addition, some landscape plants cannot tolerate herbicides. Therefore, this old hand pulling technique may be the only acceptable weed control method available. Carefully remove the weeds without disturbing the roots of the landscape plants.

Herbicides

The name herbicide describes the group of chemicals that prevent weed germination or kill actively growing weeds. An "herb" is a plant, and "cide" means to kill; thus, herbicides are plant killers. However, the term herbicide usually means weed killer, not plant killer. Herbicides can be divided into several major groups, depending on how and when they kill weeds. Always read and follow all label directions when using herbicides.

18-21. Apply granular herbicide to the shrub bed using a hand-held applicator.

Apply *pre-emergent* herbicides before the weed seeds germinate. This group of herbicides stops or prevents seed germination. The following herbicides belong to this group: Preen, Dacthal, Siduron, Galary, and Diclomec. Applying pre-emergent herbicides to actively growing weeds will waste both time and money.

Post-emergent herbicides kill actively growing weeds. Post-emergent herbicides require careful application to prevent the killing of desirable plants. During the 1980s and 90s, Roundup dominated the post-emergent group of herbicides. Wipe or spray a small amount of Roundup on the leaves of a weed and the entire weed will die.

REVIEWING

MAIN IDEAS

The best designed landscapes must have constant maintenance if they are to look their best throughout the year. Good maintenance includes proper pruning. Never prune trees and shrubs just to be pruning. Have specific goals, such as removing dangerous hanging branches or increasing light penetration to flower beds, before trimming trees. Current practices discourage the painting of tree wounds made when pruning tree branches.

Prune small shrub branches with hand pruners, but use loppers when pruning large shrub canes. Old flowering shrubs can stop flowering if not properly maintained. Rejuvenate shrubs in the late winter by cutting all the canes within 6 inches of the ground. Renewal pruning is more gradual and may take three to four years to complete, but will improve the shrub's flower production.

Both organic and inorganic mulch materials are used in landscape beds to reduce weed growth and present a beautiful planting bed. A variety of materials are available throughout the United States. Check with local landscapers to determine the best materials to use.

Fertilize shrubs in the late winter using a 5-10-10 fertilizer analysis. Landscape maintenance workers fertilize shade trees in the early spring by injecting a liquid solution into the root zone using a high pressure sprayer and feeding needle. Homeowners may drive fertilizer spikes around the dripline of shade trees to provide fertilizer nutrients to the root system.

Pests can severely injure or kill landscape plants. Both trees and shrubs can be damaged from chewing and sucking insects. Check with your local Extension Office to identify insects damaging trees and shrubs and find out the best control techniques. Control weeds in landscape beds by using one of the following methods—landscape fabric, mulch, mechanical control, or herbicides. Each method has advantages and disadvantages when compared to the other techniques. When using pesticides, always read and follow all label directions.

QUESTIONS

Answer the following questions. Use complete sentences and correct spelling.

1. What is a tree canopy, crotch, and suckers?
2. What special protective equipment should be used when running a chain saw?
3. What is the study of arboriculture?

4. Why do most landscape maintenance people discourage tree topping?

5. When during the year should trees not be pruned?

6. What is the difference between rejuvenation and renewal pruning of shrubs?

7. Homeowners usually prune evergreens using what tool?

8. What is a dead zone in an evergreen, and why is it important not to trim into this zone?

9. What is the analysis of a good tree fertilizer?

10. Describe how shade trees can be fertilized?

11. Name three sucking and two chewing insects.

12. Name the major groups of plant diseases.

13. Why mulch trees and shrubs?

14. Why should solid sheets of black plastic not be used in landscape beds?

EXPLORING

1. Visit a landscape maintenance company and watch them prune trees.

2. Practice pruning shrubs using the renewal pruning technique around your house.

3. Use fertilizer spikes to fertilize trees on the school grounds.

4. Collect and identify 10 different insects from trees and shrubs.

19

WARM AND COOL SEASON GRASSES

Beautiful, green grass is appealing and useful in many ways. All people like lawns, playing fields, parks, and other places with turfgrass. Turf is an integral part of the landscape.

Having good turf is more than throwing out a few grass seeds. Turf requires the same care as other plants. The soil must be tested and fertilized, the seedbed prepared and properly planted and watered, pests must be controlled, and the turf maintained.

Southern and northern turfgrasses need many of the same general practices. The differences are in the species used and the cultural practices in growing each. People also have different objectives for turfgrass in different locations.

19-1. A beautiful lawn or playing field is a complex community of plants.

OBJECTIVES

This chapter covers the general principles of turf as well as includes specific detail on warm and cool season turfgrasses. It has the following objectives:

1 Distinguish between warm and cool season grasses

2 Describe turfgrass functions and quality

3 Describe how to select a turfgrass

4 Identify important warm season turfgrasses and briefly describe each

5 Identify important cool season turfgrasses and briefly describe each

TERMS

cool season turfgrass
functional turf quality
ornamental turf
sports turf
transition zone

turfgrass
turf quality
utility turf
visual turf inspection
warm season turfgrass

TURFGRASS FUNCTIONS, QUALITY, AND SELECTION

The beautiful lawn or playing field is a complex community of plants. Important relationships exist between the plants, soil, and other factors in the environment. *Turfgrass* is a collection of grass plants that form a ground cover. It is an interconnecting community of plants that provide a ground cover. Areas of turf may have one species or may contain a mixture of grasses. Turf represents a high level of ecological organization. The ecological system includes plants, soil, and organisms in the soil.

In the past twenty years, major changes have occurred in the turfgrass industry. New types of equipment, new grass cultivars, and new pesticides have revolutionized the turf industry. There are over 50 million home lawns and 15 thousand golf courses in the United States. In this country, more money is spent on fertilizing turfgrass than on fertilizing field corn crops.

TURF FUNCTIONS

Turf is useful in many ways. People enjoy its beauty. Conservationists appreciate its positive effect on the environment. Athletes like turf because of the surface it provides on playing fields. Three purposes and functions of turf are—utility, ornamentation, and sports.

19-2. A well-managed turf area is an integral part of the landscape.

19-3. Turf used to bring beauty to an arid area.

The utility functions of turf are how it helps the environment. *Utility turf* has many useful functions. A number of species and cultivars can be used. It stabilizes the soil and reduces erosion. It has a cooling effect on the environment in hot weather. It helps clean the air by removing toxic emissions.

The ornamentation functions of turf are easily seen. *Ornamental turf* enhances areas around homes, businesses, in parks, and other places. Besides the utility functions, turf is a form of decoration. It brings beauty to areas that might otherwise be unattractive.

19-4. Turf creates a useful, attractive sports playing field.

Sports turf includes all types of playing fields. The turf helps both the participants and the observers. Turf provides a surface that reduces injuries to players. The surface can be specially groomed to provide the kind of area needed for football, baseball, soccer, polo, golf, bowling, tennis, and other sports. Much of the research in creating better turfgrasses was carried out to develop improved sports fields. Developing turf for golf courses has been a high priority. In addition to improved turf, the technology of establishing and maintaining turf also came from interest in sports fields.

TURF QUALITY

Turf quality is the excellence of the turf. It is closely related to the functions of turf. Turf quality is based on utility, appearance, and/or playability.

Visual Quality

Visual turf inspection is often used to assess quality. The common visual factors are density, texture, uniformity, color, growth habit, and smoothness.

- Density is the number of aerial shoots per unit area. How many blades (leaves) are present?

- Texture is based on the width of the leaf blades. Wide blades form a turf with a coarse texture. Narrow blades produce a fine texture.

19-5. Turf with outstanding visual quality. (Courtesy, Jasper S. Lee)

■ Uniformity is the evenness of distribution of the turf on a site. It involves the mass of aerial shoots that form the visible surface. This gives the surface a smooth, even appearance.

■ Color is a measure of the light reflected by the turf. Most turf should be a rich green color appropriate for the particular species.

■ Growth habit is the type of shoot growth: bunch, rhizomatous, and stoloniferous. Grain or horizontal growth patterns of the grass is a problem in stoloniferous turfgrasses.

■ Smoothness is a surface feature that affects visual quality and playability. Preparation of the soil prior to planting the turf is important in smoothness.

*F*unctional Quality

Functional turf quality is how well a turf achieves its purpose. All turf has a particular purpose. The functional factors in turf quality are rigidity, elasticity, resiliency, and recuperative potential.

■ Rigidity involves the resistance of the turf leaves to compression and is related to wear resistance. It is influenced by chemical composition of plant tissue, water content, temperature, plant size, and density.

■ Elasticity is the tendency of leaves to bounce back once a compressing force is removed.

■ Resiliency is the capacity of a turf to absorb shock without altering surface characteristics. Growth medium (organic and inorganic) is a factor.

■ Recuperative potential is the capacity of turfgrasses to recover from damage (infectious and noninfectious).

TURFGRASS SELECTION

A good turf begins with selecting the right turfgrass. Climate, use, maintenance needs, and characteristics of southern and northern turfgrasses are important in selecting the kind to grow.

*C*LIMATE AND TURFGRASS LIFE CYCLE

Turfgrasses are often placed in two groups—southern and northern. These groups are based on the ability of a species to grow and serve a useful purpose in the climate. The *cool season turfgrasses* grow best at an opti-

19-6. The desert regions in the southwest present additional problems in seedbed preparation and turfgrass selection. This shows established turf next to desert sand in Cathedral City, California.

mum temperature range of 60–75°F, and the **warm season turfgrasses** grow best at an optimum range of 80–95°F.

If cool season turfgrasses go into summer dormancy, they may not be able to survive. Irrigation and syringing (mist for surface cooling) help to minimize summer dormancy. Warm season turfgrasses have winter dormancy unless they are found in tropical climates where they remain green all year. The cool season turfgrasses are limited by heat and drought, and the warm season turfgrasses are limited by cold weather.

TRANSITION ZONE

A **transition zone** is an area between definite climate zones. The transition zone in the east is a zone separating temperate and subtropical zones in the eastern United States. It goes east and west at 37° north latitude and is 200 miles wide. Certain turfgrasses meet the limits of their southern and northern exposures.

The climate in the transition zone favors growth of some warm and some cool season turfgrasses, but is not really optimum for either. The transition zone is not ideal for any high intensity turfgrasses. However, several warm season and cool season grasses can be grown in the transition zone.

In the east, Zoysia is the most cold tolerant of the warm season turfgrasses. Bermudagrasses can be grown in the transition zone with some winter-kill. Tall fescue and ryegrasses are popular turfgrasses for the transition zone.

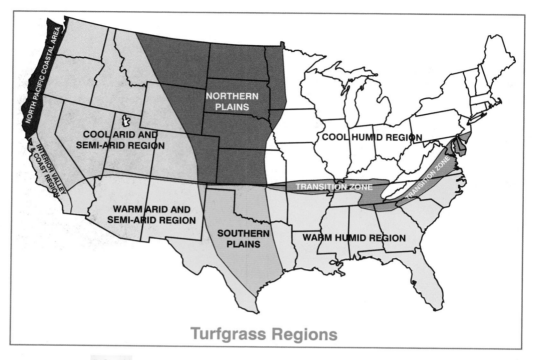

Turfgrass Regions

19-7. A map illustrating turfgrass regions and transition zones.

SCIENTIFIC CLASSIFICATION (NOMENCLATURE)

The family of grasses is the Gramineae or Poaceae. There are six subfamilies, 25 tribes, 600 genera, and 7,500 species. Of these, fewer than 50 are planted as turfgrasses throughout the world.

The trinomial nomenclature system is used in identifying turfgrasses, such as *Stenotaphrum secundatum* cv. Seville (Seville St. Augustinegrass). Here is an example:

Classification: Trinomial (*Cynodon dactylon* cv. Cheyenne)

Kingdom:	Plantae	(plant kingdom)
Division:	Embryophyta	(embryo plants)
Subdivision:	Phanaerogama	(seed plants)
Branch:	Angiospermae	(seeds enclosed in an ovary)
Class:	Monocotyledoneae	(monocotyledons)
Subclass:	Glumiflorae	(having chaffy flowers)

Order:	Poales	(grasses and sedges)
Family:	Poaceae	(grass family)
Subfamily:	Eragrostoidae	(lovegrass)
Tribe:	Chlorideae	(green)
Genus:	Cynodon	(bermudagrasses) (toothlike)
Species:	dactylon	(fingerlike)
Cultivar:	Cheyenne	(cultivated variety)

Cultivars are named after people, places, use, plant characteristics, or any combination. For example, "Tifway" Bermudagrass is named after the city of Tifton, Georgia, and a golf course fairway.

WARM SEASON TURFGRASSES

The selection of turfgrass cultivars is the most important decision in establishing turfgrass. Most warm season turfgrasses are vegetatively propagated by sod, plugs, or sprigs. Some are seeded, such as centipedegrass and zoysia. Common bermudagrass is seeded, with several cultivars being available.

The warm season turfgrasses include bahiagrass, the bermudagrasses, buffalograss, the carpetgrasses, centipedegrass, the gramas, kikuyugrass, seashore paspalum, St. Augustinegrass, weeping lovegrass, and the zoysias.

BAHIAGRASS

Bahiagrass (*Paspalum notatum*) is a tough, coarse-textured grass adapted to warm climates. Bahiagrass is primarily grown in the subtropics and tropics. It is adapted to a wide range of soil conditions, but does well on sandy, slightly acid, infertile soils.

Bahiagrass is propagated by seed and forms a wear resistant, open turf for utility and lawn purposes. It grows with rhizomes and stolons. Bahiagrass has low cultural intensity, and should be mowed at a height of 2.5 to 3.5 inches. Mowing should be with a rotary mower to remove the unsightly seedheads that may develop. Fertilization is usually at a rate of 1 to 4 pounds active ingredient nitrogen per thousand square feet each year. Bahiagrass has minimal irrigation needs. Argentine is the primary cultivar used on lawns, especially in Florida. This cultivar has good density, color, disease resistance, and response to fertilization. Pensacola is another cultivar primarily used for utility turfs.

THE BERMUDAGRASSES

The bermudagrasses are popular turfgrasses with distinct differences in color, texture, density, vigor, and environmental adaptation. Growth is by rhizomes and stolons. Bermudagrasses are adapted to a wide range of soil conditions, but not well adapted to shaded sites.

Common bermudagrass (*Cynodon dactylon*) is seeded and has a medium to coarse texture. It is used as utility and ornamental turf with little or no irrigation. Improved cultivars of common bermudagrass include Arizona and Cheyenne.

The Tifton hybrid bermudagrasses were developed at the Coastal Plain Experiment Station in Tifton, Georgia. Two Tifton hybrid bermudagrasses (*Cynodon dactylon* x *Cynodon transvaalensis*) are available: Tifway (Tifton and fairway) or Tifton 419 (used on fairways, athletic fields, and home lawns) and Tifway II (Tifton 419 II) which is a newer release of Tifway with improved cold tolerance, drought resistance, and insect, nematode, and disease resistance.

Tifgreen (Tifton and golf green) or Tifton 328 is an excellent turf for golf greens. A newer release is Tifgreen II (Tifton 328 II) which has improved cold tolerance, drought tolerance, and insect, nematode, and disease resistance. Tifdwarf is a natural mutation of Tifgreen that was first found on golf greens in Charlotte, North Carolina, and Thomasville, Georgia. Samples were taken from these courses and studied. The best cultivar was released as

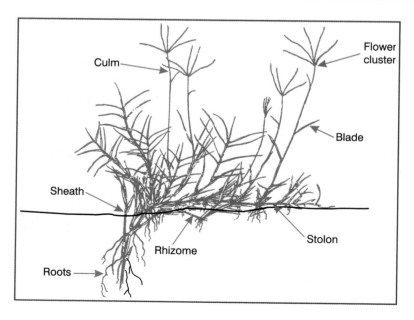

19-8. Major parts of a Bermudagrass plant.

19-9. Mowing a bermudagrass golf green in early morning.

Tifdwarf, without a number designation. Tiffine and Tiflawn are other cultivars that have been replaced by Tifway and Tifgreen.

Another bermudagrass cultivar is Midiron, which is used on lawns and athletic fields. It has good cold tolerance. Midiron has a medium texture. Texturf 1F and Texturf 10 are also used on sports fields and lawns. All bermudagrasses, except for the common varieties, are vegetatively propagated.

Bermudagrass golf greens are mowed at a height of ³/₁₆ to ⁵/₁₆ inch. Fertilizer is provided at the rate of 0.5 to 1.2 pounds nitrogen per thousand square feet per growing month. Bermudagrass tees are mowed at a height of ³/₁₀ to ³/₅ inch. Fertilizer is used on tees at the rate of 0.5 to 1.2 pounds nitrogen per thousand square feet per growing month. Bermudagrass fairways are mowed at a height of ½ to 1 inch.

Bermudagrass in lawns should be mowed at a height of ½ to 1½ inches. Fertilizer should be used at the rate of 0.25 to 0.75 pounds active ingredient nitrogen per thousand square feet per growing month. Seeded common bermudagrass lawns are mowed at 1 to 1½ inches and receive less nitrogen.

BUFFALOGRASS

Buffalograss (*Buchloe dactyloides*) is a fine-textured, grayish-green, turfgrass used with little or no irrigation. It is adapted to a wide range of soil conditions, but is best adapted to fine-textured alkaline soils. Buffalograss has excellent drought and temperature extreme hardiness. Improved cultivars include Texolea and Prairie.

Studies are under way at Texas A&M University and the University of Nebraska with buffalograss. Fairways at The Boulders Club in Atlanta and the turfgrass nursery at Abraham Baldwin Agricultural College in Tifton, Georgia, have established buffalograss.

Buffalograss has a low cultural intensity. The height of mowing is 0.5 to 1.2 inches for lawns and fairways. The fertilization rate is 0.5 to 1.2 pounds nitrogen per thousand square feet per year.

THE CARPETGRASSES

Common carpetgrass (*Axonopus affinis*) is a coarse-textured, low growing, light green, turfgrass. It is propagated by stolons or seed. Carpetgrass is found on moist, sandy, acid soils of low fertility. It is often used for lawns and utility turfs.

Carpetgrass grows more open and outward than some species. Common carpetgrass has a low cultural intensity. It should be mowed at a height of 1 to 2 inches. Mowing should be with a rotary mower to remove the unsightly seedheads. Fertilizer is used at the rate of 2 pounds nitrogen per thousand square feet per year. Irrigation is often needed on well drained soils. No cultivars are available.

Tropical carpetgrass (*Axonopus compressus*) is similar to common carpetgrass in appearance and environmental adaptation, but is less cold tolerant. It should be used in the warm subtropical and tropical climates.

19-10. Carpetgrass.

CENTIPEDEGRASS

Centipedegrass (*Eremochloa ophiuroides*) is a medium-textured, slow growing, light green, turfgrass. It is used as a utility turf and for lawns.

Centipedegrass is adapted to a wide range of soil conditions, but prefers moist, acid, sandy soils of low fertility. It is often called the "poor person's grass" because of the minimal efforts needed for maintenance.

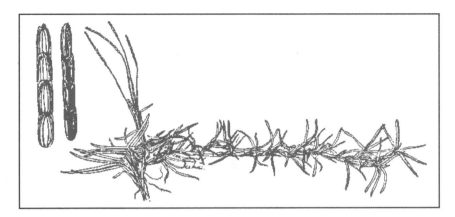

19-11. Centipedegrass stolon.

The cultural intensity of centipedegrass is low. It should be mowed at a height of 1 to 2 inches. Most homeowners make the mistake of mowing to ¾ inch and lower. This scalps the lawn and allows weeds to invade. The fertilization rate is 1 to 2 pounds nitrogen per thousand square feet per year. Skipping every third year of scheduled fertilization benefits a lawn. Centipedegrass often shows an iron deficiency. Iron applications may be necessary to correct the chlorosis.

Oklawn is a cultivar tolerant to drought and temperature extremes. Other cultivars include Tennessee Hardy and Tennessee Tough. AU Centennial is a dwarf form for use in home lawns. Research is underway to expand the gene pool for centipedegrass.

SAINT AUGUSTINEGRASS

Saint Augustinegrass (*Stenotaphrum secundatum*) is a coarse- textured, aggressive turfgrass. It is widely used on home lawns in the warm subtropical and tropical climates. Saint Augustinegrass is adapted to a wide range of soil conditions. It prefers moist, well drained, sandy, slightly acid soils of medium to high fertility. It is not very cold tolerant and is limited to coastal and piedmont areas with mild winters. It will grow well in shaded areas.

Some cultivars are susceptible to chinch bugs and the virus, which cause the disease St. Augustine Decline. Cultural intensity is moderate. It should be mowed at a height of 2½ to 3½ inches. Fertilizer should be used at the rate of 3 to 6 pounds nitrogen per thousand square feet per year.

19-12. Saint Augustinegrass shoot and leaf parts.

Cultivars offer differing characteristics. Bitter Blue, Floratine, Raleigh, and Seville offer improved density and finer texture. Seville is a dwarf form, and Raleigh is the most cold tolerant. Floratam offers chinch bug and St. Augustine Decline resistance, but is a coarse-textured cultivar. Floralawn offers chinch bug and drought tolerance.

Zoysia

Japanese lawngrass (*Zoysia japonica*) is a medium-textured, slow growing, turfgrass. It has outstanding cold tolerance. Zoysia is adapted to a wide range of soil conditions, but is best on well drained, slightly acidic soils of moderate fertility. It is slow to break dormancy in the spring, and will enter dormancy early in the fall when the temperature is 50 to 55°F. Zoysia is straw-colored in the winter.

Zoysia does not grow fast and it takes a while to establish a lawn. The cultural intensity is moderate. Mowing should be with a reel mower due to the density and toughness of the shoots. Mowing heights of ½ to 1 inch are preferred. Fertilizer is used at the rate of 1.5 to 3 pounds nitrogen per thou-

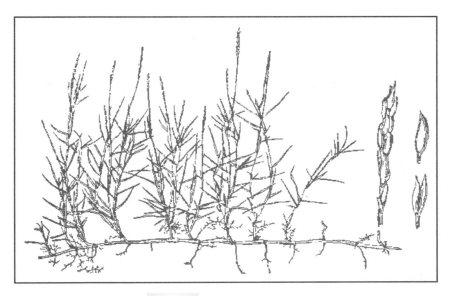

19-13. Manilagrass stolon.

sand square feet per year. Irrigation is often needed to maintain color during dry times of the summer. Improved cultivars include Meyer and El Toro.

Manilagrass (*Zoysia matrella*) is a fine-textured, dense, slower growing species that lacks the cold tolerance of Japanese lawngrass. It is used for lawns in warm subtropical and tropical climates. Mascarenegrass (*Zoysia tenuifolia*) is the finest-textured and slowest growing of the zoysias. It has very little to no cold tolerance and is used as an unmowed ground cover in the warm tropical and subtropical climates.

Interspecific hybrids of *Zoysia japonica* × *Zoysia tenuifolia* offer the two cultivars Emerald and Cashmere. They form a very dense, dark green grass but lack the cold tolerance of Meyer or El Toro.

COOL SEASON TURFGRASSES

The selection of turfgrass cultivars is the most important decision in establishing turfgrass. Although there are hundreds of different species of grasses, only a few will produce quality turf in the cooler regions of the United States. Most cool season turfgrasses are propagated by seed or sod. The principal cool season turfgrasses include bluegrass, bentgrass, fescue,

---

COMMERCIAL VARIETIES

A. Kentucky Bluegrass (Poa pratensis L.)

Adelphi

Baron

Delta

Fylking

Glade

Merion

Parade

Victa

B. Creeping Bentgrass (Argrostis stolonifera L.)

Penncross

Penneagle

Cohansey (C-7)

Pennlinks

SR 10, 20

C. Fine fescues (Festuca rubra)

Jamestown

Pennlawn

Ruby

D. Tall Fescue (Festuca arundinacea)

Jaguar

Kentucky 31

Olympic

Rebel

E. Perennial Ryegrass (Lolium perenne)

Blazer

Derby

Diplomat

Fiesta

Manhattan II

Pennfine

Regal

---

and ryegrass. A number of commercial varieties of cool season grasses are available.

KENTUCKY BLUEGRASS

Kentucky bluegrass (*Poa pratensis*) takes its name from its widespread occurrence in the fertile soils of Kentucky. It also grows across the northern two-thirds of the United States wherever the moisture supply is adequate. Kentucky bluegrass is not a native grass of the United States but was imported from Europe by the early settlers. Most bluegrass lawns are established from seed. It is a good species for sunny lawn areas in the northern cool climates. It is very aggressive grown under favorable soil conditions and can spread rapidly by creeping underground stems called rhizomes. Kentucky bluegrass is the most popular lawn grass grown in the cool humid area of the United States. Most older lawns have common Kentucky bluegrass while lawns started after 1970 have new improved cultivars of bluegrass. Kentucky bluegrass prefers fertile soil, with a soil pH between 6.0 to 7.5. However, it is not well adapted to shady locations. Kentucky bluegrass plants are identified by a boat-shaped tipped leaf and the leaves are folded in the bud.

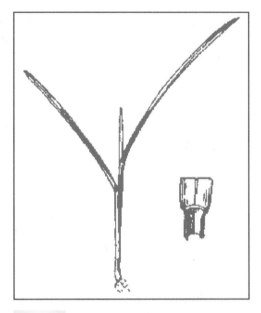

19-14. Kentucky bluegrass shoot and leaf parts.

Common Kentucky bluegrass are more erect and produce a tough dense turfgrass when mowed tall 3 to 3 ½ inches. It will go dormant in periods of heat and drought. Because of the underground creeping stems that are protected from weather extremes, the sod usually will recover from droughts or high temperatures when favorable weather returns. Common bluegrass seeds are very slow to germinate, usually taking three or four weeks. When fertilizing common Kentucky bluegrass lawns, use a complete fertilizer in the early spring and early fall and apply one pound of nitrogen per 1,000 sq. Ft.

New improved cultivars developed after the 1970s are more disease resistant, require higher levels of maintenance, and can be mowed at a 1 to 2½ inch height of cut. Seed of the improved cultivar will usually germinate in one to two weeks. The following are names of improved high-quality cultivars—Adelphi, Baron, Glade, Parade, and Touchdown. Many golf course fairways are seeded with a blend of two or three of these cultivars. New cultivars require three or four pounds of nitrogen per year. Apply one pound of nitrogen per 1,000 sq. ft. using a complete fertilizer in April, May, August, and November. Improved cultivars usually don't go dormant in hot weather and will require irrigation when rainfall is less than one inch per week.

Rough stalked bluegrass (*Poa trivialis*) is a relative of Kentucky bluegrass, but differs in several important respects. It is well adapted to moist lawn areas with moderate shade. Rough stalked bluegrass grows with a low level of maintenance. It should be cut 3 inches high and fertilized in the fall with one pound of nitrogen per 1,000 sq. ft. Saber is a new cultivar with improved heat tolerance and a dense growth habit.

Bentgrass

Creeping bentgrass (*Agrostis palustris*) is grown on golf course putting greens, lawn tennis, lawn bowling, and croquet courts. However, it is not a desirable turf for home lawns.. This species got its odd name from the characteristic position of the basal part of each stem when plants are growing in thin stands. These stems, instead of growing upright, tend to follow the ground for a short distance and then bend upward, thus bentgrass. Early settlers in this country brought with them (German mixed bent) seed which is the foundation of most new cultivars.

Creeping bent is not well adapted to shady conditions, but is tolerant of a wide range of soil conditions. Best growth occurs on soils with good water-holding capacity with a pH between 5.5 and 7.0. Bentgrass turf is also very susceptible to many turf diseases. Golf course superintendents spray fungicides to prevent and cure these disease attacks. Many new cultivars of creeping bent have been developed since the early 1900s. The United States Golf Association Greens Section developed and named many vegetatively propagated cultivars. Cohansey (C-7), Toronto (C-15), and Washington (C-50) are no longer sold but can still be found on putting greens of many older golf courses in the cool humid region of the United States.

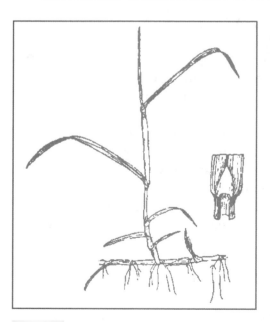

19-15. Creeping bentgrass shoot and leaf parts.

Breeding programs at state universities and private seed companies have produced improved strains of creeping bent that are propagated by seed. Perhaps the best known of these is Penncross. It is vigorous, uniform in color and texture, and has considerable disease resistance. Penneagle, SR 10, and SR 20 are other new cultivars of bentgrass that are available from seed. Since the early 1990s, many golf courses in the warm humid south have replaced Bermuda putting green turf with new bentgrass cultivars.

Bentgrass greens are mowed daily at 1/8 to 3/16 inch high with special walking or ridding mowers. Bentgrass is also planted on golf tees and fairways and cut ½ to ¾ inch high. In the past, greens were fertilized monthly to keep the grass growing. Today, many golf courses apply a total of two to three pounds of nitrogen per 1,000 sq. ft. per year to putting greens. Superintendents apply ¼ to ½ pound of nitrogen each month during the growing season. Lower rates of nitrogen can increase green speed and reduce annual bluegrass weed competition.

Fine Fescue

The bristle-like character of the leaves makes fine fescue (*Festuca rubra*) turf tough and resistant to cutting. This grass is tolerant of drought and infertile soils and grows very well in moderate shade. Fine or red fescue is the major shade grass in the northern cool humid region of the United States. Fine fescue has a blue-green color and often grows in clumps because of poor rhizome development. It is best to mix shade tolerant Kentucky bluegrass cultivars with fine fescue cultivars for establishing turf in shady areas.

Chewing fescue, hard fescue, and sheep fescue are related species that are used for soil stabilization and erosion control. They all will grow with a very low level of maintenance but usually will not provide a quality lawn turf.

Checker, Dawson, Jamestown, and Pennlawn are new improved cultivars of fine fescue that have growth habits similar to perennial ryegrass and Kentucky bluegrass. Use the same fertilizer program on fine fescue as used on Kentucky bluegrass lawns. Fine fescue is slow growing and does not require the frequent mowing that bluegrass lawns need. Cut fine fescue 2 to 3 inches tall when needed.

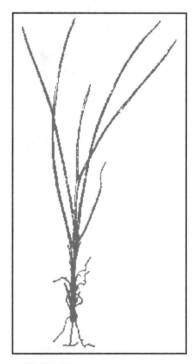

19-16. Fine fescue shoot.

Tall Fescue

Tall fescue (*Festuca arundinacea*) is a close relative of meadow fescue, but considerably larger in all respects. It is a useful grass on large areas that are

19-17. Week-old fescue with dew on the tips of the blades.

not maintained as lawn. Tall fescue is a deep rooted, strong tufted perennial grass with broad, flat leaves. It tolerates both doughty and wet soils, and will grow in sunny and moderately shady areas. It does not flourish when mowed shorter than 3 to 4 inches, and may be killed by repeated close cutting. When adequately fertilized, it makes an excellent grass cover over a wide range of unfavorable soil conditions. Tall fescue is a very popular grass in the transition zone between the cool and warm humid areas of the country.

The old coarse-leafed cultivar, Kentucky 31 is seldom used today except on roadsides and other very low maintenance turf areas. A new group of turf-type tall fescue cultivars, which are denser and more attractive, were introduced in the 1980s and hold great promise for the future.

Jaguar, Olympic, and Rebel are turf-type tall fescue cultivars that work well on athletic fields and golf course roughs. These new cultivars provide an acceptable turf that needs a lower level of maintenance than Kentucky bluegrass. They have a medium leaf blade width that is similar to Kentucky bluegrass and tolerate mowing at 2 to 3 inch height of cut. In the future, new turf-type tall fescue cultivars may become popular for home lawns.

PERENNIAL RYEGRASS

Perennial ryegrass (*Lolium perenne*) takes its name from a European name ("Rai" grass) and is not related to rye grain. Its quick germination time and relatively cheap price have made it a popular seed to include in most home lawn grass seed mixtures. It is a non-creeping, bunch-type grass that develops a turf with medium density. A major problem with ryegrass is the general appearance of the lawn after mowing. Dull rotary mower blades tend to tear and shred the grass blades and cause the lawn to turn brown the day after mowing. Pythium disease can kill large turf areas during the hot humid Midwest summers.

Since 1970, many new improved turf-type perennial ryegrass cultivars have been developed. These new cultivars, such as Blazer, Manhattan II, Fiesta, and Citation, have a similar growth habit to Kentucky bluegrass and will blend together to make a uniform lawn. Most home lawns in the northern cool humid region are seeded with a 50–50 mixture of Kentucky bluegrass and perennial ryegrass. Some golf courses have only perennial ryegrass fairways. Follow similar fertility programs for perennial ryegrass turf as for Kentucky bluegrass lawns, three to four pounds of nitrogen per 1,000 sq. ft. per season applied in the early spring, early fall, and late fall. Always keep mower blades sharp to reduce leaf shredding common to perennial ryegrass cultivars.

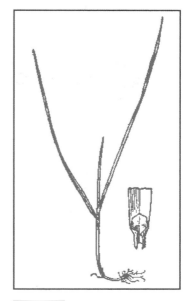

19-18. Perennial ryegrass shoot and leaf parts.

REVIEWING

*M*AIN IDEAS

The purposes and functions of turfgrasses include utility, ornamentation, and sports turf. The evolution of turfgrass science and technology is largely from the golf course industry.

The cool season turfgrasses grow best at an optimum temperature range of 60 to 75°F. Warm season turfgrasses grow best at an optimum range of 80 to 95°F. Cultivars are named after people, places, use, plant characteristics, or any combination. The warm season turfgrasses include bahiagrass, bermudagrasses, buffalograss, carpetgrasses, centipedegrass, gramas, kikuyugrass, seashore paspalum, St. Augustinegrass, weeping lovegrass, and the zoysias.

The specific selection of turfgrass cultivars for planting is one of the most important decisions made in turfgrass culture. Most warm season turfgrasses are vegetatively propagated by sod, plugs, or sprigs. However, centipedegrass is frequently seeded, and zoysia is beginning to gain precedence by seed propagation. Common bermudagrass is seeded.

QUESTIONS

Answer the following questions. Use complete sentences and correct spelling.

1. What is the difference between turf, turfgrass, and sod?
2. What are the three functions of turf? Explain each.
3. What is turf quality? What are the criteria for visual quality and functional quality?
4. Distinguish between warm season and cool season turfgrasses.
5. What is a transition zone?
6. What warm season turfgrasses are used? Select any two and provide a brief description of the grasses.
7. What cool season turfgrasses are used? Select any two and provide a brief description of the grasses.

EXPLORING

1. Select a golf course in your community and scout the course observing all features and playing areas. Review all the turfgrasses, and develop a written plan to include each somewhere on the golf course. Be able to justify and support your selections.

2. Scout a golf course in your area. Use the visual and functional factors impacting quality to rate the course. Be able to support your ratings.

20

TURFGRASS ESTABLISHMENT AND CARE

Several methods are used to establish turfgrass. The methods used depend on the kind of grass and the conditions where it will be grown. Vegetative propagation involves using parts of a growing plant to establish a turf. Common vegetative methods include using sod, sprigs, and plugs. Establishing turfgrass with seed is sexual propagation. Seeding is the most common method used with cool season grasses.

Turfgrass care is keeping a stand of grass in an attractive and healthy condition. Maintenance involves using appropriate turfgrass management practices. These include activities to keep the turf at the desired quality. It is not uncommon to find that management practices are either put into effect at the wrong time of the year or that the practices used are incorrect. Helpful management techniques and practices are presented in this chapter.

20-1. Golf courses require many different types of turf maintenance practices.

491

OBJECTIVES

This chapter provides a good foundation in turfgrass establishment and maintenance. It has the following objectives:

1   Explain the ways turfgrass can be established

2   Compare a turfgrass blend to a turfgrass mixture

3   List the steps in good seedbed preparation

4   Discuss planting techniques for seed establishment

5   Describe sod laying techniques

6   Explain how to maintain turfgrass

7   Identify correct fertilizing techniques and timing

8   Discuss proper mowing practices for maintaining turf

9   Identify weed control techniques for turfgrass

10 Explain how to control thatch in turfgrass

TERMS

acid soil

aerifying machines

armyworm

grubs

over-seeding

plug

seed purity

seed viability

sod

sprig

turfgrass blend

turfgrass maintenance

turfgrass mixture

ESTABLISHING TURF

Several methods are used to establish turfgrass. The methods used depend on the kind of grass and the conditions where it will be grown. Seeding is the most common method with cool season grasses. Sodding, plugging, or sprigging are usually used to establish warm season turfgrass species.

Table 20-1. Turfgrass Propagation Methods, Mowing Heights, and Recommended Type of Mower		Mowing Height		Type of Mower
Species	**Propagation**	**Inches**	**Centimeters**	**Mower**
Bahiagrass	Seed	2–4	5.0–10.2	Rotary
Bermudagrass				
Common	Seed	05–1.5	1.3–2.8	Reel or Rotary
Improved	Vegetative	0.25–1.0	0.6–2.5	Reel
Blue gramagrass	Seed	2.0–2.5	5.0–6.4	Rotary
Buffalograss	Seed & Vegetative	0.7–2.0	1.8–5.0	Rotary
Carpetgrass	Seed & Vegetative	1.0–2.0	2.5–5.0	Rotary
Centipedegrass	Seed & Vegetative	1.0–2.0	2.5–5.0	Reel or Rotary
St. Augustinegrass	Vegetative	1.5–3.0	3.8–7.6	Rotary
Zoysia				
Japanese lawngrass	Vegetative	0.5–2.0	1.3–5.0	Reel
Manilagrass	Vegetative	0.5–2.0	1.3–5.0	Reel
Bentgrass				
Colonial	Seed	0.5–1.0	1.3–2.5	Reel
Creeping	Seed (Veg. for a few)	0.2–0.5	0.5–1.3	Reel
Bluegrass				
Kentucky	Seed	2.5–3.0	3.8–6.4	Reel or Rotary
Rough	Seed	3.0	6.4	Reel or Rotary
Fine fescue	Seed	2.0–3.0	5.0–6.4	Reel or Rotary

TURFGRASS ESTABLISHMENT BY SEED

The advantages of seeding a lawn are desired species or cultivars can be used, the plants develop in the environment in which they must ultimately survive, and establishment usually costs less than sodding or plugging.

20-2. The site must be properly prepared before the grass seed is planted. (Courtesy, Bush Hog, Division of Allied Products Corporation, Alabama)

Early spring, late summer, or early fall weather conditions provide the best time for good grass seed germination. In the fall, weed competition is reduced and temperatures are appropriate for rapid turf growth and development. Fall planted turfgrass seedlings have plenty of time to establish a good root system before the following summer's heat. Spring establishment can also be successful. However, weed competition, especially annual weeds, can be a problem. It is also important that adequate irrigation be available during the summer's heat following spring establishment.

BLENDS, MIXTURES, AND GENETIC ENGINEERING

In selecting a turfgrass, choose cultivars that have shown resistance to diseases. Turfgrass may be of one cultivar or a blend. A *turfgrass blend* is a combination of different cultivars of the same species. Blends should have cultivars that are similar in appearance and competitive ability. At least one cultivar should be well adapted to the planting site. Blends usually have at least three cultivars. *Turfgrass mixtures* are combinations of two or more different species.

Through genetic engineering, seed companies have developed endophyte-enhanced turfgrass cultivars. Endophytes are microscopic fungi that live within the host grass plant. These fungi produce chemicals that are toxic to certain turf insects and diseases. Turfgrass cultivars containing endophytes require fewer pesticide applications. This saves time, money, and helps reduce contamination to the environment.

Seed QUALITY AND LABELING

Good seed that will grow should be planted. All seed containers have labels. The labels provide important information about the seed. ***Seed purity*** is the percentage of pure seed of an identified species or cultivar present in a particular lot of seed. ***Seed viability*** is the percentage of seed that is alive and will germinate under standard conditions. Pure live seed is the percent purity multiplied by the percent germination. For example, if a seed package is 90 percent pure and 80 percent germination rate, then the percent pure live seed is 72 percent.

Differences in seed test results and planted results are important. The expected seedling mortal-

```
APM PERENNIAL RYEGRASS
                       LOT # B29-3-36APM
                             PURITY   GERMINATION   ORIGIN
    PURE SEED                97.20%       90%          OR.

    OTHER CROP SEED           2.36%
    INERT MATTER              0.44%
    WEED SEED                 0.00%
    NOXIOUS WEED SEEDS:   NONE

    TEST DATE   7-94
                                      NET WT.  50 LBS.
                                                        PMA 185
                           MEDALIST AMERICA
          1490 INDUSTRIAL WAY S.W., ALBANY, OR., 97321

                NOTICE  READ CAREFULLY BEFORE OPENING
    1. NOTICE OF REQUIRED ARBITRATION
       Under the laws of some states (including Idaho), arbitration is required as a pre-condition of maintaining certain legal actions, counterclaims or defenses
       against a seller of seed. The buyer must file a complaint along with the filing fee within such time as to permit inspection of the crops, plants or trees and
       notify seller of complaint by certified mail. Information about this requirement, where applicable, may be obtained from a State's chief agricultural official.
    2. NOTICE OF EXCLUSION OF WARRANTIES AND LIMITATION OF DAMAGES AND REMEDY.
       The labeler warrants that this seed conforms to the label description, as required by federal and state seed laws. WE MAKE NO OTHER WARRANTIES,
       EXPRESS OR IMPLIED, OF MARKETABILITY, FITNESS FOR A PARTICULAR PURPOSE, OR OTHERWISE CONCERNING THE PERFORMANCE OF THIS
       SEED.
       The liability for damages for any cause including, but not limited to, breach of contract or breach of warranty or negligence with respect to this sale of
       seed is limited to a refund of the purchase price of this seed. This remedy is exclusive. IN NO EVENT SHALL THE LABELER BE LIABLE FOR ANY INCI-
       DENTAL OR CONSEQUENTIAL DAMAGES, INCLUDING LOSS OF PROFITS.
```

20-3. An example of a seed label.

ity rate is the difference between the laboratory pure live seed and the field pure live seed content. There is a difference because of loss of viability during storage. Also, some seed will not grow because they are planted too deep or other unfavorable conditions. Seedling mortality is due to unfavorable temperatures or inadequate moisture. Seed certification is used to assure that seed production in fields and seed lots meets standards of genetic purity.

Site PREPARATION, PLANTING, CARE, AND FIRST MOWING

After the type of grass seed has been selected, the planting site must be properly prepared. The seed must then be planted and properly managed for seed germination and plant growth. After successful establishment, the grass plants will need to be mowed for the first time.

Seedbed Preparation

Proper preparation of the planting site can reduce many drainage, aeration, pH, and fertility problems that may not become evident until after the lawn is established. It is much more difficult to correct these problems after the turf is established.

Proper site preparation steps include:

1. Control weeds at the planting site.

2. Rough grade the site and remove debris, including rocks and wood.

3. Conduct a soil test.

4. Apply soil amendments as necessary.

5. Thoroughly mix the amendments into the soil to a depth of 6 inches.

6. Fine or finish grade the site.

Eliminating weeds, especially perennial grassy weeds, will reduce competition with developing turfgrasses. Herbicides are used to eliminate weeds prior to lawn establishment. After the weeds have been eliminated, rough grade the area to facilitate surface drainage. Generally, a 1 to 2 percent slope away from buildings is adequate. This is a drop of 1 to 2 feet for every 100 feet of run. Remove all debris such as tree roots, stones, and other materials brought to the surface by rough grading.

20-4. The "Harley Rake" on this skid-steer loader is mixing soil amendments with native soil.

A good seedbed is essential for quality turf. The soil should be tested for needed nutrients. It is important to make soil improvements before planting grass seed. It is almost impossible to alter the soil after the grass is planted. Incorporate lime, organic matter, and fertilizer 6 to 8 inches deep before planting. Preparing a suitable soil seedbed for the turfgrass roots should be the landscaper's goal. A soil-testing laboratory can determine the exact quantity of nutrients needed from the soil sample. Add a complete fertilizer, high in phosphate, during seedbed preparation.

The soil must be tilled. This may include excavating earth to prepare a smooth surface. Plows and rotary tillers may be used. Plants growing on the area must be destroyed. If not, they may live and grow to be weeds in the new turf. In some cases, herbicides may be needed to kill native vegetation. Tilling the seedbed in several directions will completely mix

20-5. Installing an underground irrigation system. (Courtesy, Jasper S. Lee)

the soil amendments with the native soil. After mixing, roll the seedbed lightly to compact the soil. Complete the seedbed preparation by establishing the final grade. Apply 5 to 10 pounds of 12-25-13 starter fertilizer per 1,000 square feet of soil surface.

If topsoil is needed at the planting site, it should be incorporated into the existing soil during rough grading. Preparing the seedbed may also include the installation of drainage and irrigation systems. Heavy soils may need drain tiles or tubes to carry away excess water. Irrigation systems are installed before seeding, if possible.

Winter Over-seeding

Some lawns are over-seeded in the winter. *Over-seeding* is planting one grass in another established grass without destroying the established grass. It is often used to plant cool tolerant grasses into established dormant grasses.

20-6. A Florida lawn over-seeded with winter grass.

Ryegrass is often over-seeded with the Bermudagrasses. This gives green color in the winter while the permanent lawn is dormant.

In selecting a cool season turfgrass for over-seeding, consider the turfgrass characteristics and the microclimate of the site. Turfgrass mixtures are preferred to insure better quality during the short winter season.

*P*lanting

When establishing turf from seed, the following factors are necessary for seed germination and growth—live seed, adequate moisture, sufficient soil warmth, and adequate aeration. If these factors are not present, the seed will not germinate or the young grass plants will be too weak to survive.

The seed label provides valuable information about the grass seed. Different turfgrasses have different seeding rates. They vary according to the size and weight of the turfgrass seed. When selecting seed, look for a high purity and germination percentage, fresh seed produced the previous year, and as low as possible weed content.

After choosing the turfgrass seed and determining the seeding rate, be sure to distribute it uniformly over the planting area. Avoid excessive seedling rates that can produce crowded, weak plants and increase seedling disease invasion. Use a broadcast or drop spreader. It is advisable to apply half the seed in one direction and the other half at a 90° angle to uniformly cover the entire area.

Table 20-2. Size and Seeding Rates		
Species	**Seeds per Pound**	**lbs./1,000 sq. ft.**
Kentucky bluegrass	1,000,000 to 2,200,000	1–3
Red fescue	350,000 to 600,000	3–5
Perennial ryegrass	250,000	5–9
Tall fescue	170,000 to 220,000	5–9
Creeping bentgrass	5,000,000 to 7,000,000	5–10.5

After the seed is in place, there are two additional steps that are crucial to successful turfgrass establishment. The first is making sure there is good seed-to-soil contact. This can be accomplished by using a rake to lightly mix the seed into the upper 1/4" of soil. After raking, follow with a light rolling to produce a firm seedbed. A light rolling can be accomplished by using a half-full water-ballast roller. Rolling not only increases seed-to-soil contact, but firms the seedbed and slows the drying of the soil. The seedbed area should be mulched to prevent or slow drying. A thin layer of clean straw is commonly used. Don't apply too heavily. The soil beneath the straw should be able to be seen. Usually one bale (35 to 50 lb.) per 1,000 sq. ft. is adequate. The straw is not removed after the seed germinates. Grass seedlings will grow up through the light straw layer and gradually cover it as the straw de-

20-7. Seeding with a slit seeder.

20-8. This recently planted area surrounding the landscape berm was strawed and is now being throughly watered.

composes. Raking off the straw would injure the young grass plants. New mulches have been developed including paper materials and mesh for hillsides.

The second activity crucial to seed establishment is to make sure adequate water is available throughout the germination process. At the time of planting, water frequently and lightly, wetting the upper ½" of soil. Continue irrigation during the germination period. Water less frequently but more thoroughly and deeply as the turfgrass plants mature.

Too much activity on a newly seeded lawn can interfere with seed germination. The amount of activity on a newly seeded lawn should be limited as much as possible for two to three weeks.

*F*irst Mowing

The new lawn will be ready to mow when the turfgrass plants are higher than the height at which they will normally be maintained. For example, if the lawn is be cut regularly at 2½ inches, then mow for the first time when it is 3 inches tall. The general rule is to never remove more than 1/3 of the turfgrass leaf blade at any one time. Avoid mowing when the grass is wet. A sharp mower blade is important too. A dull mower blade can pull the grass seedlings out of the soil instead of cutting them. This can also make the turfgrass plants more susceptible to diseases and other problems.

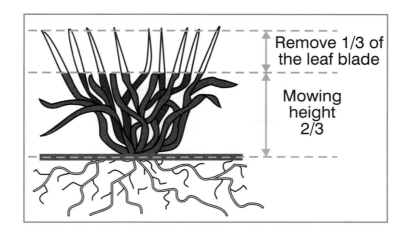

20-9. Never remove more than 1/3 of the grass leaf blade at any one time.

Remove 1/3 of the leaf blade

Mowing height 2/3

VEGETATIVE PROPAGATION

Vegetative propagation involves using parts of a growing plant to establish a turf. Common methods include using sod, sprigs, and plugs.

Sod is the surface layer of turf, including the grass plants and a thin layer of soil. Sodding can provide the home owner with an "instant lawn." Use sod containing a mixture of grass varieties that will adapt to the climatic area. While speed of establishment is usually the reason for sodding, erosion control is also a valid reason to sod a lawn. Good quality sod is uniform, pest

20-10. Newly laid dormant Bermuda grass sod. Note the checkerboarding.

20-11. Rolls of sod on a pallet.

free, sufficiently strong to hold together during handling, and capable of rooting within one to two weeks following planting.

Sod is purchased in standard rolls, macro rolls, or strips. Sod on pallets is trucked at 56 square yards per pallet and 1,008 square yards per truck under normal moisture conditions. During rainy conditions, trucks will be unable to haul 18 pallets due to the weight.

For small jobs, the pieces of sod are usually 18 inches wide, 6 feet long, and 5/8 of an inch thick. These rolls contain one square yard of sod.

The steps in site preparation for sodding are the same as for seeding a new lawn area. Improve the soil before establishing a lawn with sod. After incorporation of the necessary materials, the soil should be leveled and firmed to provide a level base for the sod. Install sod that contains suitable grass varieties for the area to be covered. For example, use shade-tolerant grasses for shaded locations and use drought-tolerant grasses for steep slopes and soils of low water holding capacity. Too often, sod is selected with little regard for the kinds of grass present in the sod.

When laying sod, care should be taken to insure that the root system of the sod contacts the seedbed. Where there is no contact, roots will be "air-pruned" and the grass plant will die. The edges of the rolls should be butted tightly to help prevent moisture loss and a place for weeds to grow. The most desirable season for laying new sod is early fall because the normal cycle of new root formation begins then. However, sod is often laid throughout the year. Sod should be laid within 24 hours after harvesting. Sod held rolled up over 24 hours will usually turn yellow and may die.

A *plug* is a small block or square of turf. Plugs are marketed by the tray and on pallets. They may be cut from sod or placed in small growing containers. Plugs are used to establish a turf without covering an entire area. The plugs are set a few inches apart in a prepared seedbed. The plants grow to cover the entire lawn. Establishing a lawn with plugs is more economical than using sod.

20-12. The square plugs shown in this photograph are being marketed by the tray.

Sprigs are used in establishing some lawns. A *sprig* is a part of the grass plant without soil. Rhizomes and stolons are used as sprigs. Sprigs are set by hand or machine in a seedbed. Using sprigs is more economical than sod or plugs; however, the sprigs must grow to provide a turf. Sprigs are delivered in bulk by trucks. Trucking includes 3,000 bushels per short dump trailer, 4,000 bushels per long flatbed trailer, and 4,500 bushels per long dump trailer. However, if a small number of bushels are needed on a specific site, then these sprigs will be packaged into cloth bags for transport.

20-13. Using sprigs to establish turf with a mechanical planter.

20-14. Hand sprigging a golf green with bermudagrass. Note the sod in the foreground.

Sprigs and sod should be bought only from reputable growers or distributors. Visiting fields prior to purchase and observing the sod will allow inspection of quality. Some sod is certified as to purity and freedom from disease. The method of delivery is also important. Keeping sod and sprigs out of the ground for a long time lowers quality.

Sod, sprigs, and plugs should be set as soon as possible after digging and delivery. If not properly cared for, the grass plants may die before being set out.

MAINTAINING TURFGRASS

Turfgrass maintenance is keeping a stand of turfgrass in an attractive and healthy condition. Maintenance involves using appropriate turfgrass management practices to keep the turf at the desired quality.

SOIL AND NUTRIENTS

Plants must have certain nutrients to grow. Soil testing should be used to determine the needed nutrients and the pH of the soil. Fertilizer, agricultural lime, and other materials may be needed for the soil to provide for the needs of plants.

20-15. Well kept turf is important to the overall beauty of the landscape.

Fertilizer and lime are applied to established turf in the spring and fall to encourage growth. Additional applications may be needed throughout the growing season. Turf in golf courses and similar uses may require fertilizer several times during the year. Turf plants are better able to survive hot summer weather if they have sufficient nutrients.

Fertilizer and lime should be evenly distributed over the turf. Spreaders are often used to assure proper application. These machines must be calibrated to put out the right amount. Too much fertilizer is wasteful. Not enough fertilizer fails to produce the desired growth results.

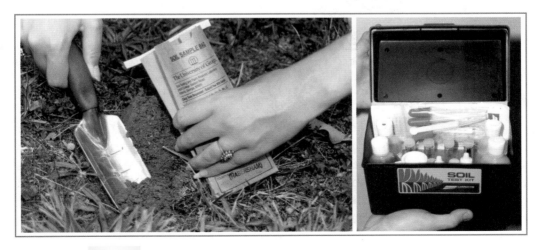

20-16. Collecting a soil sample (left) and a soil analysis kit (right).

MOWING

Mowing is used to regulate the height of the turfgrass. Each kind of grass has a recommended height for cutting. Mowers should be adjusted to provide the right cut. Two major kinds of mowers are used—rotary and reel. Rotary mowers have a blade that turns at a high rate of speed. The blade cuts grass blades, seed-heads, and other structures. Reel mowers work best on turf that is mowed very short, such as on golf course fairways and greens and on certain species of warm season turfgrasses. Reel mowers have two blades—the fixed bed knife blade and the spiral and rotating reel blade on a horizontal shaft. These blades cut grass the same way scissors cut paper. Reel mowers sometimes fail to cut tall stems from certain kinds of grass.

20-17. A large riding rotary mower is often used for mowing parks and golf course roughs.

Height of Cut

Improper mowing does more damage to home lawns than any other maintenance practice. Homeowners may set their mowers to cut the grass so short that the grass is given a setback with each mowing. Most home lawns will tolerate regular mowing at a height of 2 inches, but they will do better if cut at a height of 3 inches. Cutting grass shorter than these lengths may gradually weaken the turf grass by repeatedly defoliating the plants.

Lawnmowers, used for home lawns, should be set to cut 2½ to 3 inches in height. Permitting the grass to grow taller than this induces a stemmy erect habit of growth, rather than the desired dense prostrate growth habit. The frequency of mowing should be such as to keep grass from exceeding 3½ inches in height. In periods of rapid growth, mowing may be needed twice a week; at other times, once a week may be sufficient. In cooler periods and in dry seasons, frequency of mowing may be once in two or more weeks. The recommended cutting height is listed by species in Table 20-1. Continue mowing turf areas in the fall as long as they continue growth. Permitting long growth in fall merely encourages an increase in disease development.

Removal of Clippings

Mowing may also involve removing the grass clippings from the turf. A catcher or power bagger is used with some mowers to collect the clippings. The collected clippings may be composted to create organic material for use in flower and landscape beds.

When the amount of clippings produced by mowing the grass is such that they do not readily sift down into the turf, it is best to remove them. Such clippings are not only unsightly, but their continued presence on top of the grass is an invitation to turf diseases. Diseases prefer the high humidity conditions found under this mat of clippings. If clippings are heavy enough to require removal, this should be done promptly before fungal diseases have an opportunity to develop. The clippings that sift down through the grass to the soil are helpful since they add to its fertility. Grass clippings will decompose rapidly and do not contribute to thatch build up.

20-18. This lawn tractor is equipped with an optional power bagger to remove clippings.

*F*ERTILIZING ESTABLISHED TURF

Turfgrass growing in the cooler regions needs fertilizer applied four times each year. The best times are in early spring, late spring, early fall, and late fall. Turfgrasses will break dormancy two to three weeks earlier in the spring if fertilizer is applied in March or early April, depending on the latitude. Not only does early fertilization speed spring green-up, but it also helps the cool-season grasses to heal injuries. Thick-growing turf can prevent the invasion of summer annual weeds.

Fertilization during May will provide the turfgrass the nutrients necessary to survive the hot summer growing season. A similar fertilization in late August or in September will stimulate the production of new tillers and rhizomes for the following year. This fertilization will also help maintain a vigorous and green turf until late fall or early winter. A late fall application of fertilizer after top growth has ceased, especially with organic or slow release nitrogen, will keep the root system growing until the ground freezes. This can also improve early spring green-up of the grass.

20-19. Rotary spreader for rapid application of fertilizer to small or medium size turf areas.

*F*ertilizer Composition

Maintenance fertilizers should be high in nitrogen, but also have moderate amounts of phosphate and potash. To avoid over stimulation, immediately after fertilization, use a slow release fertilizer. Slow release fertilizers

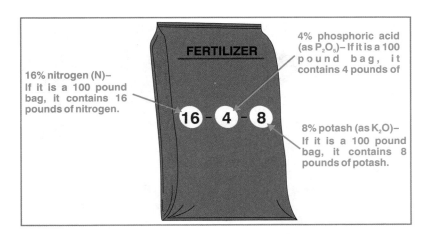

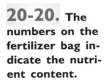

20-20. The numbers on the fertilizer bag indicate the nutrient content.

provide a sustained flow of nutrients to the grass. The exact analysis of the mixed fertilizer is not crucial; it may have a guaranteed analysis of 23-3-7, 32-4-8, or some similar content, producing equal results if properly used.

Fertilizer Rate

The amount of fertilizer should be adjusted based on a soil test, the length of the growing season, the type of grass, and the amount of traffic. Turfgrass growing on infertile soil needs extra fertilizer. The longer the growing season, the more fertilizer needed.

An average fertilizer application should contain 1 pound of actual nitrogen per 1,000 square feet 4 times each year. Calculate the pounds of actual nitrogen by multiplying the pounds of fertilizer in the bag by the percent nitrogen shown on the fertilizer label. Example: a 50 pound bag of 32-4-8 fertilizer will contain 16 pounds of actual nitrogen, 50 lb. × 32%. Apply a bag of 32-4-8 fertilizer on a 16,000 square foot lawn area. This rate is 1 pound of actual nitrogen per 1,000 square feet.

Setting the Fertilizer Spreader

All fertilizers must be spread uniformly over the turfgrass surface since fertilizers do not move laterally in the soil. Failure to provide uniform distribution will produce a very irregular response of the grass; the strips or spots that are missed will get no benefit from the fertilizer. Uniform spreading is also desirable to avoid over stimulation of grass in some areas.

20-21. Large rotary fertilizer spreader used on a golf course. (Courtesy, Danville Country Club)

Fertilizers should not burn the turfgrass if applied correctly. Adjust the fertilizer spreader to apply the desired amount in a given area. There are two major types of fertilizer spreaders. The drop-type or gravity spreader will provide a very accurate application of fertilizer; however, the application will be very slow because the spreader will only cover a 24-inch wide strip. The rotary-type or cyclone spreader will cover a 6- to 12-foot wide strip and give a uniform application. These spreaders can quickly fertilize large turf areas.

OTHER TURFGRASS MAINTENANCE PRACTICES

Watering

The natural supply of water in the soil may be inadequate for a turfgrass. Water may need to be added to the turf area. Additional water is often needed with turf that has high use or is in a location where extra quality is desired. Irrigation is artificially applying water. Automated irrigation systems may be used to assure regular and even applications of water.

In all regions receiving less than 40 inches of annual rainfall, irrigation is essential for good turf maintenance. For humid regions, irrigation is desirable during periods of prolonged drought. The turf manager should install an inexpensive rain gauge to keep track of rainfall. Such gauges should be set on a post in the open, and read each time there is an appreciable amount of rain.

20-22. A well-designed automatic sprinkler system is designed to give complete coverage.

By keeping a record of rainfall, the manager can easily figure out the need for watering. After 7 to 10 days without rain, turfgrass plants start to wilt.

Water should be added only when needed. Periodic probing of the soil with a screw driver can help the turf manager learn soil moisture content. Lightly sprinkling turf in hot weather will benefit weeds more than the desirable turf grasses. As a rule, water to wet each area to a depth of about 6 inches, moving the sprinkler from area to area as the watering system permits.

The time of day for watering is not important, except that it should be completed early enough so that the grass leaves are dry by nightfall. When grass goes into the night with wet leaves, the situation is ideal for development of fungal diseases. The claim that watering in the sun will scald grass leaves is not valid. Any such scalding is probably caused by water-logging of the soil and the resulting death of roots where drainage is inadequate.

Thatch

Thatch describes the accumulation of excess grass stems, stolons, rhizomes, and roots in the turf. In the past, turf experts believed that bagging grass clippings would prevent thatch buildup. However, grass clippings do not contribute to thatch buildup. Excessive turfgrass growth caused by over-fertilization encourages thatch accumulation. Maintaining moderate fertilization will help prevent thatch problems.

Controlling thatch requires physical removal of the excess plant growth, and adopting maintenance practices to prevent additional thatch buildup.

20-23. Thatch is the accumulated grass stems, leaves, and roots in the turf. (Courtesy, Agricultural Research Service, USDA)

Power rakes or de-thatching machines can physically remove thatch from the lawn. All the thatch must be removed from the lawn surface after de-thatching with a power rake.

A core aerifying machine removes small plugs of soil from the turf. Air and water penetrating into the thatch layer through these holes will allow soil microbes to decompose the thatch. This technique prevents thatch accumulation. Chemical thatch removers on the market today have limited value in controlling thatch problems.

Correcting Soil Acidity

Acid soils are those with a pH of below 7.0. They are universal in humid regions, unless lime has been applied. There is no danger of burning grass with limestone. To make certain that sufficient lime is being applied, a soil test should be made. If no test is done, apply 25 pounds per 1,000 square feet annually to prevent acid soil buildup.

Aerifying Turfgrass Soil

Aerifying machines can provide temporary improvement of compacted soil conditions and reduce thatch accumulation. The soil cores are usually ½ inch in diameter and the spacing of the tines on the aerifier will vary from 2 to 10 inches. The more holes made, the better the aerification job. The best type of aerifier is one that removes a soil core to depths of 2 to 3 inches. After aerification, the soil cores left at the surface can be broken up and spread over the surface by dragging with a flexible steel mat. This treatment is most effective if done before applying fertilizer or lime. These holes allow the deep penetration of fertilizers and lime into the soil.

Rolling Turf

In regions where considerable winter freezing and thawing of the soil occur, the turfgrass may appear rough in spring. Spring rolling may be necessary to smooth the turf. Use a light water filled roller to "replant" the grass plants. Repeated rolling when the soil is wet and soggy, particularly with a heavy roller, causes undesirable compaction of heavy-textured soils. Since grass roots will penetrate only into soil layers that are well aerated, spring rolling may restrict the new roots to the uppermost layers of soil if a heavy roller is used.

The freezing and thawing of heavy soils are effective natural means of producing a well-aerated soil, and unwise rolling in spring may nullify this influence. The objective in sound turf maintenance should be to roll sparingly, and thus take full advantage of soil freezing to produce soil structure favorable to grass roots. A roller should only be heavy enough to press the crowns of the grass plants into the soil and smooth the soil surface.

Seasonal Care

Turf often needs care on a seasonal basis. Warm season grasses are overseeded in the fall. Leaves are raked in the fall. Care in removing leaves is important to avoid injury to the grass plants. Tree leaves left on the turfgrass during the winter can kill the grass plants. Leaves should never be allowed to

20-24. Leaves should be removed to keep them from covering the grass and cutting off light.

accumulate. The care given depends on the kind of turf, climate, environment, and problems that develop.

CONTROLLING WEEDS AND PESTS

Established turf may have weeds, insect pests, diseases, and other problems. Regular observation of turf is needed to determine if treatments are needed. Chemicals should be used only when needed. All chemicals should be used properly for the safety of the applicator and the environment.

Controlling Weeds in Turf

People often comment that crabgrass grows like crazy in the summer while the Kentucky bluegrass growth is slowed during hot weather. In Chapter 4, the difference between C_3 and C_4 plants was explained. Understanding the difference explains these conditions. Crabgrass is a member of the C_4 group while Kentucky bluegrass is a C_3 plant. As the temperature increases, the Kentucky bluegrass growth rate slows down and may even stop. However, the growth rate of crabgrass continues to increase as the temperature rises. Therefore, crabgrass will grow faster than Kentucky bluegrass in the summer. The effective control for crabgrass is to prevent crabgrass seeds from germinating.

A good weed control strategy consists of exploiting opportunities in the following ways. (1) Natural competition makes conditions more favorable for the desired grasses and less favorable for weed growth. (2) Always plant weed-free grass seed mixtures and prevent weed seed production in the turf. (3) At-

20-25. An abundance of weeds in a turf area is evidence that conditions are not satisfactory for good turfgrass growth.

tack weeds, with herbicides, at growth stages when the weeds are most vulnerable.

From this general plan, it should be obvious that the use of herbicides without taking care of other factors is not an adequate method of weed control. An abundance of weeds in a turf area is evidence that conditions are not satisfactory for good turf grass growth. Nature controls certain types of vegetation by providing stiff competition from the desired plants, thus crowding out the undesirable kinds of plants. A healthy turf, composed of well-adapted grasses, and properly managed, will make such a dense cover that weeds will have difficulty in gaining a foothold.

Mowing Practices—Adjust mowing practices to prevent weeds from producing seed. This, combined with timely use of herbicides, will constitute good weed control practice. Mowing home lawns too short is a major reason for weedy turf. Most home lawns will have fewer weeds if the height of cut is raised to 2½ to 3 inches instead of the usual 1½ or 2 inches. Taller grass can better compete with weeds for water and nutrients. Mow tall, should be the rule, not the exception, for turfgrass maintenance.

Herbicides—Herbicides kill or prevent weed growth. Read the manufacturer's label and accompanying printed instructions as to the kinds of weeds that the herbicide will control, how to use it, and any special precautions. Herbicides are powerful compounds and may cause damage to desirable turf if improperly used. The turf manager should take appropriate care to see that spray does not drift onto other vegetation, such as flowers, shrubs, trees, and vegetables. Read the precautions and protect susceptible plants.

Treat turfgrass areas with herbicides when soil moisture and weed growth are good. The weeds are most vulnerable under those conditions. An exception to this general rule is the use of preemergence herbicides to prevent the germination of weed seeds, such as crabgrass control.

Seedling grasses are far more sensitive to herbicides than mature turf plants. It is important to apply herbicides well before the grass is re-seeded so the herbicide may be dissipated before the grass seed is planted. The time required to dissipate the herbicide is different for different chemicals. Read the label!

Today, many herbicides are mixed with turf fertilizers to make their application easy. Herbicides mixed with fertilizers may need to be applied to moist leaves to stick the herbicide on the weed leaf. Read and follow label directions After treatment at the prescribed dosage of the herbicide, injury to the weeds may take from 3 to 10 days to become evident. Do not retreat in the mistaken idea that the treatment was not effective. Over-treatment is dangerous.

20-26. Chemical methods can be used on turfgrass to control winter weeds. This shows henbit in a dormant Bermudagrass lawn before (left) and two weeks after spraying with a herbicide (right).

The pesticide manufacturers have determined the most effective rates of application and the correct time of year to apply the herbicide. Follow precautions to protect the environment and the applicator. It is in violation of federal law to use a pesticide in a way not suggested on the label. Read the label, read it again, and then use the pesticide according to label directions.

Controlling Insects and Related Pests

Insects and other pests that infest turf may damage the grass plants, produce unsightly conditions, or be obnoxious because they attack or annoy people. Grubs, cutworms, and chinch bugs are examples of insects that seriously damage turf. Earthworms and ants are objectionable because of the mounds or casts that mar the turf surface. Chiggers, ticks, bees, wasps, and yellow jackets attack people, but do little or no injury to the grass.

As discussed in Chapter 8, integrated pest management (IPM) is a pest management strategy that uses a combination of best management practices (BMPs) to reduce pest damage with the least disruption to the environment. Biological and cultural pest control are two methods for homeowners to consider in controlling turf pests.

Although sound turf best management practices minimize the damage done by disease organisms, such practices are not always effective in controlling insect pests below the economic or aesthetic injury level. Healthy turf may endure heavy infestations of some insects without visible damage, while weak turf would be injured or even killed by the same populations.

For example, 3 to 5 white grubs per square foot can kill weak turf. Vigorous growing turf can support two or three times this number of grubs without seriously injuring the grass. Proper fertilization, mowing at suitable height of cut, sound watering practices, and other good management practices are a vital part of the program of fighting insect pests. Identify turf insects before applying control measures.

Insects That Damage Turf—Grubs feed on the roots of grass. ***Grubs*** are the larval stage of a variety of beetles, such as the May, June, and Japanese beetles. Adult beetles lay eggs in green grass during the summer. These eggs hatch, in late July or early August, into tiny grubs that feed on grass roots. The grubs may be found in infested turf during the late summer and autumn. Among useful insecticides in killing grubs are *Diazinon, Oftanol, Turcam, Merit,* and *Mach II.*

Chinch bugs and false chinch bugs are different species of insects, but have similar habits. Both injure grass by sucking juice from plant stems and leaves. However, the true chinch bug is the species that usually damages lawns. Heavily infested lawns first develop yellowish spots which soon become brown dead areas. The following insecticides are effective in controlling chinch bugs: *Turcam, Dursban, Diazinon,* and *Sevin.*

Sod webworms feeding on grass leaves and stems can severely damage a lawn. The adults are small whitish or gray moths (or millers) that hide in shrubbery during the day, and in evening fly over the grass and lay their eggs. The eggs hatch to produce soft-bodied larvae that are responsible for all of the feeding on grass. Irregular brown spots in the lawn may suggest insect damage. Insecticides that are effective against sod webworms are: *Diazinon, Turcam, Dylox, Oftanol, Merit,* and *Sevin.*

Armyworms and cutworms are the larvae of moths that are brown or grayish in color. They customarily fly at night, and, therefore, may go unno-

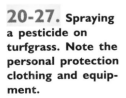

20-27. Spraying a pesticide on turfgrass. Note the personal protection clothing and equipment.

	Grubs	Chinch Bugs, False	Sod Webworms	Armyworms	Cutworms
Merit	X	X	X	X	X
Mach II	X				
Dursban		X		X	X
Dylox			X	X	X
Oftanol	X		X		
Turcam	X	X	X		
Sevin	X	X	X	X	X
Diazinon	X	X	X	X	X

Table 20-3. Insecticides and the Insects They Control

ticed. **Armyworms** are greenish worms that have black stripes along each side and down the back. The following insecticides are labeled to control armyworms and cutworms: *Diazinon, Sevin, Dursban,* and *Dylox.*

Precautions to Take When Using Pesticides—All pesticides are poisons and must be carefully used to prevent injury to humans, domestic animals, and desirable plants. When properly used, pesticides are safe and effective. Always read and follow all label directions. Pesticides should be stored properly and their containers discarded in the proper manner. Detailed safety precautions when using pesticides are listed in Chapter 8.

Controlling Disease in Turfgrass

Adopting sound management practices is a major step in controlling disease in turfgrass. Although there are many turfgrass diseases, only a few are likely to cause serious problems on turf. The selection of grasses suited to the climatic region is important. Principles of soil management and proper use of fertilizers help the turfgrass plants to develop resistance to disease and increase their ability to recover when invaded by an organism. Turfgrass mixtures are helpful because different species have different susceptibility to a specific pathogen. Avoiding close cutting, removing clippings, and avoiding excess moisture are all management practices that help prevent disease from occurring in turf.

Many diseases in turf are caused by different species of fungi. They are controlled by application of fungicides. General information on plant diseases is covered in Chapter 8. A turf management expert should be called

20-28. Some important turf diseases to be aware of are Pink snow mold (top left), Leaf spot (top center), Dollar spot (top right), Powdery mildew (bottom left), Brown patch (bottom center), and Pythium blight (bottom right).

when a disease is suspected. Some common turf diseases include—Snow mold, Leaf spot, Dollar spot, Powdery mildew, Summer patch, Brown patch, and Pythium blight.

REVIEWING

*M*AIN IDEAS

Turfgrasses are plants that form a contiguous ground cover. Turfgrass management includes a number of activities in establishing and sustaining turf at a desired quality. Turfgrass mismanagement results from selection of poorly adapted turfgrasses, improper establishment, errors in primary cultural program, errors in supplemental program, or mistakes in pesticide selection and use.

New lawns are usually established by seed, sod, plugs, or sprigs. Sod is used provide an instant lawn. Follow the steps given for proper site preparation before seeding or sodding the new lawn. Mow grass with the mower set at the proper height. Never remove more than 1/3 of the grass leaf blade at one mowing. Mow frequently when the turfgrass is actively growing, usually weekly.

Quality turfgrass needs regular fertilizer applications. Slow release types of fertilizer provide several months of sustained nutrient feeding. Always apply fertilizer according to the manufacturer's recommendations. Either a drop or rotary fertilizer spreader can be used to apply the granular fertilizer to the lawn. Do not overlap the spreader pattern when applying fertilizer.

When natural rain fall does not supply adequate water for good grass growth, artificial irrigation must be used to keep the grass vigorous. Thatch must also be controlled to have a quality lawn. Power rakes or de-thatchers remove much of the accumulated thatch from a lawn. Aerification by removing small plugs of soil will improve water and fertilizer penetration into the root zone. Both power de-thatching and aerifying should be done when the grass is actively growing, not when under stress.

Weeds can be controlled by applying the correct herbicide at the proper time. Spray lawns with insecticides to control insects that damage turfgrasses. Always read and follow all instructions on the pesticide label.

QUESTIONS

Answer the following questions using correct spelling and complete sentences.

1. What is a turfgrass blend? Turfgrass mixture? Over-seeding?
2. How is turfgrass vegetatively propagated?
3. When is the best time of the year to fertilize turf?
4. How many pounds of tall fescue should be planted per 1,000 sq. ft.?
5. What is a good fertilizer analysis for turfgrass?
6. How many pounds of actual nitrogen per fertilizer application should be applied per 1,000 sq. ft. of turf?
7. What is seed purity?
8. What is turfgrass maintenance? What activities are involved?
9. Why is aerifying turfgrass important?
10. What is thatch and how can its accumulation be prevented?
11. What conditions require rolling of turfgrass?
12. What are three important turf diseases?

EXPLORING

1. If possible, visit a sod farm in your area. Approach the site as if you were a potential customer. What questions would you ask? How would you spend your time at the farm during the visit? How would you be diplomatic in your visit to prevent being escorted from the premises? What would you recommend to future visitors to the sod farm and why?

2. Visit a golf course and talk with the superintendent about job skills necessary for employment.

3. Prepare a seedbed following the procedures discussed in this chapter.

4. Collect and identify three turf insects and 10 lawn weeds.

5. Mow three different turfgrass plots at school, each at a different height, and evaluate the turf response.

21

OLERICULTURE— VEGETABLE PRODUCTION

Vegetables have become the food of choice for many consumers. Three to five servings of vegetables per day are recommended by nutrition experts for a healthy diet. Information regarding how vegetables can improve health and the quality of life has been included in this chapter.

So where do vegetables come from and how do they grow? Olericulture is the branch of horticulture dealing with the production, storage, processing, and marketing of vegetables. Many people practice olericulture in home vegetable gardens, while others earn a living as "olericulturists."

While gardening may be fun, recreational, and provide fresh "veggies" at times, it is the commercial vegetable industry that puts vegetables on most of our tables. From roadside markets to the Jolly Green Giant, the production of vegetable crops is a major horticultural industry.

21-1. Supermarkets display numerous varieties of fresh vegetables. Vegetables are good sources of nutrients.

OBJECTIVES

Olericulture is important to the home vegetable gardener and the commercial producer. This chapter provides information on all phases of vegetable production and has the following objectives:

1   Describe the economic importance of vegetables to U.S. Agriculture

2   Explain the role of vegetables in human nutrition

3   Name the important classes of vegetables based on edible parts

4   Explain how the environment affects vegetable growth

5   Describe the important soil factors that influence vegetable growth and production

6   Describe the important techniques in producing a vegetable crop

7   Distinguish between traditional and non-traditional production methods

8   Explain how pests affect vegetable growth and production

9   Identify the major groups of vegetables and which ones belong to each group

10 Name the major components of the commercial vegetable industry

11 Explain how biotechnological advances can aid in increasing vegetable production

12 Describe the elements of home gardening

TERMS

base temperature
best stand
bolting
conservation tillage
day-neutral plant
Dicotyledonae
fresh market
frost-free period
growing degree days

hybrid
Monocotyledonae
organic production
photoperiod
plant spacing
plasticulture
processed vegetables
quality
seedbed

soil pH
strip tillage
sustainable
   horticulture
transplanting
variety
yield

IMPORTANCE OF VEGETABLES

Commercial vegetable production is usually divided between two categories. *Fresh market* vegetables are those that are grown and sold in the produce market without any processing other than washing and cleaning. *Processed vegetables* are those that are canned, dried, or frozen and sold from the grocer's shelf or the frozen foods section.

21-2. Processed vegetables may be canned and made available to consumers on supermarket shelves.

Vegetable production not only plays a significant role in our daily nutrition, but also in the agricultural economy of our nation. Although vegetables are produced on only about one percent of the total cropland in the United States, they accounted for $14.3 billion of the $202.3 billion in U.S. farm receipts according to the most recent statistics. The leading fresh market vegetable producing states are California, Florida, Arizona, Georgia, and Texas. California alone produces almost half of the vegetables grown in the United States. Table 21-1 shows the recent production of fresh market vegetable crops in the United States.

Processing vegetables allows us to have a convenient and stable source of vegetables and products made from vegetables, such as catsup and soup. California also leads the U.S. in production of vegetables for processing. Although total vegetable acreage has not changed much over the last 25 years, the total production of vegetables has risen. This has been due to more efficient production resulting in more vegetables produced per acre of land.

Table 21-1. Area Planted and Value of Principal Vegetables for Fresh Market by State and for the United States[1]

State	Area Planted		Value	
	1997	1998	1997	1998
	Acres		1,000 Dollars	
Alabama	12,300	9,400	12,362	9,592
Arizona	117,600	121,500	515,152	620,630
Arkansas	4,200	4,000	9,879	15,666
California	815,300	830,100	4,429,023	4,095,788
Colorado	39,600	37,600	107,923	131,697
Connecticut	5,400	5,400	6,768	7,911
Delaware	2,300	2,100	4,545	6,783
Florida	203,980	201,450	1,143,971	1,171,888
Georgia	107,900	108,600	255,075	243,032
Hawaii	2,420	2,470	13,791	14,445
Idaho	8,400	8,200	42,412	53,424
Illinois	9,200	8,600	13,242	12,258
Indiana	19,380	18,400	39,030	44,684
Louisiana	3,470	3,100	7,008	4,910
Maine	3,000	2,900	4,356	4,191
Maryland	11,570	12,750	25,207	27,884
Massachusetts	8,710	8,600	17,642	20,090
Michigan	59,900	60,700	123,086	148,459
Montana	2,030	1,310	5,332	13,017
Mississippi	5,000	4,700	1,464	1,527
Missouri	4,800	6,200	5,505	6,106
Nevada	1,800	2,100	13,311	12,396
New Hampshire	2,400	2,200	4,104	3,876
New Jersey	36,300	34,800	112,630	125,652
New Mexico	10,300	11,600	65,554	49,975
New York	67,650	71,600	172,054	209,164
North Carolina	45,200	48,000	76,241	75,840
North Dakota[2]				
Ohio	26,350	26,470	74,383	73,295
Oklahoma	10,000	10,000	5,200	5,076

(Continued)

Table 21-1. Area Planted and Value of Principal Vegetables for Fresh Market by State and for the United States[1]

State	Area Planted		Value	
	1997	1998	1997	1998
	Acres		1,000 Dollars	
Oregon	24,940	24,410	107,907	126,271
Pennsylvania	27,200	27,000	44,148	51,845
Rhode Island	1,200	1,100	2,079	2,640
South Carolina	17,700	18,600	37,561	37,466
Tennessee	13,800	13,700	32,020	39,161
Texas	109,800	98,500	225,542	337,117
Utah	2,400	2,500	8,451	9,050
Vermont	1,600	1,400	2,548	2,184
Virginia	20,500	20,800	65,425	65,016
Washington	49,100	49,950	185,901	192,582
Wisconsin	16,100	15,600	22,012	25,525
United States	1,931,050	1,937,630	8,040,744	8,096,114

[1]Includes processing total for dual usage crops (asparagus, broccoli, and cauliflower).

[2]Estimate discontinued.

NUTRITION

As the world population has become more health conscious, the demand for vegetables has increased. As the health benefits from some vegetables are realized, consumption of those vegetables has increased. Therefore, there have been shifts in consumption trends among vegetables. For instance, as the nutritional benefits of cauliflower and broccoli have been reported, consumption of these vegetables has increased.

The most important factor for increased consumption of vegetables is their nutritional benefit. Vegetables are important sources of several vitamins and nutrients. Many are also good sources of fiber, which is considered an important dietary ingredient. Vegetables are also generally low in fat and sodium. Vegetables differ in the amounts of each vitamin and nutrient they

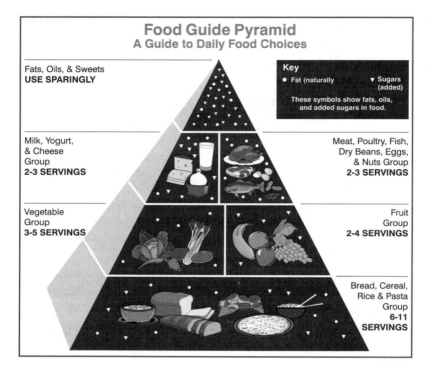

21-3. USDA recommended food chart. Note that three to five servings of vegetables should be consumed daily.

provide and in their fiber, fat, and carbohydrate content. The composition and vitamin content of selected vegetables are shown in Appendixes E and F.

Increased vegetable consumption has been shown to have many benefits. Among these are reductions in the risk of cancer and heart disease. Adequate vegetable consumption has also been shown to lower the effects of stress, insomnia, and aging.

ORIGIN AND CLASSIFICATION OF VEGETABLES

The first humans were thought to be hunters and gatherers, eating primarily wild fruits and animals. Then came the cultivation of crops. No doubt, vegetables were among those crops. As time has elapsed, improved varieties of these crops have been adapted through plant breeding and production and handling practices have improved. Therefore, most of the vegetable crops we grow today do not resemble the wild species from which those crops originally came.

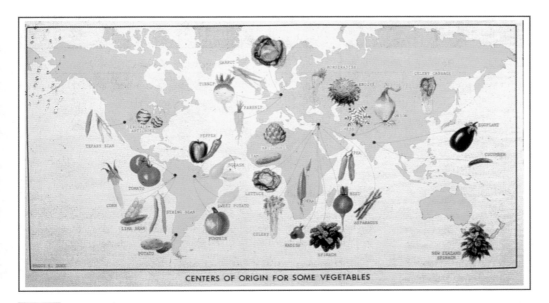

CENTERS OF ORIGIN FOR SOME VEGETABLES

21-4. Common vegetables originated in many parts of the world. (Courtesy, Agricultural Research Service, USDA)

All of the vegetable crops we know today were not originally present in North America nor all over the world. Each crop or group of crops had its "centers of origin" where it is thought to have originated. As people have traveled, they have spread these crops across the globe.

These centers of origin have been broken down into 8 regions around the world. Each of these centers has produced a number of important vegetable species. These centers and their important contributions are listed in Table 21-2. As you can see, some of the crops are thought to have originated in more than one area.

CLASSIFICATION

Vegetables can be classified based on several different criteria. Some of the categories that vegetables are classified in are (1) botanical classification—based on flower type, structure, and genetic makeup, (2) optimum growing temperature, (3) part of the plant used for food, and (4) number of seasons a plant may live. Other classifications are based on storage temperature, optimum soil conditions, and water requirements.

Botanical classifications are used by botanists and horticulturists mostly for scientific purposes. Vegetable plants all are members of the plant *king-*

Table 21-2. Centers of Origin of Selected Vegetable Crops	
Region	**Vegetable Crops**
Chinese Center—Mountains of central and western China	Radish, Chinese cabbage, onion, cucumber, and Chinese yam
Indian–Malaysian Center (Assam and Burma)	Mung bean, cowpea, eggplant, taro, cucumber, yam
Central Asiatic Center (India and Afghanistan)	Pea, mung bean, mustard, onion, garlic, spinach, carrot
Near Eastern Center	Lentil and lupine
Mediterranean Center	Pea, beat, cabbage, turnip, lettuce, celery, chicory, asparagus, parsnip, rhubarb, cauliflower, broccoli, horseradish, chard, mushroom, leek, globe artichoke
Ethiopian (Abyssinian) Center (Ethiopia and Somalia)	Cowpea, garden cress, okra
Mexican and Central American Center	Corn (maize), common bean, Lima bean, malabar gourd, winter pumpkin, chayote, sweet potato, arrowroot, pepper
South American Centers (Peru, Ecuador, Bolivia)	Andean potato, white potato, starchy maize, Lima bean, common bean, edible canna, pepino, tomato, ground cherry, pumpkin, pepper
Chiloe Center	White potato
Brazilian–Paraguayan Center	Cassava

Sources: Knott's Handbook for Vegetable Grouwers, Third Edition. Vegetables: Characteristics, Production and Marketing.

dom, the *division* Spermatophya, and the *class* Angiosperm. From that point, they may differ into *subclasses* of Monocotyledonae and Dicotyledonae and various *orders, families, genuses, species,* and *cultivars* (varieties). **Monocotyledonae** (monocot) plants characterized by one cotyledon (seed leaf) in their seedling stage, flower parts in threes or multiples thereof, and parallel leaf venation. **Dicotyledonae** (dicot) plants characterized by two cotyledons (seed leaves) in their seedling stage, usually by flower parts in fours or fives or multiples of these numbers, and reticulate leaf venation.

Classification based on the number of years a plant may live generally uses three categories. Annuals are those that complete their life cycle in one season, biennials complete their life cycle in two seasons, and perennials are those whose life cycle lasts more than two seasons. Most vegetables are

21-5. Tomatoes are a perennial but are grown as an annual. (Courtesy, Texas Department of Agriculture)

grown as annuals even though they may have a longer life cycle. Some examples include: annuals—sweet corn, biennial—cauliflower (grown as an annual), perennials—asparagus, tomato (grown as an annual).

Classification by optimum growing temperature can be broken into warm and cool season plants or temperate and tropical plants. Examples of cool-season crops are—asparagus, broccoli, cabbage, celery, garlic, kale, onion, pea, spinach, carrot, cauliflower, mustard, lettuce, turnip, and white potato. Warm-season examples are—cucumber, eggplant, Lima bean, okra, melons, squash, pepper, snap bean, sweet corn, sweet potato, and tomato.

Classification by edible parts or classification by a combination of botany and use are the most used classifications. Classification by use may be con-

21-6. Classification by edible part is one way of grouping vegetables. Edible parts may be flowers, such as in cauliflower (left), or roots, such as in turnips (right).

fusing since some vegetables may be used in more than one way. Generally, these classifications are divided into—potherbs and greens (spinach, collards), salad crops (celery), cole crops (cabbage, cauliflower), root, and tuber crops (potatoes, beets, carrots, radishes), bulb crops (onions, leeks), legumes (peas, beans), cucurbits (melons, squash), solanaceous crops (tomatoes, peppers), and sweet corn.

As would be expected, classification based on edible parts is based on that part of the plant that is normally consumed. Appendix D provides detailed information for common vegetable crops by classification. Examples of classification based on edible parts include:

Ground Parts

■ Root—beet, carrot, horseradish, sweet potato, turnip, parsnip.

■ Tuber—potato, Jerusalem artichoke.

■ Corm—taro.

■ Bulb—onion, leek, garlic.

Above Ground Parts

■ Stem—asparagus, kohlrabi.

■ Petiole—rhubarb, celery.

■ Leaf—lettuce, spinach, cabbage, mustard greens, brussels sprouts, kale, collards, garlic chive, Swiss chard, endive, chicory, parsley.

■ Flower—globe artichoke, broccoli, cauliflower.

■ Fruit—pepper, eggplant, snap bean, okra, squash, melon, cucumber, tomato.

■ Seed—dried beans, green peas, lima beans, sweet corn.

VEGETABLE CLIMATE AND ENVIRONMENT

Every living thing must have a suitable environment in which to live. There are many climatic and environmental factors that influence vegetable growth and production. Among these are temperature, light, air, water, and soil factors. The influence of soil factors on the vegetable plant are probably the most integral and yet most variable. Each of these factors can have a profound impact on the *yield* (quantity) and *quality* (degree of excellence) of vegetables.

Climatic factors generally refer to temperature and precipitation. These vary according to geographical region. This is one reason why vegetable plants have their specific points of origin. However, some climatic factors can be altered and create a suitable climate in a geographical area that was not naturally suitable.

TEMPERATURE

There are many factors that influence temperature. The distance between the earth and the sun is probably the most significant. There are seasonal changes in temperature as the earth orbits around the sun. Additionally, ocean and air currents can also dictate weather patterns, which influence temperature. Altitude and distance from large bodies of water also influence temperature. Then, of course, there are changes in temperature from day to night.

Each plant species has a temperature range at which optimum growth occurs. This is generally what differentiates cool season from warm season crops. Vegetable crops also differ in their susceptibility to frost and freeze injury. The greatest single factor in determining a crop's suitability for a particular climate is if it will have a long enough growing season to mature before it is damaged by frost or freezing temperatures. This is generally referred to as

21-7. Temperature has a great effect on vegetable growth. Squash thrives in warmer climates as do most warm-season crops. However, excessive temperatures can reduce pollination and fruit set of squash.

the "frost-free period." A region's average *frost-free period* is the average number of days from the last spring frost to the first autumn frost.

The time it takes a vegetable crop to mature can be measured in one of two ways. One way is counting the number of days from planting to maturity. Seed packages usually state a number of days to maturity. This is the average number of days it will take the crop to mature. However, a crop's rate of maturity is heavily dependent upon temperature and rainfall. Another means of measuring time to maturity is through heat units (degree days). *Growing degree days* are calculated by adding the daily high and daily low temperature together and then dividing the answer by 2. The base temperature of the crop is then subtracted from that answer to find growing degree days. The *base temperature* is the lowest or minimum temperature at which growth occurs for that crop. The minimum temperature for various vegetable crops can be found in Appendix D. Since a crop matures faster as temperature increases, the more accurate measurement of time to maturity is heat units.

The effect of temperature on crops varies according to species. Freezing or chilling injury can occur in warm season crops at temperatures of 50°F or lower. Although cool season crops can survive under much lower temperatures (40–45°F), they can also suffer from chilling or freeze injury. This type of injury can result in stress to the plants, decreased growth and production, and even death of the plant.

High temperatures can cause injury and death to plants as well. Maximum temperatures for warm season (80–95°F) and cool season crops (75–85°F) vary as well. Excessively high temperatures can cause decreased growth and production of vegetables. High temperatures result in increased respiration of plants. Excessive respiration can cause loss of yield and quality. Cool season crops that have been exposed to colder temperatures and then exposed to high temperatures may flower prematurely causing substantial yield loss. This premature flowering is called *bolting*. Yield may be decreased from excessively hot weather due to decreased pollination and fruit set. Flavor of crops with higher sugar content may be decreased as well.

MOISTURE

Moisture is also an important climatological factor for vegetable production since water is the major constituent of all vegetable crops. The plant must have water to live. Vegetable crop production is heavily dependent upon a consistent supply of water. Moisture may be present as precipitation, irriga-

21-8. Moisture is required for all vegetable crop growth. Without proper moisture yield and quality is reduced and fruit set is poor. Pole beans require good moisture during pod set to yield a good crop.

tion, or as humidity in the atmosphere. High humidity for prolonged periods can cause disease problems since this creates a favorable environment for the growth and spread of pathogens.

Moisture is most important in the soil environment. Plants must have water to absorb nutrients. Water also acts as a cooling agent for the plant either through transpiration (evaporation of water through plant tissue) or moisture on the plant surface. Water is also needed to maintain turgidity in plant cells.

Excessive moisture can result in drowning of plants if the soil stays wet enough to prevent air movement into the roots. On the other hand, insufficient moisture can result in decreased growth and production due to increased respiration, loss of turgidity, decreased nutrients or decreased pollination, and fruit set.

LIGHT

Light is a major factor in the growth and development of all vegetable crops. The most important role of light in the plant environment is to provide energy for photosynthesis. Without photosynthesis the plant cannot grow and produce. Photosynthesis is decreased under low-light conditions and the growth of the plant is slowed.

Vegetables are usually classified as long-day or short-day vegetables depending on their required ***photoperiod*** (length of day or night). Long-day vegetables require day length above a certain minimum to develop normally, while short-day plants require a night length of a certain minimum. Radish is an example of a long-day plant, sweet potato is a short-day plant, and tomatoes are day-neutral. *Day-neutral plants* are not affected by photoperiod.

AIR MOVEMENT

Air movement and wind are also important to the plant. Wind can be damaging at high speeds and cause crop injury. However, air movement is needed by the plant for several reasons. Air movement can increase or decrease humidity. Air movement may decrease the humidity around the plant and help cool the plant or it may bring in more humid air.

Movement of air can also spread diseases if airborne pathogens are present. However, wind and air currents may also aid in pollination of vegetables. Pollen may be carried from one plant to another for cross-pollinated plants. Movement of gases in the atmosphere is also important. Plants need carbon dioxide for photosynthesis, and wind currents replenish the supply of this gas. On the other hand, air pollutants, such as sulfur dioxide and ozone, may be carried into the plant environment, resulting in injury to the crop.

VEGETABLES AND THE SOIL

The soil environment is a very dynamic, complex, and very important factor in vegetable crop growth. As well as providing support for the crop, the soil also provides air, water, and nutrients to the plant. One of the most critical decisions a vegetable grower will make is where to plant the crop. This decision is largely based on the suitability of the soil for that crop.

Factors that must be considered in selecting a suitable soil are drainage, slope, soil texture, and soil structure. Slope is affected by the topography of the land. Steep sites are likely to be subject to soil erosion or have soils that are already heavily eroded. Depressions or low areas are likely to be wet.

Soil texture is defined by the relative contents of sand, silt, and clay in the soil. Light soils have more sand and silt, while heavy soils have a higher clay content. Drainage describes the soil's water-holding properties. Soils with good structure are well-aerated, loose, and friable, and do not have compacted zones (pans) that reduce root growth.

Vegetables grow best on soils that are light enough to be well drained with good aeration, but heavy enough to hold adequate nutrients and moisture. For that reason, loamy soils (approximately 20 percent clay, 40 percent silt,

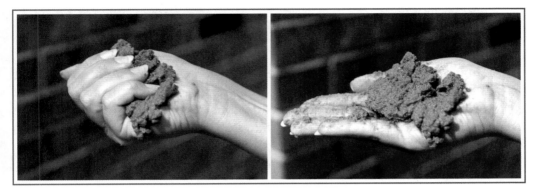

21-9. Soil high in sand content holds little water and has a gritty feel.

40 percent sand) are ideal, but a wide range of soils will produce vegetables well. Light soils are generally warmer and more uniform and are good for early spring production and production of root crops.

The soil is the source of most of the macronutrients and micronutrients needed for plant growth. Plants also get carbon (C), Hydrogen (H), and Oxygen (O) from soil gases and water.

Soil pH is a measure of the soil acidity or alkalinity. Most vegetables grow best in a pH range of 6.0 to 6.5. Soil pH can affect nutrient availability as some nutrients may become limited and some may become toxic outside the preferred range. Rainfall and fertilization tend to make soils more acid. Therefore, lime must be periodically added to the soil to adjust pH back up to a suitable range.

Probably the most important resource vegetables get from the soil is moisture. Vegetables require large amounts of water to grow and produce acceptable yields of quality produce. Soils have much pore space that contains air when the soil is dry. Although vegetables need a certain amount of air from the soil, it is from this pore space that plants get moisture. Sandy soils have larger pores which drain rapidly after they become filled with water.

Heavier soils have a greater percentage of smaller pores which hold the water longer and tighter. Therefore, moisture in sandier soils may drain more quickly, but is more available to the plant. Soils that hold large amounts of moisture for extended periods prevent air exchange around the root system, which can reduce plant growth. However, plants also will suffer stress, reduced photosynthesis, and, therefore, reduced growth before they actually wilt.

PRODUCING VEGETABLE CROPS

Although there are many components of the vegetable environment that cannot be controlled, producers can exert a great deal of modification on that environment. Crop management basically consists of the techniques that are used to increase vegetable productivity and quality. Among these techniques are irrigation, fertilization and liming, tillage, variety selection, pest management, mulching, and site selection.

VARIETY SELECTION

Another important early decision is **variety** (cultivar) selection. Varieties differ in yield, quality, horticultural characteristics, and disease resistance. Many varieties are usually suitable, but the decision as to which is the most suitable should be given careful consideration. Since varieties may perform differently in varying geographical areas, the most important criteria is to select a variety suitable to the area in which it is to be grown.

Yield and quality are the most obvious indicators of a variety's performance. However, varieties should possess resistance or tolerance to the major disease pests that affect the crop in that area. The plant growth habit, as well as product color, flavor, shape, and appearance, also must be considered.

Many crops, such as sweet corn and snap beans, are direct seeded. Others, such as tomatoes and peppers, are usually seeded in beds or in containers

21-10. Selection of varieties that are adaptable to the area in which they are to be produced is critical. Varieties of onions differ in maturity as well as in yield and quality.

and then transplanted to the field. Some, such as cabbage and squash, may be done either way. In either case, good quality material is needed. Seed should be of high quality, free of weed seed, and have good germination and vigor. Transplants should be healthy. Both should be free of disease. A **hybrid** is an improved plant developed by crossing parents of different genotype for a trait. Hybrid varieties of most vegetable crops have now been developed. Hybrid varieties generally have superior yield, quality, and disease resistance than standard varieties.

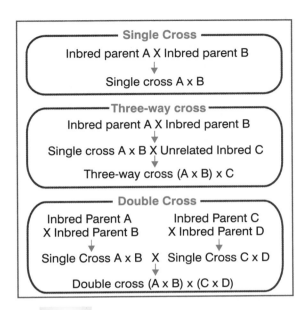

21-11. Common hybridization methods.

TILLAGE

Tillage is generally one of the earliest in-field operations performed in crop establishment. This practice improves the soil environment by increasing aeration, moisture infiltration, and weed control, as well as incorporating crop residue and creating a suitable **seedbed** (media prepared to receive seed). Most tillage operations are completed using such items as moldboard (bottom) plows and disk harrows. Where compacted zones must be broken, subsoilers may be used. Rotary cultivators may also be used to create a smooth and more uniform seedbed.

Initial tillage buries crop residue and loosens the soil to enhance root growth. This is usually followed by discing to smooth the seedbed and break up the soil. Fertilizers and weed control chemicals are often incorporated in this secondary operation. In areas that are not well-drained or where furrow irrigation is used, crops are sometimes grown on raised beds. These beds are formed by disks that hill up the soil and bed shapers that smooth and shape the bed. Improper tillage can result in poor growing conditions due to compacted soil and loss of soil structure.

Recently, the use of conservation tillage has gained in popularity. Although this is becoming a common practice in many field crops, the use of conservation tillage in vegetable crops is still non-traditional. **Conservation** *tillage,* also known as minimum tillage, no-till, and reduced tillage, involves tillage practices that leave 30 percent or more crop residues on top of the soil to prevent soil erosion. There are many methods of conservation tillage. Probably most common in vegetable production is the use of *strip tillage.* This is where strips of vegetation are left between tilled areas to prevent soil erosion.

Tillage practices generally continue after the crop has been planted. This is called cultivation. Cultivation is primarily a weed control practice. However, it is also used to incorporate fertilizer, break soil crusts, and improve infiltration of water. Vegetable crops are generally cultivated once or twice after planting, but before the crop begins to cover the row middles.

The addition of fertilizer and lime to the soil is a primary modification that can be made to the plant environment. Lime is usually applied several months before crop establishment to allow time for it to adjust soil pH.

Some mineral nutrients, such as nitrogen and potassium, are easily leached or washed out of the soil over time and are also removed during growth of previous crops. These nutrients, as well as other nutrients, must be added to the soil for proper crop growth. Complete fertilizer materials contain micronutrients, usually in sufficient quantity for the crop. Although phosphorous does not leach as easily as other nutrients, it, too, must be added to account for that removed by previous crops.

After initial fertilizer applications are made, subsequent applications are usually added during cultivation or through irrigation equipment. Fertilizers, such as nitrogen and potassium, are generally added in split applications since they are readily leached from the soil by rain and irrigation.

21-12. Fertilizer is often added to crops through fertigation. Tanks like this hold fertilizer until it is pumped to the plant through drip irrigation tubes.

MOISTURE

The addition and conservation of soil moisture is critical to proper vegetable production. The most obvious method used to increase soil moisture is through the use of irrigation. Although vegetable crops can be damaged by excessive moisture, inadequate moisture is generally of greater concern. Moisture is lost through evaporation, runoff, infiltration, and transpiration by the plant.

Crops vary as to when they are most susceptible to moisture stress. Fruiting crops are most vulnerable to injury from moisture stress at fruit set. Low soil moisture can decrease pollination during flowering. The period of fruit, tuber, head, and root enlargement is also a time of high moisture requirement. Moisture stress at this time can result in poor product quality and size.

There are several methods of irrigation used to augment the moisture supply of vegetable crops. Among these are subsurface, surface, and overhead irrigation. Probably the most familiar is overhead irrigation. The use of center pivots, solid set sprinkler systems, and traveling irrigation guns are the most noticeable irrigation systems.

Subsurface irrigation is usually used in association with a shallow water table. Water levels are adjusted as needed to supply the crop water. Surface irrigation includes both furrow and drip (trickle) irrigation. In furrow irrigation, water is siphoned into the furrow between rows and runs along the length of the row.

21-13. Water is siphoned from the ditch to irrigate this California vegetable field.

Drip irrigation involves application of water through a plastic tube that runs along the crop row. Water is pumped through the tube and emitted through small holes along the length of the tube. This type of system uses low pressure and offers the advantages of more accurate placement, lower volumes of water applied, less labor involved, and continuation of field operations during application. However, these systems require greater management and can be costly. Chapter 26 covers irrigation systems in more detail.

TEMPERATURE

The ability to modify temperature in the crop environment is limited, but not impossible. The most obvious way is through the use of greenhouse structures. Although some vegetables are produced in closed structures, the majority are field-grown. Modification of temperatures in the field is more difficult.

Recently the use of *plasticulture* (plastic mulches) to increase soil temperatures has become popular. These mulches offer several advantages, including increased weed control and moisture conservation. This type of system is often used in conjunction with drip irrigation. In such a system, a polyethylene mulch is laid on the ground over a raised bed.

Generally in plasticulture situations, the bed is fumigated for weed, disease, and nematode control, and plastic mulch and tubing is laid all in one operation. This system can be costly and involves greater management. However, it offers the advantage of producing earlier, more uniform, and greater vegetable yields.

21-14. Pepper plants can be grown traditionally on bare ground (left) or using plastic mulches and drip irrigation (right). Notice the small grain windbreaks beside the crop to prevent wind damage.

The black plastic warms the soil so that the crop can be seeded or transplanted earlier and the warmer soil enhances early growth. Fertilizers are usually applied through the drip irrigation tube throughout the growing season. This allows for more efficient use of fertilizer and water since fertilizer leaching is reduced and water does not evaporate as readily from the mulched beds.

In cooler climates, the use of row covers is a common practice. Row covers are plastic or translucent fabrics that are placed over the crop. Plastic row covers are supported by wire hoops. These covers increase the temperature in the crop environment and prevent infestation of early-season insects.

Frost and Wind

Protection from frost and wind are other production techniques that are used from time to time. Windbreaks are trees, hedgerows, or plantings of tall grasses that are spaced between vegetable plantings to reduce damage to young crops from wind. Frost protection methods include the use of row covers, as well as sprinkler irrigation and fogging. Heaters, smudge pots, fans, and wind machines are also used infrequently to reduce frost damage. These methods will not protect crops from hard freezes.

Seeding and Transplanting

Planting Density

The final step in crop establishment is planting the seed or transplant. Most vegetable crops can be planted at a variety of in-row and between-row plant spacings. The particular **plant spacing** or distance between plants in the row and the distance between plant rows depends upon the crop and its use. For instance, cabbage may be planted 12 inches apart in rows 36 inches apart for fresh market. However, this distance may be increased to as much as 20 inches or more if the crop is to be used for slaw production.

Plant spacing affects the population density of the crop. The population density affects the nutrient and water requirements of the crop. The more plants there are in a specific area, the more fertilizer and water that is needed for crop growth. Higher planting densities may also increase the chances for disease and insect infestation. However, higher densities may also crowd out weeds. If the planting density exceeds the available resources, then yield and quality of the crop will be affected.

The ideal planting density is low enough to prevent competition for resources between crop plants, but high enough to produce a canopy that covers the ground.

Direct Seeding

Most large seeded vegetables are direct seeded. With the improvement of precision seeding equipment, many smaller seeded crops are direct seeded and lower seeding rates can be used. The objective of seeding is to achieve the **best stand** (optimum population) possible. This is usually accomplished when soil moisture and temperature conditions are optimum for that crop. Under less than optimum conditions, problems may arise. For example, in cold, wet soils which occur in the spring, seed may not germinate properly and evenly and may be vulnerable to seedling disease.

Occasionally, seeding methods may need adjusting to account for suboptimal conditions. Planting seed deeper may expose them to more soil moisture in dry soils, while planting seeds more shallow may give them a warmer soil environment. Increased seeding rates may also be used to make up for lower germination conditions.

Since vegetable seeds come in various shapes and sizes, often seeds are coated to make them more uniform for use with precision seeders. Usually seeding rates are higher than the target plant density, due to the fact that all seed will not germinate. In gardens, seed tapes that have predetermined spacings are sometimes used. Seeding may often be heavier in gardens and

21-15. Proper planting densities can be achieved through various arrangements. Broccoli can be transplanted or direct-seeded. Either way an optimum planting density is critical to best yields and quality.

the crop thinned to an appropriate stand. Such techniques are rarely used in commercial vegetable production.

Transplanting

Transplanting is transferring or moving seedlings from the seedbed and setting them into the ground. Transplants are used in many crops for a couple of reasons. First, it may be the only way to assure a good stand since seeds of certain crops do not produce a suitable stand under field conditions. Secondly, the use of transplants allows extension of the growing season. This is accomplished by starting the plants indoors before conditions are suitable outside for plant growth. Then the plants are transplanted to the field after conditions are suitable for growth.

Transplants are usually grown in greenhouse structures or in covered beds. Use of the latter has declined in recent years as the use of containerized transplants has increased. Containerized transplants are grown using sterile media and plastic or polystyrene trays. These trays may have from 72 to 512 small compartments in them. A seed is placed in each compartment and the plant is grown to an appropriate size in the tray before being transplanted to the field.

The advantages of using this type of system over bed-grown transplants are several. Containerized transplants are generally more uniform and grown in a more controlled environment. They also may be held for longer periods when field conditions are not suitable for planting. Growth can be controlled more accurately by regulating applied fertilizer and water as they grow.

21-16. Modern transplant production utilizes polystyrene containers that are usually placed on rails above the ground in large plastic-covered greenhouse structures. (Courtesy, D.M. Granberry)

NON-TRADITIONAL VEGETABLE PRODUCTION

Several non-traditional methods may be used to produce vegetables. These include production practices that do not use chemical pesticides and fertilizers, utilizing integrated pest management strategies, and growing plants in a soilless environment.

ORGANIC PRODUCTION

As environmental issues have become more important to the public, so have vegetable production practices that impact the environment and food safety. U.S. vegetable products are closely monitored and have been found to be extremely safe from harmful chemicals. However, many people prefer vegetables that have been grown using *organic production* methods, without the use of chemical pesticides or fertilizers.

This type of production is more costly and generally results in lower yields and quality of vegetables. Disease and insect problems are also more common in this type of system since there are no chemical means of control and an enhanced environment is provided.

Most consumers still demand a cheap and high quality supply of produce. However, for those that are willing to pay a premium for organically produced vegetables, such production systems exist. These systems range from those

21-17. Organically produced vegetables are sometime's found at a farmer's market.

that do not use synthetically produced inputs to those that merely reduce synthetically produced inputs.

The benefits to human health are questionable since, although no chemical residues are on the product, more natural toxins may be produced, which may also be harmful. The benefits to the environment are that no chemical inputs are leached or run off into the water supply and soil.

Generally, these systems rely on the introduction of organic matter into the soil to improve soil structure and fertility. Manures, compost, and sewage sludge are added to the soils for this purpose. Nutrients are applied through manures or sludge. Generally, limestone is added to adjust soil pH. Rock sources of phosphorous and potassium may be used, although animal manures add some of each to the soil.

Organic components are added in advance of the crop, since it takes time for the nutrients to become available to the plant when applied through organic constituents. Legumes are sometimes used as rotational crops to add nitrogen to the soil.

Pest control practices include all of those mentioned in the next section except chemical controls. Sanitation, rotation, biological controls, crop resistance, and natural chemical compounds are generally important measures used.

SUSTAINABLE HORTICULTURE

Sustainable horticulture systems are those that use integrated pest management (IPM) and other best management practices (BMPs) to reduce inputs.

21-18. Sustainable horticulture often includes growing plants with mulch. These pepper plants are being transplanted into an established clover stand. (Courtesy, M. J. Bader)

This type of system does not sacrifice production or profit, but utilizes IPM, reduced tillage, and reduced chemical inputs to protect the environment.

Reduction of fertilizer and water inputs through the use of drip irrigation and plastic mulch are examples of practices used in sustainable systems. Use of improved varieties, windbreaks, strip tillage, and avoiding less productive sites are also tools of this practice.

HYDROPONICS

Systems in which vegetables are produced in the absence of soil are referred to as hydroponic systems. Nutrients, air, and water are provided to plants growing without a supporting medium (liquid soilless culture), or anchored by gravel, sand, rock wool, etc. (solid soilless culture). This type of system is generally expensive and requires intensive management.

Tomatoes and lettuce are the primary vegetable crops produced in these systems, although other crops, such as cucumbers, peppers, spinach, eggplants, and melons, can be grown successfully in this manner. Advantages of these systems are more uniform fertilization and watering.

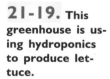

21-19. This greenhouse is using hydroponics to produce lettuce.

VEGETABLES AND THEIR PESTS

As is the case with any crop or plant, vegetables have their enemies as well. These enemies are the pests that can impair vegetable growth and production. Vegetable pests come in many forms. The most notable vegetable pests are insects, diseases, and weeds. Air pollution can also be a pest of vegetables. Some vertebrate animals can be pests as well, particularly in the home garden. Rabbits, deer, and birds can all cause production problems with vegetables, but they won't be discussed here.

DISEASES

There are three main causes of vegetable diseases. Pathogens that cause plant disease can be fungi, bacteria, or viruses. There are many different species of each of these and each can cause monumental losses in vegetable production under certain conditions. Nematodes in the soil can also result in major losses of vegetable yields. For pathogens to infect a vegetable plant the environmental conditions must be right and the pathogen must be present.

Viruses, fungi, and bacteria can be transported to vegetable plants by a number of means. Wind, water, soil, animals, insects, other plants, or seed can all be contaminated with these pathogens, and transport them to the vegetable plant.

Once the pathogen is present, certain conditions must exist before the crop is infected. First of all, the plant must be susceptible to the pathogen. Then, if environmental conditions are favorable for the pathogen, the plant may become infected and the disease may affect the growth of that plant and even spread to other nearby plants.

Diseases can be controlled or managed by a number of practices. Resistant plants, crop rotation, use of clean seed, seed treatment, use of non-

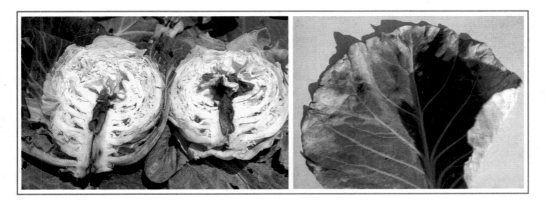

21-20. Diseases can be devastating to some vegetable crops. Black Rot in cabbage can ruin the cabbage head (left) and usually starts in the leaves (right). (Courtesy, D. M. Granberry)

infested soil, soil fumigation, control of weed hosts, control of insects that spread disease, and chemical fungicides and nematicides may be used in various combinations to manage plant diseases.

INSECTS

There are hundreds of insects that can affect the growth and yield of vegetable crops. These pests can cause damage in a number of different ways. The most obvious way insects can affect vegetables is by consuming the foliage or fruit of the plant. This can not only affect the growth of the plant, but also result in poor quality and contamination of the edible parts of the vegetable. Insects can also do damage by piercing the plant and sucking nutrients from the plant. Diseases are also spread by insects. Finally, the presence of insects on or in vegetables is generally not tolerated by the consumer and most products must be free of insect pests to be sold.

Control of insect pests can be accomplished by using a number of practices. These include the proper use of chemical or biological insecticides, avoiding infested areas, scheduling plantings to avoid times when insects are present in greatest number, use of resistant or tolerant varieties, enhancement of beneficial insect populations, and crop rotation.

21-21. Plastic mulch is used for weed control in this garden. (Courtesy, Agricultural Research Service, USDA)

WEEDS

Any plant growing out of place is a weed. In vegetable production, weeds can be a major contributing factor in reduction of yield and quality of the crop. Weeds compete with vegetables for light, nutrients, and water. As weeds rob these essentials from vegetable plants, the growth and vigor of the vegetable plant may be reduced. Weeds can also harbor insect and disease pests, which can then be spread to the vegetable crop.

Weed control is also accomplished using a variety of methods. Among these are the use of chemical herbicides, crop rotation, avoiding infested areas, and mechanical cultivation. Recently, there have been advances in the introduction of beneficial insects that attack only the weed plant and don't affect the crop.

AIR POLLUTION

The most common causes of air pollution injury to vegetable plants are ozone and sulfur dioxide. These pollutants can cause yellowing and death of the plant's leaves which in turn reduce growth, vigor, and yield of the plant. The use of fossil fuels and chlorinated hydrocarbons cause increases of these pollutants in the air and degree of damage to plants depends on environmental conditions. If the plants are susceptible to these pollutants and if they are present in sufficient quantities, they can cause problems if the air becomes stagnated enough for them to linger in the plant environment.

MAJOR VEGETABLE GROUPS

Vegetable crops are divided into cole crops, potherbs/greens, salad crops, roots and tubers, alliums, solanaceous crops, legumes, cucurbits, perennials, and miscellaneous crops. **See Appendix D for detailed information for specific vegetable crops.** Keep in mind that factors such as plant spacing may vary according to the use of the crop and the information given is general.

21-22. Carrots are grouped among the roots and tubers. Production techniques will vary according to soil type and use.

Many of the biennial or perennial crops listed are actually grown as annuals for vegetable production.

Since soil types vary widely across the world and since fertilization is largely based on soil type, nutrients removed from the soil by vegetable crops vary. A greater amount of nutrients would have to be added to the soil for proper growth and yield considering leaching and soil type. Total nutrients removed by both edible parts and other tissue are given. Crops such as beans and southern peas will remove more nitrogen from the soil than must be added, since these crops are legumes and produce some of their own nitrogen supply.

There are many different cultivars (varieties) of each of the crops listed. Varieties change rapidly as new lines are developed. Performance can also vary radically from one geographical region to another. A vegetable grower would want to consult seed catalogs and experienced farm advisors before selecting a variety.

Another item that changes rapidly is pest control recommendations. Chemical pesticides are consistently removed and added to the market as labels, registrations, and products change. Vegetable producers should consult trained farm advisors for current pest control recommendations as well.

BIOTECHNOLOGY IN VEGETABLES

The use of biotechnological techniques has far-reaching implications in vegetable production. Genetic engineering or manipulation through gene

21-23. A display used by Calgene in California to introduce the new MacGregor's Flavr-Savr Tomato (the new tomato was the first FDA approved genetically altered food product). (Courtesy, James Leising, Oklahoma)

transformation or direct gene transfer offers a wealth of potential for vegetable crop improvement. Direct introduction of DNA into the plant cell is known as direct gene transfer. These cells are then regenerated into fertile transgenic plants.

Gene transfer techniques using micropropagation make it possible to quickly and successfully transfer disease and insect resistance into existing varieties of species. Other factors that can be enhanced include yield, quality, flavor, and shelf life.

The first transgenic plant to be marketed in the United States was the Flavr-Savr Tomato. This product, introduced by Calgene, Inc., received U.S. Food and Drug Administration approval in 1994. The Flavr-Savr Tomato has the same horticultural characteristics as currently produced tomatoes, but has the advantage of having a shelf life of several weeks after harvest as opposed to several days. Many squash varieties have been altered genetically to be resistant to zucchini yellows mosaic virus and watermelon mosaic virus II and are among the latest transgenic vegetable plant to be approved by the U.S. Food and Drug Administration.

Future products will offer herbicide resistance in some crops. Herbicide resistance will allow growers to spray herbicides right over the top of vegetable crops, killing the weeds without harming the crop. These advances will make it possible to produce a greater vegetable yield and quality at a lower cost. Products will also be able to be shipped greater distances. This should make it possible for the increased health benefits of vegetables to be realized by a greater segment of the world's population.

COMMERCIAL VEGETABLE PRODUCTION

The commercial vegetable industry is a highly structured meshing of growers, buyers, shippers, wholesalers, retailers, brokers, chemical and seed industry personnel, researchers, consultants, and consumers. All of these segments work together to bring vegetables from the farm to the dining table. The commercial vegetable industry consists primarily of those produced for fresh market and those for processing.

While the number of commercial growers has declined in recent years, the size of operations has increased. Commercial growers produce vegetables to sell for a profit. The goal of each is to maximize profit while preserving

21-24. Many commercial vegetable crops are harvested mechanically. Some fresh market vegetables are still hand harvested. Often a caravan of harvesters and packers called a "mule train" is used to harvest and box the produce.

natural resources and producing safe, high quality produce. The use of some chemical pesticides is restricted to commercial operations.

The route that vegetables take in the commercial industry is generally from grower to marketing firms through brokers. From here they go to terminal markets, again through brokers. The route then goes on to wholesalers and finally to retailers and on to consumer. This route involves the intricate production practices discussed in this chapter.

Successful marketing of the finished product can depend on a number of factors. Most of those factors can be tied to supply and demand. If there are abundant sources of vegetables then prices for the product will generally be

21-25. Potatoes are one of the leading commercially grown vegetable crops. They may be sold fresh or processed into other products, such as potato chips or French fries. (Courtesy, R.D. Offutt Farms, Minnesota)

21-26. Vegetables are often transported great distances to provide consumers with fresh or processed produce. Leafy greens are often cut mechanically and hauled in large trucks to the processing facility.

low and the quality required for marketing will be high. If there are inadequate supplies, prices will generally be higher and lower quality produce may be marketed that would otherwise be discarded or sold at a lower price.

Marketing can be as simple as putting up a roadside stand. It may be through local farmers' markets or direct contract sales to wholesale distributors. The marketing process is very dynamic and is dependent on conditions and supplies in a wide geographical area. With the change to a more global economy, the geographical area that affects marketing may extend to many countries and not limited to a small area.

THE HOME VEGETABLE GARDEN

The home vegetable garden can not only provide fresh vegetables for the table. It is also a good hobby, a form of recreation and exercise, and can be a source of conversation with the neighbors. With proper maintenance, adequate space, and careful planning, the home garden can supply enough vegetables to provide fresh produce during the summer and frozen and canned produce throughout the year. However, any size garden can be a source of enjoyment and tasty produce. The types of vegetables grown will vary with personal preference.

A vegetable garden can be adapted to the available space. Some may be herb gardens in flower boxes or on patios. Dill, chives, mint, basil, parsley, and fennel are common herbs grown in flower pots. Others may consist of a

21-27. A herb garden can be grown in containers on patios or in back yards.

few vegetables in the flower beds near their homes. The garden can encompass a larger area—even large enough to be a community project or supply several large families.

Squash, tomatoes, potatoes, snap beans, sweet corn, spring onions and carrots are commonly found in home gardens. A lot of vegetables can be grown in a garden no larger than 20 feet by 50 feet (1000 square feet). The techniques used in growing the plants are similar to those with commercial vegetable production with a few exceptions.

PLANNING THE GARDEN

The most important phase of gardening comes before planting the first seed. Planning the garden involves selecting a location, deciding what to plant and when, and arranging the planting. Other decisions that must be considered in this phase are what varieties to plant, what kind of fertilizer to use, what implements are needed and what pesticides may be used, if any.

Location

The garden should be located near the home where it is readily accessible but in the open where it gets full sun. It should not be shaded by trees or buildings. The site should also be near water, have fertile soil, and not be subject to overflow from heavy rain or nearby streams. A moist, sandy loam soil with good drainage works quite well, however, many types of soil make good garden locations.

It is also good to have a site that is free of weeds and protected from animals. Fencing may be needed to keep out pets and wildlife that may damage the growing plants.

What to Plant

Deciding what to plant is an important decision. Space may be limited and proper selections should be made. A top priority is what you want to eat. Select vegetables that you like. Don't grow something you don't want!

The vegetables should be adapted to the climate where you live and the amount of space you have available. Also consider the season of the year. Cool season vegetables, such as cabbage, can be planted in the early spring. Warm season crops, such as snap beans, can be planted in late spring or summer. Some can be planted in the summer or early fall, depending on your climate. Cool season crops often grow well in the fall.

Consider the space required for a vegetable in comparison to the space available. Selecting vegetables that provide a large amount of produce in a small space are more ideal when space is a factor. Tomatoes, cabbage, broccoli, squash, carrots, radishes, onions, eggplant and beets can all produce a fair amount of produce and require little space. Other crops such as sweet corn, watermelons and snap beans grow taller or have sprawling vines that may require more space than is available.

Arranging the Planting

Arrange the garden in rows to get the best sun. Plant low-growing varieties on the side of the garden toward the sun. Plant tall-growing varieties on the side away from the sun. If tall-growing plants are planted on the side toward the sun, the low-growing plants may be shaded.

21-28. Many different vegetables can be grown in the home garden. This shows a recently planted garden with tomatoes (with stakes), corn, onions, cabbage, and potatoes. (Courtesy, Jasper S. Lee.)

Rows for small plants can be closer together than for larger plants. Squash and tomatoes need wider rows than carrots and radishes. The rows in a garden should be arranged to control soil erosion. They should be across the slope of the land. Never run rows up and down the slope. Plant perennial vegetables on one side of the garden so they will not interfere with tillage methods from one year to the next. Also, group vegetables according to maturity. This will ensure that space that is to be used for a second crop is continuous and not scattered throughout the garden.

PLANTING AND GROWING THE VEGETABLES

Planting and growing the home vegetable garden is similar to commercial vegetable production. The major difference is the size of the area to be planted. The kind of equipment used will also vary. Home gardening may involve hand tools such as shovels, hoes, and rakes. In some cases, small engine-powered rotary tillers may be used.

21-29. A rotary tiller can be used to till the soil to form a loose seedbed.

Once plans are made for the garden, the process of growing the crops begins. This process involves several steps. Here is a brief description of the major activities involved.

■ **Soil preparation**—Take a soil sample several months prior to planting to determine what soil amendments are needed. Add compost and other organic matter to the soil and till it in with any lime that is needed several months prior to planting. This can even be done in the fall to make for an earlier start come spring. Once spring arrives, add any fertilizer recommended and till the soil to form a loose seedbed.

■ **Planting**—Most home gardeners use seed or transplants. Use good seed and plant at the right depth and distance apart. A general rule of thumb is to plant seed to a depth of about three to four times the width of the seed. For example, carrots and lettuce should be planted ¼ inch deep while corn and peas are 1–2 inches deep.

Seeding rates for different vegetables are given in Appendix D. It is a good idea to over-seed by about 20% and then thin the plants to a proper stand. Using transplants is a good idea for crops such as tomatoes, pepper and eggplant which are hard to grow from seed in the ground. It also gives gardeners a head start for tender crops that cannot be planted until danger of frost has past. Transplants should be free of disease, have healthy appearance and not have blooms already on them. Set them out properly and provide water to soak the soil around each transplant.

■ **Controlling pests**—Common garden pests include insects, weeds, diseases, and animals. Insects are controlled with other insects, insecticides, and repelling plants. Some home gardeners plant marigolds around their garden to help repel the insects. Weeds in a garden are controlled by hand pulling, hoeing or chemical herbicides. Mulching with plastic sheets, straw, bark, leaves, and other materials keeps weeds from growing. Preventing disease is best accomplished by planting disease-resistant varieties as much as possible. However, the use of chemical fungicides is usually necessary to produce best yields. Chemical pesticides for the home garden are limited. Gardeners who choose to use them should always follow labeled directions and use all suggested precautions. General considerations for pest control include proper rotation of crops, destroying crop residue immediately after harvest, controlling weeds while they are small and keeping the plant free of other stresses. Refrain from garden activities when the foliage and soil are wet. It is also a good idea to learn to identify insect pests and diseases. Learn which insects are predators or parasites of the garden insect pests. Preserving these populations is an ideal way to reduce insect problems while reducing the use of chemical insecticides.

■ **Watering**—Vegetables will usually need more water than is available from the environment. This is especially true in dry areas and on lighter soils. Water is added with sprinklers, drip or trickle systems, or soaking the ground between the rows. One irrigation of three-fourths of an inch of water two times a week is better than a lighter daily irrigation. Water needs will change with the age of the crop and weather conditions. Always water early in afternoon to allow plants to dry before nightfall. This cools the plant during the hotter part of the day, prevents interference with pollinating insects and allow the foliage to dry before nightfall which reduces the spread of disease. It is a good idea to avoid spraying treated water from city water sources onto the leaves of vegetables.

■ **Fertilization**—Generally most crops will require additional fertilization as they develop. Over fertilization can be as damaging as under fertilization, however. It is best to split applications of potassium and nitrogen into three applications. The first application should be incorporated into the soil as mentioned above. Additional side dress applications should be made as the plant grows. All fertilizer should be applied by the time the plant starts to set

21-30. Tomato cage used in staking tomato plants in the home garden.

fruit. One exception would be for long season crops (less than 100 days) where the last application can be delayed a couple of weeks. Follow soil test recommendations on amounts of fertilizer applied. Always made side dress applications a couple of inches to the side of the plant row. Never apply dry fertilizer onto the leaves. This can burn the plants. It is best to water in the fertilizer after application. Lighter soils will require more fertilizer then heavier soils and more frequent application. Phosphorous can be applied all at planting since it does not tend to leach from the soil. Foliar applications are sometimes necessary for minor nutrients, however, they are seldom successful for adding major nutrients to the plant.

■ **Staking and training**—Some vegetables must be staked or trained, such as tomatoes and pole beans. If not, they grow as vines on the ground and won't produce a good crop. Staking and training also help make better use of the available space. Vegetables such as tomatoes can be staked using individual stakes and twine or by using cages. Beans can easily be staked by placing a post every ten feet, stretching a strong wire between the tops of the posts and training the vines on twine placed perpendicular to the wire every few inches. Even plants such as cucumbers can be trained onto a small fence to conserve space.

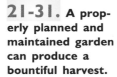

21-31. A properly planned and maintained garden can produce a bountiful harvest.

■ **Harvesting**—Harvesting is taking the desired products from the vegetable plants at the right time. With some crops, such as beets and carrots, the entire plant is harvested. With other crops, such as beans and tomatoes, the fruit is taken from the plant carefully so that the plant can produce additional fruit. Harvest as frequently as produce matures. Vegetables harvested at optimum maturity have better quality and keeping the produce harvested encourages the plant to continue producing.

■ **End of season**—When the season ends, the garden area should be cleaned. Corn stalks, tomato vines, and bean plants need to be cut and put into a compost pile or chopped and plowed into the soil. It is best to remove this debris as the crops are finished. Some areas are immediately replanted. For example, Lima beans may be planted following spring cabbage. Remove temporary trellises and cages. The gardener can then begin adding organic matter and compost and tilling it into the soil for the next season. Don't forget to take another soil test at this time as well. This is also a good time to make a notation of what plants or varieties grew well in the garden and which ones did not. This can make planning for the next garden much easier.

REGULAR OBSERVATION

Having a good garden requires continuous attention. Planting a garden and ignoring it will not produce many vegetables. It is a good idea to check a garden each day for water needs, weeds, diseases, and insect damage. Action can be taken to correct any problems before they get too big.

REVIEWING

MAIN IDEAS

Olericulture is the branch of horticulture dealing with the production, storage, processing, and marketing of vegetables. Vegetables are important economically as well as nutritionally. They are high in vitamins and minerals. Vegetable growth is influenced by temperature, moisture, light, and soil.

Although many factors affecting vegetable growth cannot be controlled, the influence producers can have on the vegetable environment is important in successful crop production. Variety selection, tillage, and influence on moisture and temperature are factors that humans can affect.

Diseases, insects, and weeds can all have serious effects on vegetable yield and quality. These pests can be controlled to an extent by the producer. Biotechnology has the potential to have a profound impact on these pests and product quality.

QUESTIONS

Answer the following questions using correct spelling and complete sentences.

1. What are the five leading U.S. vegetable producing states?
2. What nutritionally important components do vegetables have?
3. What are the eight centers of origin of vegetable crops?
4. What are the edible vegetable parts?
5. Name the major components of the vegetable climate/environment.
6. Explain how temperature can affect vegetable growth.
7. Name the factors to consider in selecting a suitable soil for vegetables.
8. Name the mineral nutrients that vegetables get from the soil.
9. What is important in the selection of a vegetable variety?
10. How does tillage affect the soil environment?
11. How can moisture be applied to vegetable crops?
12. Explain the difference in organic and traditional production.
13. What are the four main causes of vegetable diseases?
14. How can biotechnology affect vegetable production in the future?
15. What components are involved in the commercial vegetable industry?

EXPLORING

1. Visit a local farmers' market. Observe the different vegetables that come into the market. Inquire as to where they were grown and what the prices are for various vegetables. Compare these to prices in a local grocery store.

2. Make a trip to your local cooperative extension office. Find out what the local recommendations are for growing various vegetable crops in your area.

3. Prepare a vegetable garden either at your home or school. Plant and grow a variety of vegetables according to local recommendations.

4. Make a table listing each vegetable crop that would be suitable for the soils in your area and the time of year that would be best to grow them. Chart the major disease and insect pests that occur in your area that affect those crops.

22

POMOLOGY—FRUIT AND NUT PRODUCTION

Pomology is the branch of horticulture science involving the production of fruit and nuts. The word comes from the Latin word, *pomum*, meaning fruit. Pomology combines the art and science of fruit grafting, pruning, landscaping, and marketing with the science of fruit planting, fertilization, weed control, pollination, fruit growth, fruit maturation, picking, packing, storage, and shipment. Art is human skill and creativeness. Science is systematized (orderly arranged) knowledge derived from observation, study, and experimentation carried on in order to determine the nature of what is being studied.

What is a fruit? Botanically, it is the mature ovary of a plant or tree, including the seed, its envelope, and any closely connected parts. Is a tomato a fruit? Yes. However, pomology deals with perennial fruit plants, such as peaches, while olericulture (vegetable production) deals with short lived, annual fruiting crops, such as tomatoes, peppers, and eggplants.

22-1. Harvesting cranberries. (Courtesy, Agricultural Research Service, USDA)

OBJECTIVES

In this chapter, you will have an introduction to pomology which should be useful to you whether you are involved in home gardening, commercial nursery production, retail nursery sales, commercial fruit growing, or retail fruit sales. It has the following objectives:

1 Describe how cultivated fruits originated

2 Name some of the reasons for eating fruit

3 Name the types of occupations involved in fruit production, processing, and marketing

4 Describe the importance of eco-dormancy and endo-dormancy

5 Describe how a fruit is pollinated, fertilized, and grows

6 Explain what factors are needed for a good fruit growing environment

7 Explain how a fertilization program should be conducted

8 Explain the reasons for pruning fruits

9 Describe the formation of an open center and central leader fruit tree

10 Describe the importance of fruit thinning, weed control, and pest control in pomology

11 Describe how fruits change at maturity

12 Name the processes fruit go through in being packed and marketed

TERMS

advective freezes
aggregate fruit
berries
cane-pruned
controlled
  atmosphere storage
drupe
eco-dormancy
endo-dormancy

fruit spur
hydro-cooling
leaf analysis
microirrigation
multiple fruit
nuts
open center pruning
parthenocarpy
petal fall

pome
radiation freezes
rootstock
simple fruit
spur-pruned
winter chilling
  requirement

HISTORY OF POMOLOGY

Pomology is one of the oldest and most interesting branches of horticulture. The production of fruit is frequently mentioned in the Bible, Koran, and other ancient scriptures. How did fruit production begin?

ORIGINS OF FRUITS

Long before fruit was domesticated it was gathered from the wild. Fruits gathered from the wild are still very important in the diets of many people in our world. Even in the United States a great deal of fruit is gathered from the wild. What could be more fun than than picking wild blueberries, blackberries, plums, mayhaws, or persimmons on a beautiful day and enjoying the jelly and preserves on a cold winter day? Where did the various fruit crops originate? From Southeastern Europe, in the area around the Black Sea, came the apple, cherry, English walnut, European (prune type) plum, and pear. From Turkey, came the fig and grape. From China, came the apricot, chestnut, Oriental persimmon, kiwi fruit, orange, peach, and Japanese plum. From tropical America, came the avocado, cacao (chocolate), and papaya. From Africa, came the coffee tree. From tropical Asia, came the lychee, mango, banana, and coconut. From North America, came the American persimmon, American plum, blackberry, black walnut, blueberry, cranberry, mayhaw, muscadine grape, and pecan. From South America, came many of the genes for the cultivated strawberry. Even today, many groves and patches of wild fruits still exist and are being collected for use in breeding more disease-resistant and better fruit cultivars for tomorrow.

The Beginnings of Fruit Cultivation

The first known fruits brought into cultivation were the fig and grape. This occurred over 4,000 years ago. These fruits are easily propagated from hardwood (dormant) cuttings, so when an ancient horticulturist found an especially good fig or grape, he or she could bring it home and grow it in the home garden and expect the quality to be just as good or better than the plant growing in the wild. About 2,000 years ago, grafting came into use. This allowed ancient horticulturists in Greece, Rome, and China to propagate selected wild apples, pears, etc., which cannot be easily rooted from cuttings.

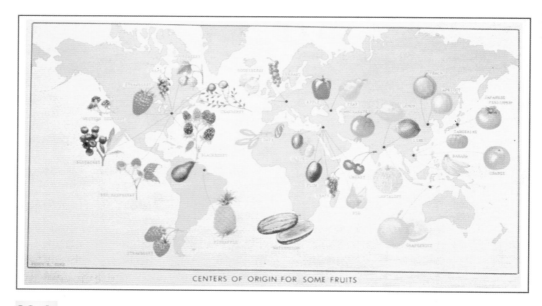

22-2. The origin of fruits can be traced to places around the world. (Courtesy, Agricultural Research Service, USDA)

When fruits are grown from seeds, they are quite variable and often not of good quality.

After the discovery of grafting, orchards became more productive. From observations, knowledge of fruit production gradually increased. During the 17, 18, and 19th centuries, many fruit species were introduced to new areas of the world, far from their centers of origin.

22-3. Grapes are one of the first known fruits brought into cultivation.

Origins of Fruit Cultivars

During the 1700s and 1800s, many excellent fruit cultivars were selected from chance seedlings and propagated by grafting and budding. Many of the apples and pears grown today are over 100 years old! Some of the grape and fig varieties we grow today are thousands of years old! In the early 1800s, the first controlled breeding work on fruit began. After 1900, this work greatly increased. Today, most of the blueberry, bramble (blackberries, dewberries, and raspberries), peach, plum, and strawberry cultivars come from controlled crosses from the many fruit breeding programs around the world.

IMPORTANCE OF FRUIT IN HUMAN HEALTH

Fruits are an important part of the healthy human diet. The actual nutrient content of the fruit varies with the species, but, in general, fruits are important sources of vitamins A, C, B_6, and folacin and the minerals potassium, magnesium, copper, and iron. Nuts are often rich sources of the

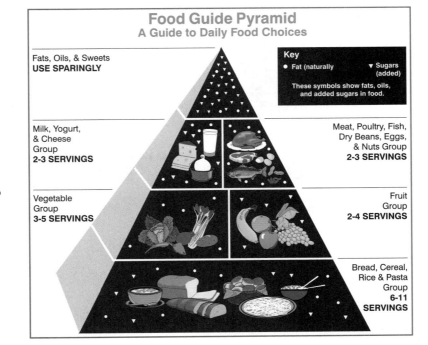

22-4. **USDA recommended food chart. Note that fruits should be consumed two to four times a day.**

Food Guide Pyramid
A Guide to Daily Food Choices

Fats, Oils, & Sweets
USE SPARINGLY

Key
● Fat (naturally   ▼ Sugars
                                    (added)
These symbols show fats, oils,
and added sugars in food.

Milk, Yogurt,
& Cheese
Group
2-3 SERVINGS

Meat, Poultry, Fish,
Dry Beans, Eggs,
& Nuts Group
2-3 SERVINGS

Vegetable
Group
3-5 SERVINGS

Fruit
Group
2-4 SERVINGS

Bread, Cereal,
Rice & Pasta
Group
6-11
SERVINGS

22-5. Small fruits such as blueberries are a rich source of anti-oxidants. Anti-oxidants help neutralize "free radicals" which cause aging and cancer.

vitamins thiamin, riboflavin, and niacin, and the minerals calcium, phosphorus, iron, and potassium. In recent years, many fruits have been discovered to contain phytochemicals which are associated with many health benefits. An example of this is elagic acid, which is abundant in strawberries and muscadine grapes. Elagic acid is associated with a decline in the risk of cancer. Blueberries have been discovered to be the highest in anti-oxidants of all fruits and vegetables tested. Anti-oxidants are compounds that neutralized bad substances call "free radicals" which cause aging and cancer. The source anti-oxidants in blueberries are the blue pigments know as anthocyanins.

Fruit pectins are also important chemicals in the control of dietary cholesterol. Pectin can trap dietary cholesterol, ultimately preventing its deposit in the linings of blood vessels. Fruits are a good source of dietary fiber which is so important in the prevention of colon cancer and constipation. In summary, fruits are more than just something delicious to eat. They are the very essence of good health.

IMPORTANCE OF FRUIT AND NUT PRODUCTION TO THE ECONOMY

Fruit production is an important part of U.S. and Canadian agriculture. In 1993, about 3.5 million acres (8,648,381 ha) of commercial fruit and nut trees or plants were growing in the United States. Canada is also a major fruit producing country.

Although the area occupied by orchards is modest compared to row crops, the value of the crops is enormous. For instance, in 1998, the total farm gate value of the U.S. fruit and nut crop was over 10 billion dollars. The United States is one of the world leaders in fruit production. The United States produces about 10 percent of all the world's apples, pears, plums, and prunes, about 20 percent of the world's peaches, about 25 percent of the world's citrus.

Hundreds of thousands of people are employed in pomology and related fields. Fruit growing is a labor-intensive business with many opportunities. It is also an agricultural enterprise in which you do not need a huge amount of land to make a living. Land which is too rolling for vegetable or row crops, and subject to erosion, can often be used for fruit production. Fruit growing can also be an excellent way to supplement the income from your regular job or farm enterprise. Fruit growing is a popular hobby, and the sale of fruit trees, bushes, and vines is a very significant part of the wholesale and retail nursery business.

22-6. Fruit growing is a labor-intensive business with many opportunities.

THE YEARLY CYCLE OF A TEMPERATE ZONE FRUIT TREE

In order to present a more complete picture of how the many aspects of plant and fruit development come together, the yearly life cycle of a fruit tree needs to be understood.

WINTER—DORMANCY

How does a fruit tree know when to bud out so it blooms will not be killed? What keeps a fruit tree from blooming during a warm spell in mid-Winter? In the fall, temperate fruit leaf and flower buds go into dormancy and will not come out until they receive winter chilling followed by warmth in the spring. The two phases dormancy are call **Endo-dormancy** or rest, which is controlled by internal physiological blocks, which are removed by winter chilling and **Eco-dormancy**, a state in which the plant is not growing and will not grow until external conditions are satisfied, usually warm spring temperatures. Endo- and eco-dormancy work together to prevent plants from breaking bud at the wrong time of the year. So a combination of chilling in the winter followed by warm temperatures in the spring is needed to release the leaf buds and flower buds from dormancy and let the temperate plant come into flower at the correct time of the year. Tropical plants to not face the danger of spring freezes so their flowering and fruit development period is usually governed by mother nature to occur at the best time of the year (rainy season, etc.) for fruit production.

Winter chilling requirement is the number of hours at and below 45°F (7.2°C) required for the plant to break dormancy and develop normally when temperatures rise in the spring. The temperature regime the plant actually responds to is more complex than just 45°F (7.2°C) and below, but this model works well for most pomology purposes. Lack of winter chilling results in

22-7. Generalized map showing approximate average winter chilling hours (at or below 45°F.) from Nov. 1 to Feb. 15 in the Southeast. Some areas of the lower Southwest also experience winters with few chill hours, but in this area mountains and oceans create microclimate effects that make chill hour accumulations variable within only a few miles.

Table 22-1. Approximate Chilling Requirements at or Below 45°F (7.2°C) to Break Winter Dormancy for Fruit and Nut Species			
Northern papaw	100–1800	Sour cherry	600–1400
American plum	700–1700	Sweet cherry	500–1300
Domestic plum	900–1700	Blueberry	150–1200
Apple	50–1700	Peach	250–1100
Raspberry	800–1700	Kiwi	600–800
Filbert	800–1700	Apricot	300–900
Pear	200–1500	Blackberry	200–400
Currant and Gooseberry	800–1500	Quince	100–400
Walnut	400–1500	Almond	100–400
Japanese plum	300–1200	Persimmon	100–400
Peach (Florida)	50–400	Grape[2]	100–1500
Peach (Texas)	350–950		
Peach (General)	800–1200	Strawberry	200–300
Pecan	300–1000	Fig[3]	0–300

Source: N. B. Childers. *Modern Fruit Science,* 9th ed. Gainesville, Florida: Horticultural Pubns. 1983.

[1]The ranges shown for each species indicate the differences between low- and high-chilling cultivars within the species.

[2]Grapes will grow with very little winter chilling, but like many species will grow much faster after longer chilling.

[3]Figs are almost evergreen, but will grow better with moderate chilling.

very poor leaf development and fruit production in the Spring. Approximate chilling requirements for fruit and nut species are listed in Table 22-1.

Why not just plant low chilling cultivars in all areas so the winter chilling requirement is always met? Planting cultivars which are too low chilling for the area results in very early bloom in late Winter or early Spring. Usually the crop is then destroyed by freezes.

SPRING—THE BEGINNING OF FLOWER, LEAF, SHOOT, AND ROOT GROWTH

Once a plant has received its winter chilling requirement and its heat unit requirement, bud break occurs. The heat unit requirement is the number of hours of warmth needed for a certain growth phase to occur. Heat units are calculated as growing degree days for most plants by averaging the high and low temperature for that day and subtracting the base temperature needed for normal growth of that species. For many fruit plants, a base temperature of 50°F (10.0°C) is used in the calculation.

Some plants have a high chilling requirement and a low heat unit requirement, such as highbush blueberries. Others have a low chilling requirement and a high heat unit requirement, such as muscadine grapes and pecans.

Buds are of two main types—mixed buds, which contain leaves and flowers, and simple buds, which contain leaves or flowers alone. Apple flower buds are mixed, peach flower buds are simple. Both types of fruits also have simple leaf buds which become the shoots.

Leaves—Leaves serve two main purposes. They manufacture photosynthates and help bring water up from below ground as they lose water during transpiration. With the water comes the essential nutrients for plant growth and hormones from the root system.

Shoots—Shoots transport the photosynthates to the roots, fruits, and other parts of the plant. They also transport hormones which control plant growth. The major plant hormones are: auxins, gibberellins, cytokinins, inhibitors, and ethylene. Auxins and gibberellins are hormones which generally promote cell expansion at low concentrations. At higher concentrations, auxins inhibit growth of side shoots. Cytokinins have a special role in helping plant cells divide. They also encourage the growth of side shoots. There are several natural inhibitors, but abscisic acid is one commonly thought to have a role in the start of plant dormancy. Ethylene is a gaseous hormone which can act as an antagonist to auxin. It is also involved in fruit ripening.

Roots—Root growth begins in the early spring when temperatures start to rise. The roots support the plant, provide moisture, assist in the uptake of nutrients and cytokinins which help regulate growth, and store carbohydrates and minerals. These stored carbohydrates in the roots and woody shoots will actually provide the fruit plant with the energy needed for most of its fruit and shoot growth for about the first 30 days. It takes this long for the shoots to grow and leaves to be mature enough to start sending photosynthates to the developing fruit.

SPRING—THE BEGINNING OF FRUIT DEVELOPMENT

Pollination and Fertilization—The transfer of pollen to the stigma is called pollination. With most fruit crops, bees of various types transfer the pollen to the stigma. Once on the stigma, the pollen germinates much like a seed.

A pollen tube forms and grows down through the stigma and style. When it reaches the ovary (female portion of the flower), the male pollen tube cell divides. One sperm nucleus fertilizes the egg cell, which will become the seed, and the other sperm nucleus fertilizes the cells which will become the endosperm. The endosperm will nourish the developing seedling during germination.

In some pomology crops, such as nuts, it is the seed that is the commercial crop. In fruits, the seed is the vital organ which causes the fruit to grow and swell into a delicious product.

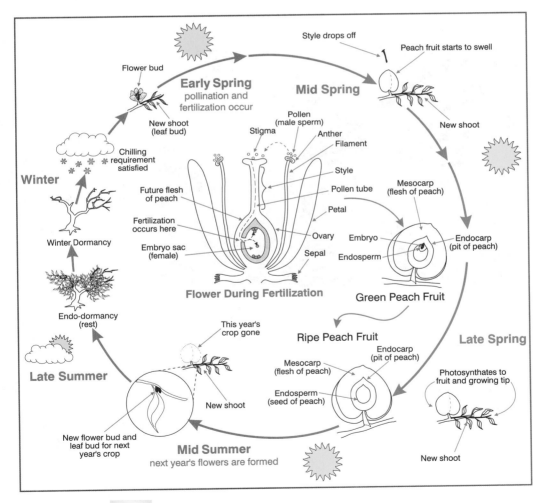

22-8. The yearly life cycle of a temperate fruit tree (peach).

LATE SPRING AND SUMMER—THE FRUIT DEVELOPS

Fruit Set and Drop—For various reasons, such as lack of pollen, incompatible pollen type, improper temperatures, competition between developing fruits, etc., fruit plants usually only set a portion of their blossoms. Often this is good because the number of blossoms the plant has is far in excess of what the plant can properly develop as fruit. Most fruits have several periods in which excess fruits are shed. The first is usually just after bloom, the second is before the final growth periods.

Some fruits also have a pre-harvest drop. Surprisingly, many larger fruits, such as apples and peaches, still fail to thin themselves enough to produce the large size the market requires. In a later section, we will discuss how farmers remove excessive amounts of fruit by thinning.

22-9. A peach about to drop naturally, due to competition from the other fruits on the shoot. There are not enough photosynthates for all four fruit so the tree aborts one fruit.

What Is Controlling Fruit Set, Drop, and Growth?

Although **parthenocarpy,** or the successful development of fruit without the presence of seeds, occurs in some fruits, such as Japanese persimmons, most fruits require seed development for fruit set. Hormones such as gibberellic acid are produced by the developing seeds of many fruit plants. Hormones produced by the developing seed probably cause photosynthates to move into the fruit and allow its development.

Besides effecting growth of the fruit, hormones have another important role in preventing excessive fruit drop. Besides accidental damage, fruits drop because of a change in a specialized layer of cells between the fruit and the

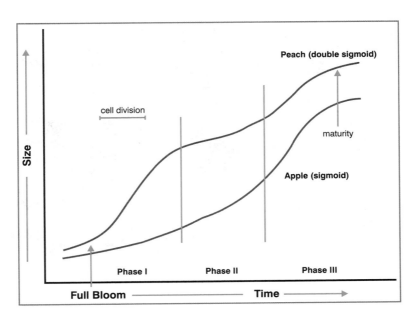

22-10. Growth in diameter of an apple and peach. The first phase is when cell division occurs. In the second and third periods, cell enlargement occurs. (Adapted from Jackson, 1986)

stem or the stem and the branch, called the abscission layer. The abscission layer is thought to be under hormonal control. Formation of the abscission layer generally occurs when the level of hormones in the fruit is low.

Fruit Growth—Following fertilization a period of cell division occurs. During this time, all the cells the fruit will ever have are created. In many fruits, this period lasts for just three to four weeks. Ultimate fruit size will be determined to some extent by the number of cells the fruit develops during this period. The first phase of fruit growth, which includes the period of cell division in most fruits, is called Phase I. Phases II and III are the primary times of cell enlargement in most fruits.

In stone fruits, which have large bony pits surrounding the seed, the growth rate slows as the pit hardens during Phase II. Phase III is also called the final swell of the fruit. In the last few days of fruit development, the growth rate slows again. When the increase in diameter is plotted out, some fruit have an S curve or sigmoid growth rate and some have an double S or double sigmoid growth rate.

MID AND LATE SUMMER—FLOWER BUDS ARE FORMED FOR NEXT YEAR'S CROP

Flower Initiation—Most deciduous fruits form their flowers one season but bloom the next spring. Usually they start to initiate flower buds in sum-

22-11. Buds of flowers of peach in the spring. Note how some became flower buds and some leaf buds.

mer after the spring season of rapid shoot is finished. Early in their growth these new buds are not differentiated and can become either leaf buds or flower buds depending on the environment and condition of the tree. It is very important that trees are well cared for late in the summer, because this will impact the number and quality of flowers for the next spring. Final development of the flower buds continues slowly during the fall and winter.

Most fruit plants have perfect flowers, or the male and female parts are both found in the same flower. A few species initiate their flowers at other times and have different flower arrangements. Hazelnut, walnut, and pecan have separate male and female flowers on the same tree, they are called monoecious. In these crops, the male flowers are formed before the female flowers. Pecan actually forms the male flowers the summer before flowering and the female flowers in the spring of the season they open.

June-bearing type strawberries form their flower buds in response to the shortening days of late summer. Most fruits and nuts do not respond directly to shortening day length by forming flower buds, but indirectly the shortening days of late summer slow the growth rate of the shoots and promote the formation of the flower buds.

FALL—THE START OF ENDO-DORMANCY (REST)

The shortening day length and dropping temperatures of late summer cause fruit plant to form terminal buds and quit elongating shoots. In late fall, the plants move into endo-dormancy and the leaves fall. Winter chilling is required to break endo-dormancy. If the plant did not have endo-dormancy, a warm spell in the fall would cause the blooms to emerge in the fall.

The yearly cycle of the fruit plant is complete. It may continue for many, many seasons. Some fruit trees, such as apples, pears, grapes, and blueberries, may live to be over 100 years old! The fruit tree you plant may be enjoyed by your great-great-great grandchildren!

TYPES OF FRUITS

There are many types of fruits with many different forms and shapes. *Simple fruits* are comprised of a single large ovary with or without some other flower parts which have developed as part of the fruit. They may be further grouped as fleshy pericarp (the wall of the ovary) and dry pericarp simple fruits.

■ **Fleshy Pericarp—***Berries* have a fleshy ovary wall and two or more carpels containing seeds. The outer wall may be hard or leathery. Examples are tomatoes, grapes, blueberries, pumpkins, and oranges. *Drupes* are single carpels with three layers and a single seed. Examples are cherries, peaches, and ol-

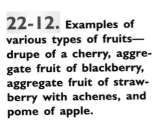

22-12. Examples of various types of fruits—drupe of a cherry, aggregate fruit of blackberry, aggregate fruit of strawberry with achenes, and pome of apple.

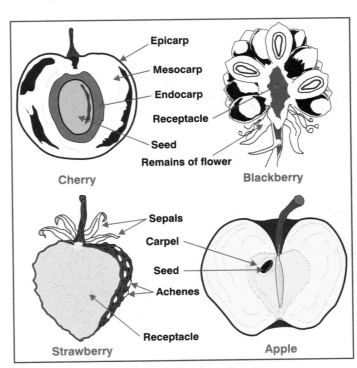

Epicarp
Mesocarp
Endocarp
Receptacle
Seed
Remains of flower

Cherry

Blackberry

Sepals
Carpel
Seed
Achenes
Receptacle

Strawberry

Apple

ives. *Pomes* are composed of several fused carpels and the flesh of the fruit is derived mostly from flower tissue outside the ovary. Examples are apples, mayhaws, and pears.

■ **Dry Pericarp**—Achenes are one-seeded, dry fruits. Strawberries are an aggregate fruit with achenes on their surface. *Nuts* are one-seeded fruits, often produced from a compound ovary. Examples are walnuts, pecans, and hazelnuts.

Aggregate fruits are comprised of a single receptacle (base of flower) with masses of similar fruitlets. Examples are blackberries and strawberries.

Multiple fruits are comprised of ovaries of many separate, but closely clustered flowers. Examples are mulberries and pineapples.

IMPORTANT ASPECTS OF FRUIT PRODUCTION

22-13. Apples and pears are grown in these Washington orchards. (Courtesy, Agricultural Research Service, USDA)

The ideal fruit environment varies with the type of fruit, but generally has some of these qualities—high quality, well-adapted cultivars and sufficient winter chilling to break the endo-dormancy requirement if it is a deciduous fruit. If it is a tropical fruit, the climate should be free from the dangers of winter freezes. The climate should be free from the dangers of late spring freezes. The weather for most of the growing season should be dry to reduce problems with fruit skin russetting, cracking, and fruit diseases. Light levels should be high. Summer temperature should be correct for the fruit being grown. Irrigation water should be clean and abundant. The soil should be well drained and free of pathogens. Land costs should be low and taxes reasonable. Labor should be abundant and inexpensive. Supplies needed for fruit growing should be easily obtainable and reasonable in price. Markets should be close and prices well above production costs.

No area of the country has all these factors in one place. However, certain areas of the country have more of these factors in one place. These are the locations where fruit is cultivated on a large scale. Examples are the Great Lakes Region, the Central Valley of California, the apple growing valleys of Washington and British Columbia, and the Fort Valley Plateau of Georgia. However, there are many successful fruit growers serving local and pick-your-own markets outside the areas where fruit is grown on a large scale.

CULTIVAR SELECTION

Cultivar selection is extremely important in successful fruit growing. The cultivar must be adapted to the climate, reasonably disease resistant, produce high quality fruit, and ripen in the correct market window. A localized breeding program is one of the keys to a successful commercial fruit industry. Many cultivars have only a relatively narrow range where they perform well.

ROOTSTOCKS IN FRUIT PRODUCTION

Different rootstocks are used to combat various soil problems and control the growth of the trees. *Rootstock* is the root system and base of the tree on which the fruiting top or scion cultivar is budded or grafted. Some rootstocks

22-14. An apple scion prepared for grafting to a special rootstock to achieve a better tree. (Courtesy, M. E. Ferree)

are propagated from cuttings, while other rootstocks are propagated by seed. Some plants, such as strawberries, blueberries, and many grapes, are grown on their own root systems and propagated by runners or cuttings.

Some rootstocks are tolerant of poorly drained soil conditions, nematodes (small parasitic soil worms), soil insects, or soil diseases. Size controlling rootstocks are widely used on apple and pear trees. These prevent the trees from growing excessively large and also cause the trees to start fruit production earlier in their life. General information on rootstocks is presented in Appendix G.

Soil

A desirable soil type is important in fruit growing. For most fruit crops the soil should be well drained, but with enough water-holding capacity to provide a good reserve of available water. Fruit crops vary greatly in their ability to stand wet or poorly oxygenated soils. Some crops, such as cranberries and mayhaws, can tolerate flooding for periods of time. Pears, pecans, and rabbiteye blueberries can tolerate damp soils. Almonds, apricots, and peaches require very well drained soils. As a general rule, the best all around soil types for fruit crops are sandy loams and loamy sands. These soils have enough nutrient and moisture holding capacity for good plant growth, but are usually well drained. Soils which are extremely fertile may cause excessive vegetative growth at the expense of fruit production. On these soils fertilizer must be applied very sparingly to minimize this problem. Some crops,

22-15. Soil type has a tremendous effect on successful fruit growing. Note that this soil has a bright colored subsoil. This is an indication of good drainage. (Courtesy, Jim Strawser)

such as highbush blueberries, require soil which is high in organic matter (over 3 percent) for best growth.

An adequate supply of water is very important in fruit production. In recent decades it has been discovered that fruit can be successfully grown on soils once thought to be too shallow, rocky, or calcareous for fruit production using fertigation. Fertigation is the injection of fertilizer in the irrigation water. All the necessary nutrients and water are supplied to the limited root system by the irrigation system.

Soil pH and Adjustment

Fruit crops in general perform best with a soil pH in the range of 6.0–6.5, with 6.5 considered near optimum. Below pH 6.0 availability of the major nutrients decreases and problems with calcium and magnesium deficiency increase. Above a pH of 7.0 micronutrients, such as iron and zinc, become less available and deficiencies of these elements become common. An exception to the general optimum pH of 6.0–6.5 are blueberries and a few other acid loving crops. They like a pH of 4.0–5.2.

Dolomitic limestone is the usual means of raising the pH of acid soils. It supplies both calcium and magnesium. Sulfur is a necessary element for plant growth, but it is also used to lower the pH of soils which are too high for optimum plant growth.

Rates of dolomitic limestone or sulfur to be applied are determined by the degree of adjustment needed and the soil characteristics. This information is determined during a soil test.

LIGHT

Light is very important in fruit production. Day length determines when some fruit crops, such as June-bearing strawberries, will form flower buds. They form daughter plants in the long days of spring and flower buds for next year's crop in the short days of late summer. Many fruit crops respond to decreasing day length in late summer and fall by slowing their growth rate. A reduction in the vegetative growth rate increases the formation of flower buds. These flower buds then emerge the following spring.

Sunlight is needed for photosynthesis. When light levels drop too low, the twig no longer produces enough photosynthates to be productive. The branch no longer initiates flower buds and will often die as the tree self-prunes to remove unproductive branches.

22-16. Most fruit trees need about 20 percent of full sunlight to form flower buds and keep fruit twigs healthy. This tree has lost much of its interior fruit zone from lack of sunlight. Proper summer pruning of watersprouts would have allowed more sunlight to reach these twigs and allowed them to remain productive.

Fruit plants, through photosynthesis, convert carbon dioxide and water into carbohydrates and oxygen. Some of the carbohydrates are stored in the woody portions of the plant. When spring comes most of the energy for the first month of growth is due to these stored reserves. With some types of plants, such as pecans, the female flowers are not formed until the following spring. If stored reserves are low, few female flowers are formed and the crop will be light that year.

TEMPERATURE

Winter Temperature

Temperature plays an important part in fruit production. Fruits can be grown in every climate, but only certain species of fruit will grow and produce well. One factor which needs to be considered is winter hardiness or how low a temperature the flower buds and plant can stand without damage.

In much of North America, winter freezes are a significant hazard in fruit production. In temperate growing areas, low winter temperatures usually first destroy the flower buds, and, if the temperature drops low enough, the tree itself may be injured or destroyed. In subtropical or tropical areas, such as Florida and California, low winter temperatures may cause severe damage to the citrus and tropical fruit orchards. Information on the approximate temperature each species can tolerate is in Appendix G. There are numerous references to the USDA hardiness zone map which can be found on page 408 (figure 16-24) of this book. By comparing the information in Appendix G and the USDA hardiness zone map, you can determine in what zones a fruit species may be adapted.

Spring Freezes and Topographic Effects

Many types of fruit plants have flowers that emerge early in the spring. Once the flowers are emerged, on most species they are subject to damage at or below approximately 28°F (−2.2°C). For this reason, site selection in relation to spring freezes is very important in fruit production. Two types of freezes occur, freezes accompanied by wind, called *advective freezes,* and freezes which occur on still nights, called *radiation freezes.*

22-17. Effect of topography on moderating the climate for fruit growing. Cold air settles in low areas during a radiation freeze. Note how the peach trees are planted on the higher ground.

Passive Freeze Protection— Site Selection in Freeze Protection

It is difficult to protect against advective freezes because of the wind. Radiation freezes, however, are more common and site selection is an important aspect of protection against this type of freeze. Because cold air is heavier than warm air, it settles to the lowest point in an area. In many cases, this can make the difference between a good crop and no crop of fruit. For this reason, orchards are usually planted on relatively high elevations. Some of these areas, such as the Fort Valley Plateau of Georgia are fairly flat, but relatively high in elevation. An interesting exception to this rule are the many western valleys, such as the San Joaquin. Here the high mountains to the east block the movement of very cold air streaming down from the Arctic. These valleys have relatively little problem with spring freezes on deciduous fruit. However, even in these valleys the warmest area are the low hills surrounding the valley. On the hills even tropical and subtropical fruit can be grown.

Bodies of water are also very important moderators of the climate in some fruit growing areas such as Florida and Southwest Michigan. Water moderates the climate, reducing the likelihood of a spring freeze. Good weed control is also important in freeze protection and will be discussed later.

Active Freeze Protection Methods

Freeze protection is an important tool in many fruit growing areas, especially against radiation freezes. Some of the most successful techniques are overhead irrigation for freeze protection, covers, and orchard heating. Overhead irrigation for freeze protection requires applying water continuously during the freeze and usually well into the next day. It works because water releases heat as it freezes and the temperature inside the ice encasing the flowers stays in the range of 31.5°F (−.3°C) which is not cold enough to kill the flowers. Covers made of translucent synthetic fabric are useful in strawberry production. They "float" on top of the crop and build up heat during the day, giving up to 6°F (3.3°C) of cold protection at night. Orchard heaters are also used in some areas, but they are not as popular as they formerly were because of rising fuel costs.

Summer Temperature

Heat tolerance is the ability of a plant to tolerate high temperatures and still produce well. Lack of heat tolerance is the reason raspberries are not widely grown in the South. Heat units are also needed to properly mature the crop. The optimum temperature for photosynthesis varies with the crop. For some crops, such as strawberries, the optimum day temperature may only be 75°F (23.9°C), but for other crops, such as Southern highbush blueberries, the optimum day temperature may be 85°F (29.4°C). To some extent, this determines where crops are best adapted. In very northern growing areas, insufficient heat units may be accumulated to properly mature some crops such as peach.

ESTABLISHING AND MANAGING THE ORCHARD

Once a suitable fruit crop has been found, an excellent cultivar selected, and a proper rootstock for the soil chosen, the plants are ordered from a nursery, normally the season before planting. A relatively spring-freeze-free site, with good soil drainage, usually on a slope or hilltop (in the Eastern United States), is chosen and soil preparation begun.

PREPARING THE SOIL AND LAYING OUT THE PLANTING

The soil pH is adjusted and dolomitic lime incorporated by plowing and discing the soil. Phosphorus is added if needed since it moves slowly in most soils. The rows are laid out according to the type of fruit crop being grown and the machinery which will be operated in the orchard. Once the plant spacing is established, a sub-soiler which will break the hard pan (hard layers) of the soil is run as deeply as possible down the row and across the row. The cross sub-soiling intersects the row at each spot where a tree is planted. The subsoiler is merely a narrow plow which can be sunk deep in the soil, usually at least 15 inches (38.1 cm).

22-18. Fruit trees are set out in patterns and spaced for better sunlight penetration.

Orchards may be set out in several possible patterns, such as a checkerboard or diamond pattern, to allow for better sunlight penetration. If possible, a north-south row orientation is used, since this allows both sides of the plant to receive sun. The east side receives sun in the morning and the west side in the afternoon as the sun moves across the sky. However, since many orchards are on sloping ground, row direction must be established to minimize erosion. Buried irrigation lines are laid to service the planting.

IRRIGATION AND WATER MANAGEMENT

Most fruits are 81 to 92 percent water and adequate water greatly increases fruit yields and quality. In the high rainfall areas of North America, many fruits can be grown without irrigation, but yields will often be disappointing in dry years. Irrigation should be scheduled based on the crop demands and soil type. Crop demands vary with species, time of the year, and plant size.

In the heat of the summer, a newly planted blueberry bush watered with a drip irrigation system needs about one gallon of water every one or two days for most rapid growth. As the plant matures to 8 to 10 feet in height, the irrigation requirement in mid-summer increases to 6 to 8 gallons per day. On a sandy soil, with a low water holding capacity, daily drip irrigation is recommended, but on clay soil with a high water holding capacity, twice the amount of irrigation water applied every other day can be satisfactory during some parts of the season. The difference is the water holding capacity of the soil. As a general rule, sandy soils can hold about 1 inch of available water per foot of top soil; loam soils, about 1 to 2 inches per foot of top soil; and clay soils, about 2 inches per foot of top soil.

In Georgia, peach **microirrigation** (irrigation by a drip system or microsprinkler system) can be scheduled based on pan evaporation (evaporation from an open pan). In mild weather, when pan evaporation is about .1 inch per day, a mature peach tree at normal spacing (109–145 trees/acre) with fruit on the tree needs about 8 gallons per tree per day. When pan evaporation is about 0.2 inch per day (warm weather), the tree needs about 16 gallons per day, and at 0.3 inch per day (hot weather), the tree needs about 24 gallons per day. The drip system is run daily, just replacing the water the tree used. When a rain is received, for instance, 1 inch, the system is turned off for four to five days, then started again.

Growers also use tensiometers and other soil water measuring devices to monitor soil water levels, especially when using sprinkler irrigation. Deter-

22-19. Micro-irrigation systems are widely used in fruit growing. Here a drip emitter is being used. Micro-sprinklers (small sprinklers) are usually considered better systems than drip because they wet a larger soil area, which helps the tree obtain more water in a very hot day. (Courtesy, Jim Strawser)

minations of when to water with a tensiometer varies with the crop, but typically an irrigation is applied when the gauge drops to 10–20 centibars; 0 centibars is saturated soil.

Farmers also use observations on the turgidity of leaves and appearance of the soil when other methods are not being employed. When using sprinkler irrigation, as a rough rule of thumb, apply 1–1½ inches (2.5–3.8 cm) of water per week during the growing season when fruit is on the tree unless rainfall occurs. Irrigation after harvest, except on young trees where rapid growth is desired, is normally limited to keeping the trees healthy.

*F*ERTILIZATION AND NUTRIENTS

Fertilization (the word fertilization is also used in reference to seed formation) of commercial fruit crops should be done in a scientific manner starting with detailed soil samples. These are followed by leaf analysis samples, which are usually collected in August. *Leaf analysis* is the laboratory analysis of the nutrient content of the leaves. Together they form a picture of what nutrient levels are in the soil and what nutrient levels are actually in the plant leaves. Based on this information, a grower can apply the correct amount of nutrients for good growth, but not overstimulate the plant. Excessive vegetative growth causes fewer flower buds to form, reducing the crop potential. It also causes excessive shading of the developing fruit, which reduces fruit red color formation.

The amount of fertilizer required varies with the crop, plant age, soil type, soil fertility level, region of the country, region of the state, etc. For this reason, the author will not attempt to define the amount of fertilizer needed.

A good source of information for both commercial and home gardeners are the Extension service bulletins written by the specialist in your state or region. This information can be obtained at your local county Extension office. However, to acquaint you with the procedure, samples of soil sample reports and leaf analysis reports are found in Figures 22-20 and 22-21.

Soil Analysis Results

Name: Gerard Krewer

Soil Testing Laboratory
Cooperative Extension Service
The University of Georgia

Date 9/21/2000

County Extension Director
Ware County
P. O. Drawer 1425
Waycross, GA 31501

Sample Identification			Soil Test Results							Recommendations				
Lab Number	Sample Number	Crop	Soil Ph	Lime Index	P	K	CA	MG	ZN	Lime-stone Tons/A	LBS per Acre N	P205	K20	See Comments
45219	1	Blueberries	4.0	7.40	7 Low	28 Low	222 Adeq	72 Med	3	.0	—	30	30	273 997 998
	Manganese Soil Test Level Is		1 Lbs/Acre			See Comment 998								
45220	2	Blueberries	4.8	7.50	8 Low	32 Low	484 Adeq	128 High	2	.0	—	30	20	273 997 998
	Manganese Soil Test Level Is		1 Lbs/Acre			See Comment 998								
45221	3	Blueberries	4.9	7.65	37 Med	36 Low	606 Adeq	97 Med	4	.0	—	20	20	273 997 998
	Manganese Soil Test Level Is		3 Lbs/Acre			See Comment 998								
45222	4	Blueberries	4.0	7.55	23 Low	66 Low	329 Adeq	76 Med	8	.0	—	20	20	273 997 998
	Manganese Soil Test Level Is		6 Lbs/Acre			See Comment 998								
45223	5	Blueberries	4.3	7.60	20 Low	62 Low	261 Adeq	38 Low	13	.0	—	20	20	273 997 998
	Manganese Soil Test Level Is		13 Lbs/Acre			See Comment 998								
45224	6	Blueberries	4.4	7.50	11 Low	29 Low	153 Low	13 Low	2	.0	—	30	30	273 997 998
	Manganese Soil Test Level Is		5 Lbs/Acre			See Comment 998								
45225	7	Blueberries	4.5	7.60	7 Low	15 Low	131 Low	13 Low	1	.0	—	30	30	273 997 998
	Manganese Soil Test Level Is		1 Lbs/Acre			See Comment 998								
45226	8	Blueberries	4.5	7.55	5 Low	10 Low	123 Low	16 Low	1	.0	—	30	30	273 997 998
	Manganese Soil Test Level Is		5 Lbs/Acre			See Comment 998								

273 If pH is greater than 5.5, consider another site or add one third cup of sulfur per planting hole 3 to 6 months before planting. Also, apply a big pinch of chelated iron at the base of the root system when planting. See Circular 713, "Commercial Blueberry Culture: for sulfur recommendations on a per acre basis.
If the soil test calcium (ca) level exceeds 900 pounds per acre or if the soil test phosphorus level is very high, the site is not well suited for blueberries.
Use liberal quantities of peat moss or pine bark mixed with the soil when planting. Following planting mulch heavily with pine bark, rotted sawdust, or pine straw if practical.
* For newly set and plants less than 4 feet high that have been mulched and settled by rain, apply 1 to 2 oz. of 10-10-10 or 12-4-8 per plant in March, May, and July using the lesser amount on small bushes. If potash and phosphorus levels are high, 1 oz of ammonium sulfate can be used. Broadcast over an area 18 inches in diameter. A minimum of 4 inches of water as rainfall or overhead irrigation should be received between fertilizer applications. Do not pile fertilizer directly at the base of the plant as this could result in death of the plant.
* For blueberry plants 4 feet and larger, apply 0.8 oz/N per plant (30 lbs. N/A) in the early spring and a similar amount after harvest. This is the equivalent of 6 oz. of 12-4-8 (225 lbs./A) applied in the early spring and repeated immediately after harvest. Apply P205 and K20 as recommended. Note: to convert P205 and K20 per acre recommendation to per plant rate, divide by 605.
* Organic or slow-release nitrogen sources such as is found in certain lawn fertilizer formulations (an example is 12-4-8) are excellent nitrogen sources for blueberries. Part of the N is readily available while part is available in small amounts over a long period. However, since such sources of N are quite expensive, it is suggested that these nitrogen sources be banded under the bushes.
997 There is no research data to determine adequate zinc levels for this crop. If zinc deficiency is suspected, follow up with plant analysis.
998 There is no research data to determine adequate manganese levels for this crop. If manganese deficiency is suspected, follow up with plant analysis.
Note:

22-20. A soil sample report on blueberries.

Plant Analysis Report

Grower: Gerard Krewer

Date: 9/22/2000

County: Research

Kit Number:

Lab Number: 1289

Sample Number: 1

Crop: Muscadine

University of Georgia

Cooperative Extension Service

Athens, GA 30602

Percentage (%)						Parts Per Million (PPM)					
N	P	K	CA	MG	S	MN	FE	AL	B	CU	ZN
1.44	0.14	0.49	1.10	0.21	??	275	99	127	16	18	23

Recommendations

The nitrogen (N) level is low. Apply 60 lbs. N/A after fruit set and 50-60 lbs. of N per acre in the spring.

The phosphorus (P) level is adequate for good growth. For next year's fertilization program, apply 40 to 50 lbs. of phosphate (P205) per acre.

The potassium (K) level is low and is generally due to low soil test K levels or a heavy fruit set. For next year's fertilization program, apply 100-125 lbs. of potash (K20) per acre.

All other elements are sufficient for good growth.

22-21. A plant analysis report on muscadine grapes.

Newly planted fruits are usually fertilized with small but frequent applications of fertilizer to produce as much vegetative growth as possible the first year.

Typically, mature fruits receive a spring application of a complete fertilizer based on the soil sample and leaf analysis reports. Later in the growing season, a second or third application of just nitrogen is often needed. Many fruit crops suffer from deficiencies of certain secondary elements, such as magnesium, or minor elements, such as zinc, boron, or copper, on some soils.

PLANT CARE—PRUNING, THINNING, AND PEST MANAGEMENT

Pruning is one of the most expensive and labor intensive aspects of pomology. The reasons for pruning are:

- Pruning increases fruit quality by allowing more sunlight and air movement through the plant. This allows the fruit to develop more red color and have fewer fungal problems. It also allow pesticides to penetrate the tree better.

- Pruning increases fruit size. Since the pruned plant has the same size root system, but a smaller top and better light penetration to the remaining shoots, the plant produces larger fruit.

- Pruning helps prevent over cropping by removing some of the flower buds.

- Pruning is also necessary to control plant size.

Some fruit trees such as plums, apples, and pears develop very short, compressed fruiting twigs called *fruit spurs.* Fruit spurs only grow an inch or less each year and allow the tree to easily maintain fruiting wood in the interior of the tree.

PRUNING—OPEN CENTER TREE FORM

Peaches, nectarines, plums, sour cherries, and pears can be pruned into many different shapes, but the most popular is the vase or *open center pruning* method. This system allows for easy access to the tree for fruit thinning and good sunlight penetration to help the fruit ripen with a red blush. At the time of planting, cut the tree back just above the point where branches are desired. Any side branches should be completely cut off so all that remains is a whip, usually 25–30 inches (63.5–76.2 cm) tall. Allow only three to five major limbs, called scaffolds, to develop. These limbs should originate up and down the trunk to increase the strength of the tree. About 12–18 (30.5–45.7 cm) inches from the trunk each scaffold will naturally branch into two or three sub-scaffolds. During the first summer or late winter start pruning the tree to produce the desired shape. Always prune back to an outward growing limb or a bud which is pointing away from the center of the tree. The severity of the pruning will depend on how much ground area you wish the tree to cover and the ultimate height you will allow the tree to grow. Remember, flat limb angles produce many watersprouts, but may be needed to make the tree spread properly if a short tree is desired. Remove any watersprouts grow-

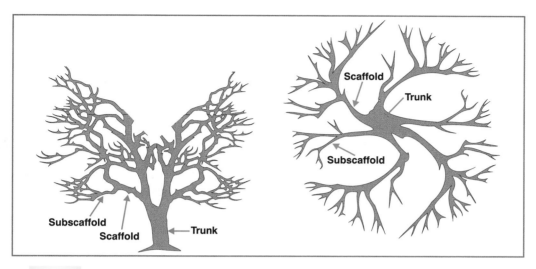

22-22. Mature open center peach tree showing basic structure from the side and top.

ing in the center of the tree. Do not allow two major limbs to develop on top of each other forming a "double deck."

Pruning in Later Years—After the second, third, and fourth growing years, continue to shape the tree, pruning during late winter. Remove low hanging limbs or head them back to an upward growing shoot. Remove crossing or broken limbs. Remove watersprouts growing in the center of the tree, leaving the smaller wood for fruit production. Watersprouts are ideally

22-23. These older trees have been pruned to promote fruiting.

pruned in summer to let more light into the center of the tree, but can also be pruned in late winter. Then make a heading cut to an outward growing limb on the subscaffold. It may be desirable to thin the fruiting twigs to about 6 inches (15.2 cm) apart measured at the middle of their length.

Pruning Mature Trees—Start with pruning low hanging limbs to upward growing shoots, removing watersprouts, and removing crossing and dead limbs. Then move to the branches on the outside of the tree and reduce the height by making bench cuts to outward growing limbs. In the Eastern United States, most peach growers prune their trees to 8–10 feet (2.4–3.1 meters) in height. In the Western United States, many growers prune the trees to 12 feet (3.7 meters) in height. The final procedure is thinning out crowded fruiting twigs to about 6 inches (15.2 cm) apart measured at the middle of their length.

PRUNING—CENTRAL LEADER TREE FORM

The central leader form of tree training is widely used on apples and sweet cherries, which have a natural growth habit, which lend themselves to this type of training. At the time of planting, head or prune back the un-branched whip to 24–30 inches (60.1–76.2 cm) from the ground. This will cause the buds just below the cut to grow and form scaffold branches. The upper most bud will become the upright central leader of the tree. If branched one or two year old trees are planted, then select four or five lateral branches with wide-angle crotches that are spaced equidistant around the trees and 4–5 inches (10.2–12.7 cm) apart vertically. The selected laterals should be no lower than 18 inches (45.7 cm) above the ground and they should be pruned back slightly by cutting off one-fourth of each limb's length.

During the first growing season, when 2 or 3 inches (5.1–7.6 cm) of growth has occurred, it is necessary to begin training the tree. Position wooden spring-type clothespins or wooden toothpicks between the main branch and the new succulent growth. The clothespin will force the new growth outward and upward thus forming the strong crotch angles needed to support the fruit load in years to come.

When the tree is one year old, a number of branches should have developed. If they have been spread with clothes pins or toothpicks, they should have good, wide crotch angles. The objective now is to develop a strong central leader and framework of scaffold branches. Only four scaffold branches spaced around the tree should be left. All of the remaining branches, as well as the central leader, are pruned back about one-fourth. Always make sure the ends of the scaffold branches are below the end of the central leader after they have been pruned back.

Two Year Old Tree—During the second growing season, develop a second layer of scaffolds 24–30 inches (61–76.2 cm) above the scaffolds you established the year before. Be sure to use the clothespins on the new succulent growth, particularly shoots that develop below the pruning cuts, so you will develop wide crotch angles. The use of limb spreaders can aid in bringing about earlier fruit production, improved tree shape, and stronger crotch angles. Spreaders can either be short pieces of wood with sharpened nails driven into each end, sharpened metal rods, or "W" clips anchored in the ground and tied to the limb with twine.

Limbs should be spread to a 45–60 degree angle, but not below a 60 degree angle from the main trunk. Limbs spread wider than 60 degrees have a tendency to produce watersprouts along the top side of the branch and may stop terminal shoot growth. The spreaders will need to remain in place for up to one year until the wood "stiffens up." Pruning consists of entirely removing undesirable limbs, and, only where necessary, reducing the length of terminal scaffolds by one-fourth. Weaker side limbs should not be pruned, unless excessively long, so they can develop flower buds. Excessive and unnecessary pruning will invigorate a tree and delay fruit production.

Pruning In Later Years—As the tree increases in age, a third layer of scaffold limbs may be added 24–30 inches (61–76.2 cm) above the second tier. Once the final height of the tree is determined, the central leader is pruned back to a small side twig each year if possible. The following year, this side twig and the previ-

22-24. A limb spreader in a tree to form a 45–60 degree angle.

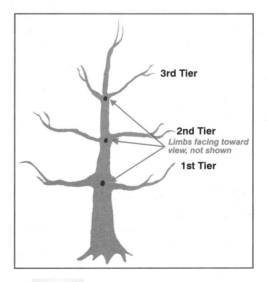

22-25. Mature central leader tree.

ous season's growth from it is removed and another side twig at about the same height is selected. Annual winter pruning on mature trees consists mainly of pruning out suckers which grow upright in the tree, removing broken branches, branches which are too low, branches which cross over each other, and thinning out branches which are too crowded. A properly trained and pruned central leader tree should conform to roughly a pyramidal (Christmas tree) shape. Much of the fruit on apples, pears, and plums is produced on spurs, which are short fruiting shoots.

Pruning Grapes

Grapes are usually grown on trellises. Many different types of trellises are used and space does not permit a discussion of trellises, but the single wire trellis and double wire vertical trellis are two of the most common types. Two basic pruning types are used. Vines which are *spur-pruned* have a permanent trunk and arms. Each winter the previous season's growth is cut back to two or four buds. This creates a short shoot which resembles a fruit tree spur, hence the name. In the spring, these buds will break, grow a short distance and produce flowers, and fruit from a lateral bud on the current season's growth. Vines which are *cane-pruned* have only a permanent trunk. Each year, new arms are placed on the trellis and all the remaining old arms are removed in pruning. Several short spurs are left near the top of the trunk to provide a place for renewal growth for the following year.

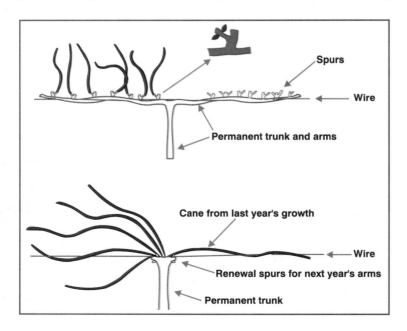

22-26. Spur (top) and cane (bottom) pruning in grapes. Left side shows appearance before pruning.

PRUNING—CANE RENEWAL OF SMALL FRUITS

Many small fruits which have a bush form are pruned by cane renewal. In this process, old canes are pruned off close to ground level after harvest. In the case of blueberries, the canes live for many years so several of the oldest, tallest canes are removed each winter. In the case of blackberries, the canes grow one year, fruit, and die the next year. After harvest, the canes which just fruited are pruned off close to ground level.

FRUIT THINNING

Fruit thinning is necessary on many tree fruits to produce large size fruit. It is also important in preventing poor flower bud production for next year's crop. Developing seeds produce hormones which not only make the fruit grow, they also prevent flower bud initiation if found in too high a concentration. Pruning accomplishes some of the thinning by removing many flower buds. In the spring, but before the fruit is very large, additional fruit is removed by chemical sprays or by hand. When finished, it is desirable to have most fresh market plums spaced 3 to 4 inches (7.6–10.2 cm) apart and most apples, nectarines, peaches, and pears about 6 to 8 inches (15.3–20.3 cm) apart. Fruit thinning should be accomplished early for greatest benefit.

22-27. After harvest, blackberry canes which just fruited are pruned off close to ground level.

WEED CONTROL

Good weed control is very important for high yields in fruit production. This is due to several factors: (1) Competition for water and nutrients (2) Improved temperatures during the spring bloom period. Bare, settled soil is able to absorb the sun's energy, then release heat at night. This can have a huge effect on the degree of freeze damage if the temperature is close to the critical mark.

22-28. Herbicide strip in a muscadine grape vineyard.

Most orchard operators now use a system called the herbicide strip system for weed control. In this system, the area under the bushes, vines, or trees is kept relatively free of weeds by mechanical cultivation and/or herbicides. The area in the aisle is planted to a cover crop, usually a grass, to help prevent erosion and supply organic matter to the soil. The cover crop in the aisle is mowed periodically during the summer. Some growers also use mulches under their berry bushes and trees, but this is normally an expensive practice so its use is limited to certain situations. Strawberries, blueberries, and raspberries are some of the crops most frequently mulched. Growers using the organic method of production also frequently use mulches. In the warmer parts of the United States, strawberries are normally grown on raised beds covered with plastic mulch. This serves as a barrier to assist with soil fumigation when the planting is established and also helps prevent moisture loss and controls weeds in later months.

Herbicides, to prevent weed growth (pre-emergent herbicides) and kill back weeds already emerged (post-emergent herbicides), are commonly applied in the early spring. Several more applications of post-emergent herbicide may be needed during the summer to kill emerged weeds. In areas where winter weeds are a problem, a fall application of pre-emergent and post-emergent herbicides may also be required.

PEST CONTROL

Some small fruits can be grown with very few pesticide applications, but many fruits require a spray of a fungicide and/or insecticide every two to four weeks while the fruit is on the plant.

22-29. Aerial application in a citrus grove. (Courtesy, Mississippi State University)

A typical spray program involves several sprays during the winter, of dormant oil, to control scale insects. During bloom, no insecticides are applied to prevent bee kill. After *petal fall*, or the point at which the petals dry up and start to fall off, a combination of an insecticide and fungicide is applied. This usually continues on a two week basis for the main part of the growing season. Near harvest, no insecticide is applied, but fungicide applications for fruit rots continue. A fungicide may also be applied in the packing house, if necessary. Organic production of some fruits is possible and is increasing each year. Unfortunately, most fresh markets demand fruit with a near perfect appearance and this is very difficult to achieve without modern fungicides. If more consumers would accept slightly blemished fruit, chemical use could be decreased.

CROP MATURITY AND HARVEST

Near the time of harvest for most fruits, sugars increase, the fruit acid levels decline, the green fruit color disappears, the red or purple color characteristic of the fruit increases, and

22-30. A new coating compound promises to extend the shelf life of fruits and vegetables. (Courtesy, Agricultural Research Service, USDA)

the fruit cell walls soften and grow thinner. Some fruit, such as peaches, nectarines, and apples, need light for good red color development, so pruning and tree training are very important. Moderate temperatures also favor good fruit color development.

Determining Maturity

Fruits are harvested at different stages of maturity for different uses and markets. For instance, peaches which will be distant-shipped must be picked at the firm mature stage. Peaches which will be sold that day on a fruit stand can be picked slightly soft. Generally, when the fruit has developed a good color and a good flavor, but is still firm enough to ship, it is ready for harvest. Fruit for processing is often picked based on acid, sugar, and color levels which can be measured with instruments.

22-31. Harvesting blueberries mechanically. Beaters inside the machine knock off the ripe fruit. It falls on to spring-loaded plates that press against the base of the bush. Conveyors carry the ripe fruit to the back of the machine where workers place it in lugs.

Harvesting and Packing

Certain types of fruit for processing are often picked mechanically, such as raspberries, blackberries, blueberries, cherries, prune plums, and cranberries.

Most large fruits are picked by hand and placed into picking sacks or buckets. The fruit is then transferred into 18–20 bushel bins for transport to the packing house. At the packing house, the fruit is poured onto a line where the fruit is sized mechanically, brushed and waxed, graded by the workers, and poured into cardboard boxes of various sizes.

Fruit Cooling and Storage

Fruit is still living when detached from the plant. They respire, using oxygen and carbohydrates and giving off carbon dioxide. For the fruit to have a long storage life, it must be cooled to slow the respiration and softening processes. Fruit may be cooled with water, called **hydrocooling,** or cooled with air. Cooling is important because the rate of respiration is approximately double for each 10°C (18°F) increase in temperature. Fruit may be held in common storage or in special cases in controlled atmosphere storage. ***Controlled atmosphere storage*** (CA storage) is storage in a cool room where the gas mixtures are carefully regulated. Usually the oxygen levels are greatly decreased from normal, and the carbon dioxide levels increased.

*M*arketing

After cooling, the fruit is loaded into large trucks and sent to various markets. Much of it is sent to the central warehouses operated by each grocery store chain. A second outlet is the terminal markets and large farmer's markets where the fruit is resold to fruitstand owners, restaurants, food service companies, and small grocery stores. Many farmers also sell directly to the consumer and fruitstand owners.

22-32. A small and a large peach packing operation.

REVIEWING

*M*AIN IDEAS

Pomology is one of the oldest branches of horticulture. Fruit production is an important part of agriculture and generates billions of dollars in farm income. Not only is fruit wonderful to eat, it's very nutritious. The fruit plant has an interesting life cycle from endo-dormancy to eco-dormancy, through spring growth and development, and summer flower bud formation. Pollination and fruit growth are one of the miracles of nature. Fruit growing is a very technical area of agriculture with many fascinating plant, soil, and climate interactions. The science and the art of pruning is another interesting and surprising aspect of pomology. We hope this introduction to pomology has been enjoyable and stimulated you to develop a better understanding of horticulture in general.

QUESTIONS

Answer the following questions using correct spelling and complete sentences.

1. What is pomology?
2. Where did the peach originate?
3. How long has fruit been in cultivation?
4. What is the approximate value of U.S. fruit production?
5. Name five jobs in pomology and related fields.
6. What are eco-dormancy and endo-dormancy?
7. How are winter chilling requirements for fruits calculated?
8. Briefly outline the function of leaves, shoots, and roots.
9. Briefly describe what happens in pollination and fertilization.
10. Why do some fruits drop off?
11. What are the three phases of fruit growth?
12. What time of the year are flowers initiated on most fruits?
13. What is a pome?
14. Describe five of the things which make an ideal fruit growing area.
15. What is a rootstock?
16. What is the optimum soil pH for most fruit crops?
17. How do advective and radiation freezes differ?
18. What is sub-soiling?
19. What are several methods of scheduling irrigation?
20. Why is pruning conducted?
21. What is CA storage?

EXPLORING

1. Using the information in the tables and charts, determine if peaches, mangos, and raspberries can be grown successfully in your area.

2. Compare your climate to the ideal climate for fruit growing. How suitable would your area be for commercial fruit production?

3. Record temperatures at the bottom and top of a hill near your house on several mornings; is there a difference on a still morning?

4. Visit the grocery store and pick out examples of several different types of fruit, such as a pome, a drupe, and a berry.

23

HAND AND POWER TOOLS

People need simple mechanical skills in their daily living. Selecting and using hand and power tools requires knowledge of the job to be done and how to properly use the tools. Tools and equipment help people do work in less time. Having good tools available also requires proper care and maintenance to keep them in good working and operating condition.

Safety with tools is important. Accidents can happen quickly, especially with power tools. Selecting the right tool and using it properly is essential to safety. Protective equipment and clothing should also be worn when using certain tools and horticultural equipment.

23-1. Leaf blowers are a common power tool used in horticulture. (Courtesy, Husqvarna Forest and Garden Company)

599

OBJECTIVES

Technology requires the use of hand and power tools for equipment maintenance and adjustment. This chapter has the following objectives:

1 Describe common wrenches and how they are used in doing work

2 Identify the four primary sources of power for power tools

3 Describe safety precautions in using power tools

4 Explain the difference between 2- and 4-cycle engines

5 List the ten steps in trouble shooting a small engine that will not start

TERMS

2-cycle engine
4-cycle engine
adjustable-jaw wrench
ball peen hammer
claw hammer
combination wrench
cordless drill
fixed-jaw wrench
grounding

hammer
hand tool
operator's manual
Phillips screwdriver
pneumatic
power tool
straight-blade screwdriver
torque wrench
wrench

HAND TOOLS

A *hand tool* is a small powerless tool. Many kinds of hand tools are used. Hand tools are selected based on the job. Each tool should be cared for properly. They should be kept clean and dry. Tools with cutting edges should be protected. Tools should be stored properly after use.

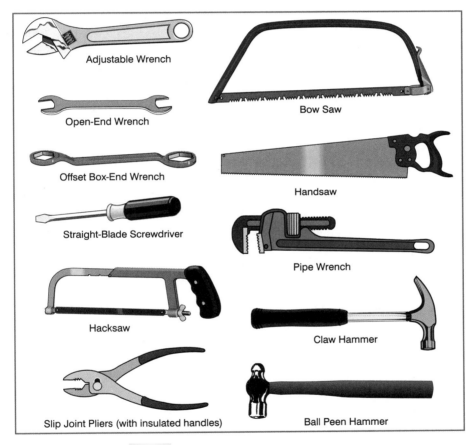

23-2. **Examples of common hand tools.**

HAMMER

A *hammer* is a tool made for driving or pounding. It has a head made of metal or another material. A handle fits into the head. Handles are made of wood, metal, or plastic. Many kinds of hammers are available.

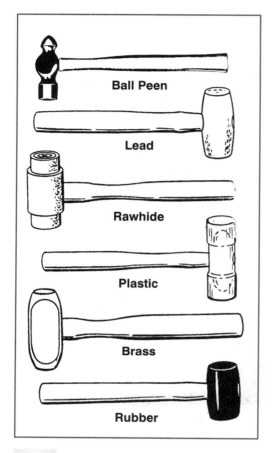

Always wear safety glasses when using any kind of hammer. Hold the hammer close to the end of the handle away from the hammer's head. The hammer face should squarely strike the surface being hit. Never use a hammer if the head is loose. Metal or wooden wedges can be inserted into the hammer's handle, head end to tighten it on the handle.

Most mechanics use the **ball peen hammer** when working on equipment. It is designed with one end of the head ball-shaped. The hardened ball peen hammer head will not be damaged when striking metal, such as chisels, punches, bolts, or other metal objects. A claw hammer's head is not made from hard steel. It should not be used for striking metal objects other than nails. A **claw hammer** also has one end of the head forked and curved for pulling nails. Other special hammers have heads made from a variety of materials, such as plastic, leather, rubber, or brass. These special hammer heads have specific purposes and prevent damage to delicate parts when used properly.

23-3. Examples of different types of hammers.

Wrenches

A **wrench** is a tool for gripping and turning bolts, nuts, and other fasteners and materials. Several different types of wrenches are used. The two main types are fixed-jaw and adjustable-jaw wrenches.

A *fixed-jaw wrench* is made for one size of fastener. It comes in many sizes in both American and metric measurements. Two of the common types of fixed-jaw wrenches are open-end and box-end. An open-end wrench has a notch of different sizes at each end. A box-end wrench has openings of different sizes at each end that completely box (encircle) the nut or bolt head. A **combination wrench** has an open-end wrench on one end and box-end wrench of the same size on the other end.

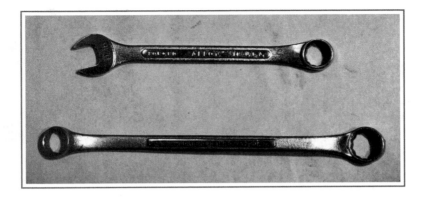

23-4. A combination wrench (top) and box-end wrench (bottom).

A socket wrench is also a form of a fix-jaw wrench. It has a removable end that boxes (encircles) the nut or bolt which is called a socket. Sockets come in a variety of sizes and are attached to a handle or lever. Various types of handles are available including ratchets, breaker bars, and braces or speed handles. Extensions of various lengths are also available which may be placed between the socket and the handle. These allow the mechanic to reach bolts and nuts that may not be otherwise accessible.

Sockets also come with a varying number point sizes—6, 8, and 12 point. Point refers to the number of notches for gripping the heads of bolts and nuts. They attach to handles or extensions with a square hole opposite their gripping ends. These square holes come in $1/4$-, $3/8$-, and $1/2$-inch sizes. They

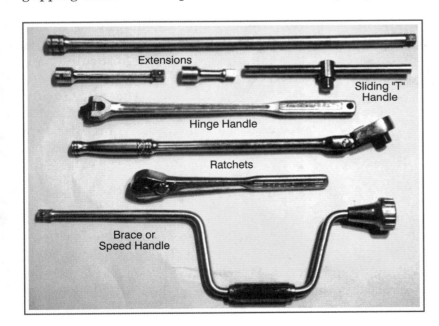

Extensions

Sliding "T" Handle

Hinge Handle

Ratchets

Brace or Speed Handle

23-5. Various handles and attachments for a socket wrench.

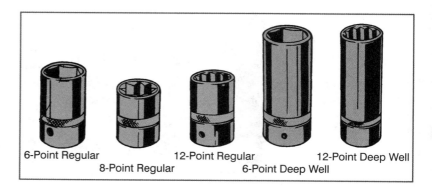

23-6. Examples of points and depths of sockets.

6-Point Regular   12-Point Regular   12-Point Deep Well
8-Point Regular   6-Point Deep Well

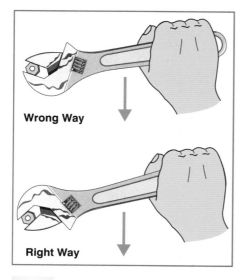

Wrong Way

Right Way

23-7. Use a crescent type adjustable wrench correctly. (Note the position of the adjustable jaw as related to the direction of pull)

are referred to as drive sizes. Sockets also come in various depths—regular and deep well.

A special handle that allows setting and/or measuring the amount of pressure applied in tightening a bolt or nut is knows as a *torque wrench*. This precise setting measured in foot pounds of pressure is used where an exact tightness called "torque" is required. This is used in engine assembly, as well as in other horticultural applications.

An *adjustable-jaw wrench* has a moveable jaw so the wrench can be changed to fit various sizes of bolts and nuts. A few wrenches may be adjustable in other ways. The common types used in horticulture are the crescent type and the vise-grip. These must be properly adjusted when used to prevent damage to the nut, bolt, or fastener. Adjustable-jaw wrenches should be used only when a fixed-jaw wrench will not work.

SCREWDRIVERS

A screwdriver is a tool with a handle on a long metal shank. Screwdrivers vary in head design, blade length, and blade width. A *straight-blade screwdriver* has a straight-edge head designed to fit into a slot. It comes in a variety of sizes of blade lengths and widths. The correct size screwdriver should be used to prevent damage to the screw. The blade should fit tightly in the slot of the screw. Never use the screwdriver as a pry tool or hit it with a ham-

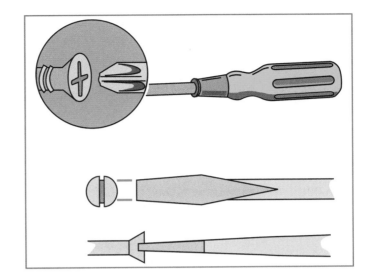

23-8. Select a screwdriver that properly fits the slot or cross-point head.

mer. A **Phillips screwdriver** has a cross-point head design. This design increases the torque that can be applied to a screw head. Phillips screwdrivers also come in a variety of head sizes from #1 to #4. Use the correct size Phillips to prevent damage to the screw head.

LEVEL

A level is used to determine whether a surface is flat (horizontal) or plumb (vertical). Levels involve using a bubble in a freeze-resistant liquid in a

23-9. Using a level in the installation of landscape timbers. (Courtesy, Jasper S. Lee)

small tube-shaped container. Markings on the tube are used to access the position of the bubble and to determine if the surface is level or plumb. Carefully handling levels helps prevent warping and protects accuracy.

Saws

A saw is a tool used in cutting materials. Most saws are made with handles attached to thin blades of metal having sharp teeth on one edge. Proper care of saws is important if they are to work properly. Rust is particularly damaging to unprotected saw blades.

Using Hand Tools Safely

Use the following safety precautions when using hand tools:

- Keep cutting tools sharp.
- Make cuts or point tools away from your body.
- Check tools regularly for damage or defects.
- Wear required personal protective equipment.
- Use the proper tool for the job.
- Use the tool as it was intended to be used.
- Carry tools in a tool box or holder.
- Keep tools clean and stored in their proper place.

POWER TOOLS

A *power tool* is any tool that has power for its operation from a source other than human force. Power tools can speed up work and get more done in less time. Power tools can do bigger jobs with less human effort. Power for power tools is from four primary sources—electric motors, fuel engines, pneumatics, and hydraulics. A number of power tools are used in horticulture.

Both portable and stationary power tools are available. Stationary power tools require connection to a fixed power source. Some may be able to be moved in a limited area if the power source is available. This is often done with extension cords. Cordless power tools are totally portable and are not

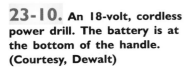

23-10. An 18-volt, cordless power drill. The battery is at the bottom of the handle. (Courtesy, Dewalt)

dependent on a power connection. They usually receive power from electric energy stored in a battery or a small gasoline engine.

USING POWER TOOLS SAFELY

A big difference between hand tools and power tools is that power tools are more dangerous. Accidents with power tools are often more serious. Safety practices should be followed and used when operating power tools.

An *owner's manual* is a written description of how to safely use and maintain a power tool. A manual usually accompanies every power tool. Most manuals also describe how to properly install and service the equipment. Parts lists are often included. These are helpful when equipment repair is needed.

Grounding is a safety method designed to protect people from electrical shock. All power tools that use electricity should be grounded properly before use. Never use an adapter to connect a three-pronged plug to an ungrounded power source. If the power tool is going to be used near water or moisture, it should be protected with a ground fault circuit interrupter (GFCI).

Since power tools often create hazards, operators should use protection appropriate for the tool. It is the operator's responsibility to do this. Common personal protection devices needed with power tools include: eye protection, hearing protection, long sleeves, gloves, masks or other respiratory protection, and leather shoes with steel toes.

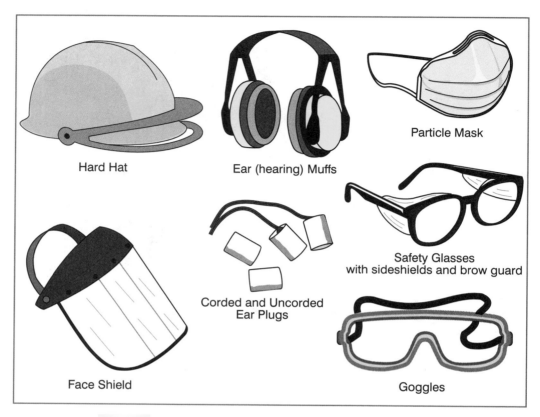

Hard Hat

Ear (hearing) Muffs

Particle Mask

Face Shield

Corded and Uncorded
Ear Plugs

Safety Glasses
with sideshields and brow guard

Goggles

23-11. Examples of common personal protective equipment.

Power tools must be kept in good, working condition. A tool used properly, but in poor condition, is dangerous. Always check the condition of a power tool before using it. Be sure all fasteners are tight and that needed adjustments are properly made. Many power tools have safety shields and other safety devices that should be in place before operation.

ELECTRIC DRILLS

All shops should have one or more electric drills. The size of electric drills relates to the maximum size drill bit they will hold. Common drill sizes are ¼, ³/8-, and ½-inch. Electric drills perform a variety of jobs in addition to drilling holes. They can drive and remove screws and remove or install nuts when used with sockets. Good electric drills have a reverse switch which allows the direction the drill turns to be reversed. A **cordless drill** is portable and contains a rechargeable battery. It is very useful when an electri-

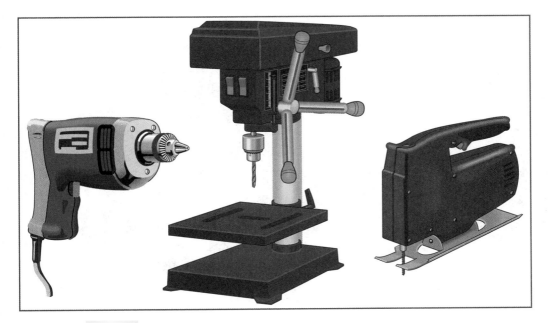

23-12. Electric drill (left), drill press (center), and sabre saw (right).

cal power source is not readily available. A drill press is a stationary drill designed for accurate and heavy drilling jobs.

Sabre Saws

A sabre saw is also known as a portable jig saw. Sabre saws are used to cut circles, curves, and holes in wood, metal, plexiglass, fiberglass, and similar materials. Different blades are used for different materials. Before using a sabre saw, be sure the blade is firmly in place. A bench vise can often be used to hold material being cut.

Grinders

A grinder is used to remove rough edges, smooth surfaces, and sharpen cutting edges of tools. Grinders may be stationary or portable. Both involve using an abrasive material that turns at high speed, known as a grinding wheel. Stationary grinders may be mounted on a bench or pedestal.

Bench-mounted grinders can be used to sharpen mower blades and other hand tools. Always wear safety glasses when grinding to protect your eyes from sparks and metal chips. Dress grinding wheels frequently with a stone

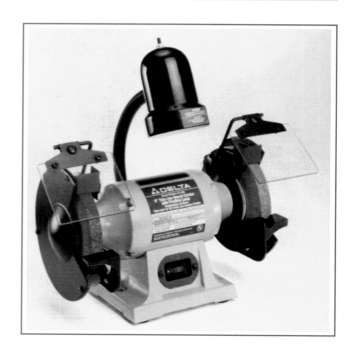

23-13. Bench-mounted grinder with enhanced lighting. (Courtesy, Delta International Machinery Corporation.)

23-14. A stationary air compressor in a horticulture maintenance shop.

dresser to restore the wheel to its original condition. Adjust the tool rest close to the grinding wheel to prevent the wheel from pulling a piece of metal into the turning stone.

AIR COMPRESSORS

Air compressors not only provide air necessary to inflate equipment tires, but are also used to generate power for pneumatic shop tools. Compressors are sold by the horse power of the electric motor used to operate the compressor. One to five horse compressors are commonly used today. *Pneumatic* (air driven) tools, such as drills, sanders, and impact drivers, provide the mechanic additional time-saving power tools.

SMALL GASOLINE ENGINES

Small gasoline engines provide the power for most landscape and turf equipment. Proper maintenance will keep this equipment running trouble

23-15. This photograph shows a small engine-powered machine grinding a stump. (Courtesy, Husqvarna Forest and Garden Company)

free for years. Read and follow all manufacturer's recommendations on type and schedule for all lubrications.

ENGINE CYCLES

Reciprocating engines may use either two-stroke or four-stroke cycles. An engine cycle is the number of strokes needed for combustion of a fuel-air mixture. A stroke is the back-and-forth motion of a piston in a cylinder.

Chain saws, weed eaters, blowers, and some lawnmowers use 2-cycle engines. A *2-cycle engine* is an internal combustion engine which completes the steps of intake, compression, power, and exhaust in two strokes, or one revolution of the crankshaft. Lubrication for this type of engine comes from oil mixed with the gasoline. These engines can operate in any position—sideways, upside down, etc. Always follow the manufacturer's recommendation when selecting and mixing the oil and gas mixture.

Many mowers, tillers, and thatching machines use 4-cycle engines. The *4-cycle engine* is an internal combustion engine with four strokes—intake,

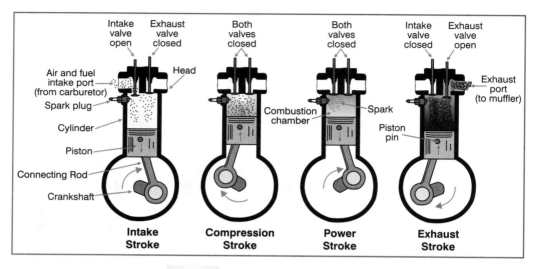

23-16. Strokes of a 4-cycle engine.

compression, power, and exhaust. It has a crankcase that holds oil necessary for engine lubrication. An important safety feature on many new engines prevents them from starting if the crankcase oil level is low. Follow the manufacturer's recommendations on oil selection and oil change interval.

TEN STEPS IN TROUBLESHOOTING ENGINES

Follow these steps when the engine will not start. Try to start the engine after each step.

1. Check the ignition switch, it must be in the on or run position. Check all safety switches for proper setup.

2. Check and fill the gas tank if needed.

3. Check the crank case oil level on 4-cycle engines and fill if needed.

4. Remove spark plug and check its condition.

5. If wet, the engine may be flooded. Leave the spark plug out of the engine, turn off ignition switch, and pull starter rope several times to clear engine flood. Replace spark plug and start engine.

6. If the spark plug is dry, the carburetor may not be working. Let a mechanic trouble shoot this problem.

7. With the spark plug grounded, turn engine over with starter rope and check for a blue spark at the spark plug.

8. If no spark occurs, have mechanic trouble shoot the problem.

9. If a good blue spark is present, replace the spark plug in the engine.

10. If the engine does not start, have a mechanic trouble shoot this problem.

AIR SYSTEMS

All engines need clean air, but this is especially important for small gasoline engines. The air intake system provides clean air for engine operation. It removes dust and dirt from the air before it is mixed with the fuel. Allowing dirty air into an engine increases the rate of engine wear and leads to engine failure. Operating landscape and turf equipment under dusty conditions requires frequent replacement or cleaning of air filters.

Three types of air cleaners are commonly used—dry element, oil foam, and oil bath. The oil bath cleaner is declining in use but is the best under very dusty conditions. A dirty dry element filter is replaced with a new, clean filter. If the area housing the filter is dirty, carefully wipe out the dust with a

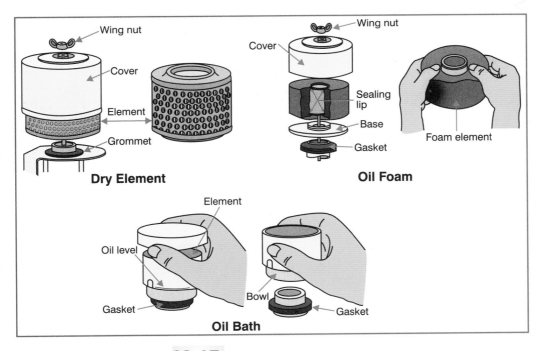

23-17. Three types of air cleaners.

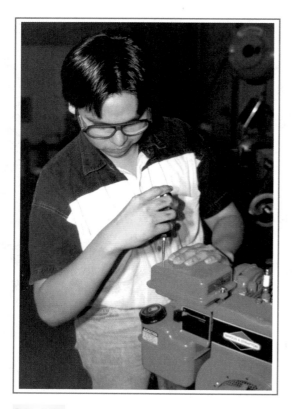

23-18. Proper maintenance will keep small engines running trouble-free for years.

clean cloth. Dry element filters should be checked each time engine oil is changed and more often in very dusty conditions. Oil foam filters can be removed and washed in a non-flammable solvent or in soapy water. Allow the filter to dry before adding engine oil to it. The amount of oil to use is no more than the filter would hold when tightly squeezed in the hand. Sometimes, buying a new foam element may be preferable to trying to wash a dirty filter. Most foam filters should be cleaned or replaced after 25 hours of use or more frequently in very dusty conditions.

Small engines have air cooling systems. The surrounding air cools the engine. Air is blown or forced over the engine to remove heat. The flywheel may be designed as a fan and have blades that create air movement. Engine blocks are designed with fins that transfer heat to the air. Maintenance involves keeping the system clean. Wipe off dust. Remove leaves, twigs, and other materials that may restrict the flow of air.

Safety precautions for gasoline-powered tools include:

- Inspect the tool before using it.
- Make sure all safety guards are in place.
- Read the operator's manual before using the equipment.
- Dress appropriately, including the use of personal protective equipment.
- Know the position of all the controls and what they do.
- Know the safety features and how they work.
- Do not allow anyone to smoke or light matches while filling the gasoline tank.
- Fill the gas tank when the engine is cold and wipe off any spills.

■ Operate and maintain the equipment according to the manufacturer's recommendation.

■ Operate in a well-ventilated area.

■ Do not operate equipment under the influence of any substance or drug.

■ Keep hands and feet away from moving parts.

■ Avoid operation near flammable materials or vapors.

■ Clear any obstructions from the work area.

■ Do not leave the tool running, unattended.

■ Turn off the engine and disconnect the spark plug wire before adjusting, repairing, or servicing.

■ Properly clean and store the tool after use.

REVIEWING

*M*AIN IDEAS

Hand tools are designed for specific uses. Both fixed- and adjustable-jaw wrenches are widely used. Properly using a torque wrench and following manufacturer's specifications assures that bolts and nuts are properly tightened. Power tools are used for many different jobs in horticultural mechanics. They may be powered by electricity, fuel engines, pneumatics, and hydraulics. Safe use of power tools is essential to prevent injury. Wearing protective equipment, being sure electrical equipment is properly grounded, and keeping tools in good condition are important parts of using them safely and efficiently.

Today, specialized equipment has replaced many manual labor tasks of the past. Equipment must be maintained in good operating condition at all times. Simple maintenance and adjustments can be made on equipment by following the operator's manual instructions. Always use the correct tools when servicing equipment and take proper safety precautions when operating equipment.

Clean air is the life of small gasoline-powered engines. Keeping air filters clean increases engine life. Clean air is important for all gasoline engines, but it is especially important for 2-cycle engines. Follow the ten step troubleshooting guide if a small gasoline-powered engine will not start.

QUESTIONS

Answer the following questions. Use complete sentences and correct spelling.

1. When should a box-end wrench or an open-end wrench be used on equipment?

2. What are the common socket wrench drive sizes?

3. When should a torque wrench be used?

4. List the different kinds of hammers used by horticulture mechanics.

5. What is another name for the jig saw?

6. What are the four primary sources of power for power tools?

7. What are the major safety considerations with power tools?

8. List 10 safety precautions for gasoline-powered tools.

9. What is the difference between a 2- and 4-cycle engine?

10. What is the first thing to check when a small gasoline-powered engine will not start?

EXPLORING

1. Tour a hardware store and study the hand and power tools that are on display. Determine the manufacturer and price for the following tools: cordless 3/8-inch drill, set of combination wrenches, torque wrench, gas-powered weed eater, and a 3/8-inch drive socket wrench set.

2. Practice using the tools described in this chapter. Be sure to use them properly and follow all safety precautions.

3. Use the trouble shooting technique discussed in this chapter to diagnose a small engine that will not start.

24

GROUNDS MAINTENANCE EQUIPMENT

The way grounds are maintained today has changed—thanks to equipment. Using equipment is exciting. However, when using grounds maintenance equipment, safety practices must be followed.

People have developed ways of getting more done with less physical effort. Mowers, trimmers, edgers, and leaf blowers make landscape care much easier for the homeowner. Other grounds equipment can help the environment through recycling, such as with a mulching mower or by composting grass clippings.

Golf courses use many different pieces of grounds maintenance equipment in maintaining fairways, greens, and other areas. Many job opportunities exist for those who can operate or maintain grounds equipment properly.

24-1. Well maintained grounds improve the economic and aesthetic value of property. (Courtesy, Church Landscapes)

617

OBJECTIVES

Safety and proper operation are essential when operating grounds maintenance equipment. This chapter helps to promote safety in equipment operation and has the following objectives:

1 Describe safety precautions for operating grounds equipment

2 Identify the four major types of mowers

3 Differentiate between mulching and standard rotary mowers

4 Demonstrate proper mower blade sharpening

5 Identify landscape maintenance equipment

6 Explain how to prevent chain saw kickback

7 Identify golf course grounds maintenance equipment

8 Explain how turf aeration improves the soil

9 Identify the major types of hydraulic systems used on grounds equipment

TERMS

aerification
back-lapping
backpack blower
bed knife blade
chain saw
core aerator
flail mower
hydraulics
kickback

kickers
mulching mower
power hedge trimmer
preventive
    maintenance
reel blade
reel mower
roller chain
rotary mower

sheave
sickle mower
sod cutter
sprocket
topiaries
v-belt
vertical mower

GROUNDS EQUIPMENT SAFETY

Safety is important when using grounds maintenance equipment. Federal law requires every push mower to be equipped with an operator-presence control system that stops mower blades within three seconds after the operator releases the safety lever. These safety devices Should never be disconnect or disable.

WALK-BEHIND MOWER SAFETY GUIDELINES

- Mow across a slope and watch your footing. Never mow a slope that is too steep to keep your balance or to control the machine.

- Always push the mower. Don't pull it toward your feet.

- Keep the mower flat. Don't lift the front end over tall grass or weeds.

- Turn off mower blades when crossing a sidewalk or drive.

- Take care when changing mowing directions to ensure that other people have not moved into your mowing path.

- Immediately stop and turn off the mower after hitting an object. Inspect the mower and repair any damage before resuming work.

24-2. Walk-behind mowers are often used by landscape maintenance firms.

RIDING MOWER OPERATION SAFETY GUIDELINES

- Never carry passengers.

- Do not mow in reverse unless absolutely necessary.

- Never leave an unattended machine running.

- Turn off blades and stop the engine before dismounting.

- Turn engine off before removing the grass catcher or unclogging the chute.

24-3. A riding mower with center-mounted mower deck and utility trailer.

■ Turn off blades and attachments when not mowing.

■ Check for traffic when crossing or operating near roadways.

■ Watch for holes, ruts, bumps, or other uneven terrain that could overturn the mower.

*S*AFETY GUIDELINES FOR REDUCING GASOLINE SPILLS

Avoid spilling and overfilling fuel tanks on all equipment, especially mowing units. The U.S. Environmental Protection Agency (EPA) estimates 17 million gallons of fuel are spilled each year during refueling.

24-4. Care should be used to avoid spilling gasoline and overfilling fuel tanks.

■ Use a gasoline container that can be easily handled and hold it securely so you can pour slowly and smoothly.

■ Use an approved gasoline container that is the proper color (red).

■ Use a spout or funnel to pour gasoline from the container into equipment.

■ Avoid overfilling power equipment gasoline tanks or allowing fuel to run over and spill.

■ Tightly close the cap and vent hole on the gasoline container after filling the tank.

■ Transport and store the gasoline container and power equipment out of direct sunlight and in a cool place.

■ Use caution when filling the gasoline container at the gasoline pump.

MOWERS

There are four major types of lawnmowers in common use today—rotary, reel, flail, and sickle. Rotary mowers are the most popular with homeowners and lawn care professionals. Reel mowers provide the very high quality cut required on golf course turf. The flail mower, while not as popular as the rotary, does provide a good cut with improved safety for both operators and other people around the mower. The sickle type mower provides a poor quality of cut and is mainly used on very low maintenance areas, such as roadsides.

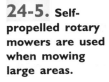

24-5. Self-propelled rotary mowers are used when mowing large areas.

Lawnmowers range in cutting width from 1 foot to more than 17 feet for the large commercial gang mower units. Most mowers are powered by a 2-cycle or a 4-cycle gasoline engine. Some homeowner mowers may have and use 120 volt electric motors for power. A few large commercial mowers may use diesel engines for power. However, most mowers are gasoline powered.

Self-propelled walk-behind mowers are available in sizes from 21 to 61 inches wide. Landscape maintenance firms often use self-propelled walk-behind mowers. They are not always able to operate these mowers under optimal conditions. Many times, especially in the spring, the mowing schedule does not allow time for the turfgrass to dry thoroughly. These conditions require a deck that is designed to lift and cut the wet turf and discharge the clippings. The quantity and degree of slopes, types of grass to be cut, soil types, and drainage conditions of residential and/or commercial lawns should be considered when selecting the size of a walk-behind mower.

RECYCLING MOWERS

Decreasing landfill space, combined with emphasis on recycling, has helped promote the use of mulching mowers. Mulching mowers are not new to the marketplace. The Bolen Company manufactured a mulching mower in the early 1960s. Landscape maintenance professionals are also using more mulching mowers in their businesses. A ***mulching mower*** recuts the clippings, returning them to the soil. Clippings return nutrients to the turf. Mulching mowers cut down on the time previously needed for raking and bagging grass clippings and leaves. Mulching reduces mowing time and eliminates the cost of clipping collection and disposal. This saves time and money for the homeowner or landscape maintenance firm. Environmental issues are also important. Mulching avoids contributing to landfill overuse.

24-6. Top and bottom view of a mulching mower. (Courtesy, Jasper S. Lee)

The Toro Company developed the recycler mower system. It consists of a special blade, kickers installed under the deck, and a new deck design. *Kickers* are the part of a recycler mower system that causes grass clippings to oscillate between the deck and blade. The blade is designed to stand the grass up, cut it off smoothly, and then recut the clippings as they bounce up and down from the blade to the kickers and back down to the blade. After cutting the grass three to four times, the clippings are forced by the kickers' air flow down beneath the surface of the grass.

Mulching mowers work best when the grass is not too long. If the grass is very tall, removing more than one inch will create more clippings than can be mulched into the lawn. Use a bag to catch the clippings for very tall grass and in the future mow more frequently. For best results, mow the lawn with a mulching mower every four to five days during the growing season. Sometimes a blower may be needed to spread piles of clippings on the lawn. However bagging and removal of clippings may be necessary on finely manicured residential lawns to retain the desired appearance.

Types of Mowers

Rotary Mower

A *rotary mower* has one or more blades that are parallel to the ground which are attached to a vertical shaft. The mowing height is adjusted by raising or lowering the wheels of the mower or the mower's deck. Rotary mowers

24-7. A rotary mower with a rear bagger for collecting grass clippings.

24-8. Underside of a front-mounted rotary mower deck.

are very popular with homeowners and lawn care professionals. Homeowner mowers are available in cutting width from 16 to 42 inches. They may be self-propelled or hand-pushed. Most rotary mowers have a 2-cycle or 4-cycle engine for power. Some smaller units may have an electric motor for power. Rotary mowers may also have a bag for collecting grass clippings. Some rotary mowers may mulch the grass clippings.

24-9. Reel mowers are commonly used on golf courses to cut the putting greens.

Reel Mower

A *reel mower* has a fixed blade and a set of rotating spiral blades on a horizontal shaft. These blades cut grass much the same way scissors cut paper. The mowing height is adjusted by setting the height of the front roller. Reel mowers are primarily used to cut turf on golf courses, especially putting greens. Reel mowers are also used to maintain certain warm season turfgrass species on southern lawns. Reel mowers give a very high quality of cut. They are best used for turfgrass species

24-10. Flail mowers are used for cutting course and tall grass. Side- and under-view are shown.

that are to be closely cut. A reel mower requires a high degree of maintenance through the season to keep it in top operating condition.

Flail Mower

Flail mowers are good for cutting coarse or tall grass along highways or parks. *Flail mowers* have 25 to 50 free-swinging flail blades which turn on a horizontal rotor that is completely enclosed by the rotor shield. It is the safest mower on the market today. Each small blade requires individual sharpening. Therefore, substantial maintenance time is necessary to keep these mowers in top cutting condition.

Sickle Mower

Sickle mowers resemble giant hedge trimmers. They may be walk-behind units or large mowers mounted on tractors. Triangular blades oscillate back and forth over a second set of fixed blades. Sickle mowers will cut large, woody weed stems and small tree trunks. Sickle

24-11. Sickle mowers are used on very low maintenance areas. (Courtesy, Troy-Bilt)

mowers are used for very low maintenance areas that may only be cut only once or twice per year.

MOWER BLADE SHARPENING

Mower blades become dull and nicked from use. Dull mower blades injure turfgrass plants by tearing the grass blade instead of cutting it. Sharp mower blades cut the grass blade. Rotary mower blades can be sharpened using a bench grinder. Special grinding equipment is needed to sharpen reel mower blades. Mower blades should be kept sharp to properly maintain turfgrass.

Sharpening Rotary Mower Blades

Before removing a rotary mower blade, the spark plug wire should be removed from the spark plug. This will prevent the engine from accidentally starting while removing the blade. Before attempting to remove the blade, secure the shaft so it will not turn. Note the exact position of the blade before removal. This will be an aid when reinstalling the blade after sharpening. Carefully remove the blade. This is usually done by removing a bolt that holds the blade in position on the shaft.

Always wear eye protection when sharpening mower blades. A bench grinder is used to sharpen the outer two to three inches of each end of the mower blade. Begin by carefully grinding away nicks in the blade. After the nicks are removed, begin the sharpening process. When grinding, keep the

24-12. Grinding a rotary mower blade.

24-13. Determining the balance of a rotary mower blade.

sharpened angle of the blade the same as the original angle from the factory. Prevent overheating of the blade by frequently dipping the end being ground into a pail of water. Overheating the metal will remove the temper from the blade and weaken it. The metal will turn blue indicating overheating from not cooling the blade often enough in the water.

After sharpening both ends, use a blade balance to check for proper weight of each end of the blade. If one end is heavier, even the balance by grinding the bevel of the heavier end a small amount and then re-check. Repeat this process until the blade remains horizontal when placed on the balance. Unbalanced blades cause excess vibration and may damage the mower. After sharpening, reinstall the blade on the shaft of the mower in the same position that it was previously. Then, reattach the spark plug wire.

Sharpening Reel Mower Blades

Reel mowers have two blades, the bed knife blade and the reel blade. The *bed knife blade* is the fixed blade and the *reel blade* is the spiral rotating blade on the horizontal shaft. These two metal parts cut the grass much the same way scissors cut paper. Special bed knife blade and reel blade grinders are needed to sharpen them. During the mowing season these blades are not usually sharpened. However, a quick touch up of the blades can be done by back lapping the reel blade. *Back lapping* is a method of sharpening a reel mower blade by spinning the reel blade backwards while applying a gritty solution to the blade. By grinding the reel and bed knife blades together with the gritty solution, a new sharp edge is created.

24-14. A bed knife grinder for sharpening reel mower blades.

24-15. Back lapping reel mower blade.

LANDSCAPE MAINTENANCE EQUIPMENT

POWER HEDGE TRIMMERS

Shearing, a form of pruning, involves removal of soft new growth in order to encourage a plant to grow into a certain shape. Shearing is done to shape dwarf trees, hedges, or evergreen shrubs. Shearing can also be used to shape shrubs into sculptures called *topiaries*. Power hedge trimmers will save time when shearing hedges and shrubs. *Powered hedge trimmers* may be electric or gasoline-engine powered. They have blades that oscillate back and forth

24-16. **Powered hedge trimmers.**

over a second set of fixed blades. When choosing electric hedge trimmers, make sure they open wide enough for cutting heavier plant growth. Care should be used not to cut the electric power cord when operating electric hedge trimmers.

Hedge trimmers are also available in cordless models with rechargeable batteries. They eliminate the need for a long electrical power cord. Small 2-cycle gasoline engines provide the power for several different models of hedge trimmers. The gasoline-powered models are designed for heavier use. Adjust the idle control screw if the cutting blades move when the hedge trimmer's engine is at idle.

Avoid accidents by using power hedge trimmers carefully. Always read and follow precautions found in the owner's manual before operating power hedge trimmers. Be careful not to cut the electric cord when using electric hedge trimmers with a power cord. Both hands should be kept on the trimmer's handles to avoid accidentally cutting fingers or hands.

WEED EATERS

A weed eater or string trimmer is used to trim grass and weeds in areas not acces-

24-17. **Care should be used while operating a weed eater.**

sible to mowers. It allows a person to neatly trim along landscape beds, edges of walks and driveways, and in grass areas too narrow for a mower. Both electric and gasoline-powered models are available. A weed eater has a single line that cuts an8 to 12 inch swath. Models are also available that have rotating blades instead of the single line. Eye protection and proper clothing are needed when operating a weed eater.

HAND-HELD BLOWERS

Homeowners often use hand-held blowers instead of leaf rakes and brooms. They can be used to distribute piles of clippings in the lawn, blow leaves and other materials from landscape beds and hard surfaces, and blow leaves from roof valleys and gutters. Hand-held blowers operate on household electrical current with extension cords attached to the blower or with power supplied by 2-cycle small gasoline engines. Always wear eye protection when operating blowers.

CHAIN SAWS

A *chain saw* is a portable power saw with a chain that has cutting teeth. Chain saws are much easier to use than an ax. They have a variety of uses in the landscape maintenance. Tree trimming, tree removal, and wood retaining wall construction are some of the many uses for chain saws. However, if used improperly, chain saws can cause serious or fatal injuries. In an average year, there are 36,000 chain saw related injuries treated in hospital emergency rooms.

There are many types and sizes of chain saws, ranging from lightweight saws to heavy-duty stump saws. Most chain saws have a chain brake device

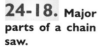

24-18. Major parts of a chain saw.

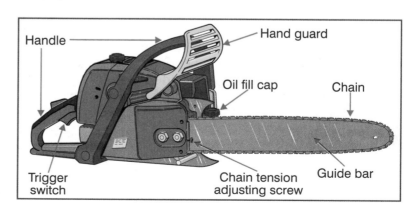

designed to stop the chain from turning in $1/10$ second or less. This helps protect the operator from kickback, the most violent accident you can have with a chain saw. *Kickback* is the violent up and back motion of the entire chain saw toward the operator. It occurs when the turning chain stalls in the upper tip of the guide bar. Kickback can send the chain saw back at the operator very fast and without warning.

Personal protective equipment is needed and safety procedures should be followed when operating a chain saw. Chain saws have two handles. Use both of them at all times when the saw is in operation. Never attempt to run a saw using only one hand. You can easily lose control of the saw. Run the saw by placing your left hand on the front handle and your right hand on the rear handle. Always wrap your thumbs under the handles, not on top. Follow safe wood cutting techniques. Plan your work before you start cutting. This will help prepare you for problems before they happen.

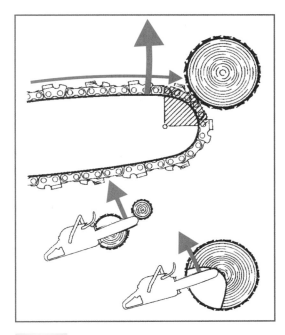

24-19. Chain saw kickback is the violent up and back motion of the entire chain saw toward the operator.

24-20. Safety is very important when operating a chain saw.

Safety Practices When Operating a Chain Saw

- ■ Helmet with ear and eye protection.
- ■ Protective pants or chaps.
- ■ Cut-resistant gloves.
- ■ Steel-toed, cut-resistant chain saw boots.

Pruning tree branches requires careful thought and constant attention. Determine which branches need to be removed and follow these guidelines.

- ■ Move your feet only when the chain completely stops turning.
- ■ Accelerate the saw before you start cutting each limb.
- ■ Do not cut with the upper tip of the guide bar (kickback portion).
- ■ Study each branch before cutting.

Even experienced wood cutters get a chain saw pinched in a tree from time to time. This is when the chain becomes lodged in the cut due to pressure. Shut the saw off and find the cause of the pinch. In most cases, it is easy to reopen the cut by removing a piece of wood or by using another piece of wood or wedge to pry the cut open.

BACKPACK BLOWER

A backpack blower is commonly used by landscape maintenance firms. It has replaced brooms and rakes and the labor and time that goes with it. Back-

24-21. A backpack blower uses air speed to move dust, dirt, leaves, and grass clippings from walkways.

pack blowers are more user-friendly. By taking a few precautions, they can be safe and trouble free to operate. Check fuel lines, air filters, and air intakes each day before operating this equipment. Goggles and hearing protection are among personal safety precautions that should be followed.

Backpack blowers use air speed to move dust, dirt, leaves, and grass clippings from walkways and to separate clumps of grass on the lawn. Backpack blowers have plenty of speed and power to do these jobs. Many operators run the blower at full throttle, when half or three-quarter throttle would do the same job with less noise. Some communities have created ordinances that restrict the use of blowers during certain times of day because of the noise pollution they create.

GOLF COURSE MAINTENANCE EQUIPMENT

AERATORS

Aerification, frequently referred to as cultivation, is the selective tillage of a turfgrass area by mechanical means. Lawns are aerified to improve the exchange of air and water between the soil below and the atmosphere above the turf. Aerification alleviates the adverse effects of soil compaction from natural forces, such as rain, equipment operation, and foot traffic.

24-22. Aerifying a golf course green with a core aerator.

A *core aerator* is an aerifying machine that has hollow metal tubes, called tines. They punch holes in the soil and remove a core of soil which is deposited on the turf's surface. The aerator tines are spaced from 2 to 12 inches apart. The closer the spacing the more holes punched and the better the aerification job.

There are many types and sizes of aerators available from different manufacturers. The walk-behind aerators range from closed coring tines types, which utilize a cam-driven piston mechanism to drive the tines into the ground, to aerators that resemble a modified rotary tiller with spoon-type or slicing tines. The larger the tine diameter and the closer the tine spacing, the more material will be removed from the turf. Several different types of pull-

24-23. Collecting cores after aerification.

24-24. Top dressing a golf green after aerification.

behind aerators are also available for use on large turf areas. After aerification, golf greens are usually top dressed to fill in the holes created by removal of the cores.

Water aerification is the newest method of aerification. The Toro Company introduced the Hydro-Jet 3000, a machine that aerates by injecting high-speed jets of water through eleven nozzles spaced across a manifold bar into the soil. It uses 150 gallons of water to aerate a 7,000 square foot golf green. This aerator was developed specifically for use on golf course putting greens during the summer.

24-25. Water aerifying is the latest method of aerification.

VERTICAL MOWERS AND POWER TRAP RAKES

Vertical mowers are wheeled power units equipped with a horizontal shaft, fitted with vertically oriented blades. The shaft spins rapidly, and the blades slice into the turf. Today, turf managers use vertical mowers for several new tasks, such as breaking up cores after aerification, opening up the turf before over-seeding in the fall, and slicing stolons to reduce horizontal turf growth.

A power trap rake is a self-propelled machine with a hydraulic rake mounted on the rear. It is used to mechanically smooth traps on golf courses. It saves many hours of hand labor which was required when traps were smoothed by hand. Always wear heavy shoes and, long pants when operating vertical mowers or power rakes.

24-26. A power rake is used to mechanically smooth traps on golf courses.

SOD CUTTER

Many homeowners do not want to wait for grass seed to germinate to establish a lawn. Laying sod can provide the homeowner with an "instant lawn." Grass seed is planted in turf farm fields and allowed to grow for one to two years. The sod is sold, cut, and rolled for landscape contractors to install. A *sod cutter* is the machine used to remove the surface layer of turf, including the plant and a thin layer of soil. Both tractor and walk-behind models are available.

24-27. Cutting bermudagrass sod at a turf farm.

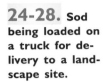

24-28. Sod being loaded on a truck for delivery to a landscape site.

Landscape contractors may need to remove existing turfgrass before installing walks, patios, or new lawns. Some golf courses also maintain their own mini turf farm as a source of sod as needed. This sod will also need to be harvested to be used. Landscape contractors and golf course workers often use a walk-behind sod cutter. It cuts the grass 12 inches wide and ¾ of an inch thick. Commercial sod farms use large tractors with sod cutting equipment attached. They cut the sod 18 inches wide and ⅝ of an inch thick. After cutting the sod, it is rolled and stacked on pallets automatically by the sod handling equipment. Fork lifts then move the pallets of sod to waiting trailers for quick delivery to golf courses and landscape contractors.

PREVENTIVE MAINTENANCE AND SERVICE

Preventive maintenance is regularly scheduled service to reduce equipment failure. Putting off regular equipment maintenance is the leading cause of premature equipment failure. "I'll change the oil tomorrow when I have more time." Tomorrow arrives with more work to do and again the oil change is delayed. Tomorrow never comes! Good preventive practices can reduce failures, save on operating costs, and keep equipment safe.

The heavier the work load, the more the equipment needs regular maintenance work. Operators never have time to do preventive maintenance, but

24-29. A maintenance schedule on an equipment decal.

they always have time to repair the broken equipment. Proper preventive maintenance will reduce down time from equipment failure.

An engine tuneup can reduce fuel consumption by 15 percent or more. Properly lubricated equipment parts are easier to turn; therefore, reduced engine power means less fuel used. Saving one dollar per day for each piece of equipment will add up to major dollars over the entire year.

Many operators try to save time by taking chances adjusting poor performing equipment while it is running. Most accidents are caused from poorly maintained or improperly adjusted equipment. Under inflated tires can make horticulture equipment harder to steer and keep under control.

KEEP RECORDS

Accurate records help insure that proper preventive maintenance practices are followed in a timely manner. Always follow the recommendations found in the operator's manual for specific service intervals. Keep the service record in a central place in the maintenance building. Most modern turf equipment contains an hour meter to show the total hours of operation indicating the time between scheduled maintenance.

EQUIPMENT OPERATION TIPS

DO:

- ✔ Read and follow operator's manual recommendations
- ✔ Perform daily maintenance procedures
- ✔ Check equipment for problems before starting
- ✔ Warm up the engine before operating the equipment
- ✔ Check and keep all bolts and nuts tight

✔ Operate at safe speeds

✔ Routinely check instrument readings while operating equipment

✔ Clean all screens around radiator in dusty conditions

✔ Fill fuel tanks after the day's work when the equipment is cool

✔ Select the proper equipment for each job

DON'T:

✘ Operate cold equipment

✘ Jam transmission into gear

✘ Pop the clutch

✘ Overload the capacity of equipment

✘ Allow engines to idle for a long time

✘ Disable safety devices or switches

✘ Remove safety covers

✘ Spin tires

✘ Leave fuel tanks empty overnight

✘ Race with other equipment operators

EQUIPMENT SERVICE INTERVALS

Recommended service intervals range from 5 hours to 1,000 hours. Hourly intervals can be easily converted to days of operation by the following table.

    5 hours (twice a day)
   10 hours (daily)
   50 hours (weekly)
  250 hours (monthly)
 1000 hours (seasonally)

Drive Belts and Chain Drives

Most turf and landscape equipment uses belts or chains to transmit engine power to the wheels and other moving parts. Proper maintenance will increase belt life and reduce down time due to belt replacement. The *v-belt* is a very common flexible rubber belt with a cross-section shape of the letter "V." It is used to connect two or more sheaves together.

Belt Drive Systems—Never check belt tension while equipment is operating. Too little belt tension can cause belt slipping and wear, while too much tension can damage bearing and over heat the belt. Squeaking belts suggest too little belt tension. Never lubricate a belt to stop belt noise; properly adjust the belt. The operator's manual will give the proper belt tension specifications and details of proper belt adjustment. Before replacing damaged belts, check sheave (pulley) alignment. A *sheave* is a grooved wheel to be installed on a shaft for belt-driven equipment. The shafts must be parallel to align belts properly. Misalignment will shorten belt life. Also check sheaves for nicks that can cut the belt. Adjust alignment and replace worn parts before replacing damaged belts.

Chain Drives—Some equipment uses a metal *roller chain* to connect two or more sprockets. It is used to transmit power from the engine to the wheels. A *sprocket* is a wheel or gear containing teeth for chain-driven equipment. Alignment is very critical on chain drives. Minor sprocket misalignment can cause drive chains to come off or sprockets to wear faster. Proper lubrication is very important for long chain life. A dry chain will increase sprocket wear and reduce chain life. Use aerosol spray chain lubricant as recommended in the operator's manual. Check the operator's manual for correct chain tension. A chain tightener (tensioner) adjusts the tension on the chain drive system. When the tension is loose, chains can come off the sprockets, while too much tension will increase wear and reduce chain life.

24-30. Check the hydraulic fluid level before operating equipment.

HYDRAULIC SYSTEMS

Hydraulics is fluid power. Hydraulic systems provide the muscles for doing work. Horticulture equipment can use one or more different hydraulic systems. Lifting the tractor bucket, turning mower reels, help in steering, and providing transport movement are several different hydraulic applications used on turf equipment.

A well maintained hydraulic system will provide years of trouble free operation. Changing hydraulic fluid and replacing hydraulic filters on the regular schedule as recommended by the equipment manufacturer will prevent most system problems. Horticulture equipment uses several different types of hydraulic fluids. Mixing two different kinds of fluid may damage the hydraulic parts.

Hydraulic systems operate best at 100 degrees above ambient (outdoor) temperature. Horticulture equipment will have some type of fluid cooler that looks like an engine radiator. Check and make sure this fluid cooler is always clean.

Common Hydraulic Fluids

There are several different types of hydraulic fluid used in turf equipment. Always read and follow the operator's manual when selecting fluid.

1. Crankcase oil, used in the transmissions of tractors, hydraulic motors, and lift cylinders.

2. Type F hydraulic fluid, used in transmissions manufactured by Ford.

3. Dexron II hydraulic fluid, used in transmissions, hydraulic motors, lift cylinders, and power steering systems.

4. Type C-3 hydraulic fluid is also used in transmissions

5. Brake fluid, a special fluid designed to be used in hydraulic brake systems. Do not use transmission hydraulic fluid in any brake system.

When replacing hydraulic fluids, always check the owner's manual for the correct fluid type for each different system. Never mix different kinds of hydraulic fluids in one system.

Hydraulic System Leaks

Leaking hydraulic fluid can injure equipment operators and/or kill plant

24-31. Hydraulic fluid leak shows on cardboard.

tissue. Hydraulic systems operate at very high pressure, over 2,000 psi. Fluid at this pressure can be injected, from a leaking hose, directly into the body through the skin. Never use your hand to find a hydraulic system leak. Instead, carefully move a piece of cardboard around each hose and other hydraulic components. The leaking fluid will discolor the cardboard and identify the location of the leak.

Hot hydraulic fluid leaking from mowing equipment can kill turfgrass plants. Washing grass plants with soap and water immediately after a hydraulic spill may prevent serious damage to the turfgrass plants. Always lower all moveable parts, such as buckets or mower units, when the equipment is not in use. Leaving equipment in the "up" position can cause extra strain on the hydraulic system. Always make sure all pressure is off the system before working on any hydraulic component.

REVIEWING

MAIN IDEAS

People are injured each year in grounds equipment related accidents. Federal law now requires every push mower to be equipped with an operator-presence control system that stops mower blades three seconds after the operator releases the safety lever. Never disconnect or disable these safety devices.

Avoid spilling and overfilling fuel tanks on all turf equipment, especially mowing units. Use a funnel and slowly pour gasoline from the container into the gas tank. Tightly close the cap and vent hole on the gasoline container after filling the tank. Keep the gasoline container out of direct sunlight and in a cool place.

There are four major types of lawnmowers in common use today—rotary, reel, flail, and sickle. Rotary mowers are very popular with homeowners and professional lawn care personnel. Reel mowers are mainly used on golf courses while flail mowers are the safest to operate. Sickle type mowers provide a poor quality of cut and are mainly used on very low maintenance areas such as roadsides. Decreasing landfill space combined with a national emphasis on recycling, has helped increase the sales of mulching mowers.

Special mower blade grinders should be used to sharpen rotary blades. Keep the mower blade cool by dipping the end into a container of cold water to prevent overheating and blade damage. Reel mowers also require special grinding equipment to sharpen the blades. Back lapping can provide temporary sharpening of reel mower blades.

Shearing is done to shape dwarf trees, hedges, or evergreens into sculptures, called topiaries. Electric shears or clippers will save time when trimming large hedges or many shrubs. Be careful not to cut the electric cord when using electric hedge trimmers. Avoid accidents by using power hedge trimmers carefully.

There are many types and sizes of chain saws, ranging from lightweight bucket saws to heavy-duty stump saws. Most saws have a chain brake device designed to stop the chain from turning in $1/10$ second or less. This will help protect the operator from kickback.

Aerify to alleviate the adverse effects of soil compaction from either natural forces, such as rain, or artificial forces, such as equipment operation or foot traffic. A core aerator has hollow tines which punch holes in the soil and remove a core of soil which is deposited on the turf's surface.

Blowers use air speed to move dust, dirt, and grass clippings from walks and to separate clumps of grass on the lawn. Backpack models have plenty of speed and power to do this job. By running the blower half or three-quarter throttle it will do the same job but make less noise.

Follow all service recommendations as to when fluids should be replaced and which types of fluids should be used. Read operation manual and follow all recommendations. Keeping accurate records of all equipment service will provide a written record for future reference. Check all v-belt and chain drives for proper tension and alignment when the equipment is not running. If a belt breaks, check the sheaves for damage before replacing the belt.

Most power equipment today uses some type of hydraulic system, either for transport or to lift or lower equipment parts. All equipment needs clean, cool hydraulic fluid. Leaking hydraulic fluid can damage or kill actively growing plants. The hydraulic fluid is under several thousand pounds of pressure; therefore, exercise care when looking for any system leaks.

QUESTIONS

Answer the following questions. Use complete sentences and correct spelling.

1. Why remove the lawnmower spark plug wire before removing the blade?

2. Why dip the mower blade into water when sharpening?

3. Name the two types of blades on a reel mower.

4. What can cause belt damage on horticulture equipment?

5. What is the best temperature for operating hydraulic systems?

6. How should a hydraulic spill on turf be "cleaned up"?

7. How should a hydraulic leak on equipment be located?

8. How wide are most homeowner rotary lawnmowers?

9. Why have mulching rotary mowers become popular?

10. Where should flail mowers be used?

11. How many gallons of fuel are spilled each year as estimated by the U.S. Environmental Protection Agency (EPA)?

12. Why are turf areas aerified?

13. What is an aerifier tine?

14. Describe how the Hydro-Jet 3000 aerifier operates.

15. What is chain saw kickback?

16. What part of the chain saw blade usually causes kickback?

17. What maintenance jobs are backpack blowers used for?

18. What is the most common accident when using electric hedge trimmers?

19. How wide is the standard strip of sod when cut at a sod farm?

EXPLORING

1. Visit a golf course and observe the different types of mowing equipment used. Prepare a report to the class identifying the type of mower and how it was being used.

2. Take a survey of your neighbors to determine the popularity of mulching mowers in your neighborhood. Make a report to the class.

3. Remove and sharpen a rotary mower blade using the proper procedure.

4. Correctly adjust the belt and chain tension on a piece of grounds maintenance equipment following the operator's manual specifications.

5. Carefully check a piece of grounds equipment for hydraulic leaks.

6. Check the operator's manual and identify the correct maintenance interval for each major maintenance item for a lawnmower.

25

ELECTRICAL CONTROLS

Today, electrical controls are used in all phases of horticulture. Computers now control greenhouse heating and cooling systems and landscape irrigation systems. Installing, maintaining, and repairing these electrical systems is an important part of horticulture technology.

Where should greenhouse thermostats be installed? What gauge wire should be used on a residential irrigation system? Where in the landscape should low-voltage power pack be installed? What is a control station? Why should a ground fault circuit interrupter be used on outdoor electrical devices? What is a mist system used for and what controls regulate it? These and many more questions about electrical controls used in horticulture will be covered in this chapter.

25-1. Electrical controls monitor the steam boiler heating system in this greenhouse.

OBJECTIVES

Many phases of horticulture use electricity and electrical controls. This chapter has the following objectives:

1   Explain where thermostats should be placed in a greenhouse

2   Discuss controlling greenhouse ventilation systems

3   Describe greenhouse mist system operation

4   Identify the major components of an irrigation systems

5   List the advantages of low-voltage lighting

TERMS

aspirated thermostat
common wire
day-night clock
electro-mechanical
   controllers
ground fault circuit
   interrupter (GFCI)
hot air heat
hot water heat
low-voltage current

master controller unit
moist-scale
percent timer clock
satellite
solenoid valve
solid-state controllers
station
steam heat
ventilation sash

GREENHOUSE CONTROLS

After labor, the cost of heat is the greatest expense a greenhouse manager has. Greenhouse growers save money by properly regulating the heating system. Coal, fuel oil, and gas provide the heat source necessary to grow plants in the greenhouse during the cold winter months. Hot water, steam, and hot air heat systems distribute this heat throughout the greenhouse. ***Hot water heat*** systems circulate 200°F water in pipes, ***steam heat*** circulates low pressure steam to produce heat, and ***hot air heat*** is produced by burning gas to heat air in the greenhouse. Greenhouse managers manually controlled greenhouse heating systems until the early 1970s. As labor costs continued to increase, greenhouse managers installed electrically controlled heating systems, replacing the expensive manual systems.

Heat small, individual, plastic-covered greenhouses (1,000 to 5,000 square feet) using gas as the heat source and hot air as the distribution system. Heat medium-sized greenhouses (5,000 to 15,000 square feet) using fuel oil or gas as the heat source and hot water as the distribution system. Using coal, fuel oil, or gas as heat sources and a steam distribution system provide warm growing conditions in large greenhouses (over 15,000 square feet).

25-2. Polyethylene tubes connected to a gas hot air furnace circulate heat throughout this greenhouse.

CONTROLLING HEATING SYSTEMS

Electrically controlled valves connected to thermostats allow the greenhouse manager automatic heat control. Most greenhouse heat control sys-

25-3. Naturally aspirated greenhouse thermostat.

25-4. Aspirated greenhouse thermostat.

tems operate on low voltage, usually 24 volts. **Aspirated thermostats** have small electric fans that provide constant air movement over the sensing devices. Most home heating systems use a naturally aspirated thermostat, one without small fans.

Place thermostats at the same level as the crop to monitor and regulate the temperature around the plants. To prevent false temperature readings, protect the thermostat from direct sunlight. Placing thermostats near the center of the greenhouse provides accurate temperature sensing of the entire greenhouse.

CONTROLLING VENTILATION

There are three major reasons for greenhouse ventilation—reducing air temperature, providing a fresh supply of carbon dioxide for photosynthesis, and reducing the relative humidity to help prevent plant diseases. The term **ventilation sash** describes glass greenhouse windows that open and allow fresh air to enter the greenhouse. This ventilation sash can be opened manually or with automated systems using electric motors. However, many new plastic-covered greenhouses use only large electric fans for ventilation.

With an automatic ventilation system, the thermostat is set at the desired greenhouse temperature. Some controls have a stepped program that opens the ventilation sash in stages, not all at once. The sash opens a few inches wide and then holds this position for several minutes. After monitoring the temperature during this time, the controller either opens or closes the sash.

Thermostats also control the electric fans used in plastic-covered greenhouses. These fan motors usually have low, medium, and high speeds. The electric control unit adjusts the fan speed to correlate with changes in the greenhouse temperature. Greenhouse cooling thermostats use a low-voltage, 24-volt system similar to the heating system.

25-5. Thermostats control electric fans in greenhouses.

WATERING SYSTEM CONTROLLERS

Greenhouse watering electrical systems operate on low-voltage, 24-volt AC, electric current. This provides greenhouse workers a degree of safety when working around wet greenhouse benches and electricity. Always turn off electrical power before working on any electric system in the greenhouse.

25-6. Solenoid valves used to automatically water greenhouse crops.

The **moist-scale,** developed by the Chapin Watermatic Company, senses the weight of a pot plant and turns on an electric water valve to irrigate an entire crop. Place the moist-scale on the bench with other similar pot plants. As the soil in the pot plant dries, it weighs less, causing the moist-scale to lift and mechanically turn on an electrical switch. A **solenoid valve** connected to the switch starts the water cycle. As the pot fills with water its weight increases until it mechanically turns off the switch on the moist-scale, stopping the watering cycle. This is a simple, but very effective, water control device.

Modern electronic time clocks are also used to control greenhouse watering systems. This system relies on turning water valves on for specific periods of time instead of sensing the amount of water delivered. Greenhouse water controllers are similar to turf irrigation controllers. Most city ordinances require the installation of a **ground fault circuit interrupter (GFCI)** on the 117-volt AC receptacle used by greenhouse watering controller. A GFCI device is designed to protect people from electrical shock caused by short-circuits.

MIST SYSTEM CONTROLLERS

Rooting of cuttings is a common method of reproducing greenhouse plants. Insert small 2- or 3-inch long pieces of stem tissue having one or two leaves into a sand rooting media. Rooting of cuttings usually takes three or four weeks to complete. To prevent wilting during this time, spray a fine mist of water over the cuttings for several seconds every couple of minutes until they are well rooted.

Fortunately, electric time clocks can provide this repetitive misting cycle. Two separate clocks control the mist cycle. The **day-night clock** turns the system on in the morning and off at night. A **percent timer clock** turns the three second mist cycle on and off every two minutes through the day.

25-7. Electronic mist control unit.

IRRIGATION SYSTEM CONTROLS

Electric operated water valves, called solenoid valves, and automatic controllers make up the modern turf irrigation system.

An electric impulse delivered through a control wire opens and closes these remote solenoid valves that allow water to flow to the sprinkler head.

RESIDENTIAL IRRIGATION SYSTEMS

All controllers have several similar features: a day-of-the-week clock to regulate which days water is applied, a twenty-four-hour electric clock to regulate time of day sprinklers operate, and a station clock to control the time each valve is turned on. Electro-mechanical and solid-state are the two major kinds of controllers used today.

Electro-mechanical controllers use electric clocks to mechanically turn switches on and off to control solenoid valves. The sound of switches clicking on and off can be heard as these units run. *Solid-state controllers* use modern electronic circuits to turn solenoid valves on and off without mechanical switches.

The term *station* describes a single electric switch in a controller that turns on a solenoid valve. Each station is similar to separate light switches in a home, one that turns on a bed room light and a different switch that turns on the kitchen light. Small controllers used in residential lawn irrigation systems have four or six separate stations that each turn on a separate solenoid valve.

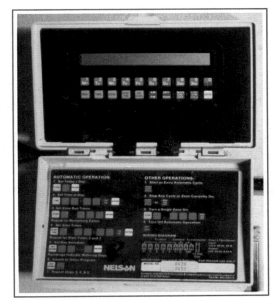

25-8. Home lawn irrigation controller.

Installing Residential Irrigation Systems

Install the controller where the sprinklers operated by the unit are visible from that location. This greatly simplifies system operation tests during installation and later during normal maintenance. However, most residential sprinkler controllers are located out of sight in the garage area. Most city ordinances require the installation of a ground fault circuit interrupter (GFCI) on the 117-volt AC receptacle used by the irrigation controller. This additional safety device will prevent accidental electrical shocks from the irrigation system.

Between the controller and the electric solenoid valves that feed the sprinklers, a network of valve control wires exists. Connect each valve to the controller with two wires, its own individual power or control wire, and the "common" or "ground" wire. The **common wire** (neutral wire) is connected to and shared by all the valves and completes the circuit back to the controller.

These wires conduct **low-voltage current**, usually 24 volts AC to energize the solenoid on the valve. A solenoid is simply a coil of copper wire that, when energized, lifts a plunger to open a control port in the valve. When the control port opens, it allows water pressure above the diaphragm in the valve's upper portion, or bonnet, to bleed off down stream. Most residential systems use a number 16-gauge control wire. Install larger size wire on large estate or light-commercial irrigation systems.

GOLF COURSE IRRIGATION SYSTEMS

The first fully automatic golf course irrigation systems were installed in California in the early 1950s. Golf course irrigation systems are complex and costly capital investments. Properly designed, installed, and operated irrigation systems provide golfers the quality turf that they want. Automatic irrigation systems consist of fixed pop-up sprinkler heads automatically activated by solenoid water valves turned on by a control device.

Golf course irrigation systems operate on a low-voltage, 24-volt system. Most city ordinances require the installation of a ground fault circuit interrupter (GFCI) on the 117-volt AC receptacle used by the irrigation controller. The automatic irrigation system may be integrated with an automatic sensing unit that monitors soil moisture levels to decide watering schedules.

The golf course superintendent establishes the specific day, time of day, and operating duration for each solenoid valve that then feeds water to one or more sprinklers. Automatic irrigation systems apply water more accurately than manual valve systems. When a green requires 10 minutes of irrigation, the controller applies 10, not 15, minutes of water to the green. This can save large quantities of water during the year.

Modern irrigation systems use remote control valves connected to one or more sprinkler heads. These valves are electrically controlled by the satellite control unit. The manner in which the automatic valve is actuated is termed either normally open or normally closed. Should an electric control line be broken in a normally closed system, the valve will remain closed and no water will flood the green. Flooding of turf can occur when normally open valve systems control lines are broken.

Valve-in-head sprinkler design installs the solenoid valve in each sprinkler head. The electric valve is incorporated in the base of the sprinkler head housing. This design allows individual control of sprinkler heads. Today, most golf course irrigation systems use valve-in-head sprinklers.

Golf Course Irrigation Controllers

Controllers have developed from simple clocks sending a signal to one valve to sophisticated computer driven units that can control hundreds of valves at once.

Electro-mechanical controllers use electric clocks to mechanically turn switches on and off to control solenoid valves. The sound of switches clicking on and off can be heard as these units run. Solid-state controllers use modern electronic circuits to turn on solenoid valves without mechanical switches.

The individual irrigation control units, called *satellites,* are similar in design to the residential controllers, but usually have 12 or more control stations. Satellites provide irrigation control for the sprinklers on one or two greens, tees, and fairways. All sprinklers controlled by the satellite should be visible from the controller location. Satellite controllers operate independently or may be controlled by a master controller unit.

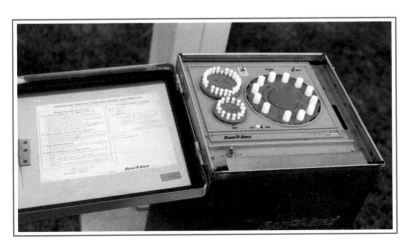

25-9. Solid-state satellite golf course irrigation controller. (Courtesy, Danville Country Club)

The *master controller unit* of an irrigation system operates each satellite, providing a single location for controlling all sprinklers on the golf course. Both electro-mechanical and solid-state satellite controllers are commonly installed today. Golf course irrigation controllers have: day-of-the-week clocks to regulate which day or days water is applied, twenty-four-hour electric clocks to regulate time of day sprinklers operate, and station clocks to control time each valve is turned on.

25-10. MAXI SYSTEM, golf course irrigation controller. (Courtesy, Rain Bird)

In the past, most satellites were hard wired to the master controller. That is, control cables run underground from the controller to each satellite. However, today, complete radio-controlled satellite systems, not hard wiring, are installed. A software program installed on a personal computer in the superintendent's office is really the master irrigation controller.

Computer control systems provide many additional features not available with simple electro-mechanical controllers. Optional sensors add an extra level of control to the irrigation system. Rain sensors, tensiometers to measure soil moisture, wind speed meters, and other sensors can be programmed into the system. The irrigation controller can register the precise quantity of any rainfall and shut off sprinklers in those areas where it is raining. If it only rains for a brief period, the controller will resume the irrigation schedule after the rain.

Installing Control Lines

Electric cable comes in a variety of gauge sizes, including 10- and 12-gauge types. The smaller the cable gauge size number the larger the cable, thus the more electricity it will carry. The long distance between controller and valve requires installing number 10-gauge wire on golf course irrigation systems. Follow all manufacturer's recommendations on cable selection.

Install control lines in the pipe trenches to one side of the pipe when possible. Placing electric control lines on top of the irrigation pipe increase the chance of damage when digging up the pipe to make plumbing repairs. Always install cable according to the engineer's and manufacturer's specifications and carefully check all connections before backfilling the trench with subsoil on the bottom and topsoil on the top. Splice all wire using special connectors that provide a watertight connection. This practice will eliminate future electrical problems from bad electrical connections.

LOW-VOLTAGE LANDSCAPE LIGHTING

Landscape designers have discovered that outdoor lighting can enhance planting and make turf, walkways, and gardens more visually pleasing at night. Besides providing beauty, safety, and security, lighting reinforces the image that property is well maintained. Many landscape contractors recommend low-voltage outdoor lighting systems because they operate on 12 volts instead of the standard 117-volt AC. Technological advances in low-voltage lighting have allowed manufacturer's to offer products equal in quality, if not better, than many high-voltage lights of just 10 years ago.

25-11. Components of a low-voltage lighting system.

ADVANTAGES OF LOW-VOLTAGE LIGHTING

Low-voltage systems have several advantages over 117-volt systems. Low voltage systems do not have to be wired by a master electrician, as required by many states for 117-volt lighting. Underground 117-volt lighting cables must be installed in rigid metal conduit pipe making it difficult to move as plants grow. Safety is another concern. Reduced voltage prevents electrical shocks even when touching the bare wires or accidentally cutting a buried cable with a garden tool.

25-12. Using low-voltage lights next to garden steps.

Low-voltage lighting systems consist of three components. The heart of any low-voltage lighting system is the power pack. Operating off a standard 117-volt electrical outlet, the power pack reduces the electrical current to a safe 12 volts for use by the lighting fixtures. Electric cables transmit the 12-volt electric current from the power pack to the individual lighting fixtures. Different lighting fixtures provide a unique lighting effect to a landscape.

POWER PACKS

Power packs are available with wattage ranging from 121 to 500 watts. Find out the correct size of the power pack by adding up the wattage of each lighting fixture. The power pack's rating should always be greater than the total lighting fixture wattage with all fixtures turned on. Controller options include manual and timer-operated models. Power packs also may have photo-cells, remote controls, and built-in motion/heat sensors.

Timer-operated power packs automatically turn the lights on and off at a predetermined time. More advanced controllers use a photocell that turns the lights on when the sun sets and turns them off at sunrise. A sophisticated motion/heat sensor power pack uses a timer that turns the lights on and off at predetermined times each day. However, if someone enters the protected area when the lights are off, the passive infrared motion sensor automatically turns the lights on for a brief period.

25-13. Low-voltage lighting power pack and controller.

*E*LECTRIC CABLE

Electric cables comes in a variety of gauge sizes, including 10- and 12-gauge types. The smaller the cable gauge size number the larger the cable, thus the more electricity it will carry. For small to medium size applications use 12-gauge cable, but install 10-gauge cable for large operations. Follow all manufacturer's recommendations on cable selection.

*I*NSTALLING LOW-VOLTAGE LIGHTING

Mount the power pack on a wall or fence post, at least one foot above the ground, next to a GFCI outdoor electrical outlet. Plug in the power pack, turn it on, and stretch the cable to the desired locations. Install the cable by following the landscape's natural contour and barriers and leave one or two feet of extra cable at each fixture site to allow for adjustments. Low-voltage cable is self-sealing and weather resistant; thus, bury it in the ground, cover it with wood chips, or hide it under bushes or foliage.

Assemble the fixtures as required and connect the fixtures to the cable using the supplied fastener. Finally, push the light into place in the ground. If

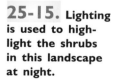 **25-14.** Installing low-voltage lighting cable.

the ground is hard, use a garden spade to dig a small hole, inset the fixture into the hole, and refill it with soil or gravel. Never use a hammer to drive the fixtures into the ground. Adjust the position of the fixtures so they create the desired lighting effect.

CARE AND MAINTENANCE

Low-voltage lighting systems require little care once installed. Occasionally, clean dirt, leaves, and other debris away from the fixtures. Periodically, check the fixtures for proper adjustment. Annually, remove the bulbs and spray the lamp sockets with a silicone-base spray. Inspect the cable yearly for damage or fraying near the connection points. If a bulb burns out, replace it immediately. The voltage that this damaged bulb no longer uses, flows to the remaining operating bulbs. This extra voltage can reduce the life of the remaining bulbs.

25-15. Lighting is used to highlight the shrubs in this landscape at night.

REVIEWING

MAIN IDEAS

Electrically controlled valves connected to thermostats allow the greenhouse manager to control the temperature inside the greenhouse. Some thermostats have small electric fans that blow air across the sensors and accurately monitor the greenhouse temperature. Naturally aspirated thermostats, similar to home heating system thermostats, do not use a fan, but rely on natural air currents to activate the control valves. Greenhouse cooling fans also use thermostats to regulate the cooling cycle in a greenhouse.

Greenhouse watering systems use a variety of control sensing devices to turn on and off the irrigation solenoid valves. Plant propagation done with a water mist system must have several clocks to control the mist cycle. The potential for electric shock always exists when water and electricity are used together. GFCI devices can help prevent electric shocks, but good personal safety practices must still be practiced.

Turf irrigation is big business. Home owners want beautiful lawns but they do not want to pull hoses and set sprinklers on the lawn. Automatic controllers allow the home owner to setup watering schedules several weeks in advance. Golf course irrigation controllers are similar to home owner control units but can handle more sprinkler heads. These controllers can also turn on irrigation pump systems before starting the watering cycle.

Low-voltage lighting systems allow the landscape designer the opportunity to install outdoor lighting without the fear of electrical shock common with the old 117-volt systems. A power pack not only turns on and off the lights, but it also changes the 117-volt electricity to a safe low-voltage rate. Once installed, low-voltage lighting systems require little, if any, maintenance, other than occasionally replacing burnt out bulbs.

QUESTIONS

Answer the following questions. Use complete sentences and correct spelling.

1. What is an aspirated thermostat?

2. Why do greenhouses need ventilation?

3. How are electronic time clocks used with automatic irrigation systems?

4. Why do mist systems need two different clocks?

5. What is a ground fault circuit interrupter and why should it be used when installing outdoor electric units?

6. What gauge wire should be used on a residential irrigation system?

7. When did golf courses start installing irrigation systems?

8. How do electro-mechanical and solid-state irrigation controllers differ?

9. What is the difference between the master controller and satellite controllers?

10. What are the advantages of using low-voltage landscape lighting?

11. What does the power pack in a landscape lighting system do?

12. What are the advantages of photo-cells over timers when controlling home landscape lights?

13. What annual maintenance should be performed on residential landscape lighting systems?

14. What is the correct mounting procedure for low-voltage power packs?

EXPLORING

1. Visit a golf course and identify the different parts of their irrigation system.

2. Visit a greenhouse and observe the different kinds of electric controllers used.

3. Practice programming a home irrigation system controller.

4. Install a low-voltage landscape light system at home or at school.

5. Identify the type of heat system used at your school and list all the electric controls used by the system.

26

IRRIGATION SYSTEMS

Water is essential to plants. Plants use water in photosynthesis and to maintain all growth processes. Plants use large quantities of water that is absorbed by plant roots. Plant wilting occurs when plants do not receive enough water. Water evaporating from plant leaves keeps plant tissue cool. Plants also require water to transport nutrients from the roots up the stem to the leaves.

How much water does a plant need? Is there a difference between rain water and artificial watering? What is irrigation? When should plants be watered? What are different ways vegetable crops can be irrigated? How does a drip irrigation system work? Answers to these questions can be found in this chapter about irrigation and watering plants.

26-1. Irrigating turfgrass at a California sod farm.

OBJECTIVES

Water is essential to plants. This chapter covers the ways plants are irrigated and has the following objectives:

1 Describe how plants use water

2 Describe ways to determine soil moisture level

3 List the major different kinds of irrigation systems

4 List automatic greenhouse irrigation systems

5 Describe a plant's drought stress symptoms

6 Discuss best times of day to water plants

7 List the different kinds of drip irrigation systems

8 Describe surface irrigation methods

TERMS

available water
backflow valve
drip irrigation
emitters
evaporation
field moisture
 capacity
footprinting
gravitational water
hand-moved watering
 system
irrigate
irrigation
lateral lines

main lines
microsprinkler
moisture meter
ooze tube watering
 system
percolation
permanent wilting
 point
point source emitters
pop-up heads
porous tubing
side-roll watering
 system
soil tension

spray stakes
surface irrigation
 system
tensiometer
trickle tape
tube watering system
turgid
turgidity
valve
water infiltration
waterlogged
wilting point

DETERMINING WATER NEEDS

The evaporation of moisture is influenced by temperature. *Evaporation* is the changing of water from a liquid to a gaseous state. In the natural environment, temperature is continually changing. Temperature varies with latitude, altitude, and topography. Moisture is naturally made available to plants by precipitation (rain and snow), water vapor (relative humidity), and dew.

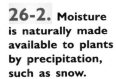

26-2. Moisture is naturally made available to plants by precipitation, such as snow.

Water is the most important requirement for plant growth. Water helps plants with many functions. Water in plants gives the cells *turgidity*. Without enough water, the cells are soft and the plant wilts. Plant cells must remain full of water or *turgid.* Water exerts pressure on the cell walls preventing the cells from collapsing or wilting. Plant wilting occurs when the rate of transpiration is greater than the rate of water uptake. All cell activity slows and will eventually stop when wilting occurs. If a wilted plant is watered, it may overcome the wilted condition if it hasn't been dry for too long.

Plants use water in photosynthesis and to maintain all growth processes. Plant roots must absorb water before photosynthesis can take place. Water transports nutrients and organic compounds throughout the plant. Water helps plants survive temperature changes. Transpiration cools the leaves in hot weather. Transpiration involves the absorption, transport, and release of water to the atmosphere by the plants. Evapo-transpiration is the rate of transpiration as affected by the rate of evaporation, known as the ET rate.

26-3. The water in the ditches between these rows of vegetables is infiltrating into the soil.

Water infiltration is movement of water into the soil. The downward movement of water through soil is *percolation.* As percolation occurs, dissolved nutrients move about in the soil. This movement is known as leaching. An ideal growth media may need approximately ½ inch of water each week. To effectively use this water, the infiltration and percolation rates must be favorable. Good soils will percolate between 3 to 6 inches of moisture per hour. *Gravitational water* is the water that moves from large pores due to the pull of gravity. This movement occurs after precipitation or irrigation.

Available water is the amount of retained water in soil that plant roots can absorb. Unavailable water is tightly held in the small pore space and is unavailable to plants. *Field moisture capacity* is when the water content of the soil fills the small pore spaces. Aeration capacity is the air content of the soil following drainage of the aeration pores. As the amount of moisture in the small pore spaces goes down or is held tightly by the soil colloids, it may become unavailable to plants. The *wilting point* is when plants cannot get enough moisture. The *permanent wilting point* is the point at which plants wilt, fail to recover turgidity, and die.

*E*VALUATING SOIL MOISTURE

When natural precipitation does not provide adequate moisture, plants must receive supplemental water. The term *irrigation* describes this supplemental watering process. In regions with less than 40 inches of annual precipitation, irrigation provides the necessary water to maintain good plant growth.

A rain gauge can be used to monitor rainfall outdoors. It should be placed in an open area and read after each rainfall. A soil moisture meter or tensiometer can also be used to measure soil moisture. A *moisture meter* is an instrument used to indirectly measure soil moisture. A *tensiometer* or electrical resistance block are other devices used to measure soil tension. *Soil tension* describes the force by which moisture is held in the soil. By keeping accurate rainfall records, you can easily figure out the need for irrigation. Whenever a period of 7 to 10 days receives less than 1 inch of recorded rainfall, the soil moisture may be exhausted to the point where plants may wilt.

Plant roots cannot grow in *waterlogged* (over-watered) soils because of poor soil aeration. Also, over-watering in warm weather can cause leaf diseases and increased weed development. Apply water only when needed by plants. Water the soil deeply and less frequently for best results. Normally, one such watering per week will suffice, except on the droughty kinds of soils or where much of the water runs off.

26-4. Precipitation is measured with rain gauges, with two examples of rain gauges shown here.

26-5. Using a moisture meter to determine soil moisture level.

Watering should be completed early enough in the day so that the plant leaves dry before nightfall. Wet plant leaves provide an ideal environment for the development of many plant diseases. Watering plants during the day will not scald the leaves but some water loss may occur due to evaporation.

GREENHOUSE WATERING

Small plants or seedlings require more frequent watering than larger plants. Plants use more water during the hot summer months than during the short days of winter. Greenhouse plants are usually fertilized while they are watered using a water-fertilizer solution.

Methods OF APPLICATION

26-6. Greenhouse irrigation systems often include the use of an injector which mixes a concentrated fertilizer solution with irrigation water.

In many greenhouses, a hand-held hose is still used to water plants. Manually watering plants is time consuming and expensive. However, most greenhouses use hand watering for a few select plants that become dry before others on the bench. A water breaker is usually attached to the end of the water hose for hand watering.

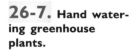

26-7. Hand watering greenhouse plants.

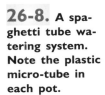

26-8. A spaghetti tube watering system. Note the plastic micro-tube in each pot.

Automated greenhouse watering systems has replaced much of the hand-held hose watering techniques. Commercial systems, such as watering tubes, ooze tubes, spray stakes, and capillary mats, provide the greenhouse manager labor-saving watering methods. Automatic systems work well with crops of the same size, in the same type of container, and planted in a like medium.

Plant crops in pots can be watered with a ***tube watering system.*** Many small plastic tubes (¹/₈ inch in diameter) are connected to a ¾ inch main wa-

26-9. A computer-controlled moving spray boom is another form of overhead irrigation. (Courtesy, Ashton Hi-Tech Seed Company, Ashton, Idaho)

ter line. The small micro-tubes are placed in individual pots. Hundreds of pot plants can be controlled by one water valve. This is also called spaghetti tube irrigation.

Carnations and chrysanthemums are often watered using an ***ooze tube watering system.*** Polyethylene tubes are used to attach a 2-inch wide by 100-foot long plastic strip by sewing along one edge. The small holes provide openings for water to slowly escape from the tube onto the medium surface. The tubes should be placed between the rows of plants. A master control valve can be used to water several benches at one time.

Spray stakes are used to water propagation benches and bedding plants. Water is applied over the canopy of the plants with spray nozzles mounted on risers. A flat circular spray of water 2 feet in diameter irrigates the plants. The spray stakes should be placed to allow spray overlap to occur in order to prevent dry spots.

A capillary mat watering system is a form of sub-irrigation in which potted plants are set on a moist, porous mat. The greenhouse benches must be level before installing this system. Water moves upward through the drainage holes in the containers into the growing medium by wick action. However, algal buildup on the mats can become a problem.

Electric solenoid water control valves allow greenhouse managers greater automation possibilities and reduced labor costs, while meeting the water needs of plants.

26-10. Irrigation in this greenhouse is controlled with electric solenoid water control valves.

WATERING TURFGRASS, FLOWERS, TREES, AND SHRUBS

Turfgrass requires 40 inches of rain annually for good uniform growth. Ideally, 1 inch of rain will fall each week during the growing season. Few places in the country have adequate rainfall. Therefore, turf managers must *irrigate* (water) to make up for the moisture shortfall from rain.

Footprinting is the first visual sign showing that turfgrass needs watering. *Footprinting* is when grass leaves do not spring-up after being walked on. Turfgrass will turn bluish in color as a second stage of water stress. Quick application of water at these early stages will prevent further damage to turf from lack of water. Once grass turns yellow from lack of water, the third stage, it can take several days for it to regain its green lush color.

WATERING TURFGRASS

When watering turf, always apply a minimum of 1 inch of water at each watering. Light frequent watering encourages shallow, poor rooting of turf. The time of day the water is applied is not as important as the water itself. Golf courses usually irrigate turf at night. The watering schedules for golf courses is dictated by golf course play and people.

Sprinkler heads distribute water from underground water lines to the turf. Turning on the water provides water pressure to force the sprinkler head out of the ground and start the irrigation process. These sprinkler heads are commonly called *pop-up heads.* Turning the water off allows the sprinkler to drop down into the ground and out of sight.

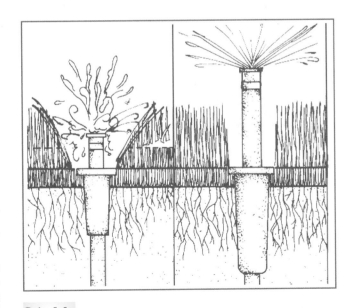

26-11. Well designed head on right will pop-up above surrounding turf; irrigation pattern from the poorly designed head on left disturbed by grass blades. (Courtesy, Rain Bird)

26-12. Proper irrigation can supplement available water for good plant growth.

Water coming out of some sprinkler nozzles cause the heads to rotate. These types of sprinklers are called rotary head sprinklers. Several different kinds of rotary sprinkler heads are available. Rotary sprinklers water turfgrass in a circular pattern. Small heads cover 3-or 4-foot diameter circles, while large heads can irrigate circles several hundred feet in diameter. Many pop-up heads can be controlled by a single water valve. This saves labor costs on golf courses and large commercial irrigation systems.

Electric solenoid water valves allow turf managers greater automation possibilities and further reduce labor costs. Chapter 25 in this book has additional information on electrical controls for automatic watering systems.

WATERING TREES AND SHRUBS

Newly transplanted trees and shrubs require a deep and thorough watering for the entire first year after planting. Homeowners may forget to water trees and shrubs regularly. A good watering technique is to thoroughly wet the soil 12 inches deep. It usually provides adequate soil moisture for most trees and shrubs.

Established trees need watering during the growing season when less than 1 inch of rain fall occurs during a single month. A lawn sprinkler can be placed under the trees. They should be watered until the soil is wet 12 inches below the surface. This may take several hours to accomplish. A lawn-type soaker hose is used to water shrub or rose beds. Wet the soil at least 6 inches deep in these beds. Other techniques include the use of a soil watering needle

26-13. This landscape technician is using a root watering needle to pump water directly to the root zone of these trees.

which supplies water directly to the root zone of plants or the use of water bags for newly transplanted trees located where watering is not easily done.

WATERING FLOWERS

Flower beds should be watered by hand using a fine stream of water. Flowers should not be let to wilt from lack of moisture. A lawn-type soaker hoses is sometimes used in flower beds. Flower beds should also be fertilized once or twice a year by applying a liquid fertilizer while watering.

26-14. Hand watering newly planted hardy chrysanthemums.

IRRIGATION SYSTEMS

MICRO OR DRIP IRRIGATION

Drip irrigation applies water directly in the root zone of plants, not to the soil between the plants. It is supplied through a thin plastic tube at a low flow rate; also called trickle irrigation. A major advantage of drip irrigation over other irrigation practices is the efficient use of limited water reserves. Many people define drip irrigation as applying small amounts of water at low pressures over long periods of time. Because drip irrigation systems usually do not wet the areas between plants, fewer weeds grow in these areas, reducing weed control problems.

Originally developed for commercial use in orchard, vine, and row crops, drip irrigation is now installed in many home landscapes to water ornamental plants. Drip irrigation provides the water necessary for many vegetable row crops such as melons, tomatoes, peanuts, and peppers. Today, drip irrigation permits orange, lemon, grapefruit, and lime trees to grow in areas where natural rainfall would not support their growth.

Drip irrigation provides water necessary for growing nut trees, such as almond and pecan. Many grape vineyards throughout America use this irrigation method for applying water for increased grape production. Ornamental tree nurseries have adapted drip irrigation techniques for watering acres of nursery stock.

System Parts

Point source emitters, low volume sprinklers placed close to root the system, and trickle tape represent the three major types of drip emitters in use today. The very popular point source emitters consist of a single outlet device that runs the water through baffles to minimize potential clogging and equalize the water output. Many newer types can automatically compensate for variation in line pressures while providing uniform water distribution.

26-15. A point source emitter is used to irrigate the root zone of this fruit tree.

Low volume sprinklers called **microsprinklers** run at pressures similar to point source emitters, but have a spray output over a very small area. Output and area coverage for microsprinklers varies with pressure, so shorter lateral and even terrain are necessary for even output. Since they cover a larger area than point source emitters, fewer are needed. Microsprinklers come in either static or spinning heads.

Trickle tape emitters are made from thin plastic lateral tubing with perforations along the entire length. Another kind of trickle tape, called **porous tubing**, emits water from openings in the entire surface area of the plastic tube. Trickle tape is very short-lived because it is easily torn and may be destroyed by sunlight. Porous tubing has a similar pattern of output, but is generally buried under the soil or mulch which increases its life.

If the water is not absolutely clean, the small orifices of a drip irrigation system will clog. A sand filter with a rinsing mechanism close to the water source may be installed for particularly dirty irrigation water. Generally, PVC pipe is installed for the main irrigation lines, running from the water source or pump. Lateral lines made of polyethylene pipe move the water from the main line into smaller pipes that contain the irrigation emitters.

Two different kinds of valves are used on drip irrigation systems. The **backflow valve**, prevents water in an irrigation system from siphoning back

26-16. Porous tubing used for irrigation in this landscape bed will be buried under mulch.

26-17. A backflow valve prevents contamination of municipal water supplies during irrigation. (Courtesy, Berry's Garden Center, Inc.)

and contaminating a municipal water supply. Manual or electric on-off valves are used to regulate the water supplied to the irrigation system.

OTHER SURFACE IRRIGATION SYSTEMS

Different kinds of irrigation systems have been developed in response to economic conditions, topography, water resources, and type of crops being grown. *Surface irrigation systems* siphon water from ditches, along the edge of the field, into furrows next to rows of plants. Surface irrigation systems provide the water needs for vegetable and orchard production in some parts of the United States.

26-18. Wells provide a clean water supply for large irrigation systems.

System Parts

Main lines, lateral lines, valves, and emitters are the major components of a drip irrigation system. *Main lines* are those that start at the source and branch into several secondary lines *(lateral lines)* which go to individual

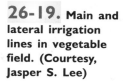

26-19. Main and lateral irrigation lines in vegetable field. (Courtesy, Jasper S. Lee)

26-20. Straw-berries watered using an aluminum pipe irrigation system.

valves or ommitters. *Valves* control the flow of water. *Emitters* are openings through which water flows in an irrigation system. Water can come from any source that can maintain at least 10 psi line pressure. Sources could include city water, which is clean but expensive, or water pumped from a well or pond, in which case filtration may be necessary.

Hand-moved watering systems have above-ground aluminum pipe laid down the rows of vegetable plants. Moving this system from one field to another requires much manual labor. A *side-roll watering system* has a series of large aluminum wheels, 5 feet in diameter, along the length of the pipe that elevate the sprinkler line above the field crops. Small motors mounted on the wheels permit moving the entire pipe system with little or no labor.

REVIEWING

MAIN IDEAS

Nutrients necessary for plant growth must be dissolved in water before the plant's root system can absorb them. Plants use water to remove heat from the leaves by transpiration, to manufacture food during photosynthesis, and to break down food to release energy during respiration.

In many parts of the United States, natural rainfall does not provide adequate moisture for good plant growth. Plant managers use irrigation practices to make up for lack of adequate rainfall. Over-watering plants can cause injury or death the same as under-watering.

Most greenhouse grown plants are automatically watered using modern irrigation systems controlled by electric clocks and control valves. A number of different systems are available.

Turfgrass requires 40 inches of water per year, either from rainfall or irrigation. Many home lawns and most golf courses use an underground irrigation pipe system with pop-up sprinkler heads. Newly transplanted trees and shrubs require careful watering during the first growing season. However, established landscape plants usually only need watering during drought periods. Most flower beds are watered using a hand-held hose.

Drip irrigation systems apply water in the soil around the plant's roots. This can conserve large quantities of water by eliminating runoff, watering only the desired plant, and reducing the amount of loss through evaporation from the soil surface. A number of different surface irrigation systems are also used to irrigate plants. Water may flow down ditches through the field or be pumped through a series of lines to irrigate plants. All irrigation systems require a good supply of clean water.

QUESTIONS

Answer the following questions. Use complete sentences and correct spelling.

1. What plant processes require water?
2. When during the day is the best time to water?
3. What greenhouse crops need the ooze tube watering system?
4. What visual signs indicate that turfgrass needs to be irrigated?
5. How many inches of water per year does turfgrass need?
6. What is a pop-up sprinkler head?
7. Why are most golf courses irrigated at night?
8. How deep should the soil around trees and shrubs be watered?
9. How is drip irrigation different from other irrigation systems?
10. Why is a backflow valve used on an irrigation system?
11. What is an irrigation emitter?
12. What horticultural crops are usually irrigated with surface irrigation?

EXPLORING

1. Visit a local greenhouse and identify the different kinds of watering systems in use.

2. Use a moisture meter to determine the moisture level in soil around trees, flowers, and turf.

3. Set up a demonstration of the water capacity of different soils. Use containers with soils high in sand, silt, and clay. Add water and observe what happens. Prepare a report on what you observe.

APPENDIXES

APPENDIX A—INTERNATIONAL SYSTEM OF MEASUREMENT

The International System of Measurement is based on units of 100, which can best be compared to our money system of dollars and cents. Different units are used for measuring length, volume, weight (mass), and temperature. Converting units within SI is extremely simple since in effect, changes are made by moving the decimal point. Initially, it is essential to learn the metric prefixes and which units are used for measuring length, volume, weight, and temperature.

Length

Length is the distance from one point to another. The SI unit of length is the meter. In making measurements, it is often more convenient to report Length in terms which signify a portion or combination of meters. The following prefixes are used with the main unit meter to specify measurements of length.

Prefix	Symbol	Meaning
kilo-	km	1,000 meters
hecto-	hm	100 meters
deca-	dam	10 meters
meter		
deci-	dm	0.1 meter
centi-	cm	0.01 meter
milli-	mm	0.001 meter

Volume

Volume is the amount of space a substance occupies and is based on measurements of length (i.e. length × width × height). The SI unit of volume is the cubic meter; however, this measurement is too

677

large for most scientific work, so scientists normally use cubic decimeters (0.1 of a meter)^3 to measure volume. One cubic decimeter (1 dm)^3 is equal to 1 liter (1). The following prefixes are used with the main unit liter to specify measurements of volume.

Prefix	Symbol	Meaning
kilo-	kl	1,000 liters
hecto-	hl	100 liters
deca-	dal	10 liters
liter		
deci-	dl	0.1 liter
centi-	cl	0.01 liter
milli-	ml	0.001 liter

Weight

Weight is a measure of the pull of gravity on an object. The SI unit of weight is the newton. Since the pull of gravity differs when you leave the earth and experiments are now conducted in space, scientists commonly measure the mass of an object which is how much matter is in something. (For example, the moon's gravity is approximately one-sixth that of the earth.) The SI unit of mass is the gram. The following prefixes are used with the main unit gram to specify measurements of mass.

Prefix	Symbol	Meaning
kilo-	kg	1,000 grams
hecto-	hg	100 grams
deca-	dag	10 grams
gram		
deci-	dg	0.1 gram
centi-	cg	0.01 gram
milli-	mg	0.001 gram

Temperature

Temperature is the amount of heat in something. The SI unit for measuring temperature is degrees Kelvin. One degree Kelvin is equal to one degree Celsius which is the common unit of measurement for the metric system. The metric system of measuring temperature is also based on 100. In this case there are 100 from the temperature at which water freezes to the temperature at which water boils. Common temperature measurements in Celsius are 18° Celsius—room temperature, 37° Celsius—body temperature.

APPENDIX B—ORDER OF INSECTS

Each insect belongs to a particular ORDER. This is the system used in grouping insects. From the description given of the orders and the illustrations shown, insects can be placed in the order in which they belong.

Thysanura (Bristletails, Silverfish)

Wings—None

Mouthparts—Chewing

Metamorphosis—None

Note—Silver-colored insects with long antennae and two or three long antennae-like appendages at the end of the abdomen. The silverfish feed on rayons, starched clothes, bookbindings and other materials having starch or glue. Can be found in feed or flour mills where starchy foods are handled or in sinks and bath tubs of homes.

Collembola (Springtails)

Wings—None

Mouthparts—Chewing

Metamorphosis—None

Note—Very small insects less than 1/5 inch long. Flip themselves into the air by means of a spring-like part under the abdomen. Found in damp places, such as under decaying vegetation, stones, and boards.

Orthoptera (Grasshoppers, Crickets, Roaches)

Wings—Two pairs—Walking sticks and camel crickets are wingless). Top pair of wings are leathery. Bottom pair are membranous and folded under top pair.

Mouthparts—Chewing

Metamorphosis—Incomplete

Note—A very common order of insects.

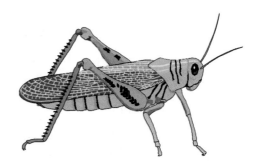

Isoptera (Termites)

Wings—Two pairs of the same length (workers are wingless).

Mouthparts—Chewing

Metamorphosis—Incomplete

Note—Kings and queens may be collected while swarming and workers may be found infesting wood. Look under wood on the ground.

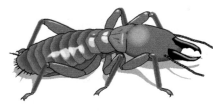

Ephemerida (Mayflies)

Wings—Two pairs—First pair much larger than second pair. Held vertically when at rest.

Mouthparts—None

Metamorphosis—Incomplete

Note—Found near water and are attracted to lights. Have two or three long antennae-like appendages at the end of the abdomen.

Plecoptera (Stoneflies)

Wings—Two pairs

Mouthparts—Chewing

Metamorphosis—Incomplete

Note—Found near running streams.

Odonata (Dragonflies, Damselflies)

Wings—Two pairs

Mouthparts—Chewing

Metamorphosis—Incomplete

Note—Feed on other insects. Usually found near water.

Neuroptera (Dobsonflies, Lacewings, Ant Lions)

Wings—Two pairs—many fine, net-like veins.

Mouthparts—Chewing

Metamorphosis—Complete

Note—Have long antennae. Found near streams, at lights, or on trees and plants.

Thysanoptera (Thrips)

Wings—Two pairs or none

Mouthparts—Rasping—sucking

Metamorphosis—Incomplete

Note—Very small insects only ⅛ inch long or less. Feeds on many plants.

Corrodentia (Book and Bark Lice)

Wings—Some wingless, some with two pairs.

Mouthparts—Chewing

Metmorphosis—Incomplete

Note—Found in old books and papers or on bark of trees or on damp stored grain.

Trichoptera (Caddisflies)

Wings—Two pairs

Mouthparts—Chewing

Metamorphosis—Complete

Note—Wings covered with short hairs and held roof-like over body when at rest. Found near water.

Mallophaga (Chewing Lice)

Wings—None

Mouthparts—Chewing

Metamorphosis—None

Note—Live on birds and to some extent on mammals. Feed on hair, feathers, scales and dried blood.

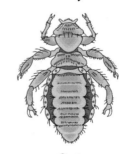

Anoplura (Sucking Lice)

Wings—None

Mouthparts—Piercing sucking

Metamorphosis—None

Note—Head narrow and long. Claws pincer-like. Feed on mammals.

Dermaptera (Earwigs)

Wings—Two pairs
Mouthparts—Chewing
Metamorphosis—Incomplete
Note—Front pair of wings like those of beetles but very
short, hind pair membranous. Have a pair of pincers
on end of abdomen. Found on plants, decayed matter
and sometimes in houses.

Hemiptera (TrueBugs)

Wings—Two pairs—Front pair is half leathery and
half membranous. Hind pair is membranous.
Mouthparts—Piercing, sucking
Metamorphosis—Incomplete
Note—Most live on land, but a few live in the water. Most
feed on plant juices, but there are some which feed on
animals and others which feed on other insects.

Homoptera (Aphids, Scales, Leafhoppers, Cicadas)

Wings—Two pairs or wingless
Mouthparts—Piercing, sucking
Metamorphosis—Incomplete
Note—All feed on plants.

Mecoptera (Scorpionflies)

Wings—Two pairs, long and narrow
Mouthparts—Chewing
Metamorphosis—Incomplete
Note—Mouthparts at the end of a long broad
snout. Found on low vegetation in dense
woods or sometimes in open fields.

Coleoptera (Beetles, Weevils)

Wings—Two pairs—Front pair—Hard and shell-like.
Hind pair—Membranous.

Mouthparts—Chewing

Metamorphosis—Complete

Note—This is one of the largest orders of insects in
the world and found almost everywhere.

Siphonaptera (Fleas)

Wings—None

Mouthparts—Piercing, sucking

Metamorphosis—Complete

Note—Live on animals. Collect them by dusting a cat or dog with pyrethrum powder and place the animal over a white cloth. Fleas will drop off on cloth.

Lepidoptera (Butterflies, Moths, Skippers)

Wings—Two pairs

Mouthparts—Siphoning

Metamorphosis—Complete

Note—Moths hide during the day and are active at night. Butterflies are active in the day and are usually brighter colored that the moths. Skippers have the tips of their antennae bent back like the handles of walking canes.

Diptera (Flies, Mosquitoes, Midges)

Wings—One pair

Mouthparts—Piercing, sucking, or sponging

Metamorphosis—Complete

Note—Found around flowers, decaying vegetation, on animals, and in houses and buildings.

Hymenoptera (Bees, Wasps, Ants)

Wings—Two pairs. Worker ants are wingless.

Mouthparts—Chewing

Metamorphosis—Complete

Note—One of the largest orders of insects and found almost everywhere.

APPENDIX C—COMMON PLANTS USED IN THE LANDSCAPE BY CLASSIFICATION

ANNUAL FLOWERS						
Common Name	Botanical Name	Height	Spacing	Exposure	Color	Use
Ageratum	*Ageratum houstonianum*	6–20"	9–12"	○	B,P,W	G/E/CF
Alyssum, Sweet	*Lobularia maritima*	2–4"	5–10"	○ or ●	W,P,V	E/GC
Aster	*Callistephus chinensis*	12–24"	10–12"	○	W,Y,Pr	B/CF
Baby's Breath	*Gypsophila elegans*	12–18"	12"	○	P,W	B/Br/CF
Balsam	*Impatiens balsamina*	12–28"	8–12"	●	P,R,W,M	B/Br
Begonia	*Begonia semperflorens*	6–10"	8–12"	●	P,R,W	B/C
Candytuft	*Iberis sp.*	6–12"	8–12"	○	W,Lc	E/B/G/GC
Carnation	*Dianthus sp.*	12–24"	12–18"	○	R,Y,W	B/E/CF
Celosia (Cockscomb)	*Celosia argentea*	6–36"	8–12"	○	R,Y,O	B/Br/CF
Coleus	*Coleus blumei*	12–24"	12"	●	R,Y,P	E/B/C
Cornflower	*Centaurea cyanus*	12–36"	12–18"	○ or ◗	B,P,W,R	B/G/CF
Cosmos	*Cosmos bipinnatus*	24–48"	12–18"	○	R,Y,O, P,W	CF/Br
Dahlia	*Dahlia sp.*	12–40"	12–24"	○	R,Y,P,M	B/E/CF
Dianthus	(See Pink)					

(Continued)

KEY	
Use	**Color**

Use

A – Accent	G – Garden
B – Bed	GC – Ground Cover
Bg – Bedding	H – Hedge
Bk – Background	P – Potpourri
Br – Border	Ps – Poisonous
C – Container	S – Seeds
CF – Cut Flower	Sc – Screen
DS – Dried Seed	Sh – Shade
E – Edging	Sp – Specimen
F – Foundation	VS – Vine Screen

Texture

C – Coarse	MC – Medium Coarse
F – Fine	MF – Medium Fine
M – Medium	

Color

B – Blue	P – Pink
B-G – Blue Green	Pr – Purple
Bk-G – Black Green	R – Red
C – Crimson	Rs – Rose
G – Emerald Green	R-G – Red Green
Gr – Green	S – Silver
Gy – Gray	Sc – Scarlet
Gy-G – Gray Green	V – Violet
L – Lavender	W – White
Lc – Lilac	Y – Yellow
M – Multi	Y-G – Yellow Green
O – Orange	

Exposure

○ – Sun   ● – Shade   ◗ – Partial Shade

APPENDIX C (Continued)

ANNUAL FLOWERS (Continued)

Common Name	Botanical Name	Height	Spacing	Exposure	Color	Use
Dusty Miller	*Cineraria maritima*	8–24"	8–12"	○	Gy	B/Br
Forget-me-not	*Myosotis sp.*	6–12"	6–8"	○ or ●	P,B	E/B/CF
Four-o'clock	*Mirabilis jalapa*	20–36"	12–18"	○	R,Y,W	Br
Gaillardia	*Gaillardia pulchella*	12–30"	9–12"	○	R with Y	Br/CF
Geranium	*Pelargonium x hortorum*	6–30"	12–15"	○	P,R,W	Bg/C
Globe Amaranth	*Gomphrena globosa*	6–24"	6–12"	○	W,P,R,O	Br/B/CF
Hollyhock	*Alcea rosea*	18–72"	12–18"	○ or ◗	R,P,Y,W,L	Sc/Br
Impatiens	*I. walleriana*	6–15"	8–12"	●	R,P,W,M	B/C
Lantana	*Lantana sp.*	12–36"	2–15"	○	R,Y,B	B/C
Larkspur	*Delphinium sp.*	18–48"	12–18"	○	B,P,W,R	Sc/Br/CF
Lobelia	*Lobelia sp.*	3–6"	4–6"	◗	B,P,W	B/C
Lupine	*Lupinus sp.*	12–36"	8–12"	○	P,Y,W,M	B/CF
Marigold	*Tagetes sp.*	6–30"	6–12"	○	R,Y,P	B/E
Nasturtium	*Tropaeolum sp.*	12–15"	9–12"	○	M	B/E
Pansy	*Viola sp.*	6–8"	6–8"	○ or ◗	M	E/B/C/CF
Petunia	*Petunia x hybrida*	8–24"	12–15"	○	B,P,R,W	B/C
Phlox	*Phlox drummondi*	6–15"	6–15"	○	M	B/E
Pink	*Dianthus chinensis*	8–15"	6–12"	○	P,R,W,M	B/E
Poppy	*Papaver sp.*	18–36"	6–12"	○	R,Y,W	Br/B/CF
Rose Moss	*Portulaca grandiflora*	4–6"	6–12"	○	P,R,W, Y,Sc	B/E/G
Rudbeckia	*Rudbeckia sp.*	18–24"	18"	○	Y	B/Br
Salvia (Scarlet Sage)	*Salvia splendens*	10–36"	8–12"	○	R,W,Pr	B/Br
Snapdragon	*Antirrhinum majus*	10–36"	10–12"	○	M	B/Br/CF
Strawflower	*Helichrysum sp.*	12–36"	12–18"	○	R,Gr,W, Y,Pr	Br/CF
Sunflower, Common	*Helianthus sp.*	2–10'	2–4'	○	Y,O	Sc/B/S
Sweet Pea	*Lathyrus odoratus*	20–30"	8–12"	○	M	VS/B/Br
Verbena	*Verbena x hybrida*	9–18"	8–12"	○	R,W,P,B	B
Vinca (Periwinkle)	*Vinca rosea*	10–18"	8–12"	○	P,W,Cr	B
Zinnia	*Zinnia sp.*	6–48"	8–15"	○	M	B/Br/CF

(Continued)

APPENDIX C (Continued)

BIENNIAL FLOWERS

Common Name	Botanical Name	Zone	Height	Spacing	Exposure	Color	Use
Canterbury Bells	Campanula medium	4	3–4'	18–24"	○ or ◐	Y,W,B,P	CF/Br
Daisy, English	Bellis perennis	4	4–8"	10–12"	○	R,W,B,P	E/G
Forget-me-not, Alpine	Myosotis sylvatica	3	8–24"	12–18"	●	B	B/Br
Foxglove	Digitalis purpurea	4	2–4'	18–24"	○ or ◐	P,Y,W,Pr	B/Br/Po
Hollyhock, Old Fashion	Alcea rosea	3	4–8'	12–18"	○	R,Y,P,W	Br
Honesty (Money Plant)	Lunaria annua	4	3'	18–24"	◐	B,W	G/DS
Pansy	Viola cornuta	4	6–8"	10–12"	○ or ◐	M	E/B/Br
Sweet William	Dianthus barbatus	4	12–20"	12–18"	○	R,W,P	E/Br/CF

PERENNIAL FLOWERS

Common Name	Botanical Name	Zone	Height	Spacing	Exposure	Color	Use
Alyssum (Basket-of-Gold)	Aurinia saxatilis	4	12–15"	24"	○	Y	G/E/CF
Anchusa (Alkanet)	Anchusa azurea	3	1–4'	24"	◐	B	Br/CF
Anthemis	Anthemis tinctoria	4	30–36"	18–24"	○	Y,W	Br
Artemisia (Silver Mound)	Artemisia schmidtiana	3	8–12"	10–15"	○	Gy-Gr	B/E/Br
Aster, Hardy	Aster sp.	4	12–60"	36"	○	W	B/G/CF
Astilbe (False Spirea)	Astilbe x arendsii	4	15–40"	18–24"	●	R,W,P,Sc	Br
Baby's Breath	Gypsophila paniculata	4	2–3'	48"	●	W,P	Br/CF
Balloon Flower	Platycodon grandiflorus	3	20"	12–18"	○ or ◐	W,B,P	Br
Baptisia (False Indigo)	Baptisia australis	3	4–5'	3'	○ or ◐	W,Y,B	Br/GC
Bellflower	Campanula sp.	3	8–20"	12–24"	○ or ◐	B,W	Br/E
Bleeding Heart	Dicentra spectabilis	4	2–3'	1–3'	○ or ◐	W,P	G
Blue Fescue (Festuca)	Festuca ovina	4	8–12"	12–18"	○ or ◐	S-B	B/E/GC
Carnation	Dianthus caryophyllus	3	18–24"	12–15"	○	P,R,W,Y	B/Br/CF
Chinese Lantern	Physalis alkekengi	4	24"	36"	○	O	Br/Sp
Chrysanthemum	Chrysanthemum sp.	5	8–36"	12–30"	○	M	B/Br/E/CF
Columbine	Aquilegia sp.	3	30–36"	12–18"	○ or ◐	M	B/CF
Daisy, Painted	Chrysanthemum coccineum	4	12–24"	12–18"	○	M	Br/CF
Daisy, Shasta	Chrysanthemum maximum	4	15–36"	18–30"	○	W	B/CF

(Continued)

APPENDIX C (Continued)

PERENNIAL FLOWERS (Continued)

Common Name	Botanical Name	Zone	Height	Spacing	Exposure	Color	Use
Daylily	*Hemerocallis* sp.	4	12–48"	24–36"	○ or ◗	P,R	B/G
Delphinium	*Delphinium* sp.	4	48–60"	24"	○	B,W,Pr	B/CF
Dianthus (Pink)	*Dianthus* sp.	3	3–24"	12"	○	P	Br/E/CF
Gaillardia (Blanket Flower)	*G. x grandiflora*	3	12–30"	24"	○	Y,Sc	Br/CF
Globeflower	*Trollius europaeus*	3	30"	8–10"	●	Y,O	B/CF
Globe Thistle	*Echinops exaltatus*	4	4'	24–36"	○	B	Br/CF
Hibiscus	*Hibiscus* sp.	5	3–8'	24"	○	R,W,P,Rs	Br/B/G
Iberis (Candytuft)	*Iberis sempervirens*	4	8–12"	12–18"	○	W	G/E/GC
Lavender	*Lavandula angustifolia*	5	12–36"	12–18"	○	Pr	Br/G/P
Liatris (Gayfeather)	*Liatris* sp.	3	18–72"	18"	○ or ◗	Rs, Pr	Br/CF
Lobelia (Cardinal Flower)	*Lobelia cardinalis*	6	24–30"	24"	◗	Sc	Br/B
Lupine	*Lupinus* sp.	4	3–5'	3'	○	R,P,W,O,Y	Br/CF
Lychnis (Maltese Cross)	*Lychnis* sp.	3	1–3'	18"	○ or ◗	R	Br
Peony	*Paeonia* sp.	3	18–36"	24–36"	○ or ◗	P,R,W	B/Br/CF
Phlox, Moss	*Phlox subulata*	3	4–5"	8"	○	R,P,W,B	Br
Phlox, Summer	*Phlox paniculata*	4	36"	18–24"	○	R,P,W,B	Br/CF
Poppy, Iceland	*Papaver nudicaule*	2	12–18"	24"	○	R,P,W	Br/CF
Poppy, Oriental	*Papaver orientale*	3	24–36"	24"	○	R,P,W,O	G/Br/CF
Potentilla (Cinquefoil)	*Potentilla* sp.	4	24–36"	24–36"	○	W,Y,O,P	B/Br
Primrose	*Primula* sp.	3	6–12"	10–12"	◗ or ●	R,P,W,Y,B	G/Br
Primrose, Evening	*Oenothera* sp.	3	12–18"	8"	○ or ◗	Y	G/Br
Red-hot Poker Plant (Torch Lily)	*Kniphofia* sp.	5	3–5'	18"	○	R,W,Y	Br/Sp
Rudbeckia (Coneflower)	*Rudbeckia* sp.	4	2–4'	24–30"	○	R,P,W	B/Br/CF
Salvia (Sage)	*Salvia* sp.	4	3–4'	18–24"	○	R	Br
Snow-in-Summer	*Cerastium tomentosum*	2	6–9"	24–36"	○	W	G/GC
Speedwell	*Veronica* sp.	3	6–36"	18"	○	Pr	Br/G/CF
Thrift	*Armeria maritima*	3	18–24"	12"	○	P	E/Br/CF
Virginia Bluebells	*Mertensia virginica*	4	12–24"	12–18"	◗ or ●	B	Br
Yarrow	*Achillea* sp.	3	6–36"	36"	○	R,Y,W,P	Br/CF

(Continued)

APPENDIX C (Continued)

GROUND COVERS—Herbaceous Perennials

Common Name	Botanical Name	Zone	Height	Width	Exposure	Texture	Color
Ajuga (Bugle Flower)	Ajuga reptans	4–8	4–6"	6–12"	◗	C	R-G
Crown Vetch	Coronilla varia	3	1–2'	18–24"	○	M	B-G
Lily-of-the-Valley	Convallaria majalis	2	6–12"	6–8"	●	MC	B-G
Phlox, Creeping	Phlox subulata	2	2–6"	8–12"	○	F	G
Plantain Lily, Fragrant	Hosta plantaginea	3	18–24"	2–3'	◗ or ●	C	B-G
Sedum, Red Creeping	Sedum spurium	3	6–8"	12–18"	○	F	G
Spurge, Japanese	Pachysandra terminalis	4	6–8"	10–16"	○ or ●	M	Y-G
Thyme, Creeping	Thymus serpyllum	1–3	1–4"	12–18"	○	F	G

GROUND COVERS—Evergreens

Common Name	Botanical Name	Zone	Height	Width	Exposure	Texture	Color
Barren-strawberry	Waldsteinia ternata	4	6–12"	12"	○ or ●	MC	G
Bearberry	Arctostaphylos uva-ursi	2–6	8–12"	spreads	○ or ◗	F	G
Bearberry, Cotoneaster	Cotoneaster dammeri	5–8	12–18"	3–6'	○ or ◗	F	G
Candytuft	Iberis sempervirens	4–8	6–12"	18–24"	○ or ◗	F	G
Ivy, Baltic	Hedera helix 'Baltica'	4–9	6–8"	8–12"	◗ or ●	M	Bk-G
Euonymus, Purple Wintercreeper	Euonymus fortunei 'Coloratus'	5	6"	12–24"	○ or ●	M	G
Juniper	(See Narrowleaf Evergreens)						
Lilyturf, Creeping	Liriope spicata	5	8–12"	12–18"	○ or ●	F	G
Mahonia, Creeping	Mahonia repens	5	10"	18–24"	○ or ◗	M-C	G
Pachistima, Canby	Pachistima canbyi	5	12"	12–24"	◗ or ●	MF	G
Periwinkle, Common (Myrtle)	Vinca minor	3	3–6"	9–18"	◗ or ●	MF	G
Spring Heath	Erica carnea	5	6–12"	24–36"	●	F	G
St. Johnswort, Aaronsbeard	Hypericum calycinum	5–8	12–18"	spreads	○ or ◗	MF	G

VINES

Common Name	Botanical Name	Zone	Height	Width	Exposure	Texture	Color
Akebia, Fiveleaf	Akebia quinata	4–8	20–40'	spreads	○ or ●	MF	B-G
Bittersweet, American	Celastrus scandens	3–8	20'	spreads	○	M	G
Clematis, Hybrid	Clematis x hybrida	3–8	6–18'	4–10'	◗	M	B-G
Fig, Climbing	Ficus pumila	8–6	climbs	spreads	◗ or ●	MF	G
Honeysuckle, Trumpet	Lonicera sempervirens	4–8	10–20'	spreads	○ or ◗	M	B-G
Ivy, Boston	Parthenocissus tricuspidata	4–8	25–35'	spreads	○ or ●	MC	G
Ivy, English	Hedera helix	4–9	climbs	spreads	◗ or ●	M	Bk-G

(Continued)

APPENDIX C (Continued)

VINES (Continued)

Common Name	Botanical Name	Zone	Height	Width	Exposure	Texture	Color
Moneywort	*Lysimachia nummularia*	4	climbs	spreads	○ or ●	M	G
Morning-glory	*Ipomoea purpurea*	4	5–10'	spreads	○	M	G
Silvervine	*Polygonum aubertii*	4	25–30'	30–35'	○ or ●	M	G
Trumpet Vine	*Campsis radicans*	4–9	25–35'	6–8'	○ or ◗	M	G
Virginia Creeper	*Parthenocissus quinquefolia*	3–9	25–50'	spreads	○ or ●	M	G
Wisteria, Japanese	*Wisteria floribunda*	5	25–30'	spreads	○ or ◗	M	G

BULBS/TUBERS

Common Name	Botanical Name	Hardiness/ Zone	Flower Time	Plant Time	Planting Depth	Spacing	Height
Allium	*Allium* sp.	hardy/4	late spring	fall	4–8"	3–6"	4'
Amaryllis	*Amaryllis* sp.	tender/5	summer	spring	top to ground level	12"	2–3'
Anemone	*Anemone blanda*	†/5	very early spring	spring	4"	12"	8"
Begonia, Tuberous	*Begonia x tuberhybrida*	tender	summer	late spring	top to ground level	6–8"	7–14"
Caladium	*Caladium* sp.	tender	summer	late spring	1–4"	10–18"	15–20"
Calla Lily	*Zantedeschia* sp.	tender/9	summer	late spring	3"	12–15"	7–14"
Canna	*Canna* sp.	tender/7	summer	spring	3–4"	1–2'	24–36"
Crocus, Spring	*Crocus* sp.	hardy/4	very early spring	fall	4"	2"	2–6"
Daffodil	*Narcissus* sp.	hardy/4	mid-spring	fall	8"	6–8"	7–12"
Daffodil, Fall	*Stembergia* sp.	†/7	late summer	mid-summer	5"	3"	1'
Dahlia	*Dahlia* sp.	tender/8	summer	spring	4"	24–48"	40–48"
Eranthis	*Eranthis* sp.	hardy/4	spring	late summer	4½"	3–4"	3"
Gladiolus	*Gladiolus* sp.	tender/8	summer	spring	6"	4–6"	28–48"
Hyacinth	*Hyacinthus* sp.	hardy/4	mid-spring	fall	8"	6–8"	7–12"
Hyacinth, Grape	*Muscari* sp.	hardy/4	mid-spring	fall	4"	2–4"	1–4"
Hyacinth, Summer	*Caltonia* sp.	tender	summer	spring	4–6"	6–8"	1–6"
Iris, Bulb	*Iris reticulata*	hardy/4	summer	spring/fall	4"	12"	4–6"

(Continued)

†Tender in northern zones and hardy in southern zones.

APPENDIX C (Continued)

BULBS/TUBERS (Continued)

Common Name	Botanical Name	Hardiness/ Zone	Flower Time	Plant Time	Planting Depth	Spacing	Height
Lily	*Lilium* sp.	hardy/3	summer	spring/fall	2–8"	12–18"	2–4'
Oxalis (Wood Sorrel)	*Oxalis* sp.	†/6	summer	fall	3"	6"	—
Ranunculus	*Ranunculus* sp.	†/3	summer	late spring	2"	12"	7–18"
Tulip	*Tulipa* sp.	hardy/4	mid–late spring	fall	8"	6–8"	7–20"

NARROWLEAF EVERGREEN SHRUBS

Common Name	Botanical Name	Zone	Height	Width	Exposure	Use	Texture	Color
False Cypress	*Chamaecyparis* sp.							
Dwarf Hinoki	*C. obtusa* 'Nana'	4–8	3'	3–4'	○ or ◗	Sp/G/F	M	Bk-G
Threadleaf	*C. pisifera* 'Filifera Nana'	4	6–8'	6–8'	○	Sp/G	M	Bk-G
Juniper	*Juniperus* sp.							
Andorra	*J. horizontalis* 'Plumosa'	3	2'	8–10'	○	GC	F	B-G
Blue Pacific	*J. conferta* 'Blue Pacific'	5–9	1'	6–9'	○	GC	M	B-G
Blue Rug	*J. horizontalis* 'Wiltoni'	3	3–6"	6–8'	○	GC	MF	B-G
Blue Star	*J. squamata* 'Blue Star'	4–7	2–3'	4–5'	○	Sp/GC/B	M	Gy-G,B-G
Creeping	*J. horizontalis*	3	6–18"	4–8'	○	B/GC	F	B-G
Compact Pfitzer	*J. chinensis* 'Pfitzeriana Compacta'	4	5–6'	5–6'	○	F/G	MF	G-G
Dwarf Common	*J. communis* 'Depressa'	3	2–4'	6–8'	○ or ◗	G/F	F	B-G
Japanese Garden	*J. procumbens*	5	18–24"	6–8'	○	G/F	F	B-G
Sargent	*J. chinensis* 'Sargentii'	4	12–18'	8'	○ or ◗	GC	F	B-G
Pine	*Pinus* sp.							
Dwarf White	*P. strobus* 'Nana'	3	4–6'	4–5'	○ or ◗	G/Sp	MF	B-G
Mugo	*P. mugo* var. *mugo*	2–7	4–8'	12–20'	○ or ◗	G/F	M	G,Bk-G
Spruce	*Picea* sp.							
Bird's Nest	*P. abies* 'Nidiformis'	2	1–2'	3–5'	○	Sp	M	G
Compact Colorado	*P. purgens* 'Compacta'	2	8–10'	10–12'	○	F/Sp	M–C	Gy-G,B-G
Dwarf Alberta	*P. glauca* 'Conica'	2	3'	2'	○ or ◗	G/F	M	G
Dwarf Globe Blue	*P. pungens* 'Glauca Globosa'	2	2–3'	3–4'	○	Sp	M–C	Gy-G,B-G
Dwarf Norway	*P. abies* 'Pumila'	3	3–4'	4'	○ or ◗	Sp	M	G
Yew	*Taxus* sp.							
Dense	*T. x media densiformis*	4	4–6'	4–6'	○ or ●	F/G	M	Bk-G
Hicks	*T. x media* 'Hicksii'	5	4–6'	4–6'	○	F/G/H	M	Bk-G
Japanese	*T. cuspidata* var. *nana*	4	5–6'	5–6'	○ or ●	F/H/Sp	M	Bk-G
Spreading English	*T. baccata* 'Repandens'	6	2–3'	3–4'	○ or ◗	F/GC	M	Bk-G
Tauton Anglojap	*T. x media* 'Tauton'	4	6–10'	8–10'	○ or ●	F/G	M	Bk-G

(Continued)

APPENDIX C (Continued)

BROADLEAF EVERGREEN SHRUBS

Common Name	Botanical Name	Zone	Height	Width	Exposure	Use	Texture	Color
Aucuba, Japanese	*Aucuba japonica*	7–10	6–10'	7–9'	●	F/H	M	Bk-G
Barberry, Wintergreen	*Berberis julianae*	5–6	6–10'	4–6'	○	Sp/H	MF	G
Boxwood	*Buxus sp.*							
Common	*B. sempervirens*	6	15–20'	12–18'	○ or ◐	F/H	MF	Bk-G
Littleleaf	*B microphylla*	5	3–4'	3–4'	○ or ◐	F/H/Br	MF	G
Camellia	*Camellia sp.*							
Japanese	*C. japonica*	7–9	10–15'	6–10'	◐	Sp	M–MC	G
Sasanqua	*C. sasanqua*	7	6–10'	5–8'	○ or ●	F/Br/Sp	M–MC	G
Daphne, Rose	*Daphne cneorum*	4–7	6–12"	2'	◐ or ●	G	MF	Gy-G
Euonymus, Bigleaf	*Euonymus sp.*							
Japanese	*E. japonica*	7–9	10–15'	4–6'	○ or ●	F/Br/Sp	M	Bk-G
Wintercreeper	*E. fortunei var.* 'Vegetus'	5–8	4–5'	4–5'	○ or ●	C/F/H	MF	G
Fatsia, Japanese	*Fatsia japonica*	7	6–10'	6–10'	●	G	C	G
Firethorn	*Pyracantha sp.*							
Formosa	*P. koidzumi*	8–10	8–12'	8–12'	○ or ◐	Br/Sp	M	G
Scarlet	*P. coccinea*	6–9	6–16'	6–16'	○ or ◐	Br/Sp	M	G
Gardenia, Cape Jasmine	*Gardenia jasminoides*	8–10	4–6'	4–6'	○ or ●	Br/Sp	M	G
Holly	*Ilex sp.*							
Chinese	*I. cornuta*	7–9	8–10'	8–12'	○ or ●	F/H/Sp	M–MC	G
Dwarf Burford	*I. cornuta* 'Dwarf Burford'	7–9	5–6'	5–6'	○ or ●	F/Sp	M–MC	G
Winterberry	*I. verticillata*	3–9	6–10'	6–10'	○ or ◐	Br	M	G
Laurel, Mountain	*Kalmia latifolia*	4–9	7–15'	7–15'	◐ or ●	F/G	M	G
Leucothoe, Drooping	*Leucothoe fontanesiana*	5–8	3–6'	3–6'	◐ or ●	F/Br	M	G
Nandina	*Nandina sp.*							
Dwarf Heavenly Bamboo	*N. domestica* 'Nana'	6–9	2–4'	2–4'	○ or ●	Br	M	B-G
Heavenly Bamboo	*N. domestica*	6–9	6–8'	4–5'	○ or ◐	F/Br	M	B-G
Oleander	*Nerium oleander*	8	8–10'	6–8'	○ or ◐	Br/Sc/ Sp/Po	M	G
Oregongrapeholly	*Mahonia aquifolium*	4–8	3–6'	3–6'	◐ or ●	F/Br/Sp	M	G
Pieris, Japanese	*Pieris japonica*	5–8	9–10'	6–8'	◐ or ●	F/Sp	M	R-G,Bk-G
Privet	*Ligustrum sp.*							
Howard	*L. japonicum* 'Howard'	7–10	6–12'	6–8'	○ or ●	F/Br	M	Y-G
Japanese	*L. japonicum*	7–10	6–12'	6–8'	○ or ●	F/Sc/Sp	M	G

(Continued)

APPENDIX C (Continued)

BROADLEAF EVERGREEN SHRUBS (Continued)

Common Name	Botanical Name	Zone	Height	Width	Exposure	Use	Texture	Color
Rhododendron	Rhododendron sp.							
Carolina	R. carolinianum	5–8	3–6'	3–6'	○ or ◑	F/B/Br/Sp	M	G
Catawba	R. catawbiense	4–8	6–10'	5–8'	◑ or ●	F/B/Br/Sp	MC–C	G
Korean	R. mucronulatum	4–7	4–8'	4–8'	●	F/B/Br/Sp	MF	G
P.J.M.	R. P.J.M.	4	3–6'	3–6'	○ or ◑	A/Sp/G/B	M	G
Wilson	R. x laetivirens	4	2–4'	4–6'	○ or ◑	F/B/Br/Sp	MF	G
Skimmia Japanese	Skimmia japonica	7–8	3–4'	3–4'	○ or ◑	F/H/G	M	G
Viburnum	Viburnum sp.							
Japanese	V. japonicum	8–9	10–15'	10–15'	○	F/Sp/Sc	M	G
Laurustinus	V. tinus	9–10	6–12'	6–12'	○ or ◑	F/Br/Sc	M	G
Leatherleaf	V. rhytidophyllum	5–8	10–15'	10–15'	● or ◑	F/Br	C	G
Wayfaringtree	V. lantana	4–8	10–15'	10–15'	○ or ◑	H/Sc/Br	M	G
Yucca	Yucca sp.							
Adam's Needle	Y. filamentosa	4	3'	3–4'	○	A	C	G
Spanish Dagger	Y. gloriosa	6	6–8'	3–4'	○	A	C	G
Spanish Bayonet	Y. aloifolia	7	8–12'	5–6'	○ or ◑	A	C	G

DECIDUOUS SHRUBS

Common Name	Botanical Name	Zone	Height	Spread	Exposure	Use	Texture	Color
Abelia, Glossy	Abelia x grandiflora	6	3–6'	3–6'	○ or ◑	A/Br/F	MF	Bk-G
Azalea	Rhododendron sp.							
Exbury Hybrids	R. x exbury	5–7	8–12'	5–8'	◑	Sp/G/Br	M	G
Flame (Yellow)	R. calendulaceum	5	4–8'	4–8'	○ or ◑	Br/Sp	M	G
Glenn Dale	R. x 'Glenn Dale'	6	4–5'	4–5'	◑	F/B/E	M	G
Karens	R. kaempferi var. karens	6	3–4'	3–5'	◑	F/B/E	M	G
Kurume Hybrids	R. x obtusum	6	4–6'	3–6'	◑	F/B/Br/G	M	G
Mollis Hybrids	R. x kosterianum	5–6	3–6'	3–5'	○ or ●	F/B/Br/G	M	G
Royal	R. schlippenbachi	4–7	6–8'	6–8'	○ or ◑	B/F/A	M	G
Barberry	Berberis sp.							
Crimson Pygmy	B. thunbergii 'Crimson Pygmy'	4	2'	2'	○	A/Br	MF	R-G
Japanese	B. thunbergii	4	5–7'	4–7'	○ or ◑	H/Br/Sp	MF	G
Mentor	B. x mentorensis	5	5–6'	4–5'	○ or ◑	H/A/Sp/G	M	G
Blueberry, Highbush	Vaccinium corymbosum	4	6–12'	8–12'	○ or ◑	Sp/H	M	B-G
Buckeye, Bottlebrush	Aesculus pariflora	5	8–12'	8–15'	○ or ◑	Sp/Br	C	G
Chokeberry	Aronia sp.							
Black	A. melanocarpa	5	3–5'	3+'	○ or ◑	Br	M	G
Red	A. x arbutifolia	5	6–10'	3–5'	○ or ◑	Br	M	G
Cotoneaster	Cotoneaster sp.							
Cranberry	C. apiculatus	5	2–3'	5–8'	○ or ◑	GC/F	F	G
Creeping	C. adpressus	5	1'	4–6'	○ or ◑	G	F	G
Many-Flowered	C. divaricatus	4–8	5–6'	6–8'	○ or ◑	H/Br/Sp	F	G
Rockspray	C. horizontalis	4	2–3'	5–8'	○ or ◑	GC/B/G	F	G
Spreading	C. multiflorus	3–7	8–12'	12–15'	○ or ◑	Sp	MF	Gy-G

(Continued)

APPENDIX C (Continued)							
DECIDUOUS SHRUBS (Continued)							
Common Name	**Botanical Name**	**Zone**	**Height**	**Spread**	**Exposure**	**Use**	**Texture**	**Color**
Daphne, Fragrant	*Daphne odora*	7–9	3'	3'	○ or ◗	Sp	MF	G
Deutzia	*Deutzia* sp.							
Showy	*D. x magnifica*	5	6–10'	6–10'	○ or ◗	Sp/Bk	MF	G
Slender	*D. gracilis*	4–8	2–4'	3–4'	○ or ◗	B/G/Sp	MF	G
Dogwood	*Cornus* sp.							
Redosier	*C. sericea*	2–8	7–10'	10'	○ or ◗	H/Sc/Br	M	G
Tatarian	*C. alba*	2–7	8–10'	5–10'	○ or ◗	H/Sc/Br	M	G
Elaeagnus	*Elaeagnus* sp.							
Autumn Olive	*E. umbellata*	2–7	18'	18'	○	H/Sc	MF	Gy-G
Silver	*E. multiflora*	5–7	6–10'	6–10'	○	H/Sc/Sp	MF	Gy-G
Euonymus	*Euonymus* sp.							
Dwarf Winged	*E. alatus* 'Compactus'	3	6–8'	5–6'	○ or ●	H/B/Br	M	G
Winged	*E. alatus*	4–8	15–20'	15–20'	○ or ◗	H/Sc/Sp	M	G
(Burning Bush)								
Forsythia	*Forsythia* sp.							
Border	*F. x intermedia*	5	8–10'	8–10'	○ or ◗	H/Br/Sp	M	Y-G
Bronx	*F. viridissima* 'Bronx'	5	2'	2–4'	○ or ◗	B/A	M	G
Fothergilla	*Fothergilla* sp.							
Dwarf	*F. gardenii*	4–8	2–3'	3–4'	○ or ◗	Sp/F/G/Br	MC	G
Large	*F. major*	4–8	6–9'	6–8'	◗	G/Br	M	G
Hibiscus	*Hibiscus syriacus*	6	8–12'	6–10'	○ or ◗	H/Br/Sp	M	G
(Rose-of-Sharon)								
Honeysuckle	*Lonicera* sp.							
Amur	*L. maacki*	2–8	12–15'	12–15'	○ or ◗	H/Sc/Sp	MC	Y-G
Clavey's Dwarf	*L. x xylosteoides*	4–6	6'	4–6'	○ or ◗	H	M	Gy-G
	'Clavey's Dwarf'							
Morrow	*L. morrowi*	4–6	6–10'	8–10'	○ or ◗	H/Sc/Sp	M	B-G
Tatarian	*L. tatarica*	3–8	9–12'	10–12"	○ or ◗	Sp/H	M	B-G
Winter Fragrant	*L. fragrantissima*	4–8	6–10'	6–10'	○ or ◗	H/Sc/Sp	M	B-G
Hydrangea	*Hydrangea* sp.							
Hills-of-Snow	*H. arborescens* 'Grandiflora'	4	4–8'	5–8'	○ or ◗	F/H/Br	C	G
Oak-Leaf	*H. quercifolia*	5	4–6'	3–5'	◗ or ●	Sp/Br	C	Gy-G
Peegee	*H. paniculata* 'Grandiflora'	3–8	10–15'	6–10'	○ or ◗	Sp	C	G
Lilac	*Syringa* sp.							
Common	*S. vulgaris*	3	8–20'	4–10'	○	Sp/H/Sc	M	G
Late	*S. villosa*	2–7	6–10'	4–10'	○	Sp	M	G
Mockorange	*Philadelphus* sp.							
Minnesota	*P. virginalis*	4	8'	4–6'	○ or ◗	A/Br	M	G
Snowflake	'Minnesota Snowflake'							
Natchez	*P. x* 'Natchez'	5	8–10'	8–10'	○ or ◗	G/Br	M–MC	G

(Continued)

APPENDIX C (Continued)

DECIDUOUS SHRUBS (Continued)

Common Name	Botanical Name	Zone	Height	Spread	Exposure	Use	Texture	Color
Prunus	*Prunus* sp.							
Dwarf Flowering Almond	*P. glandulosa*	4–8	4–5'	3–4'	○	Sp/A/Br	M	Y-G
Nanking Cherry	*P. tomentosa*	3	8–10'	10–15'	○	Br/Sp	M	G
Quince	*Chaenomeles* sp.							
Flowering	*C. speciosa*	5	6–10'	8–10'	○	Sp/Br	M	R-G
Japanese	*C. japonica*	4	3–4'	4'	○	Br/F	MF	G
Spirea	*Spiraea* sp.							
Anthony Waterer	*S. bumalda 'Anthony Waterer'*	5	3–4'	3–4'	○ or ◗	Sp	MF	R-G
Bridal-Wreath	*S. prunifolia*	5	5–8'	6–8'	○ or ◗	H/Sp/Br	MF	B-G
Snowmound	*S. nipponica 'Snowmound'*	4	3–6'	3–4'	○ or ◗	Sp/Br	MF	B-G
Vanhoutte	*S. x vanhouttei*	4	6–8'	8–10'	○ or ◗	Br/Sp	M	B-G
St. Johnswort, Shrubby	*Hypericum prolifium*	3–8	1–4'	1–4'	○ or ◗	Br/G	MF	B-G
Sumac, Fragrant	*Rhus aromatica*	3	2–6'	5–8'	○	F/GC	M	R-G
Sweetshrub, Common	*Calycanthus floridus*	4–9	6–9'	6–12'	○ or ●	Br/G	M	G
Viburnum	*Viburnum* sp.							
American Cranberrybush	*V. trilobum*	2	8–12'	8–12'	○ or ◗	Br/Sc	M	G
Arrowwood	*V. dentatum*	2–8	15'	15'	○ or ◗	Sp/Br	M	G
Compact American Cranberrybush	*V. trilobum 'Compactum'*	2	5–6'	5'	○ or ◗	H/Sc/G/Br	M	G
Doublefile	*V. plicatum* var. *tomentosum*	5–8	8–10'	9–12'	○ or ◗	A/Sp	M	G
Dwarf European Cranberrybush	*V. opulus 'Nanum'*	3	2'	2'	○ or ◗	A/Sp/G	M	G
European Cranberrybush	*V. opulus*	3–8	8–12'	10–15'	○ or ◗	Br/Sc/G	M	G
Fragrant Snowball	*V. x carlcephalum*	5	6–9'	6–8'	○ or ◗	Sp/Br	M	G
Koreanspice	*V. carlesii*	5	4–8'	5–8'	○ or ◗	B/Br	M	G
Linden	*V. dilatatum*	5–7	7–10'	5–7'	○ or ●	Sp/Br/Sc	M	G
Weigela	*Weigela florida*	5	6–10'	9–12'	○ or ◗	Sp/Br	C	G
Witchhazel, Vernal	*Hamamelis vernalis*	5	6–10'	6–10'	○ or ◗	Sp/Br	M	G

FLOWERING TREES

Common Name	Botanical Name	Zone	Height	Spread	Exposure	Use	Texture	Color
Apricot, Japanese	*Prunus mume*	6–9	15–20'	15–20'	○ or ◗	Sp/Sh/Br	M	G
Cherry	*Prunus* sp.							
Dwarf Japanese	*P. serrulata 'Shogetsu'*	5–6	10–18'	10–18'	○	A/Br	M	G
Weeping	*P. subhirtella pendula*	5–6	20–30'	15–25'	○	Sp	F	G

(Continued)

APPENDIX C (Continued)

FLOWERING TREES (Continued)

Common Name	Botanical Name	Zone	Height	Spread	Exposure	Use	Texture	Color
Crabapple	*Malus sp.*							
Carmine	*M. x atrosanquinea*	4	15–20'	15–20'	○	Sp/Br	M	G
Japanese	*M. floribunda*	4	15–25'	30'	○	Sp	F	G
Crapemyrtle	*Lagerstroemia indica*	7–9	15–25'	15–25'	○	Sp	M–MF	G
Dogwood	*Cornus sp.*							
Cherokee Chief	*C. florida 'Cherokee Chief'*	5	20–30'	20–30'	○ or ●	Sp	M	G
Pagoda	*C. alternifolia*	3–7	15–25'	20–35'	◗	Br	M	G
Pink	*C. florida var. rubra*	5	20–30'	20–30'	○ or ●	Sp	M	G
White	*C. florida*	5	20–30'	20–30'	○ or ●	Sp	M	G
Goldenchain Tree	*Laburnum x watereri*	5–7	10–15'	9–12'	○ or ◗	A	MF	G
Hawthorn	*Crataegus sp.*							
Green	*C. viridis*	4	20–30'	20–30'	○ or ◗	Sp/Sc	M–F	G
Washington	*C. phaenopyrum*	4	30'	20–25'	○	Br	M–F	G
Magnolia	*Magnolia sp.*							
Big Leaf	*M. macrophylla*	5–8	30–40'	30–40'	○ or ◗	Sp	C	G
Lily	*M. liliiflora*	5–8	10–12'	8–12'	○ or ◗	F/Br/A	M–C	G
Saucer	*M. x soulangiana*	4–9	20–30'	20–30'	○ or ◗	Sp	MC	G
Star	*M. stellata*	4–8	15–20'	10–15'	○	Sp	M	G
Mimosa (Silk Tree)	*Albizia julibrissin*	5–7	25–35'	18–25'	○	Sp	F	Y-G
Pear, Bradford Callery	*Pyrus calleryana 'Bradford'*	5	30–40'	20–35'	○	Sp/Br	MF	G
Plum, Purple Leaf	*Prunus cerasifera 'Newport'*	5	15–30'	15–30'	○	A/Sp/Br	M	G
Redbud, Eastern	*Cercis canadensis*	5	20–25'	15–30'	○ or ●	Sp	MC	B-G
Serviceberry, Downy	*Amelanchier arborea*	4–9	15–25'	15–20'	○ or ●	G/Br	MF	G

DECIDUOUS TREES

Common Name	Botanical Name	Zone	Height	Spread	Exposure	Use	Texture	Color
Beech, European	*Fagus sylvatica*	5	60–80'	50–70'	○ or ●	Sp/Sh	M	G
Baldcypress	*Taxodium distichum*	5	50–70'	20–30'	○ or ◗	Sp/Br	F	G
Birch	*Betula sp.*							
Paper	*B. papyrifera*	3	50–60'	30'	○	Sp/A	MF	G
River	*B. nigra*	4	40–70'	30–60'	○ or ◗	Sp	MF	G
White	*B. pendula*	3	40–50'	20–30'	○ or ◗	Sp	MF	G
Black Gum	*Nyssa sylvatica*	5–6	30–50'	20–30'	○ or ●	Sp	MF	G
Buckeye, Ohio	*Aesculus glabra*	4	20–40'	30–40'	○ or ◗	Sp/Sh	MC	G
Corktree, Amur	*Phellodendron amurense*	4	30–45'	30–40'	○	Sp/Sh	M	G
Elm, Chinese	*Ulmus parvifolia*	5	40–50'	30'	○	Sp/Sh	F	G
Ginkgo	*Ginkgo biloba*	5	40–60'	40'	○	Sp/Br	M	G

(Continued)

APPENDIX C (Continued)

DECIDUOUS TREES (Continued)

Common Name	Botanical Name	Zone	Height	Spread	Exposure	Use	Texture	Color
Goldenraintree, Panicled	*Koelreutaria paniculata*	5	30'	20'	○	Sp	M	G
Hackberry, Common	*Celtis occidentalus*	3	50–70'	50'	○	Sh	M	G
Honeylocust, Common Thornless	*Gleditsia triacanthos* var. *inermis*	5	75'	40–50'	○	Sp	F	G
Japanese Pagoda Tree	*Sophora japonica*	5	45–70'	30'	○	Sp	MF	B-G
Linden	*Tilia* sp.							
American	*T. americana*	3	60–80'	30–50'	○	Sp/Sh	C	G
Littleleaf	*T. cordata*	3	50–70'	40'	○	Sp/Sh	M	G
Londonplane Tree	*Plantanus x acerifolia*	4–8	70–100'	65–80'	○	Sh	MC	G
Maple	*Acer* sp.							
Japanese	*A. palmatum*	5–6	15–20'	15–20'	○ or ☽	Sp	F–MF	G
Norway	*A. platanoides*	3–4	40–50'	30–50'	○	Sh	MC	Bk-G
Red	*A. rubrum*	3–6	40–60'	30–40'	○ or ●	Sp/Sh	M	G
Silver	*A. saccharinum*	3	50–70'	35–50'	○	Sh*	M	G
Sugar	*A. saccharum*	3	60–120'	50–80'	○ or ☽	Sp/Sh	M	G
Oak	*Quercus* sp.							
Northern Red	*Q. rubra*	3–8	70–90'	60–75'	○	Sp/Sh	M	G
Pin	*Q. palustris*	4–8	60–70'	25–40'	○	Sp/Sh	M	G
Southern Red	*Q. falcata*	7–9	50–60'	50–60'	○ or ☽	Sh	M	G
White	*Q. alba*	3–9	50–80'	50–80'	○	Sh	MC	G
Poplar, Lombardy	*Populus nigra* 'Italica'	3–9	90'	10–15'	○	Sc**	MC	G
Sourwood	*Oxydendrum arboreum*	5–9	25–30'	20'	○ or ☽	Sp	M	G
Sweetgum	*Liquidambar styraciflua*	4	60–90'	40'	○	Sp/Sh	M	G
Tulip Tree	*Liriodendron tulipifera*	4	60–90'	30–40'	○ or ☽	Sp/Sh	M	Y-G
Willow	*Salix* sp.							
Corkscrew	*S. matsudana* 'Tortuosa'	4	20–40'	20–30'	○ or ☽	Sp	F	G
Weeping	*S. babylonica*	6–8	30–40'	30–40'	○ or ☽	Sp	F	G

EVERGREEN TREES

Common Name	Botanical Name	Zone	Height	Exposure	Use	Texture	Color
Arborvitae	*Thuja*						
American	*T. occidentalis*	4	40–60'	○	Sc/H	MF	G
Oriental	*T. orientalis*	6	18–25'	○	Sc/H	MF	Y-G
Techny	*T. occidentalis* cv. *techny*	2	8–10'	○ or ☽	Sc/H	MF	G
Cedar, Eastern Red	*Juniperus virginiana*	3	30–40'	○	Sc/H	F	B-G (spring), R-G (summer)

(Continued)

*fast-growing, softwood
**short-lived

APPENDIX C (Continued)

EVERGREEN TREES (Continued)

Common Name	Botanical Name	Zone	Height	Exposure	Use	Texture	Color
Cryptomeria, Japanese Cedar	*Cryptomeria japonica*	5	50–60'	○	Sp	M	G,B-G
Cypress, Arizona	*Cupressus arizonica*	6	25–40'	○	Sp/Sc	F	G–GG
Douglasfir	*Pseudotsuga menziesii*	4	50–90'	○ or ◗	Sp/Sc	M	B-G,Bk-G
Fir	*Abies sp.*						
Balsam	*A. balsamea*	4	45–75'	○	Sp/G/Br	M	G
White	*A. concolor*	4	30–45'	○	Sp/G/Br	M–C	B-G,Gy-G
Holly	*Ilex sp.*						
American	*I. opaca*	6	20–40'	○ or ◗	F/H/Br	MC	Dark Y-G
Buford	*I. cornuta 'Bufordii'*	6	10–20'	○ or ●	F/H/Br	M	Bk-G
Hume	*I. x attenuata 'Hume #2'*	6	20–25'	○ or ●	F/H/Br	M	Bk-G
Yaupon	*I. vomitoria*	7	10–15'	○ or ●	F/Sp	MF	Bk-G
Laurel, Cherry	*Prunus caroliniana*	7	15–25'	○ or ◗	Sp/H	M	Bk-G
Magnolia, Southern	*Magnolia grandiflora*	7–9	60–80'	○ or ◗	Sc/H/Sh	MC	G
Pine	*Pinus sp.*						
Austrian	*P. nigra*	5	60–90'	○	Sp/Sc/Br	MC	Bk-G
Eastern White	*P. strobus*	3	60–80'	○ or ◗	Sp/H/Sc/Br	MF	B-G
Pitch	*P. rigida*	5	40–50'	○ or ◗	Sc/Br	M	Y-G,Bk-G
Scotch	*P. sylvestris*	2	30–40'	○	Sp/Sc/F	M	B-G
Spruce	*Picea sp.*						
Blue Colorado	*P. pungens f. glauca*	3	70–90'	○ or ◗	Sp/G	M–C	B-G
Colorado	*P. pungens*	3	70–90'	○ or ◗	Sp	M–C	Gy-G,B-G
Norway	*P. abies*	2–7	40–60'	○ or ◗	Sp/Sc	M	G,Bk-G
Weeping White	*P. glauca 'Pendula'*	2	40–50'	○ or ◗	Sp	M	G
White	*P. glauca*	2	30–50'	○ or ◗	Sp	M	G

APPENDIX D—COMMON VEGETABLE CROPS BY CLASSIFICATION

Cole Crops

Common Name: Cabbage

Botanical Name: *Brassica oleracea* L. Capitata group

Common Name: Cauliflower

Botanical Name: *Brassica oleracea* L. Botrytis group

Common Name: Broccoli

Botanical Name: *Brassica oleracea* L. Italica group

Common Name: Brussels Sprouts

Botanical Name: *Brassica oleracea* L. Gemmifera group

Uses/Part: Fresh market, processing, fresh cut; Leaf—Cabbage, Brussels Sprouts; Flower—Cauliflower, Broccoli

Class: Cool-season biennial

Max/Min Temp: 45–75°F, Broccoli & Brussels Sprouts: 40–75°F

Soil & pH: Well drained, pH 6.0–6.8

Planting: Direct-seeded or transplanted

Minor Nutrients: Boron, Calcium, Sulfur, Molybdenum

Spacing: in row: Cab/Broc: 12–24", 14–24" (others); between row: 24–36"

Diseases: *Fusarium* Yellows, Clubroot, Black Rot, Black Leg, Downy Mildew, *Alternaria*, Black Blight.

Insects: Cabbage Maggot, Cabbage Worms, Aphids, Cabbage Looper

Maturity: 60–120 days (cabbage); 55–80 days (broccoli); 50–125 days (cauliflower); 90–100 days (Brussels sprouts)

Harvesting: Cabbage: Cut by hand when heads are firm; Broccoli: Hand cut (head 1–4" across) with tight beads; Cauliflower: Hand cut (heads 6" across) leaves green, curd compact; Brussels Sprouts: 1–1.5" firm, bright green sprouts cut by hand or machine harvested

Storage: 32°F, Cabbage: 3–6 weeks, Broccoli: 7–10 days, Cauliflower: 3–4 weeks, Brussels Sprouts: 3–5 weeks

Potherbs/Greens

Common Name: Spinach

Botanical Name: *Spinacia oleracea* L.

Uses/Part: Fresh market, processing and fresh pack/leaf

Class: Cool-season annual

Max/Min Temp: 40–75°F

Soil & pH: Many soils, pH 6.0–7.0

Planting: Direct-seeded

Minor Nutrients: Boron

Spacing: in row: 2–6"; between row: 12–36"

Diseases: Downy Mildew, *Fusarium* Wilt, Mosaic, Curly Top, Beet Yellows, White Rust

Insects: Green Peach Aphid, Cabbage Looper, Cucumber Beetle, Leaf Miner

Maturity: 37–45 days

Harvesting: Cut just above the growing point by hand or (processing) by machine

Storage: 32°F, 10–14 days

Common Name: Kale and Collards

Botanical Name: *Brassica Oleracea* L. Acephala group

Uses/Part: Fresh market and processing/leaf

Class: Cool-season biennial

Max/Min Temp: 40–75°F

Soil & pH: Heavy to light friable loams, pH 6.0–6.5

Planting: Direct-seeded or transplanted

Spacing: in row: 12–24"; between row: 24–36"

Diseases: see cabbage

Insects: see cabbage

Maturity: 70–85 days (collard); 55 days (kale)

Harvesting: Whole plants/individual leaves cut while tender

Storage: 32°F, Collard: 10–14 days; Kale: 2–3 weeks

Salad Crops

Common Name: Lettuce
Botanical Name: *Lactuca sativa* L.
Uses/Part: Fresh market and fresh cut/leaf
Class: Cool-season annual
Max/Min Temp: 45–75°F
Soil & pH: Many soils, pH 6.5–7.0
Planting: Transplanted or direct-seeded
Spacing: in row: 10–15"; between row: 16–24"
Diseases: *Sclerotinia* Drop, *Botrytis* Rot, Downy Mildew, Powdery Mildew, Bottom Rot, Corky Root, Mosaic Viruses, Aster Yellows Big Vein
Insects: Green Peach Aphid, Six-spotted Leafhopper, Cabbage Looper, Corn Earworm, Beet Armyworm.
Maturity: 40–85 days
Harvesting: Hand cut below lowest leaf when head yields slightly to pressure
Storage: 32°F, 2–3 weeks

Common Name: Celery
Botanical Name: *Apium graveolens* L. var. *dulce*
Uses/Part: Fresh market, processing, fresh cut/petiole
Class: Cool-season biennial
Max/Min Temp: 45–75°F
Soil & pH: Mucks, well-drained mineral soils, pH 6.0–7.0
Planting: Transplanted (some direct-seeded)
Minor Nutrients: Magnesium, Boron, and Calcium foliar applied
Spacing: in row: 6–12"; between row: 18–40"
Diseases: Late Blight, Early Blight, *Fusarium* Yellows, Pink Rot, Basal Stalk Rot, Virus Diseases
Insects: Aphids, Fall Armyworm, Black Cutworm, Cabbage Looper, Celery Leaf Tier, Spider Mites
Maturity: 90–125 days
Harvesting: Machine or hand cut at 14–16"
Storage: 32°F, 2–3 months

Roots and Tubers

Common Name: Potato
Botanical Name: *Solanum tuberosum* L.
Uses/Part: Fresh market, processing, chips/tuber
Class: Cool-season perennial
Max/Min Temp: 45–75°F
Soil & pH: Well-drained mineral or organic, pH 5.0–6.5
Planting: Seed pieces planted mechanically or by hand
Spacing: in row: 6–12"; between row: 30–42"
Diseases: *Rhizoctonia*, Scab, Ring Rot, Early Blight, Late Blight, *Verticillium*, and *Fusarium* Wilts, Virus Diseases
Insects: Wireworms, Aphids, Colorado Potato Beetle, Leafhoppers, White Grubs
Maturity: 90–120 days
Harvesting: Mechanically or hand dug when tubers mature (2–3" diameter)
Storage: Fall harvested, cured @ 50–60°F for 10–14 days, temperature lowered gradually to 38–40°F, and can be stored for 5–10 months

Common Name: Sweet Potato
Botanical Name: *Ipomoea batatas* (L.) Lam.
Uses/Part: Fresh market, processing/root
Class: Very tender, warm-season perennial
Max/Min Temp: 65–95°F
Soil & pH: Light, loose friable mineral soils, pH 5.5–6.5
Planting: Transplanted
Minor Nutrients: Boron
Spacing: in row: 10–18"; between row: 36–48"
Diseases: Scurf, Black Rot, Soil Rot, *Fusarium* Wilt, *Rhizopus* Soft Rot, White Rust, Leaf Spots, Blights
Insects: Root Knot Nematode, Sweet Potato Weevil, Wireworm, Flea Beetle, Banded Cucumber Beetle, White Grubs
Maturity: 65–80 days
Harvesting: Mechanically or hand dug before frost or freeze
Storage: Cured @ 85°F for 4–7 days, stored at 55–60°F for up to 4–7 months

Common Name: Carrot
Botanical Name: *Daucus carota* L.
Uses/Part: Fresh market, processing, fresh cut/root
Class: Cool-season biennial
Max/Min Temp: 45–75°F
Soil & pH: Deep, pebble-free, well-drained light soils and mucks, pH 5.5–7.0
Planting: Direct-seeded
Minor Nutrients: Boron
Spacing: in row: 1–3"; between row: 16–30"

Diseases: Carrot Blight, Yellows, Root Rots

Insects: Rust Fly Maggot, Weevil, Carrot Caterpillar, Leafhopper

Maturity: 50–95 days

Harvesting: Machine or hand dug when shoulder 1–1.25" broad. Processing: shoulder 1.5–4" broad

Storage: 32°F, 7–9 months when mature

Common Name: Radish

Botanical Name: *Raphanus sativus* L.

Uses/Part: Fresh market, fresh cut/root

Class: Cool-season annual

Max/Min Temp: 40–75°F

Soil & pH: Well-drained, pebble-free soils, pH 6.0–6.5

Planting: Direct-seeded

Minor Nutrients: Boron

Spacing: in row: 0.5–1"; between row: 8–18"

Diseases: Black Root, Downy Mildew, Scab, *Pythium, Rhizoctonia, Fusarium* Yellows

Insects: Root Maggot, Cabbage Aphids, Flea Beetles

Maturity: 22–30 days

Harvesting: Machine/hand dug and topped at appropriate size

Storage: 32°F, 3–4 weeks

Common Name: Turnip and Rutabaga

Botanical Name: *Brassica rapa* L. Rapifera group (turnip); *Brassica napus* L. Napobrassica group (rutabaga)

Uses/Part: Fresh market, processing/root

Class: Cool-season biennial

Max/Min Temp: 40–75°F

Soil & pH: Many deep friable soils, pH 6.0–6.8

Planting: Direct-seeded

Minor Nutrients: Boron

Spacing: in row: 5–8"; between row: 18–36"

Diseases: Anthracnose, Downy Mildew, Clubroot, Turnip Mosaic Virus

Insects: Aphids, Leaf Miners, Cabbage Maggot, Flea Beetle

Maturity: 40–75 days; Rutabaga: 90 days

Harvesting: Roots mechanically or hand dug and topped at appropriate size

Storage: 32°F, 4–5 months

Considerations: Turnips leaves are also harvested for greens

Alliums

Common Name: Onion

Botanical Name: *Allium cepa* L. Cepa group

Uses/Part: Fresh market, fresh cut, processing/bulb

Class: Cool-season biennial

Max/Min Temp: 45–85°F

Soil & pH: Many well-drained soils and muck, pH 6.0–6.8

Planting: Transplanted, onion sets, or direct-seeded

Minor Nutrients: Zinc, Copper, and Sulfur, depending on soil

Spacing: in row: 1–4"; between row: 16–24"

Diseases: Smut, Downy Mildew, Pink Root, Smudge, *Botrytis* Leaf Blight, Bulb Rots

Insects: Onion Maggots, Thrips, Cutworm, Wireworm

Maturity: 90–150 days (dry); 45–60 days (green)

Harvesting: Hand or machine dug when tops collapse

Storage: 32°F, 1–8 months (dry), 3–4 weeks (green)

Common Name: Garlic

Botanical Name: *Allium sativum* L.

Uses/Part: Fresh market, processing/bulb

Class: Cool-season perennial

Max/Min Temp: 45–85°F

Soil & pH: Rich, well-drained soils, pH 6.0–6.8

Planting: Cloves placed base down

Minor Nutrients: see onion

Spacing: in row: 1–3"; between row: 12–24"

Diseases: see onion

Insects: Few insect pests

Maturity: 90–150 days

Harvesting: Pulled and dried when leaves collapse

Storage: 32°F, 6–7 months

Solanaceous Crops

Common Name: Tomato

Botanical Name: *Lycopersicon esculentum* Mill.

Uses/Part: Fresh market, processing/fruit

Class: Tender, warm-season perennial

Max/Min Temp: 65–80°F

Soil & pH: Most well-drained mineral soils, pH 6.0–6.5

Planting: Transplanted or direct-seeded

Minor Nutrients: Calcium, Magnesium

Spacing: in row: 6–36" depending on use; between row: 36–60"

Diseases: *Fusarium,* Bacterial, and *Verticillium* Wilts, Bacterial Canker, Bacterial Speck, Bacterial Spot, Early and Late Blight, *Septoria* Leaf Spot, Anthracnose, Virus Diseases

Insects: Cutworms, Flea Beetles, Aphids, Leafhoppers, Stink Bugs, Leaf Miners, Spider Mites, Colorado Potato Beetle, Tomato Hornworm, Tomato Fruitworm

Maturity: 60–90 days

Harvesting: Fresh market: hand harvested at either fully ripe stage or mature green stage; processing: mechanically harvested fully ripe

Storage: Ripe: 46–50°F, 4–7 days; Mature green: 55–70°F, 1–3 weeks

Common Name: Pepper

Botanical Name: *Capsicum annuum* var. *annuum* L.

Uses/Part: Fresh market, processing/fruit

Class: Very tender, warm-season perennial

Max/Min Temp: 65–80°F

Soil & pH: Well-drained mineral soils, pH 6.0–6.5

Planting: Transplanted or direct-seeded

Minor Nutrients: Calcium

Spacing: in row: 12–24"; between row: 18–36"

Diseases: see tomato

Insects: see tomato

Maturity: 65–80 days

Harvesting: Mostly hand picked when green fruit is full size but firm and crisp

Storage: 45–55°F, 2–3 weeks

Common Name: Eggplant

Botanical Name: *Solanum melongena* L.

Uses/Part: Fresh market, processing/fruit

Class: Very tender, warm-season perennial

Max/Min Temp: 65–95°F

Soil & pH: Well-drained mineral soils, pH 5.5–6.5

Planting: Transplanted (some direct-seeded)

Spacing: in row: 18–30"; between row: 24–48"

Diseases: see tomato, also *Phomopsis vexans*

Insects: see tomato

Maturity: 50–80 days

Harvesting: Clipped while skin glossy but not toughened

Storage: 46–54°F, 1 week

Legumes

Common Name: Common Bean (Snap Bean)

Botanical Name: *Phaseolus vulgaris* L.

Uses/Part: Fresh market, processing/fruit

Class: Tender, warm-season annual

Max/Min Temp: 50–80°F

Soil & pH: Well-drained friable soils and muck, pH 6.0–6.8

Planting: Direct-seeded

Minor Nutrients: Zinc

Spacing: in row: 2–4"; between row: 18–38"

Diseases: Root Rots (*Fusarium, Rhizoctonia, Pythium, Thelaviopsis*), Bacterial Blights, Anthracnose, Angular Leaf Spot, *Cercospera* Leaf Spot, Powdery Mildew, Downy Mildew, Gray Mold, White Mold, Mosaic Viruses, Rust

Insects: Mexican Bean Beetle, Aphids, Bean Weevil

Maturity: 48–60 days

Harvesting: Mostly mechanically harvested with pods still smooth and crisp with little or no seed bulge

Storage: 40–45°F, 7–10 days

Common Name: Garden Pea

Botanical Name: *Pisum sativum* var. *macrocarpon*

Uses/Part: Fresh market, processing/seed

Class: Cool-season annual

Max/Min Temp: 45–75°F

Soil & pH: Well-drained mineral soils, pH 6.2–6.8

Planting: Direct-seeded

Spacing: in row: 1–3"; between row: 24–48"

Diseases: Root Rots, Viruses, *Fusarium* Wilt, Bacterial Blight, Powdery Mildew, Downy Mildew

Insects: Seed Maggot, Aphid, Pea Weevil

Maturity: 56–75 days

Harvesting: Machine or hand harvested when pods are full but tender

Storage: 32°F, 1–2 weeks

Common Name: Lima Bean

Botanical Name: *Phaseolus limensis* Macfady

Uses/Part: Fresh market, processing/seed

Class: Very tender, warm-season annual

Max/Min Temp: 50–80°F

Soil & pH: see common bean

Planting: Direct-seeded

Minor Nutrients: see common bean

Spacing: in row: 3–12"; between row: 18–48"

Diseases: see common bean

Insects: see common bean

Maturity: 65–90 days

Harvesting: Mostly mechanical when pods well-filled, seeds plump and tender with green seed coat

Storage: 37–41°F, 5–7 days

Common Name: Southern Pea

Botanical Name: *Vigna unguiculata* (L.) Walp

Uses/Part: Fresh market, processing/seed

Class: Tender, warm-season annual

Max/Min Temp: 50–95°F

Soil & pH: Slightly acidic sandy loams/sandy clay loams

Planting: Direct-seeded

Spacing: in row: 3–6"; between row: 18–42"

Diseases: see common bean

Insects: Cowpea Curculio, Corn Earworm, Stinkbugs, Grubs, Wireworms, Nematodes

Maturity: 65–85 days

Harvesting: Mostly mechanical when pods slightly yellow

Storage: 40–41°F, 6–8 days

Cucurbits

Common Name: Cucumber

Botanical Name: *Cucumis sativus* L.

Uses/Part: Fresh market, processing/fruit

Class: Very tender, warm-season annual

Max/Min Temp: 60–90°F

Soil & pH: Well-drained mineral soils, pH 6.0–6.8

Planting: Direct-seeded or transplanted

Spacing: in row: 8–12"; between row: 36–72" depending on use

Diseases: Bacterial Wilt, Angular Leaf Spot, Anthracnose, Scab, Cottony Leak, Gummy Stem Blight, Powdery Mildew, Downy Mildew, Cucumber Mosaic

Insects: Cucumber Beetle, Aphids, Pickle Worm, Nematodes, Leaf Miner, Cutworm, Flea Beetle

Maturity: 48–72 days

Harvesting: Slicers: Hand picked at 2.5" maximum diameter while firm, deep green, and minimum 6" length. Picklers: Mostly mechanical, graded by size.

Storage: 50–55°F, 10–14 days

Common Name: Squash

Botanical Name: *Cucurbita pepo* var. *melopepo* (L.) Alef

Uses/Part: Fresh market, processing/fruit

Class: Very tender, warm-season annual

Max/Min Temp: 50–90°F

Soil & pH: Many well-drained mineral soils, pH 5.5–7.5

Planting: Direct-seeded or transplanted

Spacing: in row: 24–38"; between row: 36–60"

Diseases: Powdery Mildew, *Alternaria* Leaf Spot, Several Viruses, Black Rot, also see cucumber

Insects: Aphid, Cucumber Beetle, Pickleworm, Leafhopper, Spider Mite, Stem Borers, Squash Bugs

Maturity: 40–50 days

Harvesting: Fruit hand picked at 6–8" in length

Storage: 41–50°F, 1–2 weeks

Common Name: Muskmelon

Botanical Name: *Cucumis melo* L. Reticulatus group (netted)

Botanical Name: *Cucumis melo* L. Inodorous group (honeydew)

Uses/Part: Fresh market/fruit

Class: Very tender, warm-season annual

Max/Min Temp: 60–90°F

Soil & pH: Well-drained mineral soils, pH 6.0–6.8

Planting: Direct-seeded or transplanted

Spacing: in row: 9–12"; between row: 60–84"

Diseases: see cucumber, also *Fusarium* Fruit Rot, Fusarium Wilt, *Alternaria* Leaf Spot

Insects: Striped and Spotted Cucumber Beetles, Aphid, Pickleworm, Cutworm, Wireworm, Flea Beetle, Melon Worm

Maturity: 85–95 days

Harvesting: Netted: Hand picked at quarter to full slip (when stem starts to separate from vine); Honeydew: when fruit color indicates maturity

Storage: 32–36°F @ full slip, 5–14 days

Common Name: Watermelon

Botanical Name: *Citrullus lanatus* (Thunb.) Matsum & Nakai

Uses/Part: Fresh market/fruit

Class: Very tender, warm-season annual

Max/Min Temp: 65–95°F

Soil & pH: Light well-drained mineral soils, pH 6.0–6.8

Planting: Direct-seeded or transplanted

Spacing: in row: 24–36"; between row: 72–96"

Diseases: Anthracnose, Down Mildew, *Fusarium* Wilt, Gummy Stem Blight, *Alternaria,* and *Cercospera* Leaf Spots

Insects: Aphid, Cucumber Beetle, Leaf Miner, Leafhopper, Red Spider Mite, Wireworm, Cutworm

Maturity: 75–95 days

Harvesting: Hand harvested when tendril nearest fruits dies back

Storage: 50–60°F, 2–3 weeks

Perennials

Common Name: Asparagus

Botanical Name: *Asparagus officinalis* L.

Uses/Part: Fresh market, processing/stem

Class: Cool-season perennial

Max/Min Temp: 55–85°F

Soil & pH: Deep well-drained mineral soils and muck, pH 6.0–6.8

Planting: Transplanted, direct-seeded, crowns planted

Minor Nutrients: Boron

Spacing: in row: 9–15"; between row: 48–72"

Diseases: Rust, *Fusarium* Root Rot, *Fusarium* Crown Rot, *Botrytis* Blight, *Stemphylium* Leaf Spot, Viruses

Insects: Asparagus Beetles, Tarnished Plant Bug, Alfalfa Plant Bug, Asparagus Miner, Aphids, and soil larvae

Maturity: Harvest in third year and beyond

Harvesting: Hand cut spears below the ground or snap above ground when spear is 8–10" long with tips tightly closed

Storage: 32–35°F, 2–3 weeks

Miscellaneous Crops

Common Name: Okra

Botanical Name: *Abelmoschus esculentus* (L.) Moench

Uses/Part: Fresh market, processing/fruit

Class: Very tender, warm-season annual

Max/Min Temp: 65–95°F

Soil & pH: Well-drained mineral soils and muck, pH 5.8–6.5

Planting: Direct-seeded

Spacing: in row: 8–24"; between row: 42–60"

Diseases: *Fusarium/Verticillium* Wilts, Cotton Root Rot

Insects: Nematodes, Flea Beetle, Cucumber Beetle, Aphids, Leaf Miner, Corn Earworm

Maturity: 50–60 days

Harvesting: Pods hand harvested at 3–3.5" long

Storage: 45–50°F, 7–10 days

Common Name: Sweet Corn

Botanical Name: *Zea mays* var. *rugosa* Bonaf

Uses/Part: Fresh market, processing/fruit

Class: Tender, warm-season annual

Max/Min Temp: 50–95°F

Soil & pH: Many mineral soils and mucks, pH 6.0–6.8

Planting: Direct-seeded

Minor Nutrients: Boron, Zinc, Magnesium, Manganese

Spacing: in row: 8–12"; between row: 30–42"

Diseases: Seed Rots, Seedling Diseases, Stalk Rots, Northern Corn Leaf Blight, Southern Corn Leaf Blight, Yellow Leaf Blight, Anthracnose, Bacterial Leaf Spot, Stewart's Wilt, Rust, Smut, Downy Mildew, Viruses

Insects: Corn Rootworm, Wireworm, Corn Earworm, Fall Armyworm, Stem and Stalk Borers, Spider Mite

Maturity: 64–95 days

Harvesting: Hand or machine harvested when ears are at milky stage (kernels are proper color and plump)

Storage: 32°F, 5–8 days

APPENDIX E — COMPOSITION OF EDIBLE PORTION OF FRESH RAW VEGETABLES

Vegetable	Protein (g)	Fat (g)	Carb. (g)	Fiber (g)	Ca (mg)	Fe (mg)	Na (mg)
Artichoke	2.7	0.2	11.9	1.1	48	1.6	80
Asparagus	3.1	0.2	3.7	0.8	22	0.7	2
Bean, Green	1.8	0.1	7.1	1.1	37	1.0	6
Bean, Lima	6.8	0.9	20.2	1.9	34	3.1	8
Beet, Roots	1.5	0.1	10.0	0.8	16	0.9	72
Broccoli	3.0	0.4	5.2	1.1	48	0.9	27
Brussels Sp.	3.4	0.3	9.0	1.5	42	1.4	25
Cabbage	1.2	0.2	5.4	0.8	47	0.6	18
Carrot	1.0	0.2	10.1	1.0	27	0.5	35
Cauliflower	2.0	0.2	4.9	0.9	29	0.6	15
Celery	0.7	0.1	3.6	0.7	36	0.5	88
Cucumber	0.5	0.1	2.9	0.6	14	0.3	2
Eggplant	1.1	0.1	6.3	1.0	36	0.6	4
Garlic	6.4	0.5	33.1	1.5	181	1.7	17
Lettuce Crisphead	1.0	0.2	2.1	0.5	19	0.5	9
Romaine	1.6	0.2	2.4	0.7	36	1.1	8
Melon, netted	0.9	0.3	8.4	0.4	11	0.2	9
Onion (dry)	1.2	0.3	7.3	0.4	25	0.4	2
Pea, Green	5.4	0.4	14.5	2.2	25	1.5	5
Pepper, Sweet	0.9	0.5	5.3	1.2	6	1.3	3
Potato	2.1	0.1	18.0	0.4	7	0.8	6
Radish	0.6	0.5	3.6	0.5	21	0.3	24
Southern Pea	9.0	0.8	21.8	1.8	26	1.1	4
Spinach	2.9	0.4	3.5	0.9	99	2.7	79
Squash, Summer	1.2	0.2	4.4	0.6	20	0.5	2
Sweet Corn	3.2	1.2	19.0	0.7	2	0.5	15
Sweet Potato	1.7	0.3	24.3	0.9	22	0.6	13
Tomato, Ripe	0.9	0.2	4.3	0.5	7	0.5	8
Watermelon	0.5	0.2	6.4	.	7	0.5	1

Source: Knott's Handbook for Vegetable Growers, Third Edition.

APPENDIX F — VITAMIN CONTENT OF EDIBLE PORTION OF FRESH RAW VEGETABLES

Vegetable	Vitamin A (IU)	Thiamine (mg)	Riboflavin (mg)	Niacin (mg)	Ascorbic Acid (mg)	Vitamin B6 (mg)
Artichoke	185	0.08	0.06	0.76	10.8	0.11
Asparagus	897	0.11	0.12	1.14	33.0	0.15
Bean, Green	668	0.08	0.11	0.75	16.3	0.07
Bean, Lima	303	0.22	0.10	1.47	23.4	0.20
Beet, Roots	20	0.05	0.02	0.40	11.0	0.05
Broccoli	1,542	0.07	0.12	0.64	93.2	0.16
Brussels Sp.	883	0.14	0.09	0.75	85.0	0.22
Cabbage	126	0.05	0.03	0.30	47.3	0.10
Carrot	28,129	0.10	0.06	0.93	9.3	0.15
Cauliflower	16	0.08	0.06	0.63	71.5	0.23
Celery	127	0.03	0.03	0.30	6.3	0.03
Cucumber	45	0.03	0.02	0.30	4.7	0.05
Eggplant	70	0.09	0.02	0.60	1.6	0.09
Garlic	0	0.20	0.11	0.70	31.2	.
Lettuce						
Crisphead	330	0.05	0.03	0.19	3.9	0.04
Romaine	2,600	0.10	0.10	0.50	24.0	.
Melon, Netted	3,224	0.04	0.02	0.57	42.2	0.12
Onion (dry)	0	0.06	0.01	0.10	8.4	0.16
Pea, Green	640	0.27	0.13	2.09	40.0	0.17
Pepper, Sweet	530	0.09	0.05	0.55	128.0	0.16
Potato	.	0.09	0.04	1.48	19.7	0.26
Radish	8	0.01	0.05	0.30	22.8	0.07
Southern Pea	817	0.11	0.15	1.45	2.5	0.07
Spinach	6,715	0.08	0.19	0.72	28.1	0.20
Squash, Summer	196	0.06	0.04	0.55	14.8	0.11
Sweet Corn	281	0.20	0.06	1.70	6.8	0.06
Sweet Potato	20,063	0.07	0.15	0.67	22.7	0.26
Tomato, Ripe	1,133	0.06	0.05	0.60	17.6	0.05
Watermelon	590	0.03	0.03	0.20	7.0	.

Source: Knott's Handbook for Vegetable Growers, Third Edition.

APPENDIX G — COMMON
GROWN IN THE U.S.

Common Name	Botanical Name	Plant Type	Pollination	Approximate Temperature Limits
Almonds	*Prunus amygdalus*	deciduous tree	self-incompatible	flowers: 27°F (–3°C)
Apple	*Malus domestica*	deciduous tree	mostly self-incompatible	tree: –30°F (–34°C) flowers: 28°F (–2°C)
Apricot	*Prunus armeniaca*	deciduous tree	mostly self-fertile; fruits poorly in S.E.	tree: –15°F (26°C) flowers: 28°F (–2°C)
Avocado	*Persea americana*	evergreen tree	mostly self-fertile	tree: 18–30°F (–7.8 to –1.1 °C) spring growth: 32°F (0°C)
Banana	*Musa* spp.	evergreen "tree" (perennial herb)	self-fruitful, parthenocarpic	"tree": 28°F (–2.2°C)
Blackberry	*Rubus* spp.	deciduous bush or vine	mostly self-fertile	bush: –10°F (–23°C) flowers: 28°F (–2.2°C)
Blueberry, lowbush	*Vaccinium myrtilloides*, etc.	deciduous bush	self-fertile	bush: — flowers: 30° (–1°C)
Blueberry, N. highbush	*Vaccinium corymbosum*	deciduous bush	self-fertile	bush: –20°F (–29°C) flowers: 30°F (–1°C)
Blueberry, rabbiteye	*Vaccinium ashei*	deciduous bush	self-incompatible	bush: –20°F (–29°C) flowers: 30°F (–1°C)
Cherry, sour	*Prunus cerasus*	deciduous tree	self-fertile	tree: –30°F (–34°C) flowers: 28°F (–2.2°C)
Cherry, sweet	*Prunus avium*	deciduous tree	self-incompatible	tree: –15°F (–26°C) flowers: 28°F (–2.2°C)
Cranberry	*Vaccinium macrocarpon*	creeping, evergreen shrub	self-fertile	plant: –20°F (–29°C)
Currant, black and red	*Ribes nigrum and sativum*	deciduous bush	self-fertile	bush: –4°F (–20°C) flowers: 30°F (–1°C)
Fig	*Ficus carica*	deciduous tree	common type is self-fertile	tree: 10°F (–12.2°C)

FRUIT AND NUT CROPS
AND CANADA

Typical Spacing (in row × between rows)	Typical Rootstocks	Typical Pruning	Some Insects and Mites	Some Diseases and Nematodes
19.6 × 19.6 ft. (6 × 6 m)	Nemaguard peach seedlings	open center	mites, navel orange worm	blossom blight, bacterial blast
depends on scion and rootstock	M9–Dwarf, mm106–Semi-dwarf, seedlings	central leader	codling moth, apple maggot	scab, bitter rot
18 × 18 ft. (5.5 × 5.5 m)	peach seedlings	open center	plum curculio	brown rot
26 × 26 ft. (8 × 8 m)	avocado seedings	minimal central leader	scale insects, mites	root rot, scab
10 × 10 ft. (3 × 3 m)	own roots	old stalk removed after fruiting	root borers	wilt, leaf spot
3 × 10 ft. (bush type) (0.9 × 3.1 m)	own roots	cane renewal	strawberry clipper, mites	leaf spots, anthracnose
Solid mat of plants	own roots	biennial cane renewal	fruit fly	mummy berry
4 × 10 ft. (1.2 × 3.1 m)	own roots	cane renewal	fruit fly, bud mite	mummy berry, flower blight
6 × 12 ft. (1.8 × 3.7 m)	own roots	cane renewal	fruit fly, caterpillars	mummy berry, flower blight
19.6 × 19.6 ft. (6 × 6 m)	mahaleb cherry seedlings	open center	fruit fly, plum curculio	brown rot, crown gall
19.6 × 19.6 ft. (6 × 6 m)	mazzard cherry seedlings	central leader or open center	fruit fly, plum curculio	brown rot, crown gall
1 × 1 ft. (0.3 × 0.3 m)	own roots	thinning of excess runners	root worms, fire worms	viruses, twig blight
1/black–3/red × 8 ft. (3/black–1/red × 2.5 m)	own roots	cane renewal	mites	*Botrytis* sp., leaf spot
19.6 × 19.6 ft. (6 × 6 m)	own roots	open center or cane renewal	fruit beetles	rust, root knot nematodes

(Continued)

Appendix G

Common Name	Botanical Name	Plant Type	Pollination	Approximate Temperature Limits
Filbert (hazelnut)	*Corylus spp.*	deciduous tree	mostly self-incompatible	tree: −10°F (−23.3°C) flowers: 18°F (−8°C)
Gooseberry	*Ribes grossularia*	deciduous bush	self-fertile	bush: — flowers: —
Grape, American	*Vitis labrusca*	deciduous vine	self-sterile	vine: −10°F (−24°C)
Grape, European	*Vitis vinifera*	deciduous vine	self-fertile	vines: 10°F (−12°C) bloom: 31°F (−0.5°C)
Grape, muscadine	*Vitis rotundifolia*	deciduous vine	self-fertile or female	vines: 10°F (−12°C) bloom: 31°F (−0.5°C)
Grapefruit	*Citrus x paradisi*	evergreen tree	self-fertile	tree: 23°F (−5°C)
Kiwifruit	*Actinidia deliciosa*	deciduous vine	male and female vines	vines: 18°F (−8°C) spring growth: 28°F (−2°C)
Lemon	*Citrus limon*	evergreen tree	self-fertile	tree: 26°F (−3.3°C) fruit: 29°F (−1.7°C)
Lime	*Citrus aurantifolia*	evergreen tree	self-fertile	tree: 28°F (−2.2°C)
Mango	*Mangifera indica*	evergreen tree	self-fertile	tree: 30°F (−1.1°C) fruit: 40°F (4.4°C)
Mayhaw	*Crataegus opaca* or *aestivalis*	deciduous tree	self-fertile	tree: −20°F (−28.9°C) flowers: 26°F (−3.3°C)
Olive	*Olea europaea*	evergreen tree	self-fertile	tree: 12°F (11°C) green fruit: 28°F (−2.2°C)
Orange, sweet	*Citrus sinensis*	evergreen tree	self-fertile	tree: 24°F (−4.4°C) fruit: 28°F (−2.2°C)
Papaya	*Carica papaya*	evergreen tree	some self-fertile; others male and female trees	tree: 28°F (−2.2°C)
Peach and nectarine	*Prunus persica*	deciduous tree	mostly self-fertile	tree: −12°F (−24°C) flower: 28°F (−2.2°C)

(Continued)

Typical Spacing (in row × between rows)	Typical Rootstocks	Typical Pruning	Some Insects and Mites	Some Diseases and Nematodes
15 × 15 ft. (4.5 × 4.5 m)	own roots or filbert seedlings	open center or central leader	mites	—
2 × 10 ft. (0.5 × 3 m)	own roots	cane renewal	mites, scale insects	*Botrytis* sp., leaf spot
10 × 10 ft. (3 × 3 m)	own roots	cane pruned	grape root borer, grape berry moth	black rot
6 × 10 ft. (2 × 3 m)	S04, 1202	spur pruning	phylloxera, catepillars	powdery mildew, anthracnose
20 × 12 ft. (6.1 × 3.7 m)	own roots	spur pruning	grape root borer	ripe rot, bitter rot
22 × 22 ft. (6.7 × 6.7 m)	citrange seedlings, etc.	central leader, hedging later	scale insects, fruit fly	root rot, scab
16 × 16 ft. (5 × 5 m)	kiwifruit seedlings	special cane pruning	scale insects, caterpillars	root rot, *Botrytis* sp.
17 × 17 ft. (5.2 × 5.2 m)	rough lemon seedlings, etc.	central leader, hedging later	scale insects	root rot, scab
17 × 17 ft. (5.2 × 5.2 m)	rough lemon seedlings, etc.	central leader, hedging later	scale insects	root rot, wither tip
30 × 30 ft. (9.2 × 9.2 m)	mango seedlings	little pruning required	mites, scale insects	anthracnose, scab
15 × 20 ft. (4.6 × 6.1 m)	mayhaw seedlings	open center or central leader	plum curculio	rust
30 × 30 ft. (9.2 × 9.2 m)	olive seedlings or own roots	scaffolds selected, then limited pruning	scale insects	olive knot, leaf spot
17 × 17 ft. (5.2 × 5.2 m)	citrange seedlings, etc.	central leader, hedging later	scale insects	root rot, melanose
10 × 10 ft. (3 × 3 m)	own roots	none	fruit fly	viruses, root knot nematodes
19.6 × 19.6 ft. (6 × 6 m)	peach seedlings from Nema-guard, Lovell	open center	plum cucurlio, stink bugs	brown rot, leaf curl

(Continued)

Appendix G

Common Name	Botanical Name	Plant Type	Pollination	Approximate Temperature Limits
Pear, Asian	*Pyrus pyrifolia*	deciduous tree	mostly self-incompatible	tree: — flower: 29°F (−1.7°C)
Pear, European	*Pyrus communis*	deciduous tree	mostly self-incompatible	tree: −20°F (28.9°C) flowers: 28°F (−2.2°C)
Pecan	*Carya illinoinensis*	deciduous tree	mostly self-fertile	tree: −10°F (−23.3°C) spring growth: 28°F (2.2°C)
Persimmon, Oriental	*Diospyros kaki*	deciduous tree	some self-fertile, most parthenocarpic	tree: 10°F (−12.2°C) flowers: 31°F (−0.6°C)
Pineapple	*Ananas comosus*	perennial herb	self-incompatible, but parthenocarpic	plant: 32°F (0°C)
Pistachio	*Pistacia vera*	deciduous tree	separate male and female trees	tree: −15°F (−26°C)
Plum, European	*Prunus domestica*	deciduous tree	mostly self-incompatible	tree: −20°F (−29°C) flowers: 28°F (−2.2°C)
Plum, Japanese	*Prunus salicina*	deciduous tree	mostly self-incompatible	tree: −15°F (−26°C) flowers: 28°F (−2.2°C)
Pomegranate	*Punica granatum*	deciduous bush or small tree	self-fertile	tree: 10°F (−12°C)
Raspberry	*Rubus* spp.	deciduous bush	self-fertile	bush: −4°F (−20°C) flowers: 28°F (−2.2°C)
Strawberry	*Fragaria* spp.	small, evergreen perennial	mostly self-fertile	plant: −4°F (−20°C) flowers: 28°F (−2.2°C)
Walnut, English	*Juglans regia*	deciduous tree	mostly self-fertile, but two cultivars best	tree: 0°F (−18°C) spring growth: 30°F (1°C)
Quince	*Cydonia oblonga*	deciduous tree	self-fertile	tree: — flowers: —

(Continued)

Typical Spacing (in row × between rows)	Typical Rootstocks	Typical Pruning	Some Insects and Mites	Some Diseases and Nematodes
13 × 19.6 ft. (4 × 6 m)	*pyrus betulaefolia* seedlings	open center	stink bugs, scale insects	fire blight, Bot canker
13 × 19.6 ft. (4 × 6 m)	European pear seedlings	open center	scale insects, pear psylla	fire blight, fly speck
40 × 40 ft. (12.2 × 12.2 m)	pecan seedlings	central leader	weevil, nut case bearer	scab, downy spot
10 × 20 ft. (3 × 6 m)	American or Oriental persimmon seedlings	central leader	scale insects, trunk borers	leaf spot, fly speck
1 × 2 ft. (0.3 × 0.6 m) on beds—3 ft. (0.9 m) apart	own roots	none	mealy bug	root knot nematode, root rot
25 × 25 ft. (7.6 × 7.6 m)	other pistachio species	central leader or open center	—	—
19.6 × 19.6 ft. (6 × 6 m)	myrobalan plum	open center	plum curculio	brown rot
19.6 × 19.6 ft. (6 × 6 m)	usually peach seedlings	open center	plum curculio	brown rot, bacterial spot
10 × 15 ft. (3 × 4.6 m)	own roots	cane renewal	fruit beetles	leaf spot
2 × 10 ft. (0.6 × 3 m)	own roots	cane renewal	fruit beetles	leaf spot, *Botrytis* sp., fruit rot
annual system–1 × 1 ft. (0.3 × 0.3 m) on beds— 4.5 ft. (1.5 m) apart	own roots	none	mites, stink bugs	leaf spot, *Botrytis* sp., fruit rot
29.5 × 29.5 ft. (9 × 9 m)	walnut seedlings	open center	fruit worms	bacterial blight
15 × 20 ft. (4.6 × 6.1 m)	quince seedlings	open center, limited pruning needed	—	leaf spot, fire blight

APPENDIX H — WORLD WIDE WEB SITES FOR HORTICULTURE

American Horticultural Society — www.ahs.org/
American Nursery & Landscape Association — www.anla.org/
American Rose Society — www.ars.org/
The American Society for Horticultural Science — www.ashs.org/
Associated Landscape Contractors of America — www.alca.org/
Case Corporation — www.casecorp.com
Commercial Rose Growers Association — www.rosesinc.org/
Education and Work — www.balancenet.org
Floral Design Institute — www.floraldesigninstitute.com
Garden Club of America — www.gcamerica.org/
Garden Gate to the Web — www.prairienet.org/ag/garden
Garden Gate Magazine — www.augusthome.com/gardeng.htm
The Grow Zone — www.plants.org
Horticulture in Virtual Perspective, Ohio State University —
www.hcs.ohio-state.edu/hcs/hcs.html
Horticulture Net — www.garden.org/links/links.html
Husqvarna — www2.husqvarna.com/husqvarna/
Hydroponics Resource Center — www.mayhillpress.com/
International Fruit Variety Database — www.fruitdb.com/
The Irrigation Association — www.irrigation.org
National FFA Organization — www.ffa.org/
National Gardening Web Site — www.usda.gov/news/garden.htm
National Pecan Shellers Association (NPSA) — www.ilovepecans.org/
Natural Resources Conservation Service — www.nhq.nrcs.usda.gov/
Rain Bird Irrigation — www.rainbird.com
Society of Municipal Arborists — www.urban-forestry.com/
Soil and Water Conservation Society — www.swcs.org
The Southern Nurserymen's Association — www.sna.org/
Time Inc.'s Garden Site, The Virtual Garden — pathfinder.com/vg/
Toro Equipment — www.toro.com
Underwriters Laboratories, Inc. — www.ul.com/
U.S. Department of Agriculture — www.usda.gov/
The U.S. Netherlands Flower Bulb Information Center — www.bulb.com
Virtual Garden — www.vg.com/
W. Atler Burpee & Company — garden.burpee.com
Woerner Turf — www.woerner.com/

GLOSSARY

2-cycle engine—an internal combustion engine which completes the steps of intake, compression, power, and exhaust in two strokes with lubrication coming from oil mixed with gasoline.

4-cycle engines—an internal combustion engine with four strokes (intake, compression, power, and exhaust) which has a crankcase that holds oil for engine lubrication.

Abscisic acid (ABA)—the only natural growth inhibitor among plant growth regulators.

Absorption—the process by which a substance is taken into and included in another.

Acaricide—chemicals used to control ticks and spiders.

Accent planting—an area of particular beauty or interest established in a landscape.

Acclimated—accustomed.

Acute plant problems—those occurring suddenly causing immediate damage.

Achenes—one-seeded and attached to the ovary wall only.

Acid soil—a soil with a pH of less than 7.0.

Action threshold—the predetermined level at which pest control is needed.

Active ingredient—the percent nutrient that is being applied.

Acute—occurring suddenly.

Adjustable-jaw wrench—one that has a moveable jaw so the wrench can be changed to fit various sizes of bolts and nuts.

Adsorption—the increased concentration of molecules or ions at a surface, including exchangeable cations and anions on soil particles.

Advective freezes—freezes accompanied by wind.

Adventitious roots—roots beginning from a stem or a leaf.

Advertising—any message communicated to potential customers about a product or service that uses mass media.

Aeration—the exchange of air in soil with air from the atmosphere.

Aerification—the selective tillage of a turfgrass area by mechanical means.

Aerifying machines—provide temporary improvement of compacted soil conditions and reduce thatch accumulation.

Aeroponic system—involves plant roots suspended in air with a fine mist of oxygen-rich nutrient solution sprayed on them at regular intervals.

Agar—a sugar-based gel derived from certain algae.

Aggregrate fruit—comprised of a single receptacle (base of flower) with masses of similar fruitlets, such as blackberries and strawberries.

Agriculture—the production of plants and animals to meet basic human needs.

Agrostis palustris—creeping bentgrass.

Air layering—involves removing a portion of the bark on the stem of a plant and placing moist, unmilled sphagnum moss over the exposed area.

Air pollution—when harmful or degrading materials get into the air.

Algae—a group of small, primitive, filamentous, green plants that manufacture their own food and are found under very wet conditions.

Alternate leaf arrangement—when leaves and buds are alternated or staggered along the stem.

Analogous colors—hues next to each other on the color wheel.

Anchor—the medium which provides the place for the roots of a plant to grow.

Anchor support posts—sideposts providing the main structural support for a greenhouse which are spaced at regular intervals and set in concrete footings.

Anion—an ion carrying a negative charge of electricity.

Annual—a plant that germinates from seed, grows to maturity, flowers, and produces seed in one growing season.

Anti-gibberellin—a synthetic growth regulator which counteracts the effect of naturally occurring gibberellins in plant tissue.

Anti-transpirant—seals the leaf stomata and helps prevent leaf scorch and leaf burn.

Apical dominance—tip development with suppressed development of lateral buds.

Apical meristem—the primary growing point of the stem.

Arboretum—a collection of trees arranged in a naturalized fashion.

Arboriculture—the study of trees, their growth, and culture.

Armyworm—the larva of a moth that is greenish in color and has black stripes along each side and down the back.

Asexual propagation—the reproduction of new plants from the stems, leaves, or roots of a parent plant.

Aspirated thermostat—one that has small electric fans that provide constant air movement over the sensing devices.

Asymmetrical balance—a floral arrangement which, when divided by the center axis, has two halves not equal in size or shape.

Atmospheric environment—the above ground part of the environment of a terrestrial plant.

Atrimmec (Atrinal)—a plant growth regulator used after pruning or shearing to maintain the plant in the desired shape through the growing season or to suppress flowering or fruit development of certain plants.

Atrium—a special room in a building for growing plants.

Attached even-span—rafters of equal length and the end wall is attached to an existing building.

Attached greenhouse—one connected to a building.

Auxin—a growth promoting plant hormone.

Available water—the amount of retained water in soil that plant roots can absorb.

Avicide—chemicals used to control birds.

Axillary bud—a bud which will produce a new leaf or stem which is located along the side of a plant stem.

Axonopus affinis—common carpetgrass.

Azalea pot—a container slightly shorter than but with the same width as a standard pot.

B-Nine—a synthetic abscissic acid (ABA) growth inhibitor.

Backflow valve—prevents water from siphoning back and contaminating the source.

Back lapping—a method of sharpening a reel mower by spinning the reel blade backwards while applying a gritty solution to the reel blade which sharpens both the reel and bed knife blades as they come together.

Backpack blower—uses air speed to move dust, dirt, and grass clippings from walkways and separates clumps of grass on a lawn.

Bacteria—one-celled organisms that have a primitive nucleus.

Bactericide—a chemical used to control bacteria which is also known as a germicide.

Bag culture—utilizes plastic bags that are filled with substrate, such as rockwool, peatlite, or sawdust.

Balance—implied equilibrium whether symmetrical or asymmetrical; the physical or visual stability of a floral design.

Balance sheet—a financial statement that shows assets (what a business owns), liabilities (what a business owes), and net worth of a business (assets less liabilities) for a specific date.

Balled and burlapped (B&B)—trees grown in a field, dug keeping a ball of soil surrounding the root system, and covered with burlap material.

Ball peen hammer—a hammer with one end of the head ball shaped, used for striking metal objects.

Bar caps—structural members attached to the outside of a greenhouse to hold the glass in place.

Bare root (BR)—describes trees grown in a nursery or field and dug without taking soil.

Bark graft—a grafting method usually done in the early spring used to join smaller scion wood to larger diameter understock.

Bark-based mixes—a formulation of soilless media composed of aged pine bark and vermiculite or perlite.

Barrel vault—Quonset style structure with sidewalls joined together.

Base temperature—the lowest or minimum temperature which is used when calculating degree days.

Bed knife blade—the fixed blade on a reel mower that cuts when it comes in contact with the spiral reel blade.

Berries—a fleshy ovary wall and two or more carpels containing seeds, such as tomatoes, grapes, blueberries, pumpkins, and oranges.

Best management practices (BMPs)—those practices that combine scientific research with practical knowledge to optimize yields and increase crop quality while maintaining environmental integrity, includes management of surface and subsurface water runoff, erosion control, cultural control of pests, soil testing, timing and placement of fertilizers, use of controlled release fertilizers, irrigation management, biological control of pests, pesticide selection, and correct pesticide use.

Best stand—optimum population.

Biennial—a plant that completes its life cycle in two growing seasons.

Biological pest control—using living organisms that are predators to control pests.

Biomass—the amount of matter of biological origin in a given area.

Biotechnology—the management of biological systems for the benefit of humanity.

Biotic environment—the use and culture of plants by people, including the cultural practices used to help plants grow.

Blade—the flat structure of a leaf that is used to capture the greatest amount of light.

Blend—(See turfgrass blend.)

Bluegrass—the name given to a species of grasses because of the characteristic bluish color of a field when the grass plants flower and produce seed.

Bolting—occurs when a cool season plant has been exposed to cooler temperatures and is then exposed to high temperatures causing premature flowering.

Bonsai—the art of dwarfing and shaping trees and shrubs in shallow containers by pruning and controlled fertilization.

Botanical garden—a plant collection habitat.

Botanical nomenclature—the scientific classification system which identifies kingdom, division or phylum, class, order, family, genus, species, and variety or cultivar.

Botanist—a scientist who studies plants.

Botany—the study of plants, including life cycle, structure, growth, and classification.

Brake fluid—a special fluid designed to be used in hydraulic brake systems.

Branch collar—the place where bark and wood of a scaffold and trunk come together, allowing sugar to move from the trunk into the branch.

Broadleaf—plants that have flattened leaf blades.

Broadleaf evergreen—a plant that does not have needle-like leaves but holds its leaves throughout the winter.

Buchloe dactyloides—buffalograss.

Budding—a reproduction method done in the spring or fall which joins the single bred scion with a small portion of bark or wood attached with the understock to form one new plant.

Budding plant—an herbaceous plant preseeded and growing in a plastic container, peat pot, or peat pellet.

Bud scale—tiny leaf-like structures that cover the bud before it opens and begins to grow.

Bulb pan—a container half as high as its width.

Bulbs—shortened underground stems that are enclosed with fleshy leaves.

Bush planting—when an untrained person tries to landscape without a knowledge of plant materials.

Business plan—a written document that guides the operation of a business.

Callus—an undifferentiated mass of cells.

Callus tissue—white tissue that forms over the wounded area of plant.

Calyx—the outer, usually green, leaf-like parts of a flower.

Cambium—the thin layer of plant tissue between the xylem and phloem where new plant cells are formed.

Cane-prune—a method of pruning vines that are cut back to a permanent trunk with several short spurs left near the top of the trunk to provide a place for renewal growth for the following year.

Canes—individual stems developed at the crown or growing point of herbaceous plants.

Canopy—the outline of the tree top established by scaffold branches.

Capillary mat system—a form of sub-irrigation in which potted plants are set on a moist, porous synthetic mat and water moves upward through the drain holes into the growing medium by wick action.

Carbon dioxide fertilization—injecting greenhouse air with carbon dioxide via special CO_2 generators.

Career—the general direction of a person's life as related to work in the field of horticulture.

Career goal—the level of accomplishment you want to make in your work.

Carpel—one of the units composing a pistil or ovary.

Cash-flow budget—predicted cash-flow through the business for a set period.

Cash-flow statement—shows the sources of money into and out of a business for a specific period.

Cation—an ion carrying a positive charge of electricity.

Cation exchange capacity—the total quantity of cations which a soil can adsorb by cation exchange.

Central axis—an imaginary central line running through the center of a floral arrangement.

Chain saw—a portable power saw with a chain that has cutting teeth.

Chemical pest control—using pesticides to control pests.

Chemical pinching agents—those growth regulators that promote branching by either causing death to the terminal bud or temporarily stopping shoot elongation.

Chilling injury—damage to tropical plants in the 35 to 45°F temperature range.

Chlorflurenol (Maintain CF 125)—a foliar-absorbed growth regulator often used in combination with maleic hydrazide (MH) to reduce turfgrass growth.

Chlorophyll—a green pigment, contained in the chloroplast of plant cells, that must be present for photosynthesis to take place.

Chloroplast—a specialized, subcellular structure in green plants that contains chlorophyll pigments.

Chlorosis—yellowin g of green portions of a plant, particularly the leaves.

Chronic plant problems—those that develop over a long period of time.

Claw hammer—a hammer with one end of the head forked and curved, used for pulling nails.

Cleft graft—a grafting method usually done in late winter used to join small scion parts to larger diameter understock.

Clone—genetically identical to the parent.

Clutch method—a technique for wiring flowers or florets in a floral arrangement by wrapping wire around a cluster of stems.

Coastal wetland—includes tidal salt marshes, tidal freshwater marshes, and mangrove wetlands.

Coldframe—an outside propagation structure consisting of a wooden or concrete block frame with heat supplied by solar radiation through a glass or other transparent covering.

Colloid—soil particles (inorganic or organic) having small diameters ranging from 0.2 to 0.005 micron.

Combination wrench—one with an open-end wrench on one end and the same size box-end wrench on the other end.

Common wire—also called the neutral wire, connects all electrical components in a circuit together.

Complementary colors—a primary and secondary hue directly opposite each other on the color wheel.

Complete fertilizer—contains all three of the primary fertilizer nutrients (nitrogen, phosphate, and potash).

Complete flower—a flower with all four principal parts (sepals, petals, stamens, and pistil).

Complete metamorphosis—four distinct changes in an insect life cycle (egg, larvae, pupae, and adult).

Compound leaf—one that has two or more leaflets.

Coniferous—plants that bear their seeds in cones, such as pines, spruces, and cedars.

Connected greenhouses—several greenhouses joined together.

Conservation tillage—involves tillage practices that leave crop residues on top of the soil to prevent soil erosion; minimum tillage, no-till, or reduced tillage that leaves 30 percent or more of the soil surface covered with crop residue after planting.

Consumer—a person who uses goods and services.

Container—provides protection for the plant's root system and facilitates transplanting and survival.

Container bed—one that stands alone and only contains plants in containers.

Container grown—describes plants grown and sold in a container filled with a special soilless medium.

Container nursery—a business that specializes in growing plants, shrubs, and ornamental trees in containers.

Continuous flow system—involves using shallow pools with panels containing plants floating on the surface.

Controlled atmosphere storage—storage in a cool room where the gas mixtures are carefully regulated.

Controlled release fertilizer—fertilizer in which one or more of the nutrients have limited solubility so they become available to the growing plant over an extended period.

Controlling—assessing goals and objectives to see if the business is making progress.

Cool season turfgrass—those that grow best at an optimum temperature range of 60– 75°F.

Cordless drill—one that is portable and contains a rechargeable battery.

Core aerator—an aerifying machine that has hollow metal tubes, called tines, which punch holes in the soil and remove a core which is deposited on the turf's surface.

Corms—globe-shaped, fleshy underground stems.

Corner planting—plants placed at the corners-of the house in landscape beds.

Corporation—a way for people to do business by creating an artificial entity.

Corsage—a small bouquet for a woman to wear.

Cost—the amount of money spent by the business in producing the product or providing the service.

Cotyledons—seed leaves attached to a seed embryo.

Cross-benching arrangement—runs the width usually with aisles along the sidewalls.

Crotch—the junction of a scaffold branch and the tree trunk.

Cultivar—a plant variety that is cultivated and retains its features when reproduced.

Cultural pest control—using management techniques to control pests.

Curator—the person responsible for planning, obtaining, placing, and labeling all plant material throughout the arboretum, botanical, or horticultural garden.

Cuttings—a method of asexual propagation using detached portions of the plant that form missing parts to grow into complete new plants.

Cynodon bradleyi—Bradley bermudagrass.

Cynodon dactylon—common bermudagrass.

Cycocel—a synthetic abscissic acid (ABA) growth inhibitor.

Cytokinin—an organic compound that affects plants and plant tissues, including cell enlargement and division, tissue differentiation, dormancy, and retardation of leaf senescence.

Damping-off—a fungal disease which causes the stems to rot at the soil line.

Day-neutral plant—one not affected by the length of day or night.

Day-night clock—turns lights or other controllers on and off once during the day.

Dead heading—removing dead or dying flowers from annuals so that the plant will continue to live and bloom for a longer period of time.

Dead zone—that area on an evergreen 6 to 12 inches below the green needles.

Deciduous—plants that lose their leaves during a portion of the year, usually winter.

Design elements—the physical characteristics of the plant materials that a designer uses in a floral design.

Dexron II hydraulic fluid—used in transmissions (except those manufactured by Ford), hydraulic motors, lift cylinders, and power steering systems.

Dicot—a plant with two seed leaves.

Dicotyledoneae—plants characterized by two cotyledons (seed leaves) in their seedling stage, usually by flower parts in fours or fives or multiples of these numbers, and reticulate leaf venation. (See dicot.)

DIF—the mathematical difference between daytime and nighttime temperatures.

Diffusion—the movement of gases through air-filled pores from regions of higher concentrations to lower concentrations.

Directing—the process of leading and guiding employees to achieve the objectives of the business.

Direct mail—printed ad materials sent directly to consumers.

Direct markup pricing—using the actual cost per unit multiplied times a constant factor to establish the selling price.

Direct seeding—seeds planted directly into the soil outdoors.

Disbudding—removal of all lateral flower buds and only the terminal bud is allowed to develop.

Display—an exhibit of merchandise—product or service.

Division—a method of reproduction in which parts of a plant are cut into sections that will grow into new plants naturally.

DNA (deoxyribonucleic acid)—forms the basic material in the chromosomes of the cell nucleus.

Dolomitic limestone—a preplant amendment used to provide calcium and magnesium for plant growth, as well as neutralizing acidity.

Dominance—in floral design means that one design element or characteristic is more prevalent or noticeable and other elements are subordinate to the main feature.

Dormancy—the phase in the life cycle of a plant when growth is slowed or inactive.

Down lighting—positions a lamp in a tree causing the light to shine down casting an interesting shadow pattern on the ground.

Drainage—the removal of surplus surface or ground water or the manner in which water is removed.

Drip irrigation—irrigation water supplied through a thin plastic tube at a low flow rate so that the soil only in the plant's immediate vicinity is moistened; trickle irrigation.

Dripline—the outer edge of a tree branch system that creates an imaginary circle on the ground.

Drought tolerance—the ability of a plant to live and grow with low amounts of moisture.

Drupes—single carpels with three layers and a single seed, such as cherries, peaches, and olives.

Dry fruit—one that consists of seeds enclosed in a fruit wall that is hard and brittle when mature, such as peas, sunflowers, and acorns.

Economic or aesthetic injury level—the point at which plant losses due to the pests are equal to the cost of control.

Economic system—how people go about doing business.

Ecosystem—the entire functioning system of life and its environment and geographic factors, including the biotic and ambiotic environments.

Eco-dormancy—a state in which the plant is not growing and will not grow until external conditions are satisfied, usually warm temperatures; dormancy.

Edaphic environment—the soil and area where plant roots are located.

Edaphology—deals with the influence of soil and media on the growth of plants.

Electromechanical controllers—use electric clocks to mechanically turn switches on and off.

Elemental fertilizer—provides only one plant nutrient.

Element of effect—artistic features in a landscape that create moods or feelings in people.

Emitters—openings through which water flows in an irrigation system.

Endo-dormancy—dormancy of seeds and bulbs imposed by internal physiological blocks, which are removed by winter chilling; rest.

Endosperm—specialized tissue in a seed which contains stored food used by the plant in its first stage of growth and development.

Entrepreneurship—creating goods and services to meet the unique demands of consumers.

Environment—all of the factors that affect the life of living organisms.

Enzyme—a complex protein molecule that stimulates or speeds up various chemical reactions without being used up itself.

Eragrostis curvula—weeping lovegrass.

Eremochloa ophiuroides—centipedegrass.

Erosion—the wearing away of the land surface by detachment and transport of the soil through the action of moving water, wind, etc.

Espalier—a form of pruning in which plants are trained against a wall or fence.

Ethephon—a chemical growth regulator used to initiate flowers on certain plants.

Ethylene—a natural plant hormone that is a water-soluble gas produced in ripening fruits, senescent flowers, plant meristems, and at sites where plant or fruit injury occurs.

Eutrophication—when lakes or streams have too many nutrients in their water due to excess fertilizer nutrient runoff.

Evaporation—the changing from a liquid to a gaseous state.

Evergreen—a plant that holds its leaves all during the year.

Everlasting flowers—describe naturally dried flowers and weeds.

Explants—pieces of plants used to grow new plants in tissue culture.

Fan and pad cooling system—a system where large exhaust fans draw air through a moistened cellulose pad mounted on the opposite end of the structure.

Fertigation—the application of fertilizer through an irrigation system.

Fertilization—occurs when a sperm nucleus fuses with an egg cell nucleus, forming a zygote that will become a seed.

Fertilizer—any material used to provide the nutrients plants need.

Fertilizer analysis—the composition of the active ingredients in the formulation.

Fertilizer proportioner or injector—devices that are used to introduce and meter the concentration of soluble liquid fertilizer into an irrigation system.

Fertilizer ratio—the least numerical representation of the fertilizer.

Festuca arundinacea—tall fescue.

Festuca rubra—fine fescue.

Fibrous root system—a root system made up of a number of small primary and secondary roots spread out through the soil.

Field moisture capacity—when the water content of the soil fills the small pore spaces; expressed as moisture percentage.

Field nursery—grows nursery crops to marketable size in fields.

Filler flowers—small flowers that are used to add texture, color, and depth as well as fill space between the mass and line flowers in a floral design..

Filler ingredient—the carrier that allows deposition of a fertilizer; inert ingredient.

Fixed costs—the general operating expenses of running a business.

Fixed-jaw wrench—a wrench made for one size of fastener.

Flail mower—one with 25 to 50 free-swinging flail blades that turn on a horizontal rotor completely enclosed by the rotor shield.

Fleshy fruit—those comprised of a soft and fibrous material with seed or seeds enclosed, such as peach, tomato, and watermelon.

Floral design—the art of organizing the design elements inherent in plant materials, container, and accessories according to the principles of design.

Floral foam—a porous material designed to hold water and provide stability for stems within a floral design.

Floral garland—similar to a wreath but not made into a circular form.

Floral tape—a floral stem wrap made of paraffin-coated paper that is used to cover wires and stems in an unobtrusive way.

Floriculture—the production, transportation, and use of flower and foliage plants.

Florist shears—a cutting tool with short, serrated blades for cutting thick or woody stems.

Flower—the showy, colorful part of a plant that contains the reproductive organs, which then produce the fruit and the seeds of the plant.

Flower bed—one that stands alone and only contains flowers, no shrubs.

Flower border—flowers planted in front of landscape shrubs.

Flower preservatives—chemicals added to water which increase the vase life of fresh cut flowers.

Fluoridation—the addition of low concentration of fluoride into a municipal water system.

Flurprimidol (Cutless)—a growth regulator absorbed by foliage, stems, and roots which reduces vegetative growth of both cool and warm season turfgrass but is not effective on seedhead reduction.

Focalization—the creation of focal points of interest.

Focal point—the location of the center of interest; usually centered in the lower part of the floral arrangement just above the container rim.

Fog evaporative cooling system—fog is generated inside and as the minute fog droplets evaporate, heat is absorbed.

Foliage plant—one grown and sold for its beautiful colored leaves and stem.

Foliage texture—the effect created by the combination of size, light, and shadow patterns.

Foliar analysis—a leaf tissue test used to diagnose nutrient deficiency symptoms in plants.

Food guide pyramid—USDA recommended food chart guide to daily food choices.

Foot candle (fc)—a unit for measuring illumination equal to the amount of light cast by one candle on a square foot of surface.

Footprinting—describes grass leaves that do not spring-up after walking across the turf and indicates that the turf needs water.

Forced-air heaters—localized heater units that force hot air directly into a duct system.

Forcing—describes the practices that get bulbs to grow and produce flowers.

Form—the three-dimensional shape of the outline of a floral design.

Form flowers—flowers with distinctive shapes which create a striking center of interest and add uniqueness to a design.

Foundation planting—the planting along the walls or foundations of buildings.

Free enterprise—an economic system that allows people to do business with a minimum of government interference.

Freestanding even-span—separate from other buildings with rafters of equal length.

Freestanding greenhouse—one separate from other buildings or greenhouses.

Fresh market—grown and sold in the market without any processing other than washing and cleaning.

Frost-free period—the number of days from the last spring frost to the first autumn frost.

Fruit—a mature ovary of a flowering plant.

Fruit spur—a compressed branch on which the flowers of a fruit tree develop into fruit.

Functional turf quality—how well a turf achieves its purpose.

Fungi—nucleated, usually filamentous, spore-bearing organisms having no chlorophyll.

Fungicide—a chemical used to control diseases caused by fungi.

Furrow—a shallow ditch in a field between two rows of plants.

Garden center—the retail sector of the nursery industry.

Gauge—a numerical representation which describes a wire's thickness or thinness.

Genetic engineering—utilizing biotechnology by gene transfer or genetic manipulation to select and move genetic material from one plant to another.

Genetic pest control—utilizing biotechnology by gene transfer or genetic manipulation to make plants resistant to specific pests.

Genotype—the genetic makeup of an organism; allele composition.

Geotropism—plant growth in response in response to gravitational forces.

Germination—the resumption of growth by a seed embryo; occurs when the embryonic root emerges from the seed coat.

GFCI device—(See ground fault circuit interrupter.)

Gibberella fujikuroi—a fungus from which gibberellic acid is obtained for commercial use.

Gibberellin (GA)—a water-soluble chemical produced by the seed embryo which stimulates production of enzymes that break down food reserves in the endosperm.

Goal setting—describing what we want to achieve in life.

Gothic arch—styled in the shape of a pointed arch.

Grafting—the process of connecting two plants or plant parts together in such a way that they will unite and continue to grow as one plant.

Gravel culture—involves irrigating plants grown in gravel for mechanical support.

Gravitational water—the water that moves from large pores due to the pull of gravity after precipitation or irrigation.

Greenhouse—a structure that is covered with a transparent material that allows sufficient sunlight to enter for the purpose of growing and maintaining plants.

Greenhouse range—two or more greenhouses located together.

Greening—describes adding green foliage to a floral design.

Greening pins—retainers which look like hairpins with an "S" or flat top and are used to secure moss or foliage to a design.

Gro-bag—a cylinder with a porous fabric side and a plastic bottom, which prevents the formation of difficult to harvest tap roots.

Ground covers—woody or herbaceous plants that form a mat less than one foot high covering the ground.

Ground fault circuit interrupter (GFCI)—designed to protect people from electrical shock caused by short-circuits.

Grounding—a safety method designed to protect people from electrical shock.

Ground water—water within the saturated zone of earth that supplies wells and springs and is free to move under influence by gravity.

Growing degree days—a calculation made by adding the daily high and daily low temperature together and then dividing the answer by 2. The base temperature of the crop is then subtracted from that answer to find growing degree days.

Growing medium—the material in which the roots of plants grow.

Grubs—the larval stage of a variety of beetles.

Gutter-connected—several even-spans attached together, ridge-and-furrow.

Guying—attaching steel cable between metal stakes and a tree trunk.

Habitat—the place where wildlife live in nature.

Hammer—a tool made for driving or pounding.

Hand-moved watering system—a large, aboveground irrigation system using aluminum pipe laid down between the rows of crops which has to be moved by manual labor.

Hand tool—a small powerless tool.

Harden-off—placing seedlings in areas of cooler temperatures with less frequent waterings for a set period of time.

Hardiness—the ability of a plant to withstand colder temperatures.

Hard pinch—the removal of the terminal bud and more than $1/2$ inch of stem.

Hardscaping—describes permanent landscape structures, such as fences, patios, walks and driveways, water features, and retaining walls.

Hardware—the physical equipment that make up a computer.

Hardwood cuttings—cuttings taken from one-year-old wood during the dormant season from either deciduous or evergreen plants.

Hardy plants—plants that are less sensitive to temperature extremes.

Headhouse—a central building that is used for offices, storage, and work space with attached greenhouses.

Heading cuts—pruning that removes part of a shoot and directs the growth into a side limb or bud.

Headspace—the space left between the soil medium and the top edge of the container.

Heeling-in—involves temporarily placing a plant that has been dug in a field in a bed with its root system covered.

Herbaceous plant—a plant that has a soft, not woody stem that dies back to the ground each year, such as herbs, vines, and turfgrasses.

Herbicide—a chemical used to control weeds.

Holdfasts—appendages produced by certain vines that allow them to glue themselves to the support in order to climb.

Hook method—a technique for wiring flowers in a floral arrangement by pushing the wire up through the stem above the flower petals, bending the top of the wire into a hook shape, and gently pulling the hook into the flower.

Hormone—a chemical messenger substance produced in one location of an organism and then transported to another where it has a specific effect.

Horticultural garden—one that contains an assortment of plants represented by many varieties, arranged to achieve a desirable aesthetic effect.

Horticultural science—relates to the cultivation of ornamental plants, vegetables, and fruits.

Horticultural technology—applying science in horticulture production.

Horticulture—the culture of plants for food, comfort, and beauty.

Horticulture industry—all of the activities that support meeting the needs of consumers for horticulture products.

Host—a plant that provides a pest with food.

Hot air heat—produced by burning gas to heat air and blow into the greenhouse through ducts.

Hotbed—an outside propagation structure similar to a coldframe except electric or hot water thermostatically-control heating is used.

Hot water heat—a system that circulates 200°F water in pipes throughout the greenhouse for heat.

Hue—refers to the pure color without adding white or black.

Humidity—the amount of moisture in the air.

Hybrid—an improved plant developed by crossing parents of different genotype for a trait.

Hydraulics—fluid power.

Hydro-cooling—cooled with water.

HydroJet 3000—a machine that aerates by injecting high-speed jets of water through eleven nozzles spaced across a manifold bar into the soil.

Hydrologic cycle—the cycle of water in the environment.

Hydroponics—a method of growing plants in which the nutrients needed by the plant are supplied by a nutrient solution.

Ikenobo design—a Japanese floral arrangement where each flower has a specific meaning and an exact location in the arrangement.

Imperfect flower—a flower that lacks a stamen or pistil.

Income and expense summary—a listing of all revenues and expenses for a specific period.

Incomplete flower—a flower that lacks one or more of the major parts: sepals, petals, stamens, and pistil.

Incomplete metamorphosis—the gradual (simple) change in an insect's life cycle from the egg through the nymph to the adult.

Incurve—the center focal point of a corner planting.

Indirect seeding—seeds planted indoors or in a greenhouse with a germinating medium that are transplanted later to a larger container or to a permanent location outdoors.

Indoleacetic acid (IAA)—an auxin growth regulator which promotes and accelerates root formation on cuttings.

Indolebutyric acid (IBA)—a synthetic auxin growth regulator which is used to keep fruit on the trees a few extra weeks to allow further ripening and to accelerate root formation on cuttings in the greenhouse.

Inert ingredient—the carrier or filler that allows disposition.

Infiltration—the movement of water into the soil.

Inflorescence—the arrangement of flowers on a stem.

Infrared radiant heaters—individual unit heaters that produce infrared radiation that travels through the air and directly warms objects.

Inland wetland—includes freshwater marshes, northern peatlands, southern deepwater swamps, and riparian (stream) wetlands.

Inorganic—substances occurring as minerals in nature or obtainable from them by chemical means.

Inorganic fertilizers—synthetic nutrient compounds that are derived from mineral salts.

Insect—an animal with three distinct body parts, three pairs of legs, and either two, one, or no pairs of wings.

Insecticide—a chemical used to control insects.

Instar—the shedding of the external skeleton in the life cycle of insects.

Integrated pest management (IPM)—a pest management strategy that uses a combination of measures to reduce pest damage with the least disruption to the environment.

Intensive land use—involves using large fields with production practices to get top yields.

Interiorscaper—a person that designs and creates pleasing and comfortable areas inside buildings using plants.

Interiorscaping—use of foliage plants to create pleasing and comfortable areas inside buildings; to create the feeling of outdoors.

Internode—the area between two nodes.

Ion—an electrically charged particle.

Irrigate—water; provide adequate moisture.

Irrigation—the necessary watering process to maintain good plant growth when rainfall does not provide adequate moisture.

Job—specific work that a person in a horticulture occupation performs.

Job interview—a personal appearance of the job applicant with the employer.

Joint Utility Location and Information Exchange (J.U.L.I.E.)—an association that locates and marks all underground utilities for contractors.

Kickback—the violent up and back motion of the entire chain saw toward the operator which occurs when the turning chain stalls in the upper top of the guide bar.

Kickers—the part of a recycler mower system that causes grass clippings to oscillate between the deck and blade and be recut three to four times before discharge.

Landscape construction—the execution of the planting plan and hardscaping features.

Landscape architect—a person trained in engineering, graphic arts, and architectural technology to design landscapes ranging from small gardens to entire cities.

Landscape construction—the segment of landscaping that involves the installation of materials identified in the landscape design.

Landscape contractor—the person hired to install the landscape.

Landscape design—involves preparing a landscape plan for a site.

Landscape designer—a person trained in the art of design and the science of growing horticultural plants whose work is primarilyresidential landscape designs .

Landscape horticulture—deals with producing and using plants to make our outdoor environment more appealing.

Landscape maintenance—the care of established landscapes.

Landscape nursery—the area concerned with the preparation of sites for landscaping and the purchase and planting of trees, shrubs, evergreens, vines, and turfgrasses to meet the specifications of the landscape architect or designer.

Landscape planning—preparing the details of how a site will be landscaped.

Landscape symbol—shape and pattern that depicts different kinds of plants and construction features in a landscape design.

Larvae—the worm-like stage of an insect with complete metamorphosis.

Lateral bud—(See axillary bud.)

Lateral lines—secondary irrigation lines from the main irrigation line to individual valves or sprinklers.

Lathhouse—a permanent nursery structure used to protect plants from environmental factors, such as wind and solar radiation, by using lath strips on the top of a frame structure.

Layering—a method of asexual propagation in which roots are formed on a stem while it is still attached to the parent plant.

Leaching—the downward pulling of materials through the soil by percolating water.

Leader—a tree's main growing point, the tip end of the trunk.

Leaf analysis—the laboratory analysis of the nutrient content of the leaves.

Leaf apex—the tip of the leaf.

Leaf base—the part of the blade that is attached to the petiole.

Leaf-bud cutting—an asexual propagation method consisting of a leaf, petiole, and a short piece of stem with the lateral bud.

Leaf cutting—an asexual propagation method using a leaf blade or a leaf blade with a petiole attached.

Leaf margin—the outer edge of the leaf blade.

Leaf scar—a scar that is left when a leaf drops from a stem.

Lean-to—attached to the side of the building and the roof slopes away from the building.

Leggy—used to describe when plants grow too tall.

Lenticel—tiny pores located on a plant stem that allow for gas exchange between the plant and the environment.

Line—the visual movement between two points within a design.

Line planting—creates the walls of the outdoor room.

Liner—a young plant of suitable size that is ready to be planted into a larger container or field for growing into a larger plant.

Liquid feed system—uses liquid fertilizer diluted to the appropriate amount and injected into an irrigation system; fertigation.

Liquid Fertilizer—a fluid in which the plant nutrients are in true solution.

Loam—a soil textural class with particles of intermediate size that is higher in silt and lower in sand and clay.

Lolium perenne—perennial ryegrass.

Longitudinal arrangement—runs the entire length.

Lopping shears—long-handled pruning shears with a top cutting blade and a curved bottom blade for gripping used to cut medium size tree and shrub branches.

Low-voltage current—usually 24 volts AC.

Macroenvironment—the large atmosphere above a plant.

Macronutrients—those elements that are the most important to growth and must be present in large amounts.

Main lines—those that start at the water source and branch into several secondary irrigation lines.

Maleic hydrazide (MH)—a plant growth regulator for turfgrass used extensively in controlling vegetative growth and reducing mowing frequency.

Management—all of the activities needed to move a business toward its goals.

Manager—the person who is responsible for the operation of a business.

Managing—conducting or directing the operations of a business.

Marketing—providing the products and services people want.

Marketing mix—the combination of four variables that are used to reach the target customers—product, place, price, and promotion; also called the four "Ps."

Marketing plan—a written, detailed plan that outlines your specific strategies and goals to get customers to purchase your products or services.

Master controller unit—operates each satellite of an irrigation system, providing a single location for controlling all sprinklers.

Mechanically-harvested—gathered with mechanized equipment, such as tree spades.

Mechanical pest control—using tools or equipment for pest control.

Mefluidide (Embark)—a plant growth regulator absorbed primarily by the turfgrass leaf which inhibits cell elongation that results in growth suppression.

Microenvironment—the area immediately surrounding a plant which can be adapted to suit the needs of plants.

Microirrigation—one of a number of closed irrigation systems characterized by low operating pressure, small orifice size, and constructed in part from plastic materials, such as drip, microsprinkler, mister, bubbler, and fogger.

Micronutrients—trace elements required in smaller amounts for growth.

Micropropagation—(See tissue culture.)

Microsprinkler—distributes irrigation water over a very small area under shrubs or trees.

Mineral materials—matter from inorganic sources.

Mission statement—a brief (25 to 30 words) written description about the purpose of your business and the customers it serves.

Miticide—chemicals used to control mites.

Moist-scale—sense s the weight of a pot plant and turns on an electric water valve to irrigate an entire crop.

Moisture meter—an instrument used to indirectly measure soil moisture, such as a tensiometer or electrical resistance block.

Molecular biotechnology—involves changing the structure and parts of cells.

Molluscicide—chemicals used to control snails and slugs.

Monocot—a plant with one seed leaf.

Monocotyledoneae—plants characterized by one cotyledon (seed leaf) in their seedling stage, flower parts in threes or multiples thereof, and parallel leaf venation. (See monocot.)

Morphology—the classification of plants by form and structure.

Moss—tangled green mats composed of a branched, thread-like growth over the soil surface usually found in highly acidic, excessively shaded, improperly watered, or low fertility compacted conditions.

Mulching mower—recuts the turfgrass clippings returning them to the soil reducing mowing time and returning nutrients to the turf.

Multiple fruit—comprised of ovaries of many separate, but closely clustered flowers, such as mulberries and pineapples.

Naphthaleneacetic acid (NAA)—a synthetic auxin growth regulator which is used to keep fruit on the trees a few extra weeks to allow further ripening and to accelerate root formation on cuttings in the greenhouse.

Narrowleaf—plants that have owl-like, scale-like, or needle-like leaves.

Narrowleaf evergreen—one that retains its needle-like leaves through the winter.

Naturalized plant—one introduced to an area where it is not native, but has adapted so well that it appears to be native.

Nematicide—chemica ls used to control nematodes.

Nematode—an appendageless, nonsegmented worm-like invertebrate with a body cavity and complete digestive tract, including mouth, alimentary canal, and anus.

Nightlighting—the use of ornamental lighting to enhance the landscape at night.

Nitrogen cycle—the circulation of nitrogen in nature.

Node—a point along a plant stem where leaves or other stems are attached.

Nonpoint source pollution—pollution from many sources that can't be specifically identified.

Nursery—a place where plants, shrubs, and ornamental trees are started for transplanting to landscape areas.

Nursery production—growing of plants in controlled environments.

Nursery stock—nursery plants produced for resale.

Nutrient—chemical substance that support the life processes.

Nutrient film technique (NFT)—involves using a recirculating, shallow stream of nutrient solution which moves through channels in which the plants grow.

Nutrient solution—a liquid that contains water with dissolved nutrient salts.

Nuts—one-seeded fruits, usually produced from a compound ovary.

Nymph—the stage of the life cycle of an insect with incomplete metamorphosis that looks similar to the adult, only differing in appearance by size and color.

Occupation—specific work that has a title and general duties that a person in the occupation would perform.

Off-Shoot-O—a plant growth regulator used to chemically prune shrubs and hedges by destroying the meristematic tissue of the shoot apex which inhibits shoot elongation and promotes lateral branching.

Olericulture—the growing, harvesting, storing, processing, and marketing of vegetables.

Ooze tube watering system—one that drips water slowly from thin-walled plastic tubes onto the root medium surface to water greenhouse crops.

Open center pruning—the vase pruning system that allows for easy access to the tree for fruit thinning and good sunlight penetration to help the fruit ripen.

Operator's manual—a written description of how to safely use and maintain a power tool.

Opposite leaf arrangement—when the two leaves and buds are directly across from each other on the stem.

Organic—compounds of carbon other than inorganic carbonates.

Organic fertilizers—naturally occurring nutrient materials that are derived from plants or animals.

Organic matter—decayed remains of plants and animals.

Organic production—without the use of chemical pesticides or fertilizers.

Organismic biotechnology—deals with intact or complete organisms.

Organizing—the process of setting up a system for efficiently carrying out business plans.

Ornamental horticulture—describes growing and using plants for their beauty.

Ornamental turf—decorates areas around homes, businesses, in parks, and other places.

Outcurve—the sides of a corner planting.

Outdoor ceiling—the upper limit of the landscape in an outdoor room concept.

Outdoor floor—the ground covering in an outdoor room concept.

Outdoor room—the concept of designing plantings that have boundaries, ground coverings, flowers, patios, and traffic areas.

Outdoor walls—set the boundaries of an outdoor room.

Overhead—the general fixed costs of running a business.

Over-seeding—planting one grass in another established grass without destroying the established grass.

Overwintering structure—a permanent framework which is covered with polyethylene annually to prevent winter damage to nursery crops.

Paclobutrazol (TGR)—a plant growth regulator commonly used on golf courses which inhibits plant growth by interfering with the cell elongation process.

Parent material—the unconsolidated mass of rock material or peat from which the soil profile develops.

Parthenocarpy—the successful development of fruit without the presence of seeds.

Partnership—a business owned by two or more individuals.

Parts per million (ppm)—measurement stated as parts per million.

Paspalum notatum—bahiagrass.

Patch budding—the scion contains a bud with a piece of bark and is placed in an opening of the same size on the understock.

Pathegon—any microorganism or virus that can cause disease.

the physical characteristics of the plant materials; the arrangement of the leaves and petals which create many structural patterns.

Peat—partly decayed plant material of natural occurrence, composed chiefly of organic matter that contains some nitrogen of low activity.

Peat-lite mixes—a formulation of soilless media composed of one part sphagnum peat moss to one part vermiculite or perlite.

Peat moss—moss plants which grow on heath bogs, such as species of *Sphagnum* or *Polytrichum*.

Peat pellets—compressed peat moss that expands when moistened to become small peat-filled pots.

Peat pot—a container made of compressed peat used for growing seedlings which can be transplanted without removal from the pot.

Peninsular arrangement—similar to the cross-benching arrangement, except a central aisle runs the length and the benches extend to the sidewalls.

Percent germination—the percentage of seeds that will sprout and grow.

Percent timer clock—turns on the mist system nozzles for several seconds each minute to keep the plant cuttings from wilting.

Perched water table—the upper surface of a body of free ground water in a zone of saturation separated from underlying ground water by unsaturated material.

Percolation—the downward movement of water through soil.

Perennial—a plant that may be herbaceous or woody and lives for more than two seasons.

Perfect flower—the stamen and pistil in the same flower.

Perlite—a heat-treated lava rock that is lightweight with low nutrient and moisture holding capacity.

Permanent wilting poing—the point at which plants wilt, fail to recover turgidity, and die.

Personal skills—abilities of an individual to relate to other people in a productive manner.

Pest—anything that causes injury or loss to a plant.

Pesticide—a chemical used to control pests.

Petal—the usually bright colored, leaf-like part of a flower which serves to attract pollinators.

Petal fall—the point at which the petals of flowers dry up and start to fall off.

Petiole—the leaf stem or stalk.

pH—an index of the acidity of a substance, with 7.0 being neutral; numbers below 7.0 become increasingly acid and those above 7.0 become increasingly alkaline (basic).

Phenotype—the physical appearance of an organism.

Phillips screwdriver—one with a cross-point head design.

Phloem—vascular tissue that moves food from where it is manufactured to other parts of a plant.

Photoperiod—length of day or night.

Photoperiodism—th e length of darkness that influences plant growth.

Photosynthesis—the conversion of light energy into chemical energy by green plants; chlorophyll and sunlight produce carbohydrates and carbon dioxide while releasing oxygen.

Phototropism—plant response and growth toward light.

Pierce method—a technique for wiring flowers in a floral arrangement by pushing the wire through the calyx to create an artificial stem.

Pinching—simplest form of pruning used to encourage lateral branching or to remove unwanted growth.

Piscicide—chemicals used to control fishes.

Pistil—the female reproductive part of a flower.

Pith—the center portion of the plant stem where food and moisture are stored.

Planning—the process of deciding how a business will operate.

Plant breeder—the person who specializes in the development of improved cultivars.

Plant crown—the part of the plant at the soil surface from which new shoots or leaves are produced.

Plant disease—any abnormal condition in plants that interfere with its normal appearance, growth, structure, or function.

Plant environment—the above and below ground surroundings of a plant.

Plant growth—plants increasing in size by producing new leaves and stems.

Plant growth regulator (PGR)—a natural or synthetic substance that regulates or influences cell division, cell differentiation, root and shoot growth, flowering, and senescence.

Plant hardiness map—a map of the United States that identifies 11 zones by average annual minimum temperatures for each zone.

Plant heat-zone map—a map which identifies 12 different zones in the United States based on the average number of days above 86°F.

Planting plan—a landscape plan which shows the exact location for plant materials, includes a plant list, and the permanent structures to be installed.

Plant inspector—participates in activities designed to prevent the spread of pests among plants through an inspection service.

Plantlet—a plant which was created from an explant and grown by tissue culture; has tiny leaves, stems, and roots that have not yet developed into normal-sized parts of the parent plant.

Plant propagation—the reproduction of new plants from seeds and vegetative parts like leaves, stems, or roots.

Plant spacing—the distance between plants in the row and the distance between plant rows.

Plant station—an area where a plant will look better than any other object.

Plasticulture—the use of plastic mulches.

Plug—a small block or square of turf.

Pneumatic—air driven.

Poa pratensis—Kentucky bluegrass.

Point source emitters—small sprinklers placed underground close to root systems.

Point source pollution—one that comes from sources that can be readily be identified.

Pole pruner—consists of a saw mounted on a pole which allows a tree trimmer to remain on the ground and still cut tree branches 1½ inches in diameter that are 6 to 12 feet high.

Pollination—the transfer of pollen grains from the anther to the stigma.

Pollution—when harmful or degrading materials get into the environment.

Pomes—composed of several fused carpels and the flesh of the fruit is derived mostly from flower tissue outside the ovary, such as apples, mayhaws, and pears.

Pomology—the growing, harvesting, storing, processing, and marketing of fruits and nuts.

Pop-up heads—sprinkler heads that are forced out of the ground when water pressure is applied to the system.

Pore spaces—air holes between the growing medium particles which allow oxygen to reach the roots of the plant.

Porous tubing—plastic tubing with openings around its circumference and its length for dispersing irrigation water.

Postemergence—after plant emergence.

Post-emergent—(See postemergence.)

Pot-bound—plants in containers whose root systems have grown too long; root circling.

Pot-in-pot—a holder pot is placed in the ground with the lip of the pot remaining above grade and a planted container is then placed into the holder pot for the growing season.

Pour-through method—a simple method for testing media for soluble salt and range of pH, developed at Virginia Polytechnic Institute and State University.

Powered hedge trimmers—electric or gasoline-engine powered shears with blades oscillating back and forth over a second set of fixed blades.

Power tool—any tool that has power for its operation from a source other than human force.

Predacide—chemicals used to control predatory animals.

Preemergence—before plant emergence but after planting.

Pre-emergent—(See preemergence.)

Preplant—before planting.

Preventive maintenance—regularly scheduled service to reduce equipment failure.

Primary colors—red, yellow, and blue.

Primary root—the main root which grows from the seed.

Pricing strategy—determining how you are going to price your products and services.

Principles of design—general guidelines in applying art in designing landscapes; rules and guidelines to help a floral designer create a beautiful composition.

Private area—the part of the landscape that is out of the view of the general public; outdoor living area.

Processed vegetables—those that are canned, dried, or frozen.

Production schedule—the time required to produce something, such as a crop of plants to marketable size.

Profit—the amount of money the business receives after deducting all of the costs.

Profit margin—the financial return (income) in excess of direct costs and expenses.

Profit margin pricing—calculating the selling price by adding the actual cost per unit to the desired per unit percentage of profit.

Promotion—the coordination of all seller-initiated efforts to communicate with potential customers.

Propagation—involves increasing the number of plants.

Proportion—the relationship of the size of different species of plants in the landscape; the pleasing relationship in size and shape among objects or parts of objects.

Public area—the part of the landscape that will be seen from the street.

Publicity—mass media news coverage that includes the name of the business and product or service at no cost to the business.

Pupae—the transformation stage in the life cycle of an insect with complete metamorphosis.

Purlins—run the length of the structure and are attached to each truss, which adds more structural strength.

Quality—degree of excellence.

Quonset—curved roof with or without sidewalls.

Radiation freezes—freezes which occur on still (no wind) nights.

Reel blade—the rotating spiral blades on a horizontal shaft of a reel mower.

Reel mower—one with a fixed blade and a set of rotating spiral blades on a horizontal shaft allowing the mowing height to be adjusted by setting the height of the front roller.

Rejuvenation pruning—cutting all canes of a shrub within 6 inches of the ground.

Renewal pruning—a technique that gradually rejuvenates a shrub over a three or four year period by removing $1/3$ of the shrub canes at ground level each year for three consecutive years.

Reproductive phase—when a plant flowers and produces fruit.

Rhizome—underground stem that grows horizontally.

Rhythm—related, orderly organization of the design elements to create a dominant visual pathway.

Rhythm and line—a principle dealing with flow throughout a landscape that includes the shape and direction of beds and the heights of plant materials.

Ridge—the top (highest point).

Ridge-and-furrow—see gutter-connected.

Risk—the possibility of losing what has been invested.

Rodenticide—chemicals used to control rodents.

Roller chain—the chain drive which connects two or more sprockets or gears.

Root cap—the tip of the primary root that has several layers of cells which protects the root as it grows through the soil.

Root circling—(See pot-bound.)

Root cuttings—an asexual propagation method using root pieces of young plants during late winter or early spring.

Root hairs—very small roots off of primary and secondary roots that grow between soil particles.

Root pruning—the process of removing outward root tips which encourages the plant's root system to develop within a small area near the base of the plant.

Rootstock—the root system and base of the tree on which the top or scion cultivar is budded or grafted.

Rotary mower—has one or more blades that are parallel to the ground attached to a shaft with the mowing height adjusted by raising or lowering the wheels.

Safety—preventing injury or loss.

Safety policy—a brief, written definition of management's philosophy toward safety.

Sand culture—involves growing plants in sterilized sand and with individual drip irrigation.

Sash bars—structural members which hold greenhouse glass panes attached with a glazing compound.

Satellite—individual irrigation control unit.

Sawtooth—lean-to structures joined together.

Scaffold—any of the major tree branches extending from the trunk.

Scale—a part of proportion, dealing with relative size only among things, not shapes.

Scarification—breaking or softening a seed coat to allow absorption of moisture.

Science—the orderly arranged knowledge derived from observation, study, and experimentation carried on in order to determine the nature of what is being studied.

Scion—a short piece of stem with two or more buds.

Scouting—monitoring plants regularly to determine current levels of pest activity.

Secondary colors—green, violet, and orange.

Secondary root—a branch off the primary root.

Seed—the mature, fertilized ovum that contains an embryo and forms a new plant upon germination.

Seedbed—the media prepared to receive seed and promote germination and growth.

Seed coat—the tissue that surrounds the embryo and endosperm which functions to protect the seed from moisture loss, injury, or other unfavorable conditions.

Seed embryo—a plant in an arrested state of development inside of the seed coat, consisting of an axis and attached seed leaves, called cotyledons.

Seedling—a young plant grown from seed.

Seed purity—the percentage of pure seed of an identified species or cultivar present in a particular lot of seed.

Seed viability—the percentage of seed that will germinate under standard conditions.

Selling price—what the customer pays for a product or service.

Semi-hardwood cuttings—cuttings taken in the summer from new shoots that have partially matured from woody, broad-leaved plants.

Senescence—biological aging.

Sepal—the green leaf-like structure beneath the petals.

Separation—a propagation method in which natural structures are removed from the parent plant and planted to grow on their own.

Service area—the part of the landscape that is usually near the rear or the side of the house and is relatively isolated from the public and private areas.

Sexual propagation—the reproduction of plants with the use of seeds.

Shade—a graduation of a color (hue) with reference to its mixture with black.

Shadehouses—are permanent structures used to protect plants from environmental factors, such as wind, temperature, hail, heavy rain, and solar radiation by using lath strips or shade cloth.

Shape—the two-dimensional term for form of a floral design.

Shatter—when a plant drops its individual flower florets.

Sheave—a grooved wheel designed to be installed on a shaft for belt-driven equipment.

Shrub—a woody plant often with more than one stem and may branch, but is much smaller than trees.

Sickle mower—a walk-behind or tractor mounted mower with triangular blades oscillating back and forth over a second set of fixed blades.

Side-roll watering system—a series of large wheels along the length of the pipe that elevate the sprinkler line above the field crop which is easily moved.

Sidewalls—located between the concrete footings and support anchor posts of a greenhouse.

Silhouette lighting—places a lamp behind an unusually shaped plant to accent and silhouette this plant.

Silk flower—an artificial flower made from polyester fabric or new, improved plastic; permanent flower.

Simple fruit—one comprised of a single large ovary with or without some other flower parts which have developed as part of the fruit.

Simple layering—bending branches of the parent plant to the ground and covering portions of the branches with soil.

Simple leaf—one that consists of one blade per petiole.

Simplicity—designing the landscape so that it falls within the range of acceptable landscaping.

Site—an area of land having potential for development.

Site analysis—studying the site to identify its natural features and the appropriate treatment that would achieve the determined purpose.

Sitescaping—landscaping a small part of a larger area.

Sleepy—when a plant appears wilted.

Softwood cuttings—cuttings taken from soft, succulent, new spring growth of herbaceous or woody plants.

Sod—the surface layer of turf, including the grass plants and a thin layer of soil.

Sod cutter—a machine used to remove the surface layer of turf, including the plant and a thin layer of soil.

Soft pinch—removal of the terminal bud and up to $1/2$ inch of stem.

Software—the computer program instructions that run the hardware to get the worked performed as desired.

Soil—the top few inches of the earth's surface.

Soil aeration—the movement of atmospheric air into the soil.

Soil amendment—any material worked into the soil or applied to the surface to improve plant growth.

Soil ball—the soil surrounding the root system which has been balled and burlapped (B&B).

Soil biomass—the living organisms and their non-living residues that are part of the soil.

Soil compaction—when soil is compressed into a relatively dense mass.

Soilless medium—one that contains no topsoil.

Soil pH—the acidity or alkalinity of soil.

Soil porosity—the fraction of soil volume not occupied by soil particles.

Soil profile—a vertical section of soil at a particular place.

Soil salinity—the amount of salt in the soil.

Soil structure—the physical arrangement of soil particles.

Soil tension—the force by which moisture is held in the soil and does not include osmotic pressure values.

Soil tests—determine the nutrients in the soil.

Soil texture—the proportion of sand, silt, and clay in soil.

Solenoid valve—turns water off and on with electricity which allows remote clocks to time cycles.

Sole proprietorship—a business owned by one person, known as a proprietor.

Solid-state controllers—uses electronic circuits to turn on and off solenoid valves.

Spaghetti tube irrigation—irrigates individual plant containers by connecting the main water line to each pot with a small plastic tube. Water is carried to each pot by a polyethylene micro-tube or PVC pipe.

Species—the seventh division in the scientific classification system where all organisms are very similar.

Specimen plant—one that is distinctive and is used to create a focal point in the landscape.

Splinting method—a technique for wiring flowers in a floral arrangement to support weak stems by inserting the wire into the calyx and spirally wrapping the wire down the stem.

Sports turf—includes all types of playing fields.

Spray stakes—an irrigation system where water is applied over the canopy of the plants with spray nozzles mounted on risers.

Sprig—a part of the grass plant without soil, such as rhizomes and stolons.

Sprocket—a wheel or gear containing teeth for chain-driven equipment.

Spur-prune—a method of pruning vines that have a permanent trunk and arms where the previous season's growth is cut back to two or four buds each winter.

Staffing—concerned with the recruitment, selection, and training of employees.

Stamen—the male reproductive part of a flower.

Standard pot—a container equal in width and height.

Station—describes a single electric switch in a controller that turns on a solenoid valve.

Steam heat—circulates low pressure steam in pipes for heating.

Stem cuttings—asexual propagation using a portion of the plant stem that contains terminal or lateral buds.

Stem tubers—swollen tips of a rhizome.

Stenotaphrum secundatum—Saint Augustine-grass.

Stolon—above-ground stem that grows horizontally.

Stomata—pore openings in the epidermal layer of plant tissue where transpiration and respiration occur.

Straight-blade screwdriver—one with a straight-edge head designed to fit into a slot.

Stratification—placing seeds in a moist soil medium at temperatures between 32° and 50°F for a certain period of time.

Strip tillage—a method of conservation tillage where strips of vegetation are left between tilled areas to prevent soil erosion.

Stubbing—the process of cutting a seedling down to stub and allowing the plant to regenerate a new top.

Subscaffolds—the natural branches off of the major tree branches (scaffolds) extending from the trunk.

Suckers—branches that arise from latent buds near wounded areas on a tree trunk or branch; watersprouts.

Sunscald—describes a condition that causes the bark to blister from intense winter sunlight; winter burn.

Surface irrigation system—one that siphons water from ditches along the edge of the field into furrows next to rows of plants.

Surface water—water on or above the ground including rivers, lakes, canals, and reservoirs.

Surfactant—a material that helps the dispersing, spreading, wetting, or emulsifying of a formulation.

Sustainable horticulture—systems that use integrated pest management (IPM) and other best management practices (BMPs) to reduce inputs such as reduced tillage, reduced chemical inputs, drip irrigation, plastic mulch, improved varieties, wind breaks, etc.

Symmetrical balance—a floral arrangement in which each side of the central axis is equal in size and shape.

Symmetrical triangular—a floral arrangement with three equal sides or two equal sides with a different length base, and the triangle is completely filled in with flowers and foliage.

Synthetic—human-made.

T-Budding—taking buds from one plant and inserting them under the bark of the understock which has been cut in the shape of a "T".

Tap root system—one in which the primary root grows down from the stem with some small secondary roots forming on it.

Temperature—refers to how hot or cold something is and as a measure of the heat energy of molecules as they move about.

Tender plant—a plant that cannot tolerate cool weather.

Tendrils—appendages produced by certain vines that wrap around the support and allow them to climb.

Tensiometer—a device used to measure the tension with which water is held in the soil.

Terminal bud—a bud positioned at the tip of the stem that is an undeveloped leaf, stem, flower, or mixture of all.

Terrarium—a covered or closed container enclosing a miniature landscape of growing plants; bottle garden.

Texture—the surface quality and the structure or placement of the plant parts in the floral design.

Thatch—the layer of organic residue above the soil surface and between the living tissue of the host plant.

Thermoperiodic—describes plant response to changes in day and night temperature.

Thermotropism—plant growth response to temperature.

Thinning cuts—pruning that removes entire shoots.

Tint—a graduation of color (hue) with reference to its mixture with white.

Tissue culture—a method of growing pieces of plants, called explants, on an artificial medium under sterile conditions.

Topiaries—sheared shrubs that have been shaped into sculptures.

Topidary—a form of pruning in which plants are sheared into unnatural shapes.

Topography—the surface features of an area, including human-made changes.

Torque wrench—a special handle that allows setting and/or measuring the amount of pressure applied in tightening a bolt or nut; to tighten bolts and nuts to exact foot pounds of pressure.

Transgenic organism—one which carries a foreign gene that was inserted by laboratory techniques in all its cells.

Transition zone—an area between definite climate zones.

Transpiration—the loss of water from the plant through the leaves in the form of water vapor.

Transplanting—transferring or moving seedlings from the seedbed and setting them into the ground.

Treated fence posts—those that contain chemicals that prevent wood rot and therefore extend the post life.

Tree topping—reducing the total tree height by $1/3$ to $1/2$ by removing upper scaffold branches.

Trickle irrigation—(See drip irrigation.)

Trickle tape—emitters made from thin plastic lateral tubing with perforations along the entire length.

Trinexapac-ethyl (Primo)—a plant growth regulator which is absorbed directly into the leaf of turfgrass which results in a decrease of cell elongation and internode length, but does not stunt the growth of the plant in the long-term.

Trunk—the central stem of a tree.

Trusses—composed of rafters, chords, and struts which support the roof.

Tube watering system—irrigates individual plant pots by connecting a main water line to each pot with a small plastic micro-tube; see spaghetti tube irrigation.

Turf—the plants in a groundcover and the soil in which the roots grow.

Turfgrass—a collection of grass plants that form a ground cover.

Turfgrass blend—a combination of different cultivars of the same species.

Turfgrass maintenance—keeping a stand of turfgrass in an attractive and healthy condition.

Turfgrass mixture—combination of two or more different species.

Turf quality—the excellence of turf.

Turgid—describes plant cells full of water.

Turgidity—the rigidity of living cells due to pressure against the cell membrane from within by the cell contents.

Type C-3 hydraulic fluid—used in grounds maintenance hydraulic systems.

Type F hydraulic fluid—used in transmissions manufactured by Ford.

Understock—the lower portion of a graft which develops into the root system, sometimes called the rootstock.

Uneven-span—rafters of unequal lengths.

Unity—the extent to which a design is complete and whole.

Up lighting—places a lamp, which shines up, at a tree base.

Utility turf—has many useful functions.

V-belt—flexible rubber belts with a cross-section shape of the letter "V" which is used to connect two or more sheaves together.

Valve—a device that controls the flow of liquids or gases.

Variable costs—those directly associated with producing a product or providing a service.

Variety—a plant cultivar that is cultivated and retains its features when reproduced.

Vase life—the period of time after cutting flowers that they can be held and still look good.

Vegetative phase—when a plant seed germinates and grows producing leaves, stems, and roots.

Veins—tiny tubes that form patterns and move water, minerals, and nutrients in and out of the leaf blade.

Venlo—similar to gutter-connected but normally wider with twice as many ridges between supports.

Ventilation sash—describes glass greenhouse windows that open and allow fresh air to enter the greenhouse.

Ventilators—moveable units of a greenhouse to allow for natural ventilation.

Vermiculite—heat-treated mica that is light-weight and has high nutrient and moisture holding capacity.

Vertical mower—a wheel powered unit equipped with a horizontal shaft, fitted with vertically oriented blades.

Victorian design style—the pattern for early American floral arrangements brought to America by early settlers in this country through their European cultures.

Viruses—infective living agents of microorganisms, some with characteristics of nonliving matter, that can multiply only in connection with living cells and are regarded both as living organisms and as complex proteins sometimes involving nucleic acid, enzymes, etc.

Visual turf inspection—uses visual factors to assess turf quality, including density, texture, uniformity, color, growth habit, and smoothness.

Walk lighting—offers improved safety for pedestrians by illuminating walkways while providing an interesting lighting effect in the landscape.

Warm season turfgrass—grows best at an optimum temperature range of 80–95°F.

Water holding capacity—the medium's total pore space less air space after drainage.

Water infiltration—movement of water into the soil.

Waterlogged—over watered.

Water requirement—refers to the amount of water plants need to live and grow.

Water table—the upper surface of ground water.

Watersprouts—branches that arise from latent buds near wounded areas on a tree trunk or branch; suckers.

Water tubes—small, rubber-capped, plastic tubes used for holding water and a single flower or a small cluster of filler flowers.

Water zone—an area in a landscape based on the amount of water needed by the plants.

Way of doing business—how we organize and carry out free enterprise.

Wear—the physical deterioration resulting from excessive traffic.

Weed—a plant growing out of place or an unwanted plant.

Wetland—a swamp, bog, marsh, mire, pond, or other place where water often stands.

Whip-and-Tongue graft—a method of grafting a small scion and understock together that are under an inch in diameter.

Whorled leaf arrangement—when three or more leaves and buds arise from the same point on the stem.

Wildlife—plant and animal organisms that haven't been domesticated.

Williamsburg design—describes the American version of the Victorian design style using large round or globe shaped floral arrangements made by using different kinds and colors of flowers.

Wilting point—when plants cannot get enough moisture.

Window-mounted greenhouse—a prefabricated style that is available to fit many standard size windows.

Winter burn—describes a condition that causes bark to blister from intense winter sunlight; sunscald

Winter chilling requirement—calculated by how many hours of cool temperatures at or below 45°F (7.2°C) are required during the winter for the plant to break its winter resting period and develop normally when temperatures rise in the spring.

Woody plant—any shrub, tree, or certain vines which produce wood and have buds surviving above ground over the winter.

Wrench—a tool for gripping and turning bolts, nuts, and other fasteners and materials.

Xeriscaping—a form of landscaping that uses plants which require small amounts of water.

Xylem—the woody portion of the stem that conducts water and nutrients throughout a plant.

Yield—quantity.

Zeaton—a synthetic cytokinin growth regulator.

Zoysia japonica—Japanese lawngrass.

Zoysia matrella—Manilagrass.

BIBLIOGRAPHY

Acquaah, George, *Horticulture: Principles and Practices*, Prentice-Hall, Inc., Upper Saddle River, NJ, 1999.

Adams, William, and Thomas Leroy, *Growing Fruits and Nuts in the South*, Taylor Publishing Company, Dallas, TX, 1992.

Behe, Bridget K., Peter B. Pfahl, and Charles E. Hofmann, *The Retail Florist Business*, Interstate Publishers, Inc., Danville, IL, 1994.

Biondo, Ronald J. and Jasper S. Lee, *Introduction to Plant and Soil Science and Technology*, Interstate Publishers, Inc., Danville, IL, 1997.

Biondo, Ronald J. and Dianne A. Noland, *Floriculture: From Greenhouse Production to Floral Design*, Interstate Publishers, Inc., Danville, IL, 2000.

Biondo, Ronald J. and Charles B. Schroeder, *Introduction to Landscaping: Design, Construction, and Maintenance*, Interstate Publishers, Inc., Danville, IL, 2000.

Bridwell, Ferrell M., *Landscape Plants*, Delmar Publishers, Inc., Albany, NY, 1994.

Buriak, Philip, and Edward W. Osborne, *Physical Science Applications in Agriculture*, Interstate Publishers, Inc., Danville, IL, 1996.

California Fertilizer Association, *Western Fertilizer Handbook—Second Horticulture Edition*, Interstate Publishers, Inc., Danville, IL, 1998.

Childers, Norman, *Modern Fruit Science*, Horticultural Publications, Gainesville, FL, 1983.

Countryside Books, *Home Landscaping*, A. B. Morse Co., Clearwater, FL, 1995.

Crunkilton, John R., Susan L. Osborne, Michael E. Newman, Edward W. Osborne, and Jasper S. Lee, *The Earth and AgriScience*, Interstate Publishers, Inc., Danville, IL, 1995.

Davidson, H., H. Mecklenburg, and C. Peterson, *Nursery Management: Administration and Culture*, Prentice-Hall, Inc., Upper Saddle River, NJ, 1994.

731

Granberry, Darbie M., and Wayne J. McLaurin, *Gardening*, University Georgia Cooperative Extension Service Bulletin 577, 25 pp, 1990.

Hochmuth, George, Sustainable Vegetable Production, *Citrus & Vegetable Magazine*, April, pp 64-95, 1991.

Holland, I. I. and G. L. Rolfe, *Forests and Forestry*, Interstate Publishers, Inc., Danville, IL, 1997.

Krewer, Gerard, M. E. Ferree, and Steven Myers, *Bunch Grapes*, University of Georgia Cooperative Extension Service Bulletin 804, 1992.

Lee, Jasper S., and Diana L. Turner, *Introduction to World AgriScience and Technology*, Interstate Publishers, Inc., Danville, IL, 1997.

Lee, Jasper S., James G. Leising, and David E. Lawver, *AgriMarketing Technology*, Interstate Publishers, Inc., Danville, IL, 1994.

Mason, J., *Nursery Management*, Kangaroo Press Pty Ltd., Kenthurst NSW, Australia, 1994.

McKinley, William J., *The Cut Flower Companion*, Interstate Publishers, Inc., Danville, IL, 1994.

Myers, Stephen, and Gerard Krewer, *Home Garden Apples*, University of Georgia Cooperative Extension Service Circular 740, 1991.

Myers, Stephen, Gerard Krewer, and Paul Bertand, *Home Garden Peaches and Nectarines*, University of Georgia Cooperative Extension Service Circular 741, 1991.

Newman, Michael E., and Walter J. Wills, *Agribusiness Management and Entrepreneurship*, Interstate Publishers, Inc., Danville, IL, 1994.

Noland, Dianne A. and Kristen Bolin, *Perennials for the Landscape*, Interstate Publishers, Inc., Danville, IL, 2000.

Osborne, Edward W., *Biological Science Applications in Agriculture*, Interstate Publishers, Inc., Danville, IL, 1994.

Pierceall, Gregory M., *An Illustrated Guide to Landscape Design, Construction, and Management*, Interstate Publishers, Inc., Danville, IL, 1998.

Schroeder, Charles B. and Howard B. Sprague, *The Turf Management Handbook*, Interstate Publishers, Inc., Danville, IL, 1996.

Swiader, John M., George W. Ware, and J. P. McCollum, *Producing Vegetable Crops*, Interstate Publishers, Inc., Danville, IL, 1992.

USDA, National Agricultural Statistics Service, *Agricultural Statistics*, Washington, DC, 1999.

USDA, *World Agriculture Trends and Indicators 1970–1991*, Economic Research Service Bulletin No. 861, 1993.

Westwood, Melvin, *Temperate-Zone Pomology*, Timber Press, Portland, OR, 1993.

INDEX